Mathematics for Physics Students

Mathematics for Physics Students

Robert V. Steiner, Ph.D.
Project Director, Seminars on Science
National Center for Science Literacy, Education, and Technology
American Museum of Natural History

Philip A. Schmidt, Ph.D.
Director, Curriculum and Instruction
The Teachers College
Western Governors University

Schaum's Outline Series

New York Chicago San Francisco Lisbon London Madrid
Mexico City Milan New Delhi San Juan Seoul
Singapore Sydney Toronto

The **McGraw-Hill** *Companies*

ROBERT V. STEINER has a B.S. in physics from the University of California, Berkeley, and a Ph.D. in experimental elementary particle physics from Yale University. He directs the program in online science education at the American Museum of Natural History in New York City. In addition, he serves as a member of the adjunct faculty of the Department of Physics at Queens College, City University of New York.

PHILIP A. SCHMIDT has a B.S. degree (with a major in mathematics) from Brooklyn College, an M.A. in mathematics, and a Ph.D. in mathematics education from Syracuse University. He is the co-author of several Schaum's Outlines, including *College Mathematics* and *Elementary Algebra*. He is the recipient of numerous grants and awards in the fields of mathematics education, science education, and technology education.

2 3 4 5 6 7 8 9 10 ROV 1 9 8 7 6 5 4 3 2

ISBN 978-0-07-163415-1
MHID 0-07-163415-0

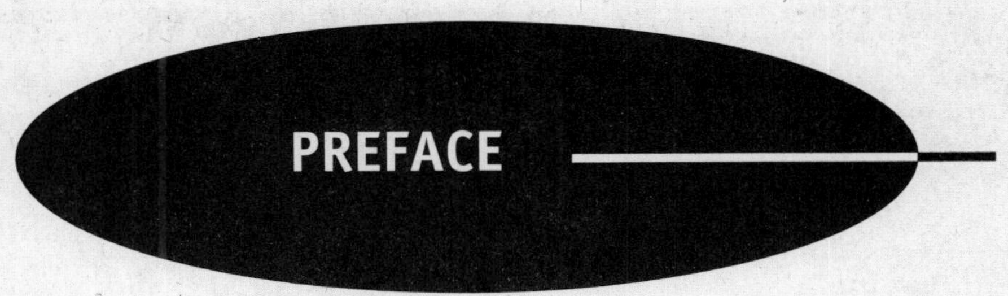

PREFACE

The idea for this book developed from our work as physics and mathematics educators. During the course of our respective careers, we have observed firsthand the problems that physics students encounter while learning mathematics. As we discussed the central challenge of providing students with an understanding of the mathematical concepts and tools essential to success in physics, we found a lack of resources that effectively address this crucial matter.

This book is intended for a wide variety of audiences: students in both algebra-based and calculus-based college physics courses, high school physics students, Advanced Placement Physics students, and physics majors. It reflects the current pedagogy in both physics and mathematics and can be used successfully as a supplement to any physics textbook. The notation used throughout the text is aligned with what one typically finds in current physics and mathematics texts. There is a special emphasis on problem-solving techniques and critical thinking throughout the book.

We express our deepest thanks to Barbara Gilson, Charles Wall, Kimberly Eaton, Stephen Smith, Sharon Rumbal, and their colleagues at McGraw-Hill. Phil thanks Janet Schnitz, Executive Director of the Teachers College at Western Governors University, for her professional support of this project. One of us (R.S.) dedicates this book to his wife Allison Steiner and to his children Russell and Samantha Steiner; he also gratefully acknowledges the intellectual inspirations of his father Eric Steiner, his uncle Herbert Steiner and Charles McClure, mathematics teacher *par excellence*. The other (P.S.) dedicates the book to his wife, Dr. Jan Zlotnik Schmidt, and his son, Reed Alexander Schmidt. As always, their inspiration has been a substantial gift.

ROBERT V. STEINER, PH.D.
PHILIP A. SCHMIDT, PH.D.

CONTENTS

PART I Algebra and Geometry 1

1. Introduction to Algebra 3
2. Functions 8
3. Graphs of Functions 13
4. Linear Equations 19
5. Simultaneous Linear Equations 23
6. Quadratic Functions and Equations 32
7. Inequalities 40
8. The Locus of an Equation 45
9. The Straight Line 51
10. Families of Straight Lines 57
11. The Circle 61

PART II Pre-Calculus and Elementary Calculus 71

12. Rational and Polynomial Functions 73
13. Trigonometric Functions 89
14. Exponential and Logarithmic Functions 134
15. Complex Numbers 149
16. The Calculus of Single-Variable Functions: A Mathematics Approach 161
17. The Calculus of Single-Variable Functions: A Physics Approach 192
18. Vectors 212

PART III Advanced Topics in Mathematics 233

19. Polar, Spherical, and Cylindrical Coordinate Systems 235
20. Multivariate Calculus 264
21. Elementary Linear Algebra 277
22. Vector Calculus: Grad, Div, and Curl 302
23. Vector Calculus: Flux and Gauss' Law 313
24. Differential Equations 319
25. Elementary Probability 345
26. Infinite Series 362

APPENDIX A Rectangular Coordinates in Space 378

APPENDIX B Units and Dimensions 390

APPENDIX C Solving Physics Problems 391

APPENDIX D	Selected Physics Formulas	394
APPENDIX E	Selected Physical Constants	395
APPENDIX F	Integration by Parts	396
APPENDIX G	The Greek Alphabet and Prefixes	399
INDEX		401

Theory and Problems of

MATHEMATICS FOR PHYSICS STUDENTS

PART I

ALGEBRA AND GEOMETRY

Introduction to Algebra

IN ARITHMETIC the numbers used are always known numbers; a typical problem is to convert 5 hours and 35 minutes to minutes. This is done by multiplying 5 by 60 and adding 35; thus, $5 \cdot 60 + 35 = 335$ minutes.

In algebra some of the numbers used may be known but others are either unknown or not specified; that is, they are represented by letters. For example, convert h hours and m minutes into minutes. This is done in precisely the same manner as in the paragraph above by multiplying h by 60 and adding m; thus, $h \cdot 60 + m = 60h + m$. We call $60h + m$ an *algebraic expression*. (See Problem 1.1.)

Since algebraic expressions are numbers, they may be added, subtracted, and so on, following the same laws that govern these operations on known numbers. For example, the sum of $5 \cdot 60 + 35$ and $2 \cdot 60 + 35$ is $(5 + 2) \cdot 60 + 2 \cdot 35$; similarly, the sum of $h \cdot 60 + m$ and $k \cdot 60 + m$ is $(h + k) \cdot 60 + 2m$. (See Problems 1.2–1.6.)

POSITIVE INTEGRAL EXPONENTS. If a is any number and n is any positive integer, the product of the n factors $a \cdot a \cdot a \cdots a$ is denoted by a^n. To distinguish between the letters, a is called the *base* and n is called the *exponent*.

If a and b are any bases and m and n are any positive integers, we have the following laws of exponents:

(1) $a^m \cdot a^n = a^{m+n}$

(2) $(a^m)^n = a^{mn}$

(3) $\dfrac{a^m}{a^n} = a^{m-n}, \quad a \neq 0, \quad m > n; \qquad \dfrac{a^m}{a^n} = \dfrac{1}{a^{n-m}}, \quad a \neq 0, \quad m < n$

(4) $(a \cdot b)^n = a^n b^n$

(5) $\left(\dfrac{a}{b}\right)^n = \dfrac{a^n}{b^n}, \qquad b \neq 0$

(See Problem 1.7.)

LET n BE A POSITIVE INTEGER and a and b be two numbers such that $b^n = a$; then b is called an nth *root of a*. Every number $a \neq 0$ has exactly n distinct nth roots.

If a is imaginary, all of its nth roots are imaginary; this case will be excluded here and treated later. (See Chapter 35.)

If a is real and n is odd, then exactly one of the nth roots of a is real. For example, 2 is the real cube root of 8, $(2^3 = 8)$, and -3 is the real fifth root of $-243[(-3)^5 = -243]$.

If a is real and n is even, then there are exactly two real nth roots of a when $a > 0$, but no real nth roots of a when $a < 0$. For example, $+3$ and -3 are the square roots of 9; $+2$ and -2 are the real sixth roots of 64.

THE PRINCIPAL n**th** **ROOT OF** a is the positive real nth root of a when a is positive and the real nth root of a, if any, when a is negative. The principal nth root of a is denoted by $\sqrt[n]{a}$, called a *radical*. The integer n is called the *index* of the radical and a is called the *radicand*. For example,

$$\sqrt{9} = 3 \qquad \sqrt[6]{64} = 2 \qquad \sqrt[5]{-243} = -3$$

(See Problem 1.8.)

ZERO, FRACTIONAL, AND NEGATIVE EXPONENTS. When s is a positive integer, r is any integer, and p is any rational number, the following extend the definition of a^n in such a way that the laws (1)-(5) are satisfied when n is any rational number.

DEFINITIONS	EXAMPLES
(6) $a^0 = 1, a \neq 0$	$2^0 = 1, \left(\frac{1}{100}\right)^0 = 1, (-8)^0 = 1$
(7) $a^{r/s} = \sqrt[s]{a^r} = (\sqrt[s]{a})^r$	$3^{1/2} = \sqrt{3}, (64)^{5/6} = \left(\sqrt[6]{64}\right)^5 = 2^5 = 32, 3^{-2/1} = 3^{-2} = \frac{1}{9}$
(8) $a^{-p} = 1/a^p, a \neq 0$	$2^{-1} = \frac{1}{2}, 3^{-1/2} = 1\sqrt{3}$

[NOTE: Without attempting to define them, we shall assume the existence of numbers such as $a^{\sqrt{2}}, a^{\pi}, \ldots$, in which the exponent is irrational. We shall also assume that these numbers have been defined in such a way that the laws (1)–(5) are satisfied.] (See Problem 1.9–1.10.)

Solved Problems

1.1 For each of the following statements, write the equivalent algebraic expressions: (*a*) the sum of x and 2, (*b*) the sum of a and $-b$, (*c*) the sum of $5a$ and $3b$, (*d*) the product of $2a$ and $3a$, (*e*) the product of $2a$ and $5b$, (*f*) the number which is 4 more than 3 times x, (*g*) the number which is 5 less than twice y, (*h*) the time required to travel 250 miles at x miles per hour, (*i*) the cost (in cents) of x eggs at 65¢ per dozen.

(*a*) $x + 2$ (*d*) $(2a)(3a) = 6a^2$ (*g*) $2y - 5$

(*b*) $a + (-b) = a - b$ (*e*) $(2a)(5b) = 10ab$ (*h*) $250/x$

(*c*) $5a + 3b$ (*f*) $3x + 4$ (*i*) $65(x/12)$

1.2 Let x be the present age of a father. (*a*) Express the present age of his son, who 2 years ago was one-third his father's age. (*b*) Express the age of his daughter, who 5 years from today will be one-fourth her father's age.

(*a*) Two years ago the father's age was $x - 2$ and the son's age was $(x - 2)/3$. Today the son's age is $2 + (x - 2)/3$.

(*b*) Five years from today the father's age will be $x + 5$ and his daughter's age will be $\frac{1}{4}(x + 5)$. Today the daughter's age is $\frac{1}{4}(x + 5) - 5$.

1.3 A pair of parentheses may be inserted or removed at will in an algebraic expression if the first parenthesis of the pair is preceded by a + sign. If, however, this sign is −, the signs of all terms within the parentheses must be changed.

(a) $5a + 3a - 6a = (5 + 3 - 6)a = 2a$ (b) $\frac{1}{2}a + \frac{1}{4}b - \frac{1}{4}a + \frac{3}{4}b = \frac{1}{4}a + b$

(c) $(13a^2 - b^2) + (-4a^2 + 3b^2) - (6a^2 - 5b^2) = 13a^2 - b^2 - 4a^2 + 3b^2 - 6a^2 + 5b^2 = 3a^2 + 7b^2$

(d) $(2ab - 3bc) - [5 - (4ab - 2bc)] = 2ab - 3bc - [5 - 4ab + 2bc]$
$= 2ab - 3bc - 5 + 4ab - 2bc = 6ab - 5bc - 5$

(e) $(2x + 5y - 4)3x = (2x)(3x) + (5y)(3x) - 4(3x) = 6x^2 + 15xy - 12x$

(f)
$$\begin{array}{rl} & 5a - 2 \\ & 3a + 4 \\ \hline & 15a^2 - 6a \\ (+) & 20a - 8 \\ \hline & 15a^2 + 14a - 8 \end{array}$$

(g)
$$\begin{array}{rl} & 2x - 3y \\ & 5x + 6y \\ \hline & 10x^2 - 15xy \\ (+) & 12xy - 18y^2 \\ \hline & 10x^2 - 3xy - 18y^2 \end{array}$$

(h)
$$\begin{array}{rl} & 3a^2 + 2a - 1 \\ & 2a - 3 \\ \hline & 6a^3 + 4a^2 - 2a \\ (+) & -9a^2 - 6a + 3 \\ \hline & 6a^3 - 5a^2 - 8a + 3 \end{array}$$

(i)
$$\begin{array}{r} x^2 + 4x - 2 \\ x - 3 \overline{)\, x^3 + x^2 - 14x + 6} \\ (-)\,x^3 - 3x^2 \\ \hline 4x^2 - 14x \\ (-)\,4x^2 - 12x \\ \hline -2x + 6 \\ (-)\,-2x + 6 \end{array}$$

(j)
$$\begin{array}{r} x^2 - 2x - 1 \\ x^2 + 3x - 2 \overline{)\, x^4 + x^3 - 9x^2 + x + 5} \\ (-)\,x^4 + 3x^3 - 2x^2 \\ \hline -2x^3 - 7x^2 + x \\ (-)\,-2x^3 - 6x^2 + 4x \\ \hline -x^2 - 3x + 5 \\ (-)\,-x^2 - 3x + 2 \\ \hline 3 \end{array}$$

$$\frac{x^3 + x^2 - 14x + 6}{x - 3} = x^2 + 4x - 2 \qquad \frac{x^4 + x^3 - 9x^2 + x + 5}{x^2 + 3x - 2} = x^2 - 2x - 1 + \frac{3}{x^2 + 3x - 2}$$

1.4 The problems below involve the following types of factoring:

$ab + ac - ad = a(b + c - d)$ $a^2 \pm 2ab + b^2 = (a \pm b)^2$

$a^3 + b^3 = (a + b)(a^2 - ab + b^2)$ $a^2 - b^2 = (a - b)(a + b)$

$acx^2 + (ad + bc)x + bd = (ax + b)(cx + d)$ $a^3 - b^3 = (a - b)(a^2 + ab + b^2)$

(a) $5x - 10y = 5(x - 2y)$

(b) $\frac{1}{2}gt^2 - \frac{1}{2}g^2t = \frac{1}{2}gt(t - g)$

(c) $x^2 + 4x + 4 = (x + 2)^2$

(d) $x^2 + 5x + 4 = (x + 1)(x + 4)$

(e) $x^2 - 3x - 4 = (x - 4)(x + 1)$

(f) $4x^2 - 12x + 9 = (2x - 3)^2$

(g) $12x^2 + 7x - 10 = (4x + 5)(3x - 2)$

(h) $x^3 - 8 = (x - 2)(x^2 + 2x + 4)$

(i) $2x^4 - 12x^3 + 10x^2 = 2x^2(x^2 - 6x + 5) = 2x^2(x - 1)(x - 5)$

1.5 Simplify.

(a) $\dfrac{8}{12x + 20} = \dfrac{4 \cdot 2}{4 \cdot 3x + 4 \cdot 5} = \dfrac{2}{3x + 5}$

(b) $\dfrac{9x^2}{12xy - 15xz} = \dfrac{3x \cdot 3x}{3x \cdot 4y - 3x \cdot 5z} = \dfrac{3x}{4y - 5z}$

(c) $\dfrac{5x - 10}{7x - 14} = \dfrac{5(x - 2)}{7(x - 2)} = \dfrac{5}{7}$

(d) $\dfrac{4x - 12}{15 - 5x} = \dfrac{4(x - 3)}{5(3 - x)} = \dfrac{4(x - 3)}{-5(x - 3)} = -\dfrac{4}{5}$

(e) $\dfrac{x^2 - x - 6}{x^2 + 7x + 10} = \dfrac{(x + 2)(x - 3)}{(x + 2)(x + 5)} = \dfrac{x - 3}{x + 5}$

(f) $\dfrac{6x^2 + 5x - 6}{2x^2 - 3x - 9} = \dfrac{(2x + 3)(3x - 2)}{(2x + 3)(x - 3)} = \dfrac{3x - 2}{x - 3}$

(g) $\dfrac{3a^2 - 11a + 6}{a^2 - a - 6} \cdot \dfrac{4 - 4a - 3a^2}{36a^2 - 16} = \dfrac{(3a - 2)(a - 3)(2 - 3a)(2 + a)}{(a - 3)(a + 2)4(3a + 2)(3a - 2)} = -\dfrac{3a - 2}{4(3a + 2)}$

1.6 Combine as indicated.

(a) $\dfrac{2a + b}{10} + \dfrac{a - 6b}{15} = \dfrac{3(2a + b) + 2(a - 6b)}{30} = \dfrac{8a - 9b}{30}$

(b) $\dfrac{2}{x} - \dfrac{3}{2x} + \dfrac{5}{4} = \dfrac{2 \cdot 4 - 3 \cdot 2 + 5 \cdot x}{4x} = \dfrac{2 + 5x}{4x}$

(c) $\dfrac{2}{3a-1}-\dfrac{3}{2a+1}=\dfrac{2(2a+1)-3(3a-1)}{(3a-1)(2a+1)}=\dfrac{5-5a}{(3a-1)(2a+1)}$

(d) $\dfrac{3}{x+y}-\dfrac{5}{x^2-y^2}=\dfrac{3}{x+y}-\dfrac{5}{(x+y)(x-y)}=\dfrac{3(x-y)-5}{(x+y)(x-y)}=\dfrac{3x-3y-5}{(x+y)(x-y)}$

(e) $\dfrac{a-2}{6a^2-5a-6}+\dfrac{2a+1}{9a^2-4}=\dfrac{a-2}{(2a-3)(3a+2)}+\dfrac{2a+1}{(3a+2)(3a-2)}$

$$=\frac{(a-2)(3a-2)+(2a+1)(2a-3)}{(2a-3)(3a+2)(3a-2)}=\frac{7a^2-12a+1}{(2a-3)(3a+2)(3a-2)}$$

1.7 Perform the indicated operations.

(a) $3^4=3\cdot3\cdot3\cdot3=81$
(b) $-3^4=-81$
(c) $(-3)^4=81$
(d) $-(-3)^4=-81$
(e) $-(-3)^3=27$
(f) $2^6\cdot2^4=2^{6+4}=2^{10}=1024$
(g) $\left(\frac{1}{2}\right)^3\left(\frac{1}{2}\right)^2=\left(\frac{1}{2}\right)^5$
(h) $a^{n+3}a^{m+2}=a^{m+n+5}$
(i) $(a^2)^5=a^{2\cdot5}=a^{10}$
(j) $(a^{2n})^3=a^{6n}$
(k) $a^{10}/a^4=a^{10-4}=a^6$
(l) $a^4/a^{10}=1/a^{10-4}=1/a^6$
(m) $(-2)^8/(-2)^5=(-2)^3=-8$
(n) $a^{2n}b^{5m}/a^{3n}b^{2m}=b^{3m}/a^n$
(o) $36^{x+3}/6^{x-1}=6^{2x+6}/6^{x-1}=6^{x+7}$

1.8 Evaluate.

(a) $81^{1/2}=\sqrt{81}=9$
(b) $81^{3/4}=\left(\sqrt[4]{81}\right)^3=3^3=27$
(c) $\left(\frac{16}{49}\right)^{3/2}=\left(\sqrt{\frac{16}{49}}\right)^3=\left(\frac{4}{7}\right)^3=\frac{64}{343}$
(d) $(-27)^{1/3}=\sqrt[3]{-27}=-3$
(e) $(-32)^{4/5}=\left(\sqrt[5]{-32}\right)^4=(-2)^4=16$
(f) $-400^{1/2}=-\sqrt{400}=-20$

1.9 Evaluate.

(a) $4^0=1$
(b) $4a^0=4\cdot1=4$
(c) $(4a)^0=1$
(d) $4(3+a)^0=4\cdot1=4$
(e) $4^{-1}=\frac{1}{4}$
(f) $5^{-2}=\left(\frac{1}{5}\right)^2=\frac{1}{25}$
(g) $125^{-1/3}=1/125^{1/3}=\frac{1}{5}$
(h) $(-125)^{-1/3}=-\frac{1}{5}$
(i) $-\left(\frac{1}{64}\right)^{5/6}=-\left(\sqrt[6]{\frac{1}{64}}\right)^5=-\left(\frac{1}{2}\right)^5=-\frac{1}{32}$
(j) $-\left(-\frac{1}{32}\right)^{4/5}=-\left(\sqrt[5]{-\frac{1}{32}}\right)^4=-\left(-\frac{1}{2}\right)^4=-\frac{1}{16}$

1.10 Perform each of the following operations and express the result without negative or zero exponents:

(a) $\left(\dfrac{81a^4}{b^8}\right)^{-1/4}=\dfrac{3^{-1}a^{-1}}{b^{-2}}=\dfrac{b^2}{3a}$ (b) $\left(a^{1/2}+a^{-1/2}\right)^2=a+2a^0+a^{-1}=a+2+\dfrac{1}{a}$

(c) $\left(a-3b^{-2}\right)\left(2a^{-1}-b^2\right)=2a^0-ab^2-6a^{-1}b^{-2}+3b^0=5-ab^2-6/ab^2$

(d) $\dfrac{a^{-2}+b^{-2}}{a^{-1}-b^{-1}}=\dfrac{(a^{-2}+b^{-2})(a^2b^2)}{(a^{-1}-b^{-1})(a^2b^2)}=\dfrac{b^2+a^2}{ab^2-a^2b}$

(e) $\left(\dfrac{a^2}{b}\right)^7\left(-\dfrac{b^2}{a^3}\right)^6=\dfrac{a^{14}\cdot b^{12}}{b^7\cdot a^{18}}=\dfrac{b^5}{a^4}$

(f) $\left(\dfrac{a^{1/2}b^{2/3}}{c^{3/4}}\right)^6\left(\dfrac{c^{1/2}}{a^{1/4}b^{1/3}}\right)^9=\dfrac{a^3b^4}{c^{9/2}}\cdot\dfrac{c^{9/2}}{a^{9/4}b^3}=a^{3/4}b$

Supplementary Problems

1.11 Combine.

(a) $2x+(3x-4y)$
(b) $5a+4b-(-2a+3b)$
(c) $[(s+2t)-(s+3t)]-[(2s+3t)-(-4s+5t)]$
(d) $8x^2y-\{3x^2y+[2xy^2+4x^2y-(3xy^2-4x^2y)]\}$

1.12 Perform the indicated operations.

(a) $4x(x-y+2)$ (c) $(5x^2-4y^2)(-x^2+3y^2)$ (e) $(2x^3+5x^2-33x+20)\div(2x-5)$

(b) $(5x+2)(3x-4)$ (d) $(x^3-3x+5)(2x-7)$ (f) $(2x^3+5x^2-22x+10)\div(2x-3)$

1.13 Factor.

(a) $8x+12y$ (e) $16a^2-8ab+b^2$ (i) $(x-y)^2+6(x-y)+5$

(b) $4ax+6ay-24az$ (f) $25x^2+30xy+9y^2$ (j) $4x^2-8x-5$

(c) a^2-4b^2 (g) $x^2-4x-12$ (k) $40a^2+ab-6b^2$

(d) $50ab^4-98a^3b^2$ (h) $a^2+23ab-50b^2$ (l) $x^4+24x^2y^2-25y^4$

1.14 Simplify.

(a) $\dfrac{a^2-b^2}{2ax+2bx}$ (d) $\dfrac{16a^2-25}{2a-10}\times\dfrac{a^2-10a+25}{4a+5}$

(b) $\dfrac{x^2+4x+3}{1-x^2}$ (e) $\dfrac{x^2+xy-6y^2}{2x^3+6x^2y}\times\dfrac{8x^2y}{x^2-5xy+6y^2}$

(c) $\dfrac{1-x-12x^2}{1+x-6x^2}$

1.15 Perform the indicated operations.

(a) $\dfrac{5x}{18}+\dfrac{4x}{18}$ (c) $\dfrac{3a}{4b}-\dfrac{4b}{3a}$ (e) $x+5-\dfrac{x^2}{x-5}$ (g) $\dfrac{2x+3}{18x^2-27x}-\dfrac{2x-3}{18x^2+27x}$

(b) $\dfrac{3a}{x}+\dfrac{5a}{2x}$ (d) $\dfrac{2a-3b}{a^2-b^2}+\dfrac{1}{a-b}$ (f) $\dfrac{a+2}{2a-6}-\dfrac{a-2}{2a+6}$

1.16 Simplify.

(a) $\dfrac{a}{2-\dfrac{3}{a}}$ (b) $\dfrac{4-\dfrac{x}{3}}{\dfrac{x}{6}}$ (c) $\dfrac{x-\dfrac{4}{x}}{1-\dfrac{2}{x}}$ (d) $\dfrac{\dfrac{1}{x}+\dfrac{1}{y}}{\dfrac{x+y}{y}+\dfrac{x+y}{x}}$

ANSWERS TO SUPPLEMENTARY PROBLEMS

1.11 (a) $5x-4y$ (b) $7a+b$ (c) $t-6s$ (d) $xy(y-3x)$

1.12 (a) $4x^2-4xy+8x$ (c) $-5x^4+19x^2y^2-12y^4$ (e) x^2+5x-4

(b) $15x^2-14x-8$ (d) $2x^3-13x^2+31x-35$ (f) $x^2+4x-5-5/(2x-3)$

1.13 (a) $4(2x+3y)$ (e) $(4a-b)^2$ (i) $(x-y+1)(x-y+5)$

(b) $2a(2x+3y-12z)$ (f) $(5x+3y)^2$ (j) $(2x+1)(2x-5)$

(c) $(a-2b)(a+2b)$ (g) $(x-6)(x+2)$ (k) $(5a+2b)(8a-3b)$

(d) $2ab^2(5b-7a)(5b+7a)$ (h) $(a+25b)(a-2b)$ (l) $(x-y)(x+y)(x^2+25y^2)$

1.14 (a) $\dfrac{a-b}{2x}$ (b) $-\dfrac{x+3}{x-1}$ (c) $\dfrac{4x-1}{2x-1}$ (d) $\frac{1}{2}(4a^2-25a+25)$ (e) $\dfrac{4y}{x-3y}$

1.15 (a) $\frac{1}{2}x$ (b) $\dfrac{11a}{2x}$ (c) $\dfrac{9a^2-16b^2}{12ab}$ (d) $\dfrac{3a-2b}{a^2-b^2}$ (e) $-\dfrac{25}{x-5}$

(f) $\dfrac{5a}{a^2-9}$ (g) $\dfrac{8}{12x^2-27}$

1.16 (a) $\dfrac{a^2}{2a-3}$ (b) $\dfrac{24-2x}{x}$ (c) $x+2$ (d) $\dfrac{1}{x+y}$

Chapter 2

Functions

A VARIABLE IS A SYMBOL selected to represent any one of a given set of numbers, here assumed to be real numbers. Should the set consist of just one number, the symbol representing it is called a *constant*.

The *range* of a variable consists of the totality of numbers of the set which it represents. For example, if x is a day in September, the range of x is the set of positive integers $1, 2, 3, \ldots, 30$; if x (ft) is the length of rope cut from a piece 50 ft long, the range of x is the set of numbers greater than 0 and less than 50.

Examples of ranges of a real variable, together with special notations and graphical representations, are given in Problem 2.1

FUNCTION. A correspondence (x, y) between two sets of numbers which pairs to an arbitrary number x of the first set exactly one number y of the second set is called a *function*. In this case, it is customary to speak of y as a *function* of x. The variable x is called the *independent variable* and y is called the *dependent variable*.

A function may be defined

(*a*) By a table of correspondents or table of values, as in Table 2.1.

Table 2.1

x	1	2	3	4	5	6	7	8	9	10
y	3	4	5	6	7	8	9	10	11	12

(*b*) By an equation or formula, as $y = x + 2$.

For each value assigned to x, the above relation yields a corresponding value for y. Note that the table above is a table of values for this function.

A FUNCTION IS CALLED *single-valued* if, to each value of y in its range, there corresponds just one value of x; otherwise, the function is called *multivalued*. For example, $y = x + 3$ defines y as a single-valued function of x while $y = x^2$ defines y as a multivalued (here, two-valued) function of x.

At times it will be more convenient to label a given function of x as $f(x)$, to be read "the f function of x" or simply "f of x." (Note carefully that this is not to be confused with "f times x.") If there are two functions, one may be labeled $f(x)$ and the other $g(x)$. Also, if $y = f(x) = x^2 - 5x + 4$, the statement "the value of the function is -2 when $x = 3$" can be replaced by "$f(3) = -2$." (See Problem 2.2.)

Let $y = f(x)$. The set of values of the independent variable x is called the *domain* of the function while the set of values of the dependent variable is called the *range of the function*. For example, $y = x^2$ defines a function whose domain consists of all (real) numbers and whose range is all nonnegative numbers, that is, zero and the positive numbers; $f(x) = 3/(x-2)$ defines a function whose domain consists of all numbers except 2 (why?) and whose range is all numbers except 0. (See Problems 2.3–2.8.)

A VARIABLE w (dependent) is said to be a function of the (independent) variables $x, y, z, \ldots$ if when a value of each of the variables $x, y, z, \ldots$ is known, there corresponds exactly one value of w. For example, the volume V of a rectangular parallelepiped of dimensions x, y, z is given by $V = xyz$. Here V is a function of three independent variables. (See Problem 2.9.)

ADDITIONAL TERMINOLOGY If the function $y = f(x)$ is such that for every y in the range there is one and only one x in the domain such that $y = f(x)$, we say that f is a one-to-one correspondence. Functions that are one-to-one correspondences are sometimes called *bijections*. Note that all functions of the form $ax + by + c = 0$ are bijections. Note that $y = x^2$ is not a bijection. Is $y = x^3$ a bijection? (Answer: Yes!)

Solved Problems

2.1 Represent graphically each of the following ranges:

(*a*) $x > -2$ (-2)

(*b*) $x < 5$ (5)

(*c*) $x \leqslant -1$

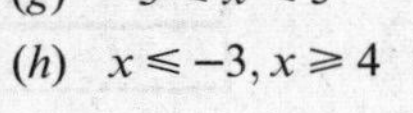

(*d*) $-3 < x < 4$ (-3, 4)

(*e*) $-2 < x < 2$ or $|x| < 2$

(*f*) $|x| > 3$

(*g*) $-3 \leqslant x \leqslant 5$

(*h*) $x \leqslant -3, x \geqslant 4$

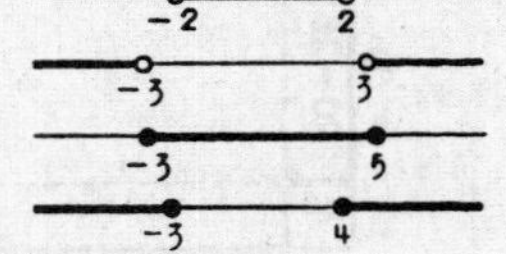

2.2 Given $f(x) = x^2 - 5x + 4$, find

(*a*) $f(0) = 0^2 - 5 \cdot 0 + 4 = 4$

(*b*) $f(2) = 2^2 - 5 \cdot 2 + 4 = -2$

(*c*) $f(-3) = (-3)^2 - 5(-3) + 4 = 28$

(*d*) $f(a) = a^2 - 5a + 4$

(*e*) $f(-x) = x^2 + 5x + 4$

(*f*) $f(b+1) = (b+1)^2 - 5(b+1) + 4 = b^2 - 3b$

(*g*) $f(3x) = (3x)^2 - 5(3x) + 4 = 9x^2 - 15x + 4$

(*h*) $f(x+a) - f(a) = [(x+a)^2 - 5(x+a) + 4] - (a^2 - 5a + 4) = x^2 + 2ax - 5x$

(*i*) $\dfrac{f(x+a) - f(x)}{a} = \dfrac{[(x+a)^2 - 5(x+a) + 4] - (x^2 - 5x + 4)}{a} = \dfrac{2ax - 5a + a^2}{a} = 2x - 5 + a$

2.3 In each of the following, state the domain of the function:

(*a*) $y = 5x$

(*b*) $y = -5x$

(*c*) $y = \dfrac{1}{x+5}$

(*d*) $y = \dfrac{x-2}{(x-3)(x+4)}$

(*e*) $y = \dfrac{1}{x}$

(*f*) $y = \sqrt{25 - x^2}$

(*g*) $y = \sqrt{x^2 - 9}$

(*h*) $y = \dfrac{1}{16 - x^2}$

(*i*) $y = \dfrac{1}{16 + x^2}$

Ans. (*a*), (*b*), all real numbers; (*c*) $x \neq -5$; (*d*) $x \neq 3, -4$; (*e*) $x \neq 0$; (*f*) $-5 \leqslant x \leqslant 5$ or $|x| \leqslant 5$; (*g*) $x \leqslant -3, x \geqslant 3$ or $|x| \geqslant 3$; (*h*) $x \neq \pm 4$; (*i*) all real numbers.

2.4 A piece of wire 30 in. long is bent to form a rectangle. If one of its dimensions is x in., express the area as a function of x.

Since the semiperimeter of the rectangle is $\frac{1}{2} \cdot 30 = 15$ in. and one dimension is x in., the other is $(15 - x)$ in. Thus, $A = x(15 - x)$.

2.5 An open box is to be formed from a rectangular sheet of tin 20×32 in. by cutting equal squares, x in. on a side, from the four corners and turning up the sides. Express the volume of the box as a function of x.

From Fig. 2-1. it is seen that the base of the box has dimensions $(20 - 2x)$ by $(32 - 2x)$ in. and the height is x in. Then

$$V = x(20 - 2x)(32 - 2x) = 4x(10 - x)(16 - x)$$

2.6 A closed box is to be formed from the sheet of tin of Problem 2.5 by cutting equal squares, x cm on a side, from two corners of the short side and two equal rectangles of width x cm from the other two corners, and folding along the dotted lines shown in Fig. 2-2. Express the volume of the box as a function of x.

One dimension of the base of the box is $(20 - 2x)$ cm; let y cm be the other. Then $2x + 2y = 32$ and $y = 16 - x$. Thus,

$$V = x(20 - 2x)(16 - x) = 2x(10 - x)(16 - x)$$

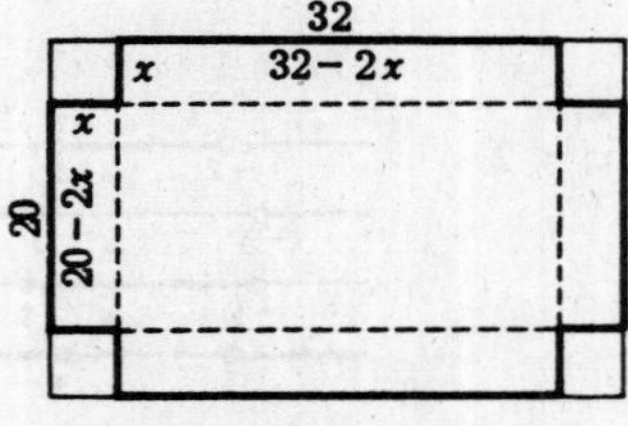

Fig. 2-1

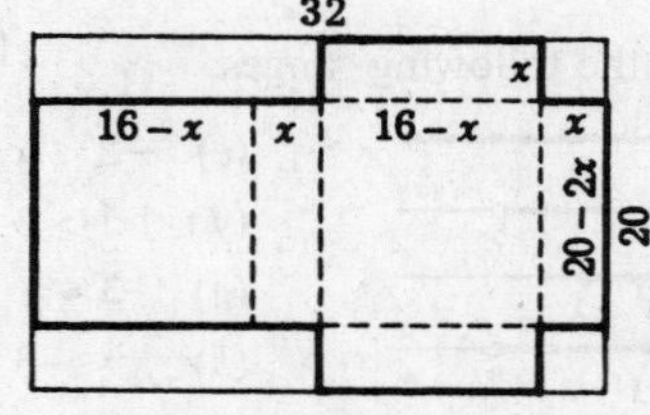

Fig. 2-2

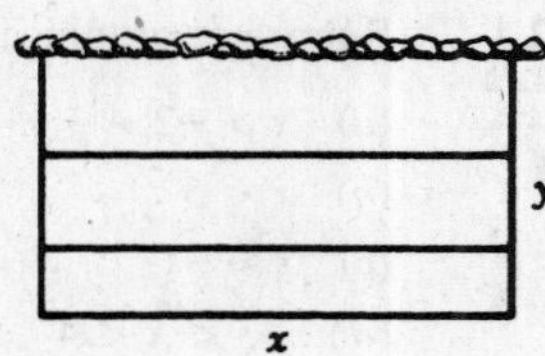

Fig. 2-3

2.7 A farmer has 600 ft of woven wire fencing available to enclose a rectangular field and to divide it into three parts by two fences parallel to one end. If x ft of stone wall is used as one side of the field, express the area enclosed as a function of x when the dividing fences are parallel to the stone wall. Refer to Fig. 2-3.

The dimensions of the field are x and y ft where $3x + 2y = 600$. Then $y = \frac{1}{2}(600 - 3x)$ and the required area is

$$A = xy = x \cdot \frac{1}{2}(600 - 3x) = \frac{3}{2}x(200 - x)$$

2.8 A right cylinder is said to be inscribed in a sphere if the circumferences of the bases of the cylinder are in the surface of the sphere. If the sphere has radius R, express the volume of the inscribed right circular cylinder as a function of the radius r of its base.

Let the height of the cylinder be denoted by $2h$. From Fig. 2-4, $h = \sqrt{R^2 - r^2}$ and the required volume is

$$V = \pi r^2 \cdot 2h = 2\pi r^2 \sqrt{R^2 - r^2}$$

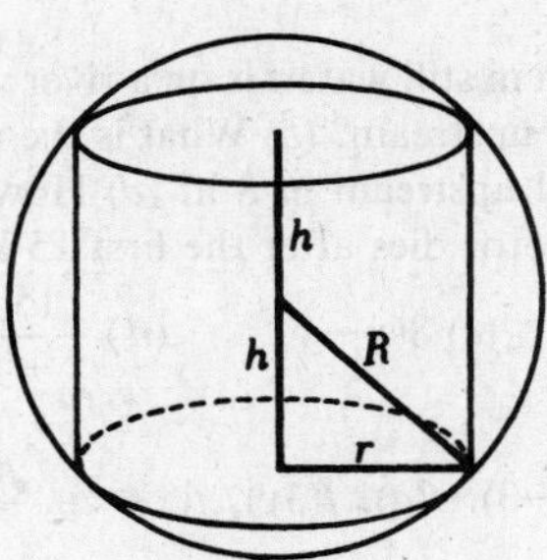

Fig. 2-4

2.9 Given $z = f(x, y) = 2x^2 + 3y^2 - 4$, find

(a) $f(0, 0) = 2(0)^2 + 3(0)^2 - 4 = -4$ (b) $f(2, -3) = 2(2)^2 + 3(-3)^2 - 4 = 31$

(c) $f(-x, -y) = 2(-x)^2 + 3(-y)^2 - 4 = 2x^2 + 3y^2 - 4 = f(x, y)$

Supplementary Problems

2.10 Represent graphically each of the following domains:

(a) $x > -3$ (c) $x \geqslant 0$ (e) $|x| < 2$ (g) $-4 \leqslant x \leqslant 4$

(b) $x < 5$ (d) $-3 < x < -1$ (f) $|x| \geqslant 0$ (h) $x < -3, x \geqslant 5$

2.11 In the three angles, A, B, C of a triangle, angle B exceeds twice angle A by 15°. Express the measure of angle C in terms of angle A.

Ans. $C = 165° - 3A$

2.12 A grocer has two grades of coffee, selling at \$9.00 and \$10.50 per pound, respectively. In making a mixture of 100 lbs, he uses x lbs of the \$10.50 coffee. (a) How many pounds of the \$9.00 coffee does he use? (b) What is the value in dollars of the mixture? (c) At what price per pound should he offer the mixture?

Ans. (a) $100 - x$ (b) $9(100 - x) + 10.5x$ (c) $9 + 0.015x$

2.13 In a purse are nickels, dimes, and quarters. The number of dimes is twice the number of quarters and the number of nickels is three less than twice the number of dimes. If there are x quarters, find the sum (in cents) in the purse.

Ans. $65x - 15$

2.14 A and B start from the same place. A walks 4 km/h and B walks 5 km/h. (a) How far (in miles) will each walk in x h? (b) How far apart will they be after x h if they leave at the same time and move in opposite directions? (c) How far apart will they be after A has walked $x > 2$ hours if they move in the same direction but B leaves 2 h after A? (d) In (c), for how many hours would B have to walk in order to overtake A?

Ans. (a) $A, 4x; B, 5x$ (b) $9x$ (c) $|4x - 5(x - 2)|$ (d) 8

2.15 A motor boat, which moves at x km/h in still water, is on a river whose current is $y < x$ km/h. (*a*) What is the rate (km/h) of the boat when moving upstream? (*b*) What is the rate of the boat when moving downstream? (*c*) How far (km) will the boat travel upstream in 8 h? (*d*) How long (hours) will it take the boat moving downstream to cover 20 km if the motor dies after the first 15 km?

Ans. (*a*) $x - y$ (*b*) $x + y$ (*c*) $8(x - y)$ (*d*) $\dfrac{15}{x+y} + \dfrac{5}{y}$

2.16 Given $f(x) = \dfrac{x-3}{x+2}$, find $f(0)$, $f(1)$, $f(-3)$, $f(a)$, $f(3y)$, $f(x+a)$, $\dfrac{f(x+a) - f(x)}{a}$.

Ans. $-\frac{3}{2}, -\frac{2}{3}, 6, 0, \dfrac{a-3}{a+2}, \dfrac{3y-3}{3y+2}, \dfrac{x+a-3}{x+a+2}, \dfrac{5}{(x+2)(x+a+2)}$

2.17 A ladder 25 ft long leans against a vertical wall with its foot on level ground 7 ft from the base of the wall. If the foot is pulled away from the wall at the rate 2 ft/s, express the distance (y ft) of the top of the ladder above the ground as a function of the time t seconds in moving.

Ans. $y = 2\sqrt{144 - 7t - t^2}$

2.18 A boat is tied to a dock by means of a cable 60 m long. If the dock is 20 m above the water and if the cable is being drawn in at the rate 10 m/min, express the distance y m of the boat from the dock after t min.

Ans. $y = 10\sqrt{t^2 - 12t + 32}$

2.19 A train leaves a station at noon and travels east at the rate 30 km/h. At 2 P.M. of the same day a second train leaves the station and travels south at the rate 25 km/h. Express the distance d (km) between the trains as a function of t (hours), the time the second train has been traveling.

Ans. $d = 5\sqrt{61t^2 + 144t + 144}$

2.20 For each function, tell whether it is a bijection:

(*a*) $y = x^4$

(*b*) $y = \sqrt{x}$

(*c*) $y = 2x^2 + 3$

Ans. (*a*) No (*b*) Yes (*c*) No

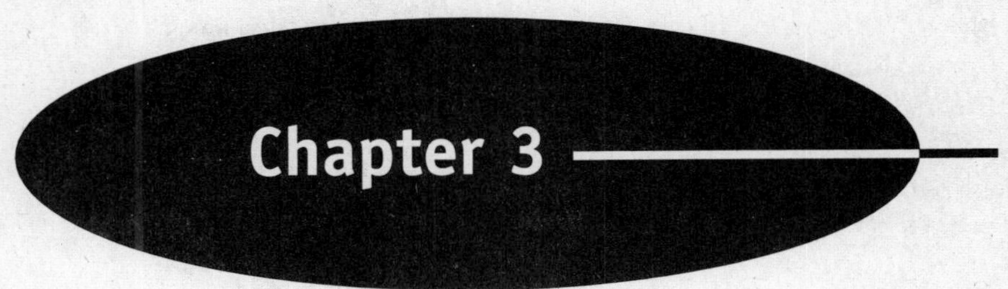

Chapter 3

Graphs of Functions

A FUNCTION $y = f(x)$, by definition, yields a collection of pairs $(x, f(x))$ or (x, y) in which x is any element in the domain of the function and $f(x)$ or y is the corresponding value of the function. These pairs are called *ordered pairs.*

EXAMPLE 1. Obtain 10 ordered pairs for the function $y = 3x - 2$.

The domain of definition of the function is the set of real numbers. We may choose at random any 10 real numbers as values of x. For one such choice, we obtain the chart in Table 3.1.

Table 3.1

x	-2	$-\frac{4}{3}$	$-\frac{1}{2}$	0	$\frac{1}{3}$	1	2	$\frac{5}{2}$	3	4
y	-8	-6	$-\frac{7}{2}$	-2	-1	1	4	$\frac{11}{2}$	7	10

(See Problem 3.1.)

THE RECTANGULAR CARTESIAN COORDINATE SYSTEM in a plane is a device by which there is established a one-to-one correspondence between the points of the plane and ordered pairs of real numbers (a, b).

Consider two real number scales intersecting at right angles in O, the origin of each (see Fig. 3-1), and having the positive direction on the horizontal scale (now called the *x axis*) directed to the right and the positive direction on the vertical scale (now called the y axis) directed upward.

Let P be any point distinct from O in the plane of the two axes and join P to O by a straight line. Let the projection of OP on the x axis be $OM = a$ and the projection of OP on the y axis be $ON = b$. Then the pair of numbers (a, b) in that order are called the plane rectangular Cartesian coordinates (briefly, the rectangular coordinates) of P. In particular, the coordinates of O, the *origin* of the coordinate system, are $(0, 0)$.

The first coordinate, giving the directed distance of P from the y axis, is called the *abscissa* of P, while the second coordinate, giving the directed distance of P from the x axis, is called the *ordinate* of P. Note carefully that the points $(3, 4)$ and $(4, 3)$ are distinct points.

The axes divide the plane into four sections, called *quadrants*. Figure 3-1 shows the customary numbering of the quadrants and the respective signs of the coordinates of a point in each quadrant. (See Problems 3.1–3.4.)

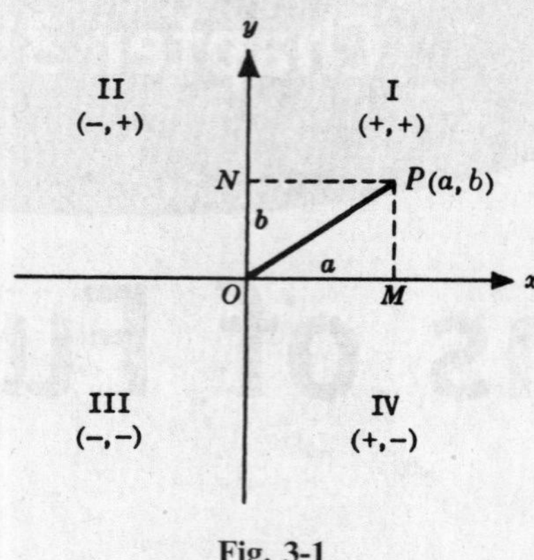

Fig. 3-1

THE GRAPH OF A FUNCTION $y = f(x)$ consists of the totality of points (x, y) whose coordinates satisfy the relation $y = f(x)$.

EXAMPLE 2. Graph the function $3x - 2$.

After plotting the points whose coordinates (x, y) are given in Table 3.1, it appears that they lie on a straight line. See Fig. 3-2. Figure 3-2 is not the complete graph since (1000, 2998) is one of its points and is not shown. Moreover, although we have joined the points by a straight line, we have not proved that every point on the line has as coordinates a number pair given by the function. These matters as well as such questions as: What values of x should be chosen? How many values of x are needed? will become clearer as we proceed with the study of functions. At present,

(1) Build a table of values.
(2) Plot the corresponding points.
(3) Pass a smooth curve through these points, moving from left to right.

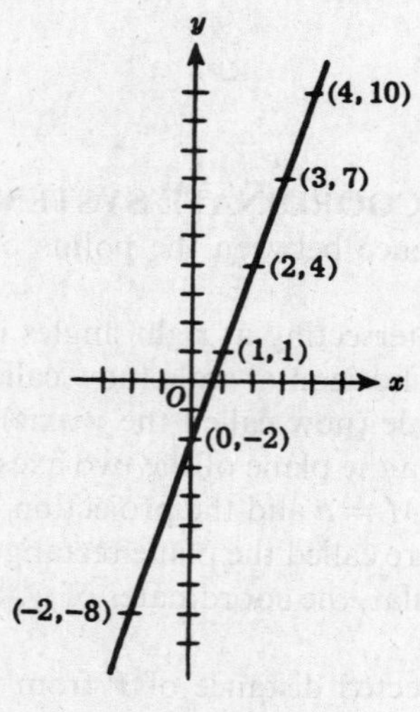

Fig. 3-2

It is helpful to picture the curve in your mind before attempting to trace it on paper. If there is doubt about the curve between two plotted points, determine other points in the interval.

ANY VALUE OF x for which the corresponding value of function $f(x)$ is zero is called a *zero* of the function. Such values of x are also called *roots* of the equation $f(x) = 0$. The real roots of an equation $f(x) = 0$ may be approximated by estimating from the graph of $f(x)$ the abscissas of its points of intersection with the x axis. (See Problems 3.9–3.11.)

Algebraic methods for finding the roots of equations will be treated in later chapters. The graphing calculator can also be used to find roots by graphing the function and observing where the graph intersects the x axis. See Appendix A.

Solved Problems

3.1 (*a*) Show that the points $A(1, 2)$, $B(0, -3)$, and $C(2, 7)$ are on the graph of $y = 5x - 3$.
(*b*) Show that the points $D(0, 0)$ and $E(-1, -2)$ are not on the graph of $y = 5x - 3$.

(*a*) The point $A(1, 2)$ is on the graph since $2 = 5(1) - 3$, $B(0, -3)$ is on the graph since $-3 = 5(0) - 3$, and $C(2, 7)$ is on the graph since $7 = 5(2) - 3$.

(*b*) The point $D(0,0)$ is not on the graph since $0 \neq 5(0) - 3$, and $E(-1, -2)$ is not on the graph since $-2 \neq 5(-1) - 3$.

3.2 Sketch the graph of the function $2x$. Refer to Table 3.2.

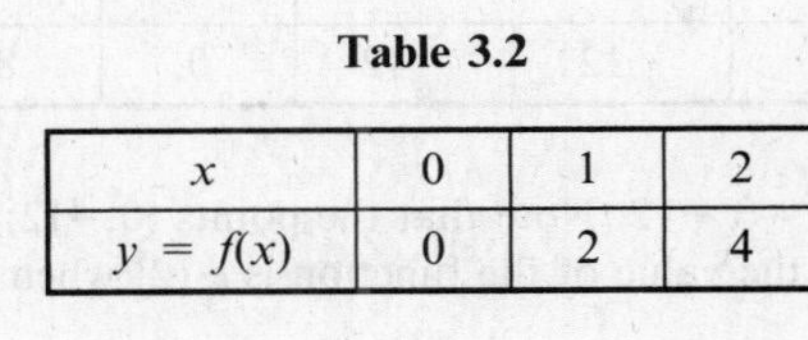

Table 3.2

x	0	1	2
$y = f(x)$	0	2	4

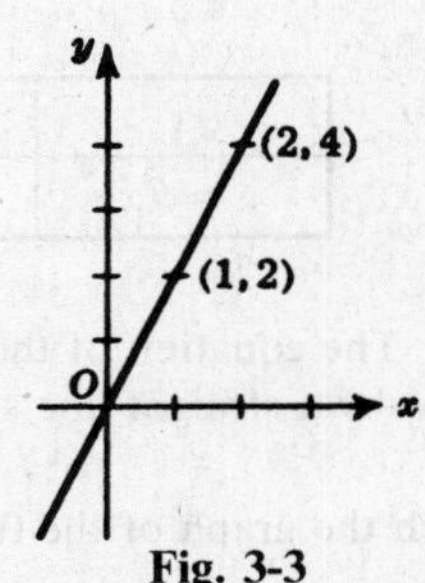

Fig. 3-3

This is a linear function and its graph is a straight line. For this graph only two points are necessary. Three points are used to provide a check. See Fig. 3-3. The equation of the line is $y = 2x$.

3.3 Sketch the graph of the function $6 - 3x$. Refer to Table 3.3.

Table 3.3

x	0	2	3
$y = f(x)$	6	0	-3

See Fig. 3-4. The equation of the line is $y = 6 - 3x$.

3.4 Sketch the graph of the function x^2. Refer to Table 3.4.

Table 3.4

x	3	1	0	−2	−3
$y = f(x)$	9	1	0	4	9

See Fig. 3-5. The equation of this graph, called a *parabola*, is $y = x^2$. Note for $x \neq 0$, $x^2 > 0$. Thus, the curve is never below the x axis. Moreover, as $|x|$ increase, x^2 increases; that is, as we move from the origin along the x axis in either direction, the curve moves farther and farther from the axis. Hence, in sketching parabolas sufficient points must be plotted so that their U shape can be seen.

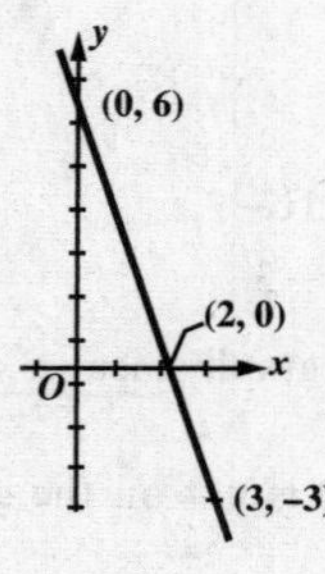

Fig. 3-4

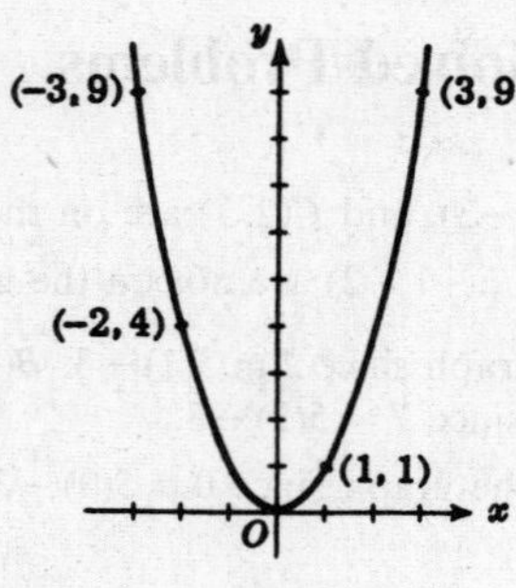

Fig. 3-5

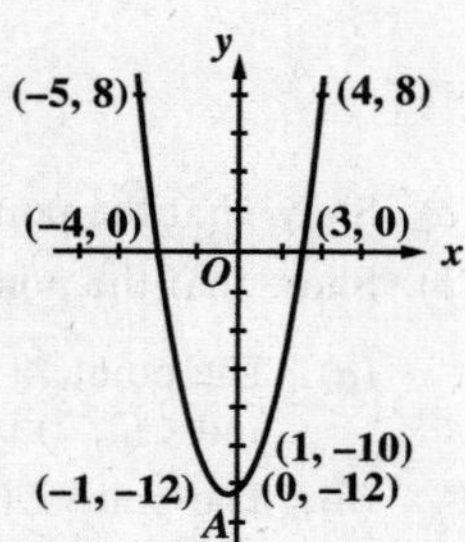

Fig. 3-6

3.5 Sketch the graph of the function $x^2 + x - 12$. Refer to Table 3.5.

Table 3.5

x	4	3	1	0	−1	−4	−5
$y = f(x)$	8	0	−10	−12	−12	0	8

The equation of the parabola is $y = x^2 + x - 12$. Note that the points $(0, -12)$ and $(-1, -12)$ are *not* joined by a straight line segment. Check that the value of the function is $-12\frac{1}{4}$ when $x = -\frac{1}{2}$. See Fig. 3-6.

3.6 Sketch the graph of the function $-2x^2 + 4x + 1$. Refer to Table 3.6.

Table 3.6

x	3	2	1	0	−1
$y = f(x)$	−5	1	3	1	−5

See Fig. 3-7.

3.7 Sketch the graph of the function $(x + 1)(x - 1)(x - 2)$. Refer to Table 3.7.

Table 3.7

x	3	2	$\frac{3}{2}$	1	0	−1	−2
$y = f(x)$	8	0	$-\frac{5}{8}$	0	2	0	−12

This is a *cubic* curve of the equation $y = (x + 1)(x - 1)(x - 2)$. It crosses the x axis where $x = -1, 1,$ and 2. See Fig. 3-8.

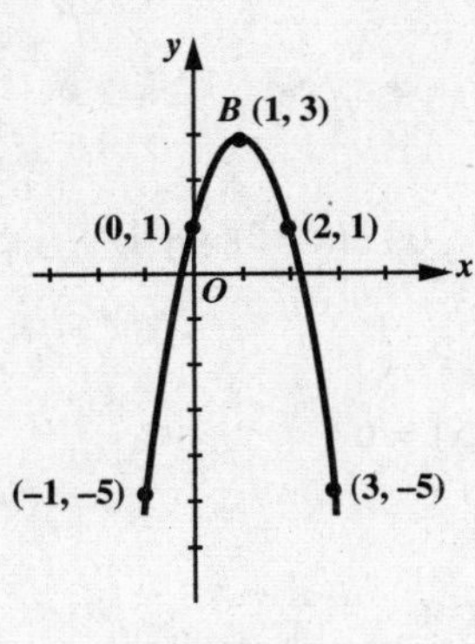

Fig. 3-7

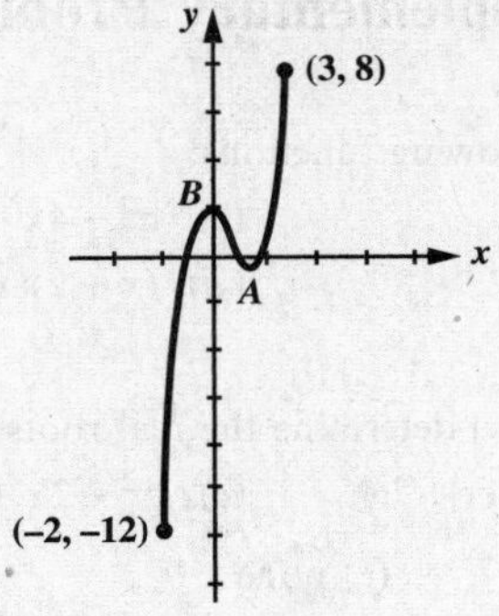

Fig. 3-8

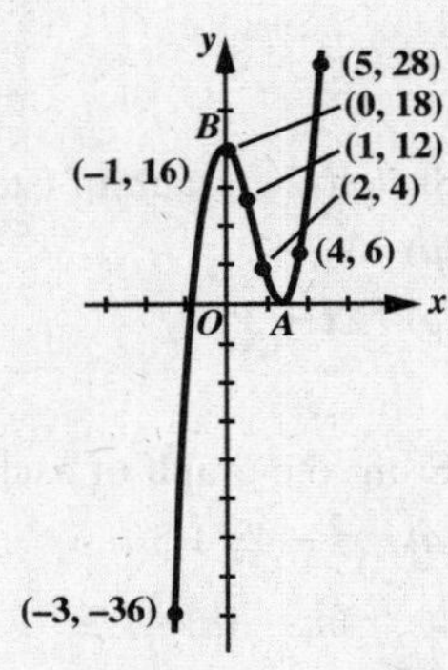

Fig. 3-9

3.8 Sketch the graph of the function $(x+2)(x-3)^2$. Refer to Table 3.8.

Table 3.8

x	5	4	$\frac{7}{2}$	3	2	1	0	−1	−2	−3
$y = f(x)$	28	6	$\frac{11}{8}$	0	4	12	18	16	0	−36

This cubic crosses the x axis where $x = -2$ and is tangent to the x axis where $x = 3$. Note that for $x > -2$, the value of the function is positive except for $x = 3$, where it is 0. Thus, to the right of $x = -2$, the curve is *never* below the x axis. See Fig. 3-9.

3.9 Sketch the graph of the function $x^2 + 2x - 5$ and by means of it determine the real roots of $x^2 + 2x - 5 = 0$. Refer to Table 3.9.

Table 3.9

x	2	1	0	−1	−2	−3	−4
$y = f(x)$	3	−2	−5	−6	−5	−2	3

The parabola cuts the x axis at a point whose abscissa is between 1 and 2 (the value of the function changes sign) and at a point whose abscissa is between −3 and −4.

Reading from the graph in Fig. 3-10, the roots are $x = 1.5$ and $x = -3.5$, approximately.

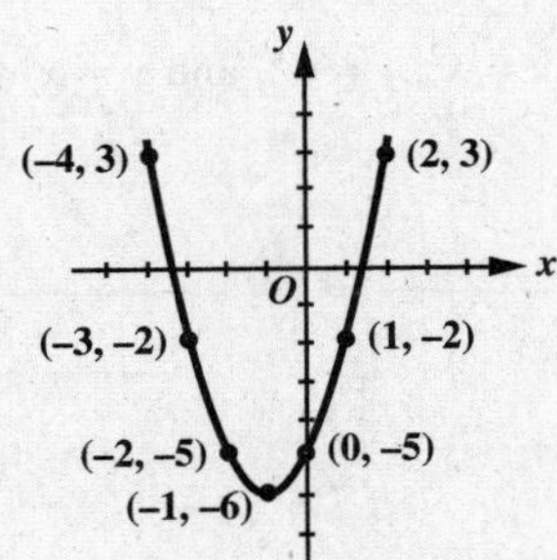

Fig. 3-10

Supplementary Problems

3.10 Sketch the graph of each of the following functions:

(*a*) $3x - 2$ (*c*) $x^2 - 1$ (*e*) $x^2 - 4x + 4$ (*g*) $(x-2)(x+1)^2$

(*b*) $2x + 3$ (*d*) $4 - x^2$ (*f*) $(x+2)(x-1)(x-3)$

3.11 From the graph of each function $f(x)$ determine the real roots, if any, of $f(x) = 0$.

(*a*) $x^2 - 4x + 3$ (*b*) $2x^2 + 4x + 1$ (*c*) $x^2 - 2x + 4$

Ans. (*a*) 1,3 (*b*) −0.3, −1.7 (*c*) none

3.12 If A is a point on the graph of $y = f(x)$, the function being restricted to the type considered in this chapter, and if all points of the graph sufficiently near A are higher than A (that is, lie above the horizontal drawn through A), then A is called a *relative minimum point* of the graph. (*a*) Verify that the origin is the relative minimum point of the graph of Problem 3.4. (*b*) Verify that the graph of Problem 3.5 has a relative minimum at a point whose abscissa is between $x = -1$ and $x = 0$ (at $x = -\frac{1}{2}$), the graph of Problem 3.7 has a relative minimum at a point whose abscissa is between $x = 1$ and $x = 2$ (approximately $x = 1.5$), and the graph of Problem 3.8 has (3, 0) as relative minimum point. Also, see Chapter 16 for a more sophisticated discussion of minima.

3.13 If B is a point on the graph of $y = f(x)$ and if all points of the graph sufficiently near B are lower than B (that is, lie below the horizontal drawn through B), then B is called a *relative maximum point* of the graph. (*a*) Verify that (1, 3) is the relative maximum point of the graph of Problem 3.6. (*b*) Verify that the graph of Problem 3.7 has a relative maximum at a point whose abscissa is between $x = -1$ and $x = 1$ (approximately $x = -0.2$), and that the graph of Problem 3.8 has a relative maximum between $x = -1$ and $x = 0$ (at $x = -\frac{1}{3}$). See Chapter 16 for additional work on extrema.

3.14 Verify that the graphs of the functions of Problem 3.11 have relative minimums at $x = 2$, $x = -1$, and $x = 1$, respectively.

3.15 From the graph of the function of Problem 2.4 in Chapter 2 read that the area of the rectangle is a relative maximum when $x = \frac{15}{2}$.

3.16 From the graph of the function of Problem 2.7 in Chapter 2 read that the area enclosed is a relative maximum when $x = 100$.

3.17 Use a graphing calculator to locate the zeros of the function $y = x^2 + 3$.

3.18 Use a graphing calculator to graph $y = x^2$, $y = x^4$, and $y = x^6$ on the same axes. What do you notice?

Linear Equations

AN EQUATION is a statement, such as (*a*) $2x - 6 = 4 - 3x$, (*b*) $y^2 + 3y = 4$, and (*c*) $2x + 3y = 4xy + 1$, that two expressions are equal. An equation is linear in an unknown if the highest degree of that unknown in the equation is 1. An equation is quadratic in an unknown if the highest degree of that unknown is 2. The first is a *linear* equation in one unknown, the second is a *quadratic* in one unknown, and the third is linear in each of the two unknowns but is of degree 2 in the two unknowns.

Any set of values of the unknowns for which the two members of an equation are equal is called a *solution* of the equation. Thus, $x = 2$ is a solution of (*a*), since $2(2) - 6 = 4 - 3(2)$; $y = 1$ and $y = -4$ are solutions of (*b*); and $x = 1, y = 1$ is a solution of (*c*). A solution of an equation in one unknown is also called a *root* of the equation.

TO SOLVE A LINEAR EQUATION in one unknown, perform the same operations on both members of the equation in order to obtain the unknown alone in the left member.

EXAMPLE 1. Solve: $2x - 6 = 4 - 3x$.

Add 6: $2x = 10 - 3x$ Check: $2(2) - 6 = 4 - 3(2)$
Add $3x$: $5x = 10$ $-2 = -2$
Divide by 5: $x = 2$

EXAMPLE 2. Solve: $\frac{1}{3}x - \frac{1}{2} = \frac{3}{4}x + \frac{5}{6}$.

Multiply by LCD = 12: $4x - 6 = 9x + 10$ Check: $\frac{1}{3}(-\frac{16}{5}) - \frac{1}{2} = \frac{3}{4}(-\frac{16}{5}) + \frac{5}{6}$
Add $6 - 9x$: $-5x = 16$
Divide by -5: $x = -\frac{16}{5}$ $-\frac{47}{30} = -\frac{47}{30}$

(See Problems 4.1–4.3.)

An equation which contains fractions having the unknown in one or more denominators may sometimes reduce to a linear equation when cleared of fractions. When the resulting equation is solved, the solution *must* be checked since it may or may not be a root of the original equation. (See Problems 4.4–4.8.)

RATIO AND PROPORTION. The ratio of two quantities is their quotient. The ratio of 1 inch to 1 foot is 1/12 or 1:12, a pure number; the ratio of 30 miles to 45 minutes is $30/45 = 2/3$ mile per minute.

The expressed equality of two ratios, as $\frac{a}{b} = \frac{c}{d}$, is called a *proportion*. (See Problem 4.11.)

VARIATION. A variable y is said to vary *directly* as another variable x (or y is proportional to x) if y is equal to some constant c times x, that is, if $y = cx$.

A variable y is said to vary *inversely* as another variable x if y varies directly as the reciprocal of x, that is, if $y = c/x$.

Solved Problems

Solve and check the following equations. The check has been omitted in certain problems.

4.1 $x - 2(1 - 3x) = 6 + 3(4 - x)$.

$$x - 2 + 6x = 6 + 12 - 3x$$
$$7x - 2 = 18 - 3x$$
$$10x = 20$$
$$x = 2$$

4.2 $ay + b = cy + d$.

$$ay - cy = d - b$$
$$(a - c)y = d - b$$
$$y = \frac{d - b}{a - c}$$

4.3 $\dfrac{3x - 2}{5} = 4 - \dfrac{1}{2}x$.

Multiply by 10: $6x - 4 = 40 - 5x$ Check: $\dfrac{3(4) - 2}{5} = 4 - \dfrac{1}{2}(4)$

$$11x = 44 \qquad 2 = 2$$
$$x = 4$$

4.4 $\dfrac{3x + 1}{3x - 1} = \dfrac{2x + 1}{2x - 3}$. Here the LCD is $(3x - 1)(2x - 3)$.

Multiply by LCD: $(3x + 1)(2x - 3) = (2x + 1)(3x - 1)$

$$6x^2 - 7x - 3 = 6x^2 + x - 1$$
$$-8x = 2$$
$$x = -\frac{1}{4}$$

Check: $\dfrac{3(-\frac{1}{4}) + 1}{3(-\frac{1}{4}) - 1} = \dfrac{2(-\frac{1}{4}) + 1}{2(-\frac{1}{4}) - 3}$,

$$\frac{-3 + 4}{-3 - 4} = \frac{-2 + 4}{-2 - 12}, \qquad -\frac{1}{7} = -\frac{1}{7}$$

4.5 $\dfrac{1}{x-3}-\dfrac{1}{x+1}=\dfrac{3x-2}{(x-3)(x+1)}$. Here the LCD is $(x-3)(x+1)$.

$$(x+1)-(x-3)=3x-2$$
$$-3x=-6$$
$$x=2$$

Check: $-1-\dfrac{1}{3}=\dfrac{6-2}{-3}, \quad -\dfrac{4}{3}=-\dfrac{4}{3}$

4.6 $\dfrac{1}{x-3}+\dfrac{1}{x-2}=\dfrac{3x-8}{(x-3)(x-2)}$. The LCD is $(x-3)(x-2)$.

$$(x-2)+(x-3)=3x-8$$
$$2x-5=3x-8$$
$$x=3$$

Check: When $x=3$, $\dfrac{1}{x-3}$ is without meaning. The given equation has no root. The value $x=3$ is called *extraneous*.

4.7 $\dfrac{x^2-2}{x-1}=x+1-\dfrac{1}{x-1}$.

$$x^2-2=(x+1)(x-1)-1$$
$$=x^2-2$$

The given equation is satisfied by all values of x except $x=1$. It is called an identical equation or *identity*.

4.8 $\dfrac{1}{x-1}+\dfrac{1}{x-3}=\dfrac{2x-5}{(x-1)(x-3)}$.

$$(x-3)+(x-1)=2x-5$$
$$2x-4=2x-5$$

There is no solution. $2x-4$ and $2x-5$ are unequal for all values of x.

4.9 One number is 5 more than another and the sum of the two is 71. Find the numbers.

Let x be the smaller number and $x+5$ be the larger. Then $x+(x+5)=71$, $2x=66$, and $x=33$. The numbers are 33 and 38.

4.10 A father is now three times as old as his son. Twelve years ago he was six times as old as his son. Find the present age of each.

Let $x=$ the age of the son and $3x=$ the age of the father. Twelve years ago, the age of the son was $x-12$ and the age of the father was $3x-12$.

Then $3x-12=6(x-12)$, $3x=60$, and $x=20$. The present age of the son is 20 and that of the father is 60.

4.11 When two pulleys are connected by a belt, their angular velocities (revolutions per minute) are *inversely* proportional to their diameters; that is, $\omega_1:\omega_2=d_2:d_1$. Find the velocity of a pulley 15 cm in diameter when it is connected to a pulley 12 cm in diameter and rotating at 100 rad/s.

Let ω_1 be the unknown velocity; then $d_1=15$, $\omega_2=100$, and $d_2=12$. The given formula becomes

$$\frac{\omega_1}{100}=\frac{12}{15} \quad \text{and} \quad \omega_1=\frac{12}{15}(100)=80 \text{ rad/s}$$

Supplementary Problems

4.12 Solve for x and check each of the following:

(a) $2x - 7 = 29 - 4x$

(b) $2(x-1) - 3(x-2) + 4(x-3) = 0$

(c) $\dfrac{x+3}{x-3} = 3$

(d) $\dfrac{4}{x-4} = \dfrac{2}{2x-5}$

(e) $\dfrac{2x+1}{4} - \dfrac{1}{x-1} = \dfrac{x}{2}$

(f) $a(x+3) + b(x-2) = c(x-1)$

Ans. (a) 6 (b) $\dfrac{8}{3}$ (c) 6 (d) 2 (e) 5 (f) $\dfrac{2b-3a-c}{a+b-c}$

4.13 A piece of wire $11\frac{2}{3}$ m long is to be divided into two parts such that one part is $\frac{2}{3}$ that of the other. Find the length of the shorter piece.

Ans. $4\frac{2}{3}$ m

4.14 A train leaves a station and travels at the rate of 40 km/h. Two hours later a second train leaves the station and travels at the rate of 60 km/h. Where will the second train overtake the first?

Ans. 240 km from the station

4.15 A tank is drained by two pipes. One pipe can empty the tank in 30 min, and the other can empty it in 25 min. If the tank is $\frac{5}{6}$ filled and both pipes are open, in what time will the tank be emptied?

Ans. $11\frac{4}{11}$ min

4.16 A man invests $\frac{1}{3}$ of his capital at 6% and the remainder at 8%. What is his capital if his total income is $4400?

Ans. $60,000

4.17 A can do a piece of work in 10 days. After he has worked 2 days, B comes to help him and together they finish it in 3 days. In how many days could B alone have done the work?

Ans. 6 days

4.18 How soon after noon are the hands of a clock together again?

Ans. 1 hr, $5\frac{5}{11}$ min

Chapter 5

Simultaneous Linear Equations

TWO LINEAR EQUATIONS IN TWO UNKNOWNS. Let the system of equations be

$$\begin{cases} a_1x + b_1y + c_1 = 0 \\ a_2x + b_2y + c_2 = 0 \end{cases}$$

Each equation has an unlimited number of solutions (x, y) corresponding to the unlimited number of points on the locus (straight line) which it represents. Our problem is to find all solutions common to the two equations or the coordinates of all points common to the two lines. There are three cases:

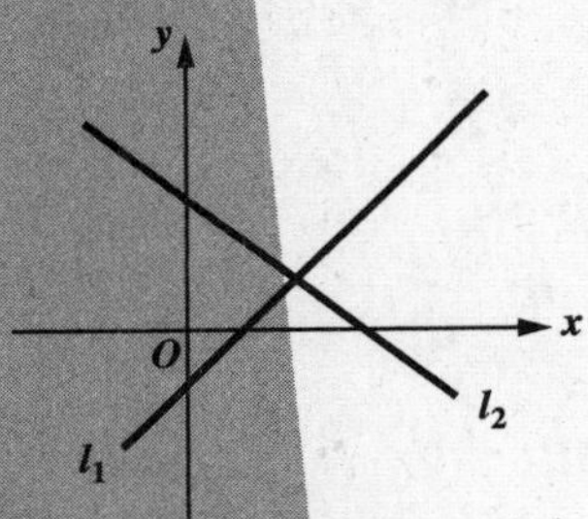

Fig. 5-1

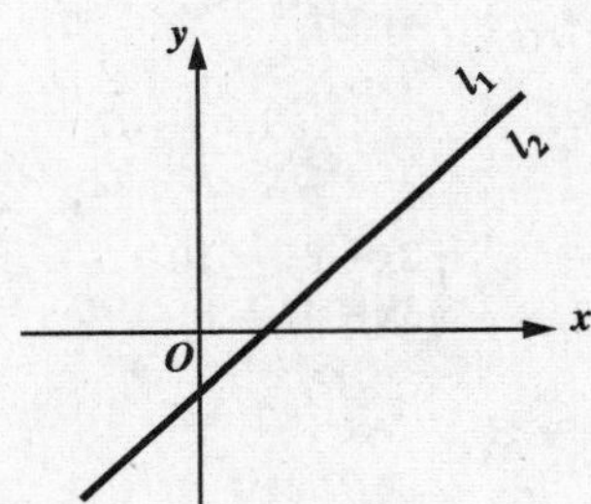

Fig. 5-2

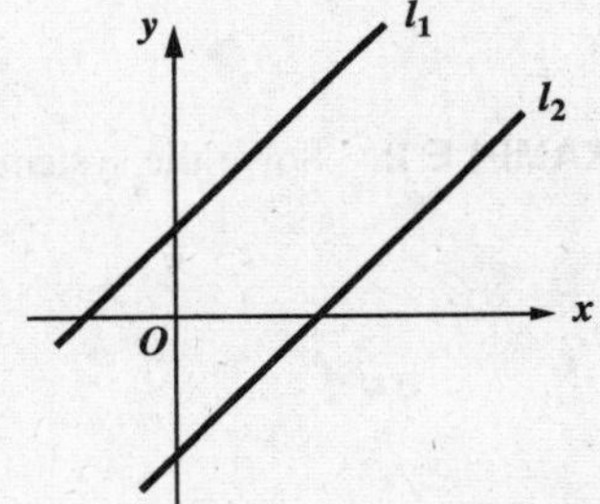

Fig. 5-3

(1) The system has one and only one solution; that is, the two lines have one and only one point in common. The equations are said to be *consistent* (have common solutions) *and independent*. See Fig. 5-1, indicating two distinct intersecting lines.

(2) The system has an unlimited number of solutions; that is, the two equations are equivalent or the two lines are coincident. The equations are said to be *consistent and dependent*. See Fig. 5-2, indicating that the two equations represent the same line.

(3) The system has no solution; that is, the two lines are parallel and distinct. The equations are said to be *inconsistent*. See Fig. 5-3, indicating that the two equations result in two parallel lines.

GRAPHICAL SOLUTION. We plot the graphs of the two equations on the same axes and scale off the coordinates of the point of intersection. The defect of this method is that, in general, only approximate solutions are obtained. (See Problem 5.1.)

ALGEBRAIC SOLUTION. A system of two consistent and independent equations in two unknowns may be solved algebraically by eliminating one of the unknowns.

EXAMPLE 1. Solve the system

$$\begin{cases} 3x - 6y = 10 & (5.1) \\ 9x + 15y = -14 & (5.2) \end{cases}$$

ELIMINATION BY SUBSTITUTION

Solve (*5.1*) for x:

$$x = \tfrac{10}{3} + 2y \qquad (5.3)$$

Substitute in (*5.2*):

$$9\left(\tfrac{10}{3} + 2y\right) + 15y = -14$$

$$30 + 18y + 15y = -14 \qquad 33y = -44 \qquad y = -\tfrac{4}{3}$$

Substitute for y in (*5.3*):

$$x = \tfrac{10}{3} + 2\left(-\tfrac{4}{3}\right) = \tfrac{2}{3}$$

Check: Using (*5.2*),

$$9\left(\tfrac{2}{3}\right) + 15\left(-\tfrac{4}{3}\right) = -14$$

EXAMPLE 2. Solve the system

$$\begin{cases} 2x - 3y = 10 & (5.4) \\ 3x - 4y = 8 & (5.5) \end{cases}$$

ELIMINATION BY ADDITION

Multiply (*5.4*) by -3 and (*5.5*) by 2:

$$\begin{aligned} -6x + 9y &= -30 \\ 6x - 8y &= 16 \\ \hline \text{Add:} \qquad y &= -14 \end{aligned}$$

Substitute for x in (*5.4*):

$$2x + 42 = 10 \quad \text{or} \quad x = -16$$

Check: Using (*5.5*),

$$3(-16) - 4(-14) = 8$$

(See Problems 5.2–5.4.)

THREE LINEAR EQUATIONS IN THREE UNKNOWNS. A system of three consistent and independent equations in three unknowns may be solved algebraically by deriving from it a system of two equations in two unknowns.

EXAMPLE 3. Solve the system

$$\begin{cases} 2x + 3y - 4z = 1 & (5.6) \\ 3x - y - 2z = 4 & (5.7) \\ 4x - 7y - 6z = -7 & (5.8) \end{cases}$$

We shall eliminate y.

$$\begin{aligned} \text{Rewrite } (5.6)\text{:} \quad & 2x + 3y - 4z = 1 \\ 3 \times (5.7)\text{:} \quad & 9x - 3y - 6z = 12 \\ \hline \text{Add:} \quad & 11x \qquad - 10z = 13 \qquad (5.9) \end{aligned}$$

$$\begin{aligned} \text{Rewrite } (5.8)\text{:} \quad & 4x - 7y - 6z = -7 \\ -7 \times (5.8)\text{:} \quad & -21x + 7y + 14z = -28 \\ \hline \text{Add:} \quad & -17x \qquad + 8z = -35 \qquad (5.10) \end{aligned}$$

Next, solve (5.9) and (5.10).

$$\begin{aligned} 4 \times (5.9)\text{:} \quad & 44x - 40z = 52 \\ 5 \times (5.10)\text{:} \quad & -85x + 40z = -175 \\ \hline \text{Add:} \quad & -41x \qquad = -123 \\ & x = 3 \end{aligned}$$

From (5.9): $11(3) - 10z = 13 \quad z = 2$
From (5.6): $2(3) + 3y - 4(2) = 1 \quad y = 1$
Check: Using (5.7),

$$3(3) - 1 - 2(2) = 4$$

(See Problems 5.5–5.6.)

SOLUTIONS OF LINEAR SYSTEMS USING DETACHED COEFFICIENTS. In Example 4 below, a variation of the method of addition and subtraction is used to solve a system of linear equations. On the left the equations themselves are used, while on the right the same moves are made on the rectangular array (called a *matrix*) of the coefficients and constant terms. The numbering (1), (2), (3), ... refers both to the equations and to the rows of the matrices.

EXAMPLE 4. Solve the system

$$\begin{cases} 2x - 3y = 2 \\ 4x + 7y = -9 \end{cases}$$

USING EQUATIONS		USING MATRICES
$2x - 3y = 2$ $4x + 7y = -9$	(1) (2)	$\left(\begin{array}{cc\|c} 2 & -3 & 2 \\ 4 & 7 & -9 \end{array}\right)$

Multiply (1) by $\frac{1}{2}$ and write as (3). Multiply (1) by -2 and add to (2) to obtain (4).

$$x - \tfrac{3}{2}y = 1 \quad (3) \qquad \left(\begin{array}{cc|c} 1 & -\tfrac{3}{2} & 1 \\ 0 & 13 & -13 \end{array}\right)$$
$$13y = -13 \quad (4)$$

Multiply (*4*) by $\frac{3}{2}/13 = \frac{3}{26}$ and add to (*3*) to obtain (*5*). Multiply (*4*) by $\frac{1}{13}$ to obtain (*6*).

$$x = -\tfrac{1}{2} \quad (5) \qquad \left(\begin{array}{cc|c} 1 & 0 & -\tfrac{1}{2} \\ 0 & 1 & -1 \end{array}\right)$$
$$y = -1 \quad (6)$$

The required solution is $x = -\frac{1}{2}$, $y = -1$.

EXAMPLE 5. Solve, using matrices, the system

$$\begin{cases} 2x - 3y + 2z = 14 \\ 4x + 4y - 3z = 6 \\ 3x + 2y - 3z = -2 \end{cases}$$

The matrix of the system

$$\left(\begin{array}{ccc|c} 2 & -3 & 2 & 14 \\ 4 & 4 & -3 & 6 \\ 3 & 2 & -3 & -2 \end{array}\right)$$

is formed by writing in order the coefficients of x, y, z and the constant terms.

There are, in essence, three moves:

(*a*) Multiply the elements of a row by a nonzero number. This move is used only to obtain an element 1 in a prescribed position.
(*b*) Multiply the elements of a row by a nonzero number and add to the corresponding elements of another row. This move is used to obtain an element 0 in a prescribed position.
(*c*) Exchange two rows when required.

The first attack must be planned to yield a matrix of the form

$$\left(\begin{array}{ccc|c} 1 & * & * & * \\ 0 & * & * & * \\ 0 & * & * & * \end{array}\right)$$

in which only the elements of the first column are prescribed.

Multiply first row by $\frac{1}{2}$:
Multiply first row by -2 and add to second row:
Multiply first row by $-\frac{3}{2}$ and add to third row:

$$\left(\begin{array}{ccc|c} 1 & -\tfrac{3}{2} & 1 & 7 \\ 0 & 10 & -7 & -22 \\ 0 & \tfrac{13}{2} & -6 & -23 \end{array}\right)$$

The second attack must be planned to yield a matrix of the form

$$\left(\begin{array}{ccc|c} 1 & 0 & * & * \\ 0 & 1 & * & * \\ 0 & 0 & * & * \end{array}\right)$$

in which the elements of the first two columns are prescribed.

Multiply second row by $\frac{3}{20}$ and add to first row:
Multiply second row by $\frac{1}{10}$:
Multiply second row by $-\frac{13}{20}$ and add to third row:

$$\left(\begin{array}{ccc|c} 1 & 0 & -\tfrac{1}{20} & \tfrac{37}{10} \\ 0 & 1 & -\tfrac{7}{10} & -\tfrac{11}{5} \\ 0 & 0 & -\tfrac{29}{20} & -\tfrac{87}{10} \end{array}\right)$$

The final attack must be planned to yield a matrix of the form

$$\left(\begin{array}{ccc|c} 1 & 0 & 0 & * \\ 0 & 1 & 0 & * \\ 0 & 0 & 1 & * \end{array}\right)$$

in which the elements of the first three columns are prescribed.

Multiply third row by $-\frac{1}{29}$ and add to first row:
Multiply third row by $-\frac{14}{29}$ and add to second row:
Multiply third row by $-\frac{20}{29}$:

$$\left(\begin{array}{ccc|c} 1 & 0 & 0 & 4 \\ 0 & 1 & 0 & 2 \\ 0 & 0 & 1 & 6 \end{array}\right)$$

The solution is $x = 4$, $y = 2$, $z = 6$.

SOLUTIONS USING THE GRAPHING CALCULATOR. Systems of equations such as

$$\begin{cases} 2x - 3y = 2 \\ 4x + 7y = -9 \end{cases}$$

are easily solved using the graphing calculator. Also, software packages, such as Maple, provide broad capabilities for this topic.

Solved Problems

5.1 Solve graphically the systems

(a) $\begin{cases} x + 2y = 5 \\ 3x - y = 1 \end{cases}$, (b) $\begin{cases} x + y = 1 \\ 2x + 3y = 0 \end{cases}$, (c) $\begin{cases} 3x - 6y = 10 \\ 9x + 15y = -14 \end{cases}$.

(a) $x = 1$, $y = 2$ (b) $x = 3$, $y = -2$ (c) $x = 0.7$, $y = -1.3$ (See Fig. 5-4.)

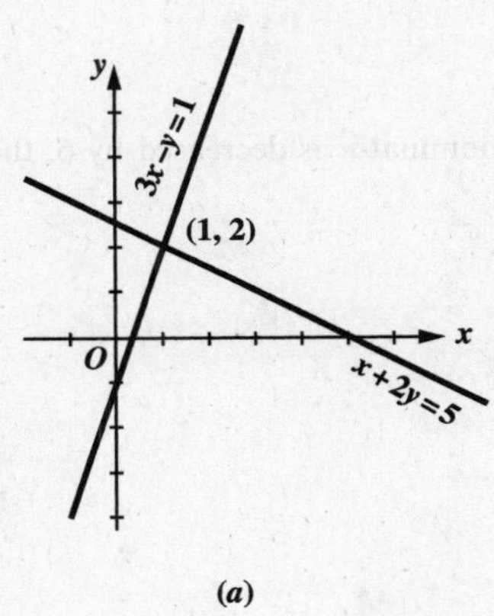

(a)

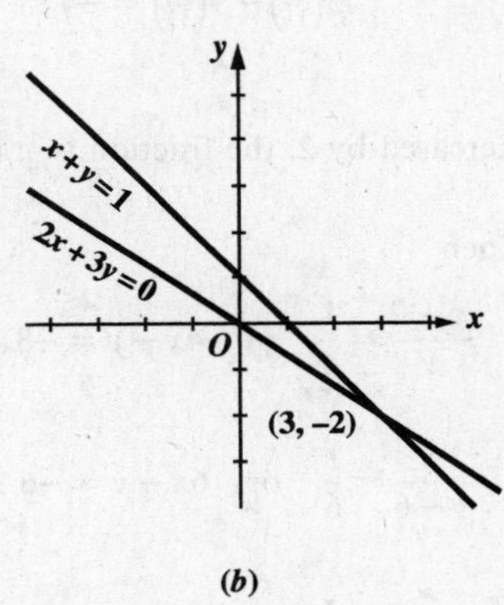

(b)

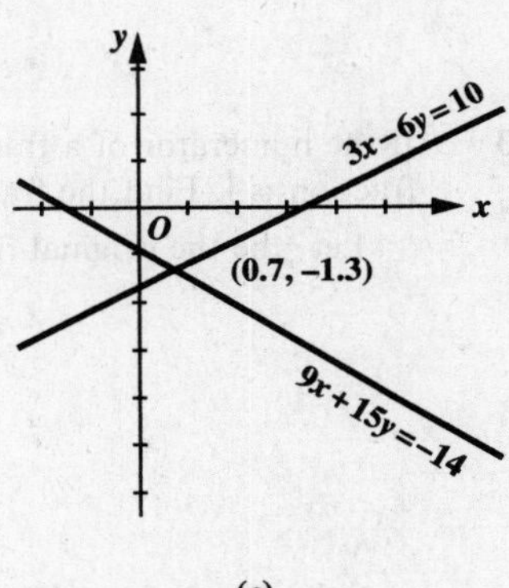

(c)

Fig. 5-4

5.2 Solve algebraically:

(a) $\begin{cases} x + 2y = 5 & (1) \\ 3x - y = 1 & (2) \end{cases}$ (b) $\begin{cases} 3x + 2y = 2 & (1) \\ 5x + 6y = 4 & (2) \end{cases}$ (c) $\begin{cases} 2x + 3y = 3 & (1) \\ 5x - 9y = -4 & (2) \end{cases}$

(a) Rewrite (1): $x + 2y = 5$
Multiply (2) by 2: $6x - 2y = 2$
Add: $7x = 7$
$x = 1$

Substitute for x in (1):

$$1 + 2y = 5, \quad y = 2$$

Check: Using (2),

$$3(1) - 2 = 1$$

(*b*) Multiply (*1*) by -5: $-15x - 10y = -10$

Multiply (*2*) by 3: $15x + 18y = 12$

Add: $8y = 2$

$y = \frac{1}{4}$

Substitute for y in (*1*):

$$3x + 2\left(\tfrac{1}{4}\right) = 2, \quad x = \tfrac{1}{2}$$

Check: Using (*2*),

$$5\left(\tfrac{1}{2}\right) + 6\left(\tfrac{1}{4}\right) = 4$$

(*c*) Multiply (*1*) by 3: $6x + 9y = 9$

Rewrite (*2*): $5x - 9y = -4$

Add: $11x = 5$

$x = \frac{5}{11}$

Substitute in (*1*):

$$3y = 3 - 2\left(\tfrac{5}{11}\right) = \tfrac{23}{11}, \quad y = \tfrac{23}{33}$$

Check: Using (*2*),

$$5\left(\tfrac{5}{11}\right) - 9\left(\tfrac{23}{33}\right) = -4$$

5.3 If the numerator of a fraction is increased by 2, the fraction is $\frac{1}{4}$; if the denominator is decreased by 6, the fraction is $\frac{1}{6}$. Find the fraction.

Let $\frac{x}{y}$ be the original fraction. Then

$$\frac{x+2}{y} = \frac{1}{4} \quad \text{or} \quad 4x - y = -8 \quad (1)$$

$$\frac{x}{y-6} = \frac{1}{6} \quad \text{or} \quad 6x - y = -6 \quad (2)$$

Subtract (*1*) from (*2*):

$$2x = 2 \quad \text{and} \quad x = 1$$

Substitute $x = 1$ in (*1*):

$$4 - y = -8 \quad \text{and} \quad y = 12$$

The fraction is $\frac{1}{12}$.

5.4 Solve the system

$$\begin{cases} x - 5y + 3z = 9 & (1) \\ 2x - y + 4z = 6 & (2) \\ 3x - 2y + z = 2 & (3) \end{cases}$$

Eliminate z.

Rewrite (*1*): $x - 5y + 3z = 9$

Multiply (*3*) by -3: $-9x + 6y - 3z = -6$

Add: $-8x + y = 3$ (*4*)

$$\begin{array}{rl} \text{Rewrite } (2)\text{:} & 2x - y + 4z = 6 \\ \text{Multiply } (3) \text{ by } -4\text{:} & \underline{-12x + 8y - 4z = -8} \\ \text{Add:} & -10x + 7y \qquad = -2 \quad (5) \end{array}$$

$$\begin{array}{rl} \text{Multiply } (4) \text{ by } -7\text{:} & 56x - 7y = -21 \\ \text{Rewrite } (5)\text{:} & \underline{-10x + 7y = -2} \\ \text{Add:} & 46x \qquad = -23 \\ & x = -\frac{1}{2} \end{array}$$

Substitute $x = -\frac{1}{2}$ in (4):

$$-8(-\tfrac{1}{2}) + y = 3 \quad \text{and} \quad y = -1$$

Substitute $x = -\frac{1}{2}$, $y = -1$ in (1):

$$-\tfrac{1}{2} - 5(-1) + 3z = 9 \quad \text{and} \quad z = \tfrac{3}{2}$$

Check: Using (2), $2(-\frac{1}{2}) - (-1) + 4(\frac{3}{2}) = -1 + 1 + 6 = 6$.

5.5 A parabola $y = ax^2 + bx + c$ passes through the points (1,0), (2,2), and (3,10). Determine its equation.

Since (1,0) is on the parabola:

$$a + b + c = 0 \quad (1)$$

Since (2,2) is on the parabola:

$$4a + 2b + c = 2 \quad (2)$$

Since (3,10) is on the parabola:

$$9a + 3b + c = 10 \quad (3)$$

Subtract (1) from (2):

$$3a + b = 2 \quad (4)$$

Subtract (1) from (3):

$$8a + 2b = 10 \quad (5)$$

Multiply (4) by -2 and add to (5):

$$2a = 6 \quad \text{and} \quad a = 3$$

Substitute $a = 3$ in (4):

$$3(3) + b = 2 \quad \text{and} \quad b = -7$$

Substitute $a = 3$, $b = -7$ in (1):

$$3 - 7 + c = 0 \quad \text{and} \quad c = 4$$

The equation of the parabola is $y = 3x^2 - 7x + 4$.

5.6 Solve, using matrices, the system

$$\begin{cases} x - 5y + 3z = 9 \\ 2x - y + 4z = 6 \\ 3x - 2y + z = 2 \end{cases} \qquad \text{(See Problem 5.4.)}$$

Begin with the matrix:

$$\begin{pmatrix} 1 & -5 & 3 & | & 9 \\ 2 & -1 & 4 & | & 6 \\ 3 & -2 & 1 & | & 2 \end{pmatrix}$$

Rewrite first row (since first element is 1):
Multiply first row by -2 and add to second row:
Multiply first row by -3 and add to third row:

$$\begin{pmatrix} 1 & -5 & 3 & | & 9 \\ 0 & 9 & -2 & | & -12 \\ 0 & 13 & -8 & | & -25 \end{pmatrix}$$

Multiply second row by $\frac{5}{9}$ and add to first row:
Multiply second row by $\frac{1}{9}$.
Multiply second row by $-\frac{13}{9}$ and add to third row:

$$\begin{pmatrix} 1 & 0 & \frac{17}{9} & | & \frac{7}{3} \\ 0 & 1 & -\frac{2}{9} & | & -\frac{4}{3} \\ 0 & 0 & -\frac{46}{9} & | & -\frac{23}{3} \end{pmatrix}$$

Multiply third row by $\frac{17}{46}$ and add to first row:
Multiply third row by $-\frac{1}{23}$ and add to second row:
Multiply third row by $-\frac{9}{46}$:

$$\begin{pmatrix} 1 & 0 & 0 & | & -\frac{1}{2} \\ 0 & 1 & 0 & | & -1 \\ 0 & 0 & 1 & | & \frac{3}{2} \end{pmatrix}$$

The solution is $x = -\frac{1}{2}$, $y = -1$, $z = \frac{3}{2}$.

Supplementary Problems

5.7 Solve graphically the systems

(*a*) $\begin{cases} x + y = 5 \\ 2x - y = 1 \end{cases}$ (*b*) $\begin{cases} x - 3y = 1 \\ x - 2y = 0 \end{cases}$ (*c*) $\begin{cases} x + y = -1 \\ 3x - y = 3 \end{cases}$

Ans. (*a*) (2, 3) (*b*) (−2, −1) (*c*) $(\frac{1}{2}, -\frac{3}{2})$

5.8 Solve algebraically the systems

(*a*) $\begin{cases} 3x + 2y = 2 \\ x - y = 9 \end{cases}$ (*c*) $\begin{cases} 3x - y = 1 \\ 2x + 5y = 41 \end{cases}$ (*e*) $\begin{cases} 1/x + 2/y = 2 \\ 2/x - 2/y = 1 \end{cases}$

(*b*) $\begin{cases} 3x - 5y = 5 \\ 7x + y = 75 \end{cases}$ (*d*) $\begin{cases} x + ay = b \\ 2x - by = a \end{cases}$ (*f*) $\begin{cases} 1/4x + 7/2y = \frac{5}{4} \\ 1/2x - 3/y = -\frac{5}{14} \end{cases}$

Hint: In (*e*) and (*f*) solve first for $1/x$ and $1/y$.

Ans. (*a*) $x = 4$, $y = -5$ (*c*) $x = \frac{46}{17}$, $y = \frac{121}{17}$ (*e*) $x = 1$, $y = 2$

(*b*) $x = 10$, $y = 5$ (*d*) $x = \dfrac{b^2 + a^2}{2a + b}$, $y = \dfrac{2b - a}{2a + b}$ (*f*) $x = 1$, $y = \frac{7}{2}$

5.9 *A* and *B* are 30 km apart. If they leave at the same time and travel in the same direction, *A* overtakes *B* in 60 hours. If they travel toward each other, they meet in 5 hours. What are their rates?

Ans. A, $3\frac{1}{4}$ km/h; *B*, $2\frac{3}{4}$ km/h

5.10 Two trains, each 400 ft long, run on parallel tracks. When running in the same direction, they pass in 20 s; when running in the opposite direction, they pass in 5 s. Find the speed of each train.

Ans. 100 ft/s, 60 ft/s

5.11 One alloy contain 3 times as much copper as silver; another contains 5 times as much silver as copper. How much of each alloy must be used to make 14 kg in which there is twice as much copper as silver?

Ans. 12 kg of first, 2 kg of second

5.12 If a field is enlarged by making it 10 m longer and 5 m wider, its area is increased by 1050 square meters. If its length is decreased by 5 m and its width is decreased by 10 m, its area is decreased by 1050 square meters. Find the original dimensions of the field.

Ans. 80 m × 60 m

5.13 Solve each of the following systems:

(*a*) $\begin{cases} x+y+z=3 \\ 2x+y-z=-6 \\ 3x-y+z=11 \end{cases}$ (*c*) $\begin{cases} 3x+y+4z=6 \\ 2x-3y-5z=2 \\ 3x-4y+3z=8 \end{cases}$ (*e*) $\begin{cases} 4x-3y+3z=8 \\ 2x+3y+24z=1 \\ 6x-y+6z=-1 \end{cases}$

(*b*) $\begin{cases} 4x+4y-3z=3 \\ 2x+3y+2z=-4 \\ 3x-y+4z=4 \end{cases}$ (*d*) $\begin{cases} 2x-3y-3z=9 \\ x+3y-2z=3 \\ 3x-4y-z=4 \end{cases}$ (*f*) $\begin{cases} 6x+2y+4z=2 \\ 4x-y+2z=-3 \\ 7x-2y-3z=5 \end{cases}$

Ans. (*a*) $x=1,\ y=-3,\ z=5$ (*c*) $x=\frac{3}{2},\ y=-\frac{1}{2},\ z=\frac{1}{2}$ (*e*) $x=-\frac{3}{2},\ y=-4,\ z=\frac{2}{3}$

(*b*) $x=2,\ y=-2,\ z=-1$ (*d*) $x=3,\ y=2,\ z=-3$ (*f*) $x=\frac{2}{3},\ y=\frac{2}{3},\ z=-\frac{5}{3}$

5.14 Find the equation of the parabola $y=ax^2+bx+c$ which passes through the points (1,6), (4,0), (3,4). Check your result using a graphing calculator.

Ans. $y=-x^2+3x+4$

5.15 Solve the systems in Problem 5.13 above using a computer software package such as Maple.

5.16 Repeat Problems 5.7 and 5.8, solving the systems using a graphing calculator.

Chapter 6

Quadratic Functions and Equations

THE GRAPH OF THE QUADRATIC FUNCTION $y = ax^2 + bx + c,\ a \neq 0$, is a parabola. If $a > 0$, the parabola opens upward (Fig. 6-1); if $a < 0$, the parabola opens downward (Fig. 6-2). The lowest point of the parabola of Fig. 6-1 and the highest point of the parabola of Fig. 6-2 are called *vertices*. The abscissa of the vertex is given by $x = -b/2a$. (See Problem 6.1.).

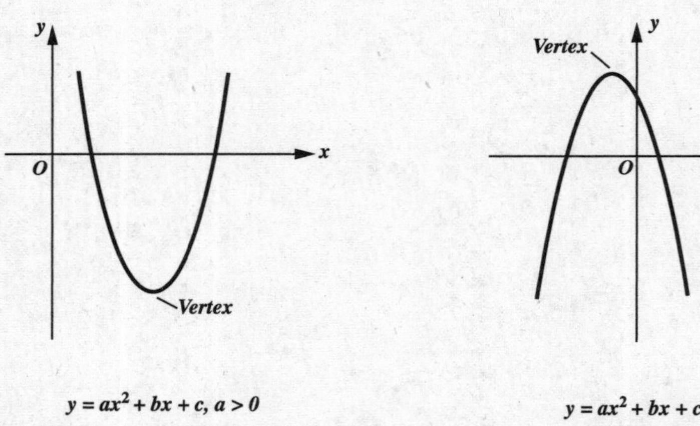

$y = ax^2 + bx + c, a > 0$

Fig. 6-1

$y = ax^2 + bx + c, a < 0$

Fig. 6-2

A QUADRATIC EQUATION in one unknown x is of the form

$$ax^2 + bx + c = 0 \qquad a \neq 0 \tag{6.1}$$

Frequently a quadratic equation may be solved by *factoring*. (See Problem 6.2.)

Every quadratic equation (*6.1*) can be solved by the following process, known as *completing the square*:

(*a*) Substract the constant term c from both members.
(*b*) Divide both members by a, the coefficient of x^2.

(*c*) Add to each member the square of one-half the coefficient of the term in x.
(*d*) Set the square root of the left member (a perfect square) equal to $\pm$ the square root of the right member and solve for x.

EXAMPLE 1. Solve $3x^2 - 8x - 4 = 0$ by completing the square.

(*a*) $3x^2 - 8x = 4,$ (*b*) $x^2 - \frac{8}{3}x = \frac{4}{3},$

(*c*) $x^2 - \frac{8}{3}x + \frac{16}{9} = \frac{4}{3} + \frac{16}{9} = \frac{28}{9},\quad \left[\frac{1}{2}\left(-\frac{8}{3}\right)\right]^2 = \left(-\frac{4}{3}\right)^2 = \frac{16}{9},$

(*d*) $x - \frac{4}{3} = \pm\frac{2\sqrt{7}}{3}.$ Then $x = \frac{4}{3} \pm \frac{2\sqrt{7}}{3} = \frac{4 \pm 2\sqrt{7}}{3}.$

(See Problem 6.3.)

Every quadratic equation (*6.1*) can be solved by means of the quadratic formula

$$x = \frac{-b \pm \sqrt{b^2 - 4ac}}{2a}$$

(See Problems 6.4–6.5.)

It should be noted that it is possible that the roots may be complex numbers.

EQUATIONS IN QUADRATIC FORM. An equation is in *quadratic form* if it is quadratic in some function of the unknown. For example, if the unknown is z, the equation might be quadratic in z^2 or in z^3.

EXAMPLE 2. Solve $x^4 + x^2 - 12 = 0$. This is a quadratic in x^2.

Factor:

$$x^4 + x^2 - 12 = (x^2 - 3)(x^2 + 4) = 0$$

Then

$$x^2 - 3 = 0 \qquad x^3 + 4 = 0$$
$$x = \pm\sqrt{3} \qquad x = \pm 2i$$

(See Problems 6.11–6.12).

EQUATIONS INVOLVING RADICALS may sometimes reduce to quadratic equations after squaring to remove the radicals. All solutions of this quadratic equation *must* be tested since some may be extraneous. (See Problems 6.13–6.16.)

THE DISCRIMINANT of the quadratic equation (*6.1*) is, by definition, the quantity $b^2 - 4ac$. When a, b, c are rational numbers, the roots of the equation are

Real and unequal if and only if $b^2 - 4ac > 0$.
Real and equal if and only if $b^2 - 4ac = 0$.
Complex if and only if $b^2 - 4ac < 0$. (See Chapter 15.)

(See Problems 6.17–6.18.)

SUM AND PRODUCT OF THE ROOTS. If x_1 and x_2 are the roots of the quadratic equation (*6.1*), then $x_1 + x_2 = -b/a$ and $x_1 \cdot x_2 = c/a$.

A quadratic equation whose roots are x_1 and x_2 may be written in the form

$$x^2 - (x_1 + x_2)x + x_1 \cdot x_2 = 0$$

THE GRAPHING CALCULATOR. The graphing calculator can easily be used to find the roots of a quadratic equation.

Solved Problems

6.1. Sketch the parabolas: (*a*) $y = x^2 - 2x - 8$, (*b*) $y = 2x^2 + 9x - 9$. Determine the coordinates of the vertex $V(x, y)$ of each.

Vertex:

(*a*) $$x = -\frac{b}{2a} = -\frac{-2}{2 \cdot 1} = 1 \qquad y = 1^2 - 2 \cdot 1 - 8 = -9$$

hence, $V(1, -9)$. See Fig. 6-3(*a*).

(*b*) $$x = -\frac{9}{2(-2)} = \frac{9}{4} \qquad y = -2\left(\frac{9}{4}\right)^2 + 9\left(\frac{9}{4}\right) - 9 = \frac{9}{8}$$

hence, $V\left(\frac{9}{4}, \frac{9}{8}\right)$. See Fig. 6-3(*b*).

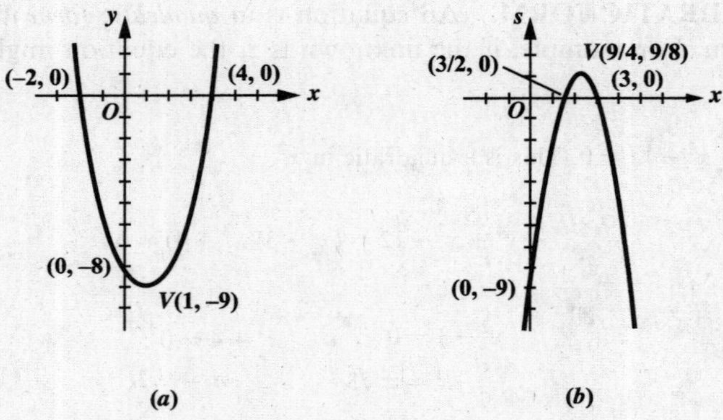

Fig. 6-3

6.2. Solve by factoring:

(*a*) $4x^2 - 5x = x(4x - 5) = 0$

(*b*) $4x^2 - 9 = (2x - 3)(2x + 3) = 0$

(*c*) $x^2 - 4x + 3 = (x - 1)(x - 3) = 0$

(*d*) $x^2 - 6x + 9 = (x - 3)(x - 3) = 0$

(*e*) $4x^2 + 20x + 25 = (2x + 5)(2x + 5) = 0$

(*f*) $6x^2 + 13x + 6 = (3x + 2)(2x + 3) = 0$

(*g*) $3x^2 + 8ax - 3a^2 = (3x - a)(x + 3a) = 0$

(*h*) $10ax^2 + (15 - 8a^2)x - 12a = (2ax + 3)(5x - 4a) = 0$

Ans. (*a*) $0, \frac{5}{4}$ (*c*) $1, 3$ (*e*) $-\frac{5}{2}, -\frac{5}{2}$ (*g*) $a/3, -3a$

(*b*) $\frac{3}{2}, -\frac{3}{2}$ (*d*) $3, 3$ (*f*) $-\frac{2}{3}, -\frac{3}{2}$ (*h*) $-\frac{3}{2}a,\ 4a/5$

6.3. Solve by completing the square: (*a*) $x^2 - 2x - 1 = 0$, (*b*) $3x^2 + 8x + 7 = 0$.

(*a*) $x^2 - 2x = 1;\quad x^2 - 2x + 1 = 1 + 1 = 2;\quad x - 1 = \pm\sqrt{2};\quad x = 1 \pm \sqrt{2}$.

(*b*) $3x^2 + 8x = -7;\quad x^2 + \frac{8}{3}x = -\frac{7}{3};\quad x^2 + \frac{8}{3}x + \frac{16}{9} = -\frac{7}{3} + \frac{16}{9} = -\frac{5}{9};$

$$x + \frac{4}{3} = \pm\sqrt{\frac{-5}{9}} = \pm\frac{i\sqrt{5}}{3};\quad x = \frac{-4 \pm i\sqrt{5}}{3}.$$

6.4. Solve $ax^2 + bx + c = 0,\ a \neq 0$, by completing the square.

Proceeding as in Problem 6.3, we have

$$x^2 + \frac{b}{a}x = -\frac{c}{a}, \quad x^2 + \frac{b}{a}x + \frac{b^2}{4a^2} = \frac{b^2}{4a^2} - \frac{c}{a} = \frac{b^2 - 4ac}{4a^2},$$

$$x + \frac{b}{2a} = \pm\sqrt{\frac{b^2 - 4ac}{4a^2}} = \pm\frac{\sqrt{b^2 - 4ac}}{2a}, \quad \text{and} \quad x = \frac{-b \pm \sqrt{b^2 - 4ac}}{2a}$$

6.5. Solve the equations of Problem 6.3 using the quadratic formula.

(*a*) $$x = \frac{-b \pm \sqrt{b^2 - 4ac}}{2a} = \frac{-(-2) \pm \sqrt{(-2)^2 - 4(1)(-1)}}{2 \cdot 1} = \frac{2 \pm \sqrt{4+4}}{2} = \frac{2 \pm 2\sqrt{2}}{2} = 1 \pm \sqrt{2}$$

(*b*) $$x = \frac{-(8) \pm \sqrt{8^2 - 4 \cdot 3 \cdot 7}}{2 \cdot 3} = \frac{-8 \pm \sqrt{64 - 84}}{6} = \frac{-8 \pm \sqrt{-20}}{6} = \frac{-8 \pm 2i\sqrt{5}}{6} = \frac{-4 \pm i\sqrt{5}}{3}$$

6.6. An open box containing 24 cm^3 is to be made from a square piece of tin by cutting 2 cm squares from each corner and turning up the sides. Find the dimension of the piece of tin required.

Let $x =$ the required dimension. The resulting box will have dimensions $(x - 4)$ by $(x - 4)$ by 2, and its volume will be $2(x - 4)(x - 4)$. See Fig. 6-4. Then

$$2(x-4)^2 = 24, \qquad x - 4 = \pm 2\sqrt{3}, \qquad \text{and} \qquad x = 4 \pm 2\sqrt{3} = 7.464,\ 0.536$$

The required square of tin is 7.464 cm on a side.

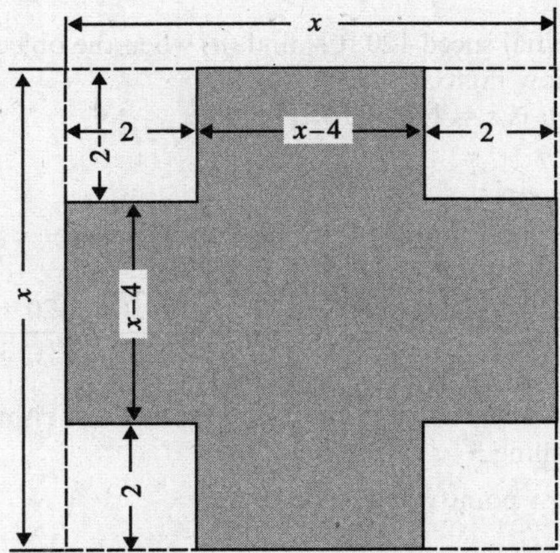

Fig. 6-4

6.7. Two pipes together can fill a reservoir in 6 h 40 min. Find the time each alone will take to fill the reservoir if one of the pipes can fill it in 3 h less time than the other.

Let $x =$ time (hours) required by smaller pipe, $x - 3 =$ time required by larger pipe. Then

$$\frac{1}{x} = \text{part filled in 1 h by smaller pipe} \qquad \frac{1}{x-3} = \text{part filled in 1 h by larger pipe}$$

Since the two pipes together fill $\frac{1}{\frac{20}{3}} = \frac{3}{20}$ of the reservoir in 1 h,

$$\frac{1}{x} + \frac{1}{x-3} = \frac{3}{20}, \qquad 20(x-3) + 20x = 3x(x-3),$$

$$3x^2 - 49x + 60 = (3x-4)(x-15) = 0, \quad \text{and} \quad x = \tfrac{4}{3}, 15$$

The smaller pipe will fill the reservoir in 15 h and the larger pipe in 12 h.

6.8. Express each of the following in the form $a(x-h)^2 \pm b(y-k)^2 = c$.

(*a*) $x^2 + y^2 - 6x - 9y + 2 = 0$.

$$(x^2 - 6x) + (y^2 - 9y) = -2 \qquad (x^2 - 6x + 9) + (y^2 - 9y + \tfrac{81}{4}) = -2 + 9 + \tfrac{81}{4} = \tfrac{109}{4}$$

$$(x-3)^2 + (y - \tfrac{9}{2})^2 = \tfrac{109}{4}$$

(*b*) $3x^2 + 4y^2 + 6x - 16y - 21 = 0$.

$$3(x^2 + 2x) + 4(y^2 - 4y) = 21 \qquad 3(x^2 + 2x + 1) + 4(y^2 - 4y + 4) = 21 + 3(1) + 4(4) = 40$$

$$3(x+1)^2 + 4(y-2)^2 = 40$$

6.9. Transform each of the following into the form $a\sqrt{(x-h)^2 + k}$ or $a\sqrt{k - (x-h)^2}$.

(*a*) $\sqrt{4x^2 - 8x + 9} = 2\sqrt{x^2 - 2x + \tfrac{9}{4}} = 2\sqrt{(x^2 - 2x + 1) + \tfrac{5}{4}} = 2\sqrt{(x-1)^2 + \tfrac{5}{4}}$

(*b*) $\sqrt{8x - x^2} = \sqrt{16 - (x^2 - 8x + 16)} = \sqrt{16 - (x-4)^2}$

(*c*) $\sqrt{3 - 4x - 2x^2} = \sqrt{2} \cdot \sqrt{\tfrac{3}{2} - 2x - x^2} = \sqrt{2} \cdot \sqrt{\tfrac{5}{2} - (x^2 + 2x + 1)} = \sqrt{2} \cdot \sqrt{\tfrac{5}{2} - (x+1)^2}$

6.10. If an object is thrown directly upward with initial speed v ft/s, its distance s ft above the ground after t s is given by

$$s = vt - \tfrac{1}{2}gt^2$$

Taking $g = 32.2\,\text{ft/s}^2$ and initial speed 120 ft/s, find (*a*) when the object is 60 ft above the ground, (*b*) when it is highest in its path and how high.

The equation of motion is $s = 120t - 16.1t^2$.

(*a*) When $s = 60$:

$$60 = 120t - 16.1t^2 \quad \text{or} \quad 16.1t^2 - 120t + 60 = 0$$

$$t = \frac{120 \pm \sqrt{(120)^2 - 4(16.1)60}}{32.2} = \frac{120 \pm \sqrt{10536}}{32.2} = \frac{120 \pm 102.64}{32.2} = 6.91,\ 0.54$$

After $t = 0.54$ s the object is 60 ft above the ground and rising. After $t = 6.91$ s, the object is 60 ft above the ground and falling.

(*b*) The object is at its highest point when

$$t = \frac{-b}{2a} = \frac{-(-120)}{2(16.1)} = 3.73\,\text{s}. \text{ Its height is given by } 120t - 16.1t^2 = 120(3.73) - 16.1(3.73)^2 = 224\,\text{ft}.$$

6.11. Solve $9x^4 - 10x^2 + 1 = 0$.

Factor: $(x^2 - 1)(9x^2 - 1) = 0$. Then $x^2 - 1 = 0$, $9x^2 - 1 = 0$; $x = \pm 1$, $x = \pm\tfrac{1}{3}$.

6.12. Solve $x^4 - 6x^3 + 12x^2 - 9x + 2 = 0$.

Complete the square on the first two terms: $(x^4 - 6x^3 + 9x^2) + 3x^2 - 9x + 2 = 0$

or $(x^2 - 3x)^2 + 3(x^2 - 3x) + 2 = 0$

Factor: $[(x^2 - 3x) + 2][(x^2 - 3x) + 1] = 0$

Then $x^2 - 3x + 2 = (x-2)(x-1) = 0$ and $x = 1, 2$

$$x^2 - 3x + 1 = 0 \quad \text{and} \quad x = \frac{3 \pm \sqrt{9-4}}{2} = \frac{3 \pm \sqrt{5}}{2}$$

6.13. Solve $\sqrt{5x-1} - \sqrt{x} = 1$.

Transpose one of the radicals: $\sqrt{5x-1} = \sqrt{x} + 1$

Square: $5x - 1 = x + 2\sqrt{x} + 1$

Collect terms: $4x - 2 = 2\sqrt{x}$ or $2x - 1 = \sqrt{x}$

Square:

$$4x^2 - 4x + 1 = x, \quad 4x^2 - 5x + 1 = (4x-1)(x-1) = 0, \quad \text{and} \quad x = \tfrac{1}{4}, 1$$

For $x = \frac{1}{4}$: $\sqrt{5(\frac{1}{4}) - 1} - \sqrt{\frac{1}{4}} = 0 \neq 1$.

For $x = 1$: $\sqrt{5(1) - 1} - \sqrt{1} = 1$. The root is $x = 1$.

6.14. Solve $\sqrt{6x+7} - \sqrt{3x+3} = 1$.

Transpose one of the radicals: $\sqrt{6x+7} = 1 + \sqrt{3x+3}$

Square: $6x + 7 = 1 + 2\sqrt{3x+3} + 3x + 3$

Collect terms: $3x + 3 = 2\sqrt{3x+3}$

Square:

$$9x^2 + 18x + 9 = 4(3x+3) = 12x + 12, \quad 9x^2 + 6x - 3 = 3(3x-1)(x+1) = 0, \quad \text{and} \quad x = \tfrac{1}{3}, -1$$

For $x = \frac{1}{3}$: $\sqrt{6(\frac{1}{3}) + 7} - \sqrt{3(\frac{1}{3}) + 3} = 3 - 2 = 1$.

For $x = -1$: $1 - 0 = 1$. The roots are $x = \frac{1}{3}, -1$.

6.15. Solve $\dfrac{\sqrt{x+1} + \sqrt{x-1}}{\sqrt{x+1} - \sqrt{x-1}} = 3$.

Multiply the numerator and denominator of the fraction by $(\sqrt{x+1} + \sqrt{x-1})$:

$$\frac{(\sqrt{x+1} + \sqrt{x-1})(\sqrt{x+1} + \sqrt{x-1})}{(\sqrt{x+1} - \sqrt{x-1})(\sqrt{x+1} + \sqrt{x-1})} = \frac{(x+1) + 2\sqrt{x^2-1} + (x-1)}{(x+1) - (x-1)} = x + \sqrt{x^2-1} = 3$$

Then $x - 3 = -\sqrt{x^2-1}$, $x^2 - 6x + 9 = x^2 - 1$, and $x = \dfrac{5}{3}$

Check:
$$\frac{\sqrt{\frac{8}{3}} + \sqrt{\frac{2}{3}}}{\sqrt{\frac{8}{3}} - \sqrt{\frac{2}{3}}} = \frac{2\sqrt{\frac{2}{3}} + \sqrt{\frac{2}{3}}}{2\sqrt{\frac{2}{3}} - \sqrt{\frac{2}{3}}} = \frac{3\sqrt{\frac{2}{3}}}{\sqrt{\frac{2}{3}}} = 3$$

6.16. Solve $3x^2 - 5x + \sqrt{3x^2 - 5x + 4} = 16$.

Note that the unknown appears in the same polynomials in both the expressions free of radicals and under the radical. Add 4 to both sides:

$$3x^2 - 5x + 4 + \sqrt{3x^2 - 5x + 4} = 20$$

Let $y = \sqrt{3x^2 - 5x + 4}$. Then

$$y^2 + y - 20 = (y+5)(y-4) = 0 \qquad \text{and} \qquad y = 4, -5$$

Now $\sqrt{3x^2 - 5x + 4} = -5$ is impossible. From $\sqrt{3x^2 - 5x + 4} = 4$ we have

$$3x^2 - 5x + 4 = 16, \qquad 3x^2 - 5x - 12 = (3x+4)(x-3) = 0, \qquad \text{and} \qquad x = 3, -\tfrac{4}{3}$$

The reader will show that both $x = 3$ and $x = -\frac{4}{3}$ are solutions.

6.17. Without solving determine the character of the roots of

(*a*) $x^2 - 8x + 9 = 0$. Here $b^2 - 4ac = 28$; the roots are irrational and unequal.
(*b*) $3x^2 - 8x + 9 = 0$. Here $b^2 - 4ac = -44$; the roots are imaginary and unequal.
(*c*) $6x^2 - 5x - 6 = 0$. Here $b^2 - 4ac = 169$; the roots are rational and unequal.
(*d*) $4x^2 - 4\sqrt{3}x + 3 = 0$. Here $b^2 - 4ac = 0$; the roots are real and equal.

(NOTE: Although the discriminant is the square of a rational number, the roots $\frac{1}{2}\sqrt{3}, \frac{1}{2}\sqrt{3}$ are not rational. Why?)

6.18. Without sketching, state whether the graph of each of the following functions crosses the x axis, is tangent to it, or lies wholly above or below it.

(*a*) $3x^2 + 5x - 2$. $b^2 - 4ac = 25 + 24 > 0$; the graph crosses the x axis.
(*b*) $2x^2 + 5x + 4$. $b^2 - 4ac = 25 - 32 < 0$ and the graph is either wholly above the wholly below the x axis. Since $f(0) > 0$ (the value of the function for any other value of x would do equally well), the graph lies wholly above the x axis.
(*c*) $4x^2 - 20x + 25$. $b^2 - 4ac = 400 - 400 = 0$; the graph is tangent to the x axis.
(*d*) $2x - 9 - 4x^2$. $b^2 - 4ac = 4 - 144 < 0$ and $f(0) < 0$; the graph lies wholly below the x axis.

6.19. Find the sum and product of the roots of

(*a*) $x^2 + 5x - 8 = 0$. *Ans.* Sum $= -\dfrac{b}{a} = -5$, product $= \dfrac{c}{a} = -8$.
(*b*) $8x^2 - x - 2 = 0$ or $x^2 + \frac{1}{8}x - \frac{1}{4} = 0$. *Ans.* Sum $= \frac{1}{8}$, product $= -\frac{1}{4}$.
(*c*) $5 - 10x - 3x^2 = 0$ or $x^2 + \frac{10}{3}x - \frac{5}{3} = 0$. *Ans.* Sum $= -\frac{10}{3}$, product $= -\frac{5}{3}$.

Supplementary Problems

6.20. Locate the vertex of each of the parabolas of Problem 3.11. Compare the results with those of Problem 3.14.

6.21. Solve for x by factoring.

(*a*) $3x^2 + 4x = 0$ (*c*) $x^2 + 2x - 3 = 0$ (*e*) $10x^2 - 9x + 2 = 0$
(*b*) $16x^2 - 25 = 0$ (*d*) $2x^2 + 9x - 5 = 0$ (*f*) $2x^2 - (a + 4b)x + 2ab = 0$

Ans. (*a*) $0, -\frac{4}{3}$ (*b*) $\pm\frac{5}{4}$ (*c*) $1, -3$ (*d*) $\frac{1}{2}, -5$ (*e*) $\frac{1}{2}, \frac{2}{5}$ (*f*) $\frac{1}{2}a, 2b$

6.22. Solve for x by completing the square.

(*a*) $2x^2 + x - 5 = 0$ (*c*) $3x^2 + 2x - 2 = 0$ (*e*) $15x^2 - (16m - 14)x + 4m^2 - 8m + 3 = 0$
(*b*) $2x^2 - 4x - 3 = 0$ (*d*) $5x^2 - 4x + 2 = 0$

Ans. (*a*) $\frac{1}{4}(-1 \pm \sqrt{41})$ (*b*) $\frac{1}{2}(2 \pm \sqrt{10})$ (*c*) $\frac{1}{3}(-1 \pm \sqrt{7})$ (*d*) $\frac{1}{5}(2 \pm i\sqrt{6})$
(*e*) $\frac{1}{3}(2m-1), \frac{1}{5}(2m-3)$

6.23. Solve the equations of Problem 6.24 using the quadratic formula.

6.24. Solve $6x^2 + 5xy - 6y^2 + x + 8y - 2 = 0$ for (*a*) y in terms of x, (*b*) x in terms of y.

Ans. (*a*) $\frac{1}{2}(3x+2), \frac{1}{3}(1-2x)$ (*b*) $\frac{1}{2}(1-3y), \frac{2}{3}(y-1)$

6.25. Solve.

(*a*) $x^4 - 29x^2 + 100 = 0$ (*c*) $1 - \dfrac{2}{2x^2 - x} = \dfrac{3}{(2x^2-x)^2}$ (*e*) $\sqrt{2x+3} - \sqrt{4-x} = 2$

(*b*) $\dfrac{21}{x+2} - \dfrac{1}{x-4} = 2$ (*d*) $\sqrt{4x+1} - \sqrt{3x-2} = 5$ (*f*) $\sqrt{3x-2} - \sqrt{x-2} = 2$

Ans. (*a*) $\pm 2, \pm 5$ (*b*) 5, 7 (*c*) $-1, \frac{3}{2}, \frac{1}{4}(1 \pm i\sqrt{7})$ (*d*) 342 (*e*) 3 (*f*) 2, 6

6.26. Form the quadratic equation whose roots are

(*a*) The negative of the roots of $3x^2 + 5x - 8 = 0$.
(*b*) Twice the roots of $2x^2 - 5x + 2 = 0$.
(*c*) One-half the roots of $2x^2 - 5x - 3 = 0$.

Ans. (*a*) $3x^2 - 5x - 8 = 0$ (*b*) $x^2 - 5x + 4 = 0$ (*c*) $8x^2 - 10x - 3 = 0$

6.27. The length of a rectangle is 7 cm more than its width; its area is 228 cm^2. What are its dimensions?

Ans. 12 cm × 19 cm

6.28. A rectangular garden plot 16 m × 24 m is to be bordered by a strip of uniform width x meters so as to double the area. Find x.

Ans. 4 m

6.29. The interior of a cubical box is lined with insulating material $\frac{1}{2}$ cm thick. Find the original interior dimensions if the volume is thereby decreased by 271 cm^3.

Ans. 10 cm

Chapter 7

Inequalities

AN INEQUALITY is a statement that one (real) number is greater than or less than another; for example, $3 > -2, -10 < -5$.

Two inequalities are said to have the *same sense* if their signs of inequality point in the same direction. Thus, $3 > -2$ and $-5 > -10$ have the same sense; $3 > -2$ and $-10 < -5$ have opposite senses.

The sense of an equality is *not* changed:

(*a*) If the same number is added to or subtracted from both sides
(*b*) If both sides are multiplied or divided by the same *positive* number

The sense of an equality *is* changed if both sides are multiplied or divided by the same negative number. (See Problems 7.1–7.3.)

AN ABSOLUTE INEQUALITY is one which is true for all real values of the letters involved; for example, $x^2 + 1 > 0$ is an absolute inequality.

A CONDITIONAL INEQUALITY is one which is true for certain values of the letters involved; for example, $x + 2 > 5$ is a conditional inequality, since it is true for $x = 4$ but not for $x = 1$.

SOLUTION OF CONDITIONAL INEQUALITIES. The solution of a conditional inequality in one letter, say x, consists of all values of x for which the inequality is true. These values lie on one or more intervals of the real number scale as illustrated in the examples below.

To solve a linear inequality, proceed as in solving a linear equality, keeping in mind the rules for keeping or reversing the sense.

EXAMPLE 1. Solve the inequality $5x + 4 > 2x + 6$.

Subtract $2x$ from each member; Subtract 4 from each member: $3x > 2$

Divide by 3: $x > \frac{2}{3}$

Graphical representation: (See Fig. 7-1.)

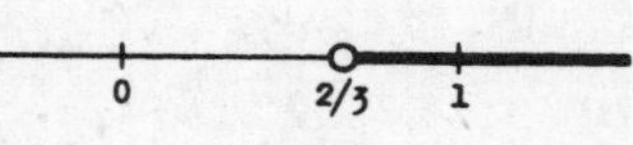

Fig. 7-1

(See Problems 7.5–7.6.)

To solve a quadratic inequality, $f(x) = ax^2 + bx + c > 0$, solve the equality $f(x) = 0$, locate the roots r_1 and r_2 on a number scale, and determine the sign of $f(x)$ on each of the resulting intervals.

EXAMPLE 2. Solve the inequality $3x^2 - 8x + 7 > 2x^2 - 3x + 1$.
Subtract $2x - 3x + 1$ from each member:

$$x^2 - 5x + 6 > 0$$

Solve the equality $x^2 - 5x + 6 = 0$:

$$x = 2, \qquad x = 3$$

Locate the roots on a number scale (see Fig. 7-2).

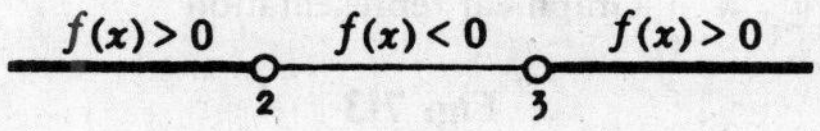

$f(x) > 0$ $f(x) < 0$ $f(x) > 0$
2 3

Fig. 7-2

Determine the sign of $f(x) = x^2 - 5x + 6$:

On the interval $x < 2$:	$f(0) = 6 > 0$
On the interval $2 < x < 3$:	$f(\frac{5}{2}) = \frac{25}{4} - \frac{25}{2} + 6 < 0$
On the interval $x > 3$:	$f(4) = 16 - 20 + 6 > 0$

The given inequality is satisfied (see darkened portions of the scale) when $x < 2$ and $x > 3$. (See Problems 7.7–7.11.)

Solved Problems

7.1 Given the inequality $-3 < 4$, write the result when (*a*) 5 is added to both sides, (*b*) 2 is subtracted from both sides, (*c*) -3 is subtracted from both sides, (*d*) both sides are doubled, (*e*) both sides are divided by -2.

Ans. (*a*) $2 < 9$, (*b*) $-5 < 2$, (*c*) $0 < 7$, (*d*) $-6 < 8$, (*e*) $\frac{3}{2} > -2$

7.2 Square each of the inequalities: (*a*) $-3 < 4$, (*b*) $-3 > -4$

Ans. (*a*) $9 < 16$, (*b*) $9 < 16$

7.3 If $a > 0$, $b > 0$, prove that $a^2 > b^2$ if and only if $a > b$.

Suppose $a > b$. Since $a > 0$, $a^2 > ab$ and, since $b > 0$, $ab > b^2$. Hence, $a^2 > ab > b^2$ and $a^2 > b^2$.
Suppose $a^2 > b^2$. Then $a^2 - b^2 = (a - b)(a + b) > 0$. Dividing by $a + b > 0$, we have $a - b > 0$ and $a > b$.

7.4 Prove $\dfrac{a}{b^2} + \dfrac{b}{a^2} > \dfrac{1}{a} + \dfrac{1}{b}$ if $a > 0$, $b > 0$, and $a \neq b$.

Suppose $a > b$; then $a^2 > b^2$ and $a - b > 0$. Now $a^2(a - b) > b^2(a - b)$ or $a^3 - a^2b > ab^2 - b^3$ and $a^3 + b^3 > ab^2 + a^2b$. Since $a^2b^2 > 0$,

$$\frac{a^3 + b^3}{a^2b^2} > \frac{ab^2 + a^2b}{a^2b^2} \quad \text{and} \quad \frac{a}{b^2} + \frac{b}{a^2} > \frac{1}{a} + \frac{1}{b}$$

Why is it necessary that $a > 0$ and $b > 0$? **Hint:** See Problem 7.3.

7.5 Solve $3x + 4 > 5x + 2$.

Subtract $5x + 4$ from each member:

$$-2x > -2$$

Divide by -2:

$$x < 1$$

See Fig. 7-3.

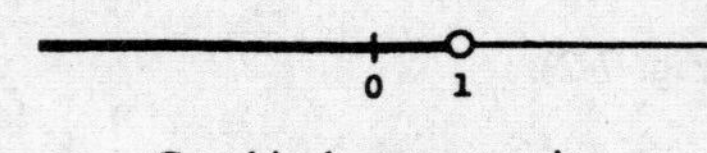

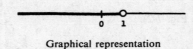

Graphical representation

Fig. 7-3

7.6 Solve $2x - 8 < 7x + 12$.

Subtract $7x - 8$ from each member:

$$-5x < 20$$

Divide by -5:

$$x > -4$$

See Fig. 7-4.

Graphical representation

Fig. 7-4

7.7 Solve $x^2 > 4x + 5$.

Subtract $4x + 5$ from each member:

$$x^2 - 4x - 5 > 0$$

Solve the equality $f(x) = x^2 - 4x - 5 = 0$:

$$x = -1, 5$$

Locate the roots on a number scale.

Determine the sign of $f(x)$

On the interval $x < -1$: $f(-2) = 4 + 8 - 5 > 0$
On the interval $-1 < x < 5$: $f(0) = -5 < 0$
On the interval $x > 5$: $f(6) = 36 - 24 - 5 > 0$

The inequality is satisfied when $x < -1$ and $x > 5$. See Fig. 7-5.

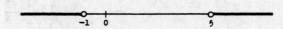

Fig. 7-5

7.8 Solve $3x^2 + 2x + 2 < 2x^2 + x + 4$.

Subtract $2x^2 + x + 4$ from each member:

$$x^2 + x - 2 < 0$$

Solve $f(x) = x^2 + x - 2 = 0$:

$$x = -2, 1$$

Locate the roots on a number scale.

Determine the sign of $f(x)$

On the interval $x < -2$: $f(-3) = 9 - 3 - 2 > 0$
On the interval $-2 < x < 1$: $f(0) = -2 < 0$
On the interval $x > 1$: $f(2) = 4 + 2 - 2 > 0$

The inequality is satisfied when $-2 < x < 1$. See Fig. 7-6.

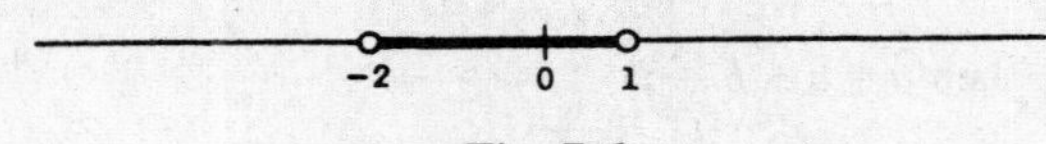

Fig. 7-6

7.9 Solve $(x + 5)(x - 1)(x - 2) < 0$.

Solve the equality $f(x) = (x + 5)(x - 1)(x - 2) = 0$. $x = 1, 2, -5$

Locate the roots on a number scale.

Determine the sign of $f(x)$

On the interval $x < -5$: $f(-6) = (-1)(-7)(-8) < 0$
On the interval $-5 < x < 1$: $f(0) = 5(-1)(-2) > 0$
On the interval $1 < x < 2$: $f(\frac{3}{2}) = (\frac{13}{2})(\frac{1}{2})(-\frac{1}{2}) < 0$
On the interval $x > 2$: $f(3) = 8 \cdot 2 \cdot 1 > 0$

The inequality is satisfied when $x < -5$ and $1 < x < 2$. See Fig. 7-7.

Fig. 7-7

7.10 Solve $(x - 2)^2(x - 5) > 0$.

Solve the equality $f(x) = (x - 2)^2(x - 5) = 0$: $x = 2, 2, 5$

Locate the roots on a number scale.

Determine the sign of $f(x)$

On the interval $x < 2$: $f(0) = (+)(-) < 0$
On the interval $2 < x < 5$: $f(3) = (+)(-) < 0$
On the interval $x > 5$: $f(6) = (+)(+) > 0$

The inequality is satisfied when $x > 5$. See Fig. 7-8.

Fig. 7-8

(NOTE: The inequality $(x - 2)^2(x - 5) < 0$ is satisfied when $x < 2$ and $2 < x < 5$.
The inequality $(x - 2)^2(x - 5) \geq 0$ is satisfied when $x \geq 5$ and $x = 2$.)

7.11 Determine the values of k so that $3x^2 + kx + 4 = 0$ will have real roots.

The discriminant $b^2 - 4ac = k^2 - 48 = (k - 4\sqrt{3})(k + 4\sqrt{3}) \geq 0$. The roots will be real when $k \geq 4\sqrt{3}$ and when $k \leq -4\sqrt{3}$, that is, when $|k| \geq 4\sqrt{3}$.

Supplementary Problems

7.12 If $2y^2 + 4xy - 3x = 0$, determine the range of values of x for which the corresponding y roots are real.

Ans. Here

$$y = \frac{-4x \pm \sqrt{16x^2 + 24x}}{4} = \frac{-2x \pm \sqrt{4x^2 + 6x}}{2}$$

will be real provided $4x^2 + 6x \geq 0$. Thus, y will be real for $x \leq -\frac{3}{2}$ and for $x \geq 0$.

7.13 Prove: If $a > b$ and $c > d$, then $a + c > b + d$.

Hint: $(a - b) + (c - d) = (a + c) - (b + d) > 0$.

7.14 Prove: If $a \neq b$ are real numbers, then $a^2 + b^2 > 2ab$.

Hint: $(a - b)^2 > 0$.

7.15 Prove: If $a \neq b \neq c$ are real numebrs, then $a^2 + b^2 + c^2 > ab + bc + ca$.

7.16 Prove: If $a > 0$, $b > 0$, and $a \neq b$, then $a/b + b/a > 2$.

7.17 Prove: If $a^2 + b^2 = 1$ and $c^2 + d^2 = 1$, then $ac + bd \leq 1$.

7.18 Solve: (*a*) $x - 4 > -2x + 5$ (*c*) $x^2 - 16 > 0$ (*e*) $x^2 - 6x > -5$
(*b*) $4 + 3x < 2x + 24$ (*d*) $x^2 - 4x < 5$ (*f*) $5x^2 + 5x - 8 \leq 3x^2 + 4$

Ans. (*a*) $x > 3$ (*b*) $x < 20$ (*c*) $|x| > 4$ (*d*) $-1 < x < 5$
(*e*) $x < 1, x > 5$ (*f*) $-4 \leq x \leq \frac{3}{2}$

7.19 Solve: (*a*) $(x + 1)(x - 2)(x + 4) > 0$ (*b*) $(x - 1)^3(x - 3)(x + 2) < 0$
(*c*) $(x + 3)(x - 2)^2(x - 5)^3 < 0$

Ans. (*a*) $-4 < x < -1,\ x > 2$ (*b*) $x < -2,\ 1 < x < 3$ (*c*) $-3 < x < 2,\ 2 < x < 5$

7.20 In each of the following determine the domain of x for which y will be real:

(*a*) $y = \sqrt{2x^2 - 7x + 3}$ (*c*) $y = \sqrt{61x^2 + 144x + 144}$ (*e*) $xy^2 + 3xy + 3x - 4y - 4 = 0$
(*b*) $y = \sqrt{6 - 5x - 4x^2}$ (*d*) $y^2 + 2xy + 4y + x + 14 = 0$ (*f*) $6x^2 + 5xy - 6y^2 + x + 8y - 2 = 0$

Ans. (*a*) $x \leq \frac{1}{2}, x \geq 3$ (*c*) all values of x (*e*) $-4 \leq x \leq \frac{4}{3}$
(*b*) $-2 \leq x \leq \frac{3}{4}$ (*d*) $x \leq -5, x \geq 2$ (*f*) all values of x

7.21 Prove that if $a < b$, $a > 0$, and $b > 0$, then $a^2 < b^2$. Is the converse true?

7.22 Solve Problem 7.12 using a graphing calculator.

7.23 Solve Problem 7.12 using a computer sofware package.

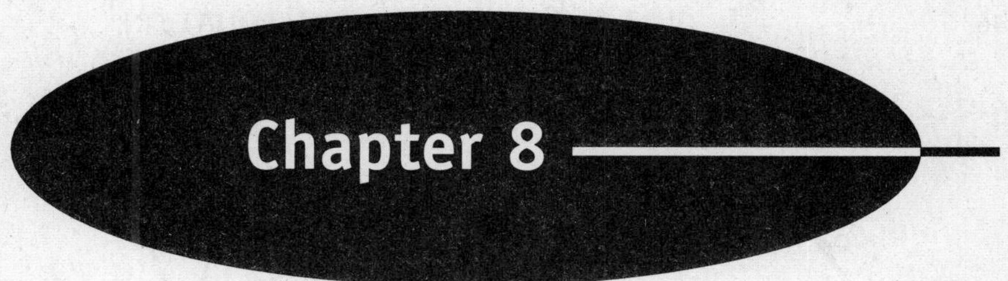

The Locus of an Equation

WHAT IS A LOCUS? *Locus*, in Latin, means location. The plural of locus is loci. A locus of points is the set of points, and only those points, that satisfy conditions.

For example, the locus of points in a plane that are 2 m from a given point P is the set of points 2 m from P. These points lie on the circle with P as center and radius 2 m.

To determine a locus:

(1) State what is given and the condition to be satisfied.
(2) Find several points satisfying the condition which indicate the shape of the locus.
(3) Connect the points and describe the locus fully. For example, the locus $x(y - x) = 0$ consists of the lines $x = 0$ and $y - x = 0$. The reader can easily sketch several points and the locus.

DEGENERATE LOCI. The locus of an equation $f(x, y) = 0$ is called *degenerate* if $f(x, y)$ is the product of two or more real factors $g(x, y), h(x, y), \ldots$. The locus of $f(x, y) = 0$ then consists of the loci of $g(x, y) = 0$, $h(x, y) = 0, \ldots$. (See Problem 8.1.)

INTERCEPTS. The intercepts on the coordinate axes of a locus are the directed distances from the origin to the points of intersection of the locus and the coordinate axes.

To find the x intercepts, set $y = 0$ in the equation of the locus and solve for x; to find the y intercepts, set $x = 0$ and solve for y. (See Problem 8.2.)

SYMMETRY. Two points P and Q are said to be symmetric with respect to a point R if R is the midpoint of the segment PQ (see Fig. 8-1). Each of the points is called the symmetric point of the other with respect to the point R, the *center of symmetry*.

Two points P and Q are said to be symmetric with respect to a line l if l is the perpendicular bisector of the segment PQ (see Fig. 8-2). Each of the points P, Q is called the symmetric point of the other with respect to l, the *axis of symmetry*.

A locus is said to be symmetric with respect to a point R or to a line l if the symmetric point with respect to R or l of every point of the locus is also a point of the locus (see Figs. 8-3 and 8-4).

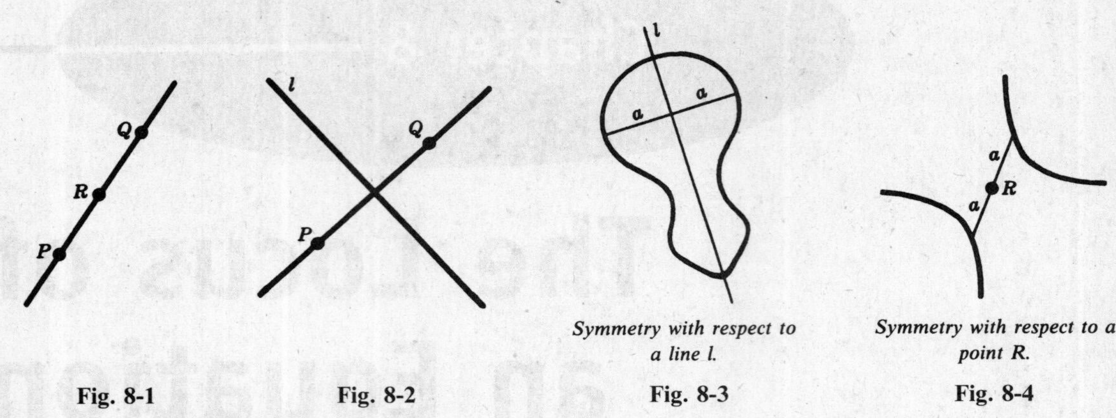

Fig. 8-1 Fig. 8-2

Symmetry with respect to a line l.

Fig. 8-3

Symmetry with respect to a point R.

Fig. 8-4

SYMMETRY OF A LOCUS. The locus of a given equation $f(x, y) = 0$ is symmetric with respect to the x axis if an equivalent equation is obtained when y is replaced by $-y$, is symmetric with respect to the y axis if an equivalent equation is obtained when x is replaced by $-x$, and is symmetric with respect to the origin if an equivalent equation is obtained when x is replaced by $-x$ and y is replaced by $-y$ simultaneously.

An equation whose graph is symmetric with respect to the y axis is called *even*; one whose graph is symmetric with respect to the x axis is *odd*.

EXAMPLE 1. Examine $x^2 + 2y^2 + x = 0$ for symmetry with respect to the coordinate axes and the origin.

When y is replaced by $-y$, we have $x^2 + 2y^2 + x = 0$; the locus is symmetric with respect to the x axis.

When x is replaced by $-x$, we have $x^2 + 2y^2 - x = 0$; the locus is not symmetric with respect to the y axis.

When x is replaced by $-x$ and y by $-y$, we have $x^2 + 2y^2 - x = 0$; the locus is not symmetric with respect to the origin. (See Problem 8.3.)

ASYMPTOTES. The line $x = a$ is a vertical asymptote to the graph of an equation (locus) if, as x gets arbitrarily close to a, the locus gets arbitrarily large.

The line $y = b$ is a horizontal asymptote to a locus if the locus gets arbitrarily close to b as x gets arbitrarily large. See Fig. 8-5. (In calculus terms, if $\lim_{x \to \infty} f(x) = a$, then $y = a$ is a horizontal asymptote to $f(x)$. If $\lim_{x \to b} f(x) = \infty$, then $x = b$ is a vertical asymptote.

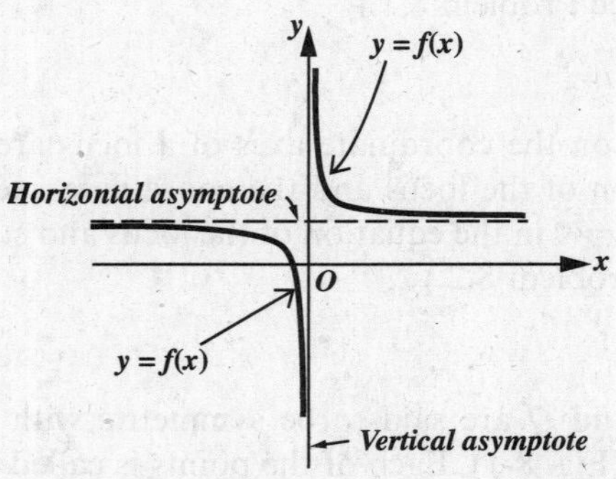

Fig. 8-5

EXAMPLE 2. Show that the line $y - 2 = 0$ is a horizontal asymptote of the curve $xy - 2x - 1 = 0$.

The locus exists for all $x \neq 0$ and for all $y \neq 2$. Since $y > 0$ for $x > 0$, there is a branch of the locus in the first quadrant. On this branch choose a point, say, $A(1, 3)$, and let a point $P(x, y)$, moving to the right from A, trace the locus.

From Table 8.1, which lists a few of the positions assumed by P, it is clear that as P moves to the right, (*a*) its ordinate y remains greater than 2 and (*b*) the difference $y-2$ may be made as small as we please by taking x sufficiently large. For example, if we wish $y-2<1/10^{12}$, we take $x>10^{12}$; if we wish $y-2<1/10^{999}$, we take $x>10^{999}$, and so on. The line $y-2=0$ is therefore a horizontal asymptote.

Table 8.1

x	y
1	3
10	2.1
100	2.01
1000	2.001
10 000	2.0001

To find the horizontal asymptotes: Solve the equation of the locus for x and set each *real* linear factor of the denominator, if any, equal to zero.

To find the vertical asymptotes: Solve the equation of the locus for y and set each *real* linear factor of the denominator, if any, equal to zero.

EXAMPLE 3. The locus of $x^2+4y^2=4$ has neither horizontal nor vertical asymptotes. The locus of $x=(y+6)/(y+3)$ has $y+3=0$ as a horizontal asymptote and $x-1=0$ as a vertical asymptote.

Solved Problems

8.1 (*a*) The locus $xy+x^2=x(y+x)=0$ consists of the lines $x=0$ and $y+x=0$.

(*b*) The locus $y^2+xy^2-xy-x^2=(x+y)(y^2-x)=0$ consists of the line $x+y=0$ and the parabola $y^2-x=0$.

(*c*) The locus $y^4+y^2-x^2-x=(y^2-x)(y^2+x+1)=0$ consists of the parabolas $y^2-x=0$ and $y^2+x+1=0$.

8.2 Examine for intercepts.

(*a*) $4x^2-9y^2=36$.

Set $y=0$: $4x^2=36$, $x^2=9$; the x intercepts are ± 3.
Set $x=0$: $-9y^2=36$, $y^2=-4$; the y intercepts are imaginary.

(*b*) $x^2-2x=y^2-4y+3$.

Set $y=0$: $x^2-2x-3=(x-3)(x+1)=0$; the x intercepts are -1 and 3.
Set $x=0$: $y^2-4y+3=(y-1)(y-3)=0$; the y intercepts are 1 and 3.

8.3 Examine for symmetry with respect to the coordinate axes and the origin.

(*a*) $4x^2-9y^2=36$.

Replacing y by $-y$, we have $4x^2-9y^2=36$; the locus is symmetric with respect to the x axis.
Replacing x by $-x$, we have $4x^2-9y^2=36$; the locus is symmetric with respect to the y axis.
Replacing x by $-x$ and y by $-y$, we have $4x^2-9y^2=36$; the locus is symmetric with respect to the origin.
Note that if a locus is symmetric with respect to the coordinate axes, it is automatically symmetric with respect to the origin. It will be shown in the next problem that the converse is not true.

(*b*) $x^3-x^2y+y^3=0$.

Replacing y by $-y$, we have $x^3+x^2y-y^3=0$; the locus is not symmetric with respect to the x axis.
Replacing x by $-x$, we have $-x^3-x^2y+y^3=0$; the locus is not symmetric with respect to the y axis.

Replacing x by $-x$ and y by $-y$, we have $-x^3 + x^2y - y^3 = -(x^3 - x^2y + y^3) = 0$; the locus is symmetric with respect to the origin.

(c) $x^2 - 4y^2 - 2x + 8y = 0$.

Replacing y by $-y$, we have $x^2 - 4y^2 - 2x - 8y = 0$, replacing x by $-x$, we have $x^2 - 4y^2 + 2x + 8y = 0$, and replacing x by $-x$ and y by $-y$, we have $x^2 - 4y^2 + 2x - 8y = 0$; the locus is not symmetric with respect to either axis or the origin.

8.4 Investigate for horizontal and vertical asymptotes.

(a) $3x + 4y - 12 = 0$.

Solve for y: $y = \dfrac{12 - 3x}{4}$. Solve for x: $x = \dfrac{12 - 4y}{3}$.

Since the denominators do not involve the variables, there are neither horizontal nor vertical asymptotes.

(b) $xy = 8$.

Solve for y: $y = 8/x$. Solve for x: $x = 8/y$. Set each denominator equal to zero:

$x = 0$ is the vertical asymptote,
$y = 0$ is the horizontal asymptote.

(c) $xy - y - x - 2 = 0$.

Solve for y: $y = \dfrac{x + 2}{x - 1}$. Solve for x: $x = \dfrac{y + 2}{y - 1}$.

Set each denominator equal to zero:

$x = 1$ is the vertical asymptote.
$y = 1$ is the horizontal asymptote.

(d) $x^2y - x - 4y = 0$.

Solve for y: $y = \dfrac{x}{x^2 - 4}$. Solve for x: $x = \dfrac{1 \pm \sqrt{1 + 16y^2}}{2y}$.

Then $x = 2$ and $x = -2$ are vertical asymptotes and $y = 0$ is the horizontal asymptote.

(e) $x^2y - x^2 + 4y = 0$.

Solve for y: $y = \dfrac{x^2}{x^2 + 4}$. Solve for x: $x = \pm 2\sqrt{\dfrac{y}{1 - y}}$.

There are no vertical asymptotes since when $x^2 + 4 = 0$, x is imaginary. The horizontal asymptote is $y = 1$.

Discuss the following equations and sketch their loci.

8.5 $y^2 = -8x$.

Intercepts: When $y = 0, x = 0$ (x intercept); when $x = 0, y = 0$ (y intercept).

Symmetry: When y is replaced by $-y$, the equation is unchanged; the locus is symmetric with respect to the x axis.

The locus is a parabola with vertex at (0, 0). It may be sketched after locating the following points: $(-1, \pm 2\sqrt{2}), (-2, \pm 4)$, and $(-3, \pm 2\sqrt{6})$. See Fig. 8-6.

8.6 $x^2 - 4x + 4y + 8 = 0$.

Intercepts: When $y = 0$, x is imaginary; when $x = 0, y = -2$ (y intercept).

Symmetry: The locus is not symmetric with respect to the coordinate axes or the origin.

The locus is a parabola with vertex at $(2, -1)$. Other points on the locus are: $(-2, -5)$, $(4, -2)$, and $(6, -5)$. See Fig. 8-7.

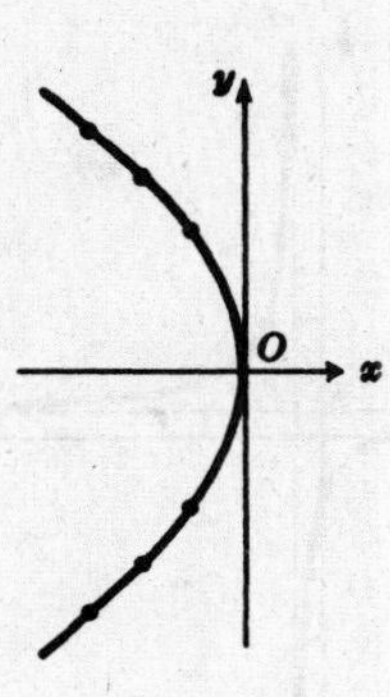

Fig. 8-6

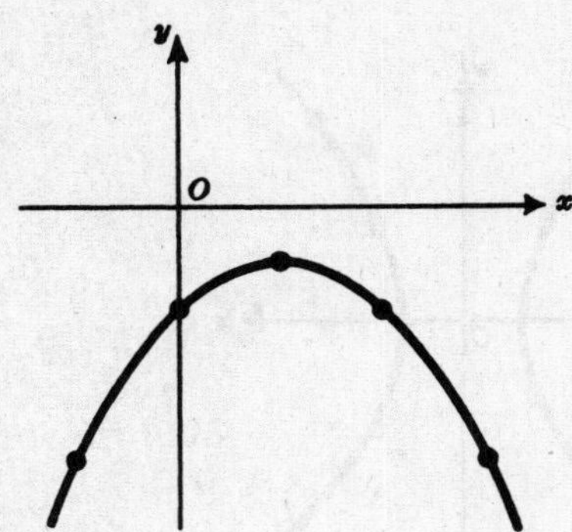

Fig. 8-7

8.7 $x^2 + y^2 - 4x + 6y - 23 = 0$.

Intercepts: When $y = 0$, $x = \dfrac{4 \pm \sqrt{16+92}}{2} = 2 \pm 3\sqrt{3}$ (*x* intercepts); when $x = 0$, $y = \dfrac{-6 \pm \sqrt{36+92}}{2}$ $= -3 \pm 4\sqrt{2}$ (*y* intercepts).

Symmetry: There is no symmetry with respect to the coordinate axes or the origin.

Completing the squares, we have

$$(x^2 - 4x + 4) + (y^2 + 6y + 9) = 23 + 4 + 9 = 36 \quad \text{or} \quad (x-2)^2 + (y+3)^2 = 36$$

the equation of a circle having center at $C(2, -3)$ and radius 6. See Fig. 8-8.

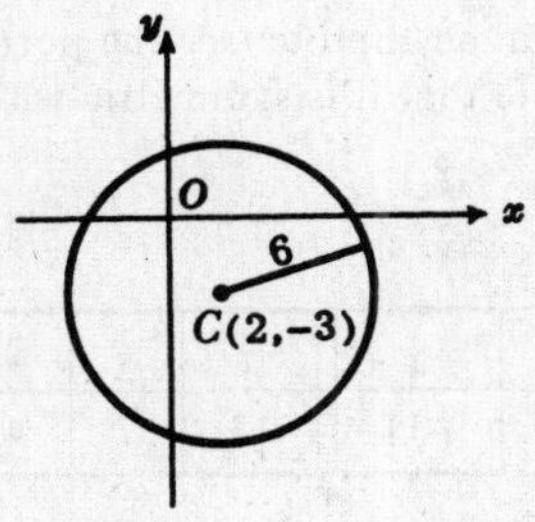

Fig. 8-8

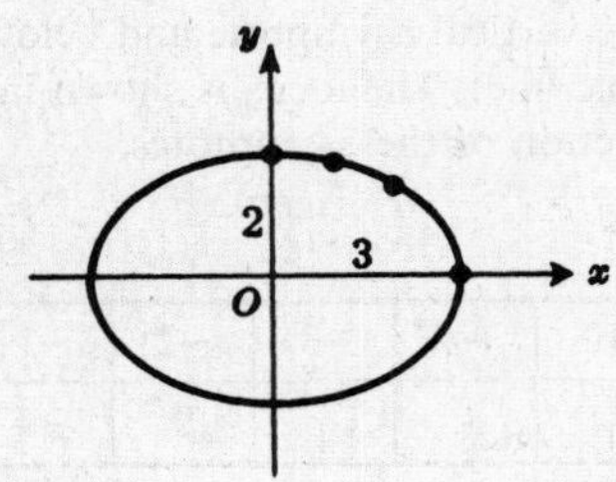

Fig. 8-9

8.8 $4x^2 + 9y^2 = 36$.

Intercepts: When $y = 0, x = \pm 3$ (*x* intercepts); when $x = 0, y = \pm 2$ (*y* intercepts).
Symmetry: The locus is symmetric with respect to the coordinate axes and the origin.

Since the locus is symmetric with respect to both the axes, only sufficient points to sketch the portion of the locus in the first quadrant are needed. Two such points are $(1, 4\sqrt{2}/3)$ and $(2, 2\sqrt{5}/3)$. The locus is called an *ellipse*. See Fig. 8-9.

8.9 $9x^2 - 4y^2 = 36$.

Intercepts: When $y = 0, x = \pm 2$ (*x* intercepts); when $x = 0, y$ is imaginary.
Symmetry: The locus is symmetric with respect to the coordinate axes and the origin.

The locus consists of two separate pieces and is not closed. The portion in the first quadrant has been sketched using the points $(3, 3\sqrt{5}/2)$, $(4, 3\sqrt{3})$, and $(5, 3\sqrt{21}/2)$. The locus is called a *hyperbola*. See Fig. 8-10.

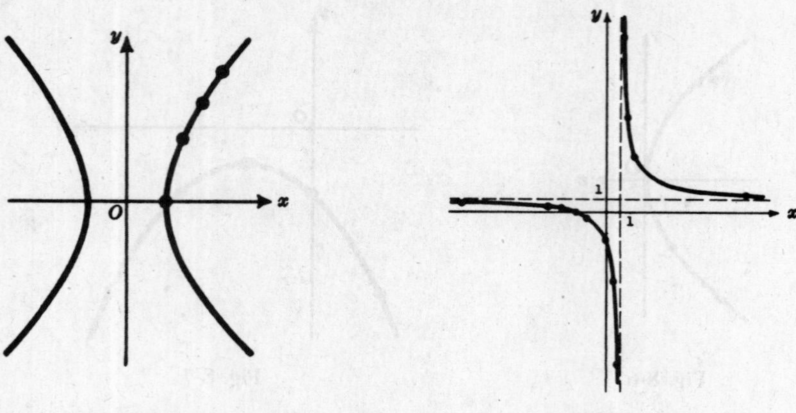

Fig. 8-10 Fig. 8-11

8.10 $xy - y - x - 2 = 0$.

Intercepts: The x intercept is -2; the y intercept is -2.
Symmetry: There is no symmetry with respect to the coordinate axes or the origin.
Asymptotes: $x = 1$, $y = 1$.

To sketch the locus, first draw in the asymptotes $x = 1$ and $y = 1$ (dotted lines). While the asymptotes are *not* a part of the locus, they serve as very convenient guide lines. Since the locus does not exist for $x = 1$ and $y = 1$, it does not cross the asymptotes. Since there is one value of y for each value of $x \neq 1$, that is, since y is single-valued, the locus appears in only two of the four regions into which the plane is separated by the asymptotes.

From Table 8.2 it is evident that the locus lies in the region to the right of the vertical asymptote and above the horizontal asymptote (see the portion of the table to the right of the double line) and in the region to the left of the vertical asymptote and below the horizontal asymptote (see the portion of the table to the left of the double line). The locus is shown in Fig. 8-11; note that it is symmetric with respect to (1, 1), the point of intersection of the asymptotes.

Table 8.2

x	-10	-4	-3	-2	-1	0	$\frac{1}{2}$	$\frac{3}{4}$	$\frac{5}{4}$	$\frac{3}{2}$	2	10
y	$\frac{8}{11}$	$\frac{2}{5}$	$\frac{1}{4}$	0	$-\frac{1}{2}$	-2	-5	-11	13	7	4	$\frac{4}{3}$

Supplementary Problems

8.11 Discuss and sketch.

(*a*) $x^2 - 4y^2 = 0$
(*b*) $x^2 + 2xy + y^2 = 4$
(*c*) $y = 9x^2$
(*d*) $y^2 = 6x - 3$
(*e*) $y^2 = 4 - 2x$
(*f*) $x^2 + y^2 = 16$
(*g*) $x^2 + y^2 = 0$
(*h*) $4x^2 + 9y^2 = 36$
(*i*) $9x^2 - 4y^2 = 36$
(*j*) $9x^2 - 4y^2 + 36 = 0$
(*k*) $xy = 4$
(*l*) $xy = -4$
(*m*) $xy^2 = -9$
(*n*) $y^3 + xy^2 = 2xy - 2x^2$
(*o*) $y = x^3$
(*p*) $y = -x^3$
(*q*) $xy - 3x - y = 0$
(*r*) $x^2 + xy + y - 2 = 0$
(*s*) $x^2y - x - 4y = 0$
(*t*) $x^2y - 4xy + 3y - x - 2 = 0$
(*u*) $x^2y - x^2 + xy + 3x - 2 = 0$
(*v*) $x^3 + xy^2 - y^2 = 0$
(*w*) $x^2 + y^2 = -4$

8.12 Use a graphing calculator to sketch the following loci:

(*a*) $x^2 - 9y^2 = 0$
(*b*) $xy = 16$
(*c*) $9x^2 + 4y^2 = 36$

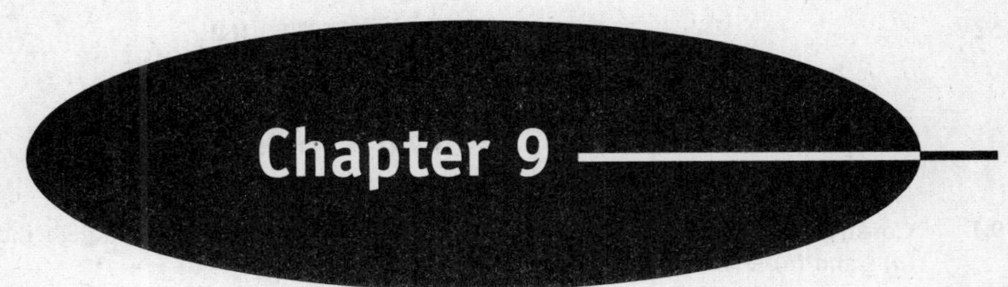

Chapter 9

The Straight Line

THE EQUATION OF THE STRAIGHT LINE parallel to the y axis at a distance a from that axis is $x = a$.

The equation of the straight line having slope m and passing through the point (x_1, y_1) is

$$y - y_1 = m(x - x_1) \qquad \textit{(Point-slope form)}$$

(See Problem 9.1.)

The equation of the line having slope m and y intercept b is

$$y = mx + b \qquad \textit{(Slope-intercept form)}$$

(See Problem 9.2.)

The equation of the line passing through the points $(x_1,\ y_1)$ and $(x_2,\ y_2)$, where $x_1 \neq x_2$, is

$$y - y_1 = \frac{y_2 - y_1}{x_2 - x_1}(x - x_1) \qquad \textit{(Two-point form)}$$

(See Problem 9.3.)

The equation of the line whose x intercept is a and whose y intercept is b, where $ab \neq 0$, is

$$\frac{x}{a} + \frac{y}{b} = 1 \qquad \textit{(Intercept form)}$$

(See Problem 9.4.)

The length of the line segment with end points (x_1, y_1) and (x_2, y_2) is

$$d = \sqrt{(x_2 - x_1)^2 + (y_2 - y_1)^2}$$

THE GENERAL EQUATION of the straight line is $Ax + By + C = 0$, where A, B, C are arbitrary constants except that not both A and B are zero.

If $C = 0$, the line passes through the origin. If $B = 0$, the line is vertical; if $A = 0$, the line is horizontal. Otherwise, the line has slope $m = -A/B$ and y intercept $b = -C/B$.

If two nonvertical lines are parallel, their slopes are equal. Thus, the lines $Ax + By + C = 0$ and $Ax + By + D = 0$ are parallel.

If two lines are perpendicular, the slope of one is the negative reciprocal of the slope of the other. If m_1 and m_2 are the slopes of two perpendicular lines, the $m_1 = -1/m_2$ or $m_1 m_2 = -1$. Thus, $Ax + By + C = 0$ and $Bx - Ay + D = 0$, where $AB \neq 0$, are perpendicular lines. (See Problems 9.5–9.8.)

Solved Problems

9.1 Construct and find the equation of the straight line which passes though the point $(-1,-2)$ with slope (*a*) $\frac{3}{4}$ and (*b*) $-\frac{4}{5}$.

(*a*) A point tracing the line *rises* (slope is positive) 3 units as it moves a horizontal distance of 4 units to the right. Thus, after locating the point $A(-1,-2)$, move 4 units to the right and 3 up to the point $B(3,1)$. The required line is AB. [See Fig. 9-1(*a*).] Using $y-y_1=m(x-x_1)$, the equation is

$$y+2=\frac{3}{4}(x+1) \quad \text{or} \quad 3x-4y-5=0$$

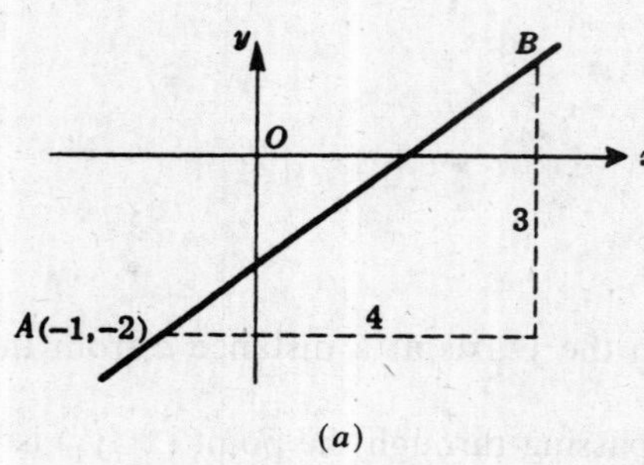

(*a*)

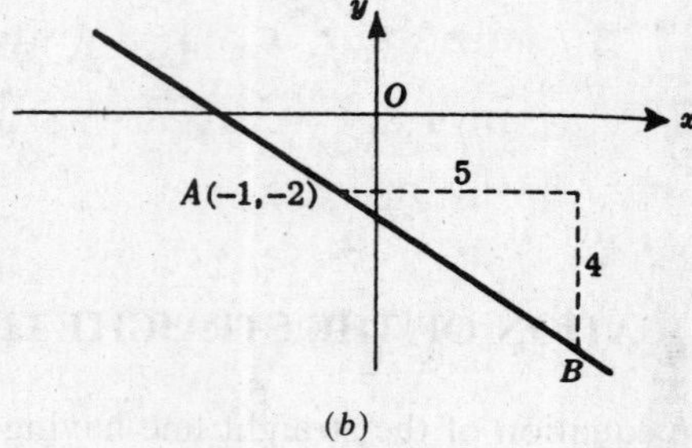

(*b*)

Fig. 9-1

(*b*) A point tracing the line falls (slope is negative) 4 units as it moves a horizontal distance of 5 units to the right. Thus, after locating the point $A(-1,-2)$, move 5 units to the right and 4 units down to the point $B(4,-6)$. The required line is AB as shown in Fig. 9-1(*b*). Its equation is

$$y+2=-\frac{4}{5}(x+1) \quad \text{or} \quad 4x+5y+14=0$$

9.2 Determine the slope m and y intercept b of the following lines. Sketch each.

(*a*) $y=\frac{3}{2}x-2$

Ans. $m=\frac{3}{2};\ b=-2$

To sketch the locus, locate the point $(0,-2)$. Then move 2 units to the right and 3 units up to another point on the required line. See Fig. 9-2(*a*).

(*b*) $y=-3x+\frac{5}{2}$

Ans. $m=-3;\ b=\frac{5}{2}$

To sketch the locus, locate the point $(0,\frac{5}{2})$. Then move 1 unit to the right and 3 units down to another point on the line. See Fig. 9-2(*b*).

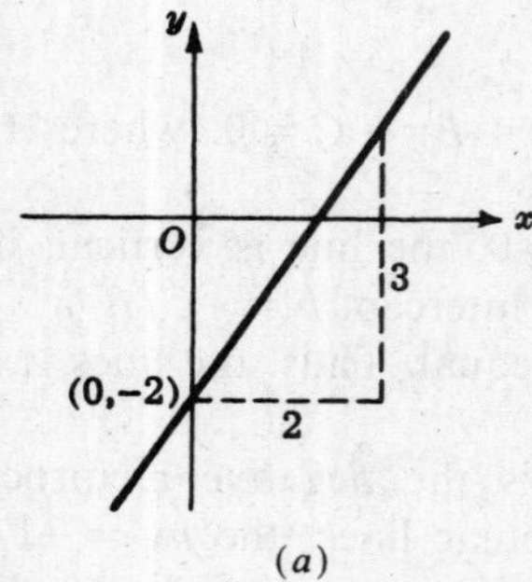

(*a*)

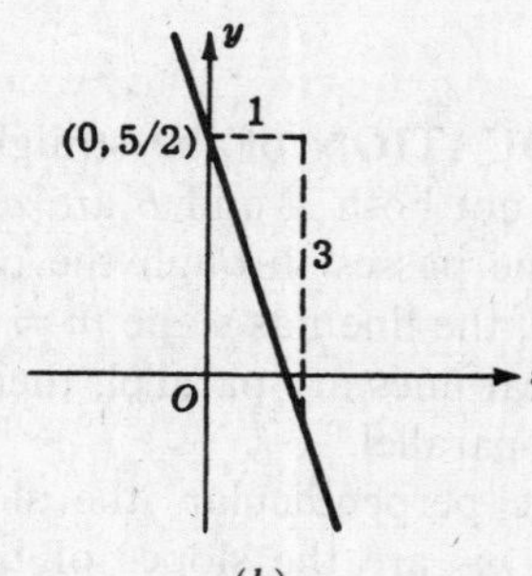

(*b*)

Fig. 9-2

9.3 Write the equation of the straight lines: (*a*) through (2, 3) and (−1, 4), (*b*) through (−7, −2) and (−2, −5), and (*c*) through (3, 3) and (3, 6).

We use $y - y_1 = \dfrac{y_2 - y_1}{x_2 - x_1}(x - x_1)$ and label each pair of points P_1 and P_2 in the order given.

(*a*) The equation is $y - 3 = \dfrac{4-3}{-1-2}(x-2) = -\dfrac{1}{3}(x-2)$ or $x + 3y - 11 = 0$.

(*b*) The equation is $y + 2 = \dfrac{-5+2}{-2+7}(x+7) = -\dfrac{3}{5}(x+7)$ or $3x + 5y + 31 = 0$.

(*c*) Here $x_1 = x_2 = 3$. The required equation is $x - 3 = 0$.

9.4 Determine the x intercept a and the y intercept b of the following lines. Sketch each.

(*a*) $3x - 2y - 4 = 0$

When $y = 0, 3x - 4 = 0$ and $x = \frac{4}{3}$; the x intercept is $a = \frac{4}{3}$.

When $x = 0, -2y - 4 = 0$ and $y = -2$; the y intercept is $b = -2$.

To obtain the locus, join the points $(\frac{4}{3}, 0)$ and $(0, -2)$ by a straight line. See Fig. 9-3(*a*).

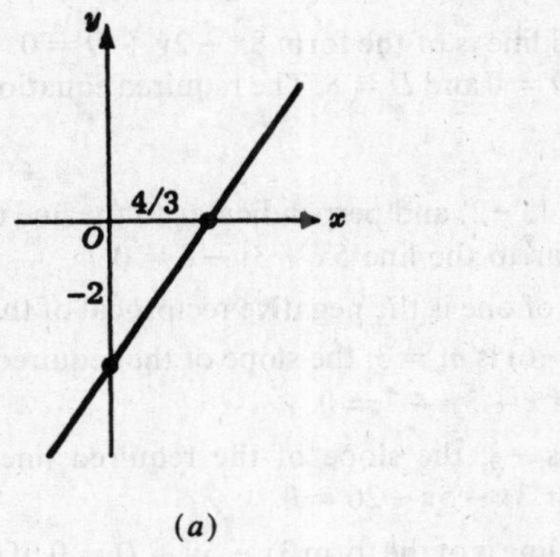

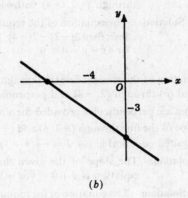

(*a*) (*b*)

Fig. 9-3

(*b*) $3x + 4y + 12 = 0$

When $y = 0, 3x + 12 = 0$ and $x = -4$; the x intercept is $a = -4$.

When $x = 0, 4y + 12 = 0$ and $y = -3$; the y intercept is $b = -3$.

The locus is the straight line joining the point (−4, 0) and (0, −3). See Fig. 9-3(*b*).

9.5 Prove: If two oblique lines l_1 and l_2 of slope m_1 and m_2, respectively, are mutually perpendicular, then $m_1 = -1/m_2$.

Let $m_2 = \tan\theta_2$, where θ_2 is the inclination of l_2. The inclination of l_1 is $\theta_1 = \theta_2 \pm 90°$ according as θ_2 is less than or greater than 90° [see Figs. 9-4(*a*) and (*b*)]. Then

$$m_1 = \tan\theta_1 = \tan(\theta_2 \pm 90°) = -\cot\theta_2 - \frac{1}{\tan\theta_2} = -\frac{1}{m_2}$$

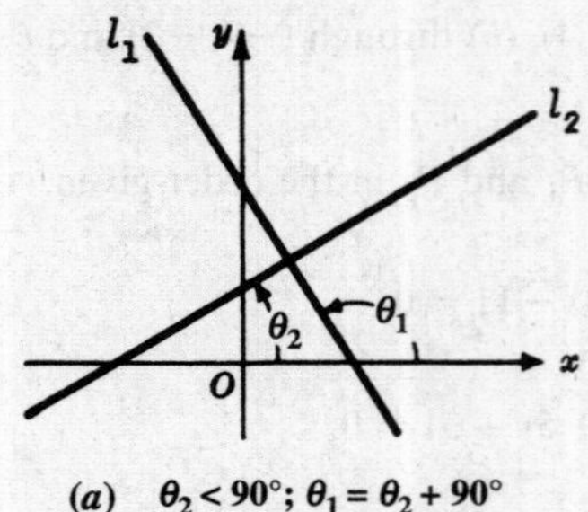

(a) $\theta_2 < 90°$; $\theta_1 = \theta_2 + 90°$

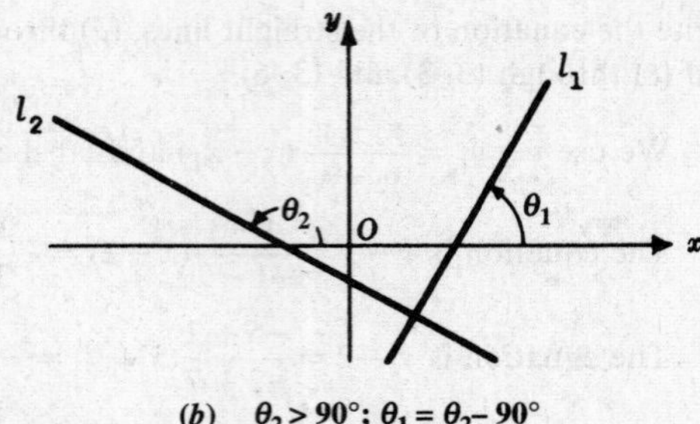

(b) $\theta_2 > 90°$; $\theta_1 = \theta_2 - 90°$

Fig. 9-4

9.6 Find the equation of the straight line (*a*) through (3, 1) and parallel to the line through (3, −2) and (−6, 5) and (*b*) through (−2, −4) and parallel to the line $8x - 2y + 3 = 0$.

Two lines are parallel provided their slopes are equal.

(*a*) The slope of the line through (3, −2) and (−6, 5) is $m = \dfrac{y_2 - y_1}{x_2 - x_1} = \dfrac{5+2}{-6-3} = -\dfrac{7}{9}$. The equation of the line through (3, 1) with slope $-\frac{7}{9}$ is $y - 1 = -\frac{7}{9}(x - 3)$ or $7x + 9y - 30 = 0$.

(*b*) **First Solution.** From $y = 4x + \frac{3}{2}$, the slope of the given line is $m = 4$. The equation of the line through (−2, −4) with slope 4 is $y + 4 = 4(x + 2)$ or $4x - y + 4 = 0$.

Second Solution. The equation of the required line is of the form $8x - 2y + D = 0$. If (−2,−4) is on the line, then $8(-2) - 2(-4) + D = 0$ and $D = 8$. The required equation is $8x - 2y + 8 = 0$ or $4x - y + 4 = 0$.

9.7 Find the equation of the straight line (*a*) through (−1, −2) and perpendicular to the line through (−2, 3) and (−5, −6) and (*b*) through (2, −4) and perpendicular to the line $5x + 3y - 8 = 0$.

Two lines are perpendicular provided the slope of one is the negative reciprocal of the slope of the other.

(*a*) The slope of the line through (−2, 3) and (−5, −6) is $m = 3$; the slope of the required line is $-1/m = -\frac{1}{3}$. The required equation is $y + 2 = -\frac{1}{3}(x + 1)$ or $x + 3y + 7 = 0$.

(*b*) **First Solution.** The slope of the given line is $-\frac{5}{3}$; the slope of the required line is $\frac{3}{5}$. The required equation is $y + 4 = \frac{3}{5}(x - 2)$ or $3x - 5y - 26 = 0$.

Second Solution. The equation of the required line is of the form $3x - 5y + D = 0$. If (2,−4) is on the line, then $3(2) - 5(-4) + D = 0$ and $D = -26$. The required equation is $3x - 5y - 26 = 0$.

9.8 Given the vertices *A*(7, 9), *B*(−5, −7), and *C*(12, −3) of the triangle *ABC* (see Fig. 9-5), find

(*a*) The equation of the side *AB*
(*b*) The equation of the median through *A*
(*c*) The equation of the altitude through *B*
(*d*) The equation of the perpendicular bisector of the side *AB*
(*e*) The equation of the line through *C* with slope that of *AB*
(*f*) The equation of the line through *C* with slope the reciprocal of that of *AB*

(*a*) $y + 7 = \dfrac{9+7}{7+5}(x + 5) = \dfrac{4}{3}(x + 5)$ or $4x - 3y - 1 = 0$

(*b*) The median through a vertex bisects the opposite side. The midpoint of *BC* is $L(\frac{7}{2}, -5)$. The equation of the median through *A* is

$$y - 9 = \frac{-5-9}{\frac{7}{2} - 7}(x - 7) = 4(x - 7) \quad \text{or} \quad 4x - y - 19 = 0$$

(c) The altitude through B is perpendicular to CA. The slope of CA is $-\frac{12}{5}$ and its negative reciprocal is $\frac{5}{12}$. The equation is $y+7=\frac{5}{12}(x+5)$ or $5x-12y-59=0$.

(d) The perpendicular bisector of AB passes through the midpoint $N(1, 1)$ and has slope $-\frac{3}{4}$. The equation is $y-1=-\frac{3}{4}(x-1)$ or $3x+4y-7=0$.

(e) $y+3=\frac{4}{3}(x-12)$ or $4x-3y-57=0$. (f) $y+3=\frac{3}{4}(x-12)$ or $3x-4y-48=0$.

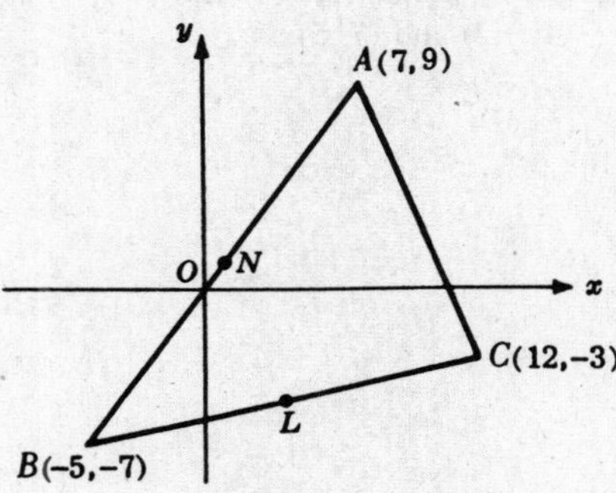

Fig. 9-5

Supplementary Problems

9.9 Determine the slope and the intercepts on the coordinate axes of each of the lines (a) $4x-5y+20=0$ and (b) $2x+3y-12=0$.

Ans. (a) $m=\frac{4}{5}$, $a=-5$, $b=4$ (b) $m=-\frac{2}{3}$, $a=6$, $b=4$

9.10 Write the equation of each of the straight lines:

(a) With slope 3 and y intercept -2. *Ans.* $y=3x-2$

(b) Through (5, 4) and parallel to $2x+3y-12=0$. *Ans.* $2x+3y-22=0$

(c) Through $(-3, 3)$ and with y intercept 6. *Ans.* $x-y+6=0$

(d) Through $(-3, 3)$ and with x intercept 4. *Ans.* $3x+7y-12=0$

(e) Through (5, 4) and $(-3, 3)$. *Ans.* $x-8y+27=0$

(f) With $a=3$ and $b=-5$. *Ans.* $5x-3y-15=0$

(g) Through (2, 3) and $(2, -5)$ *Ans.* $x-2=0$

9.11 Find the value of k such that the line $(k-1)x+(k+1)y-7=0$ is parallel to the line $3x+5y+7=0$.

Ans. $k=4$

9.12 Given the triangle whose vertices are $A(-3, 2)$, $B(5, 6)$, $C(1, -4)$. Find the equations of

(a) The sides. *Ans.* $x-2y+7=0$, $5x-2y-13=0$, $3x+2y+5=0$

(b) The medians. *Ans.* $x+6y-9=0$, $7x-6y+1=0, x=1$

(c) The altitudes. *Ans.* $2x+5y-4=0$, $2x-3y+8=0, 2x+y+2=0$

(d) The perpendicular bisectors of the sides. *Ans.* $2x+y-6=0$, $2x+5y-11=0, 2x-3y-1=0$

9.13 For the triangle of Problem 9.12: (*a*) Find the coordinates of the centroid G (intersection of the medians), the orthocenter H (intersection of the altitudes), and circumcenter C (intersection of the perpendicular bisectors of the sides). (*b*) Show that G lies on the line joining H and C and divides the segment HC in the ratio 2:1.

Ans. (*a*) $G(1, \frac{4}{3})$, $H(-\frac{7}{4}, \frac{3}{2})$, $C(\frac{19}{8}, \frac{5}{4})$

9.14 Show that the line passing through the points $(5, -4)$ and $(-2, 7)$ is the perpendicular bisector of the line segment whose end points are $(-4, -2)$ and $(7, 5)$.

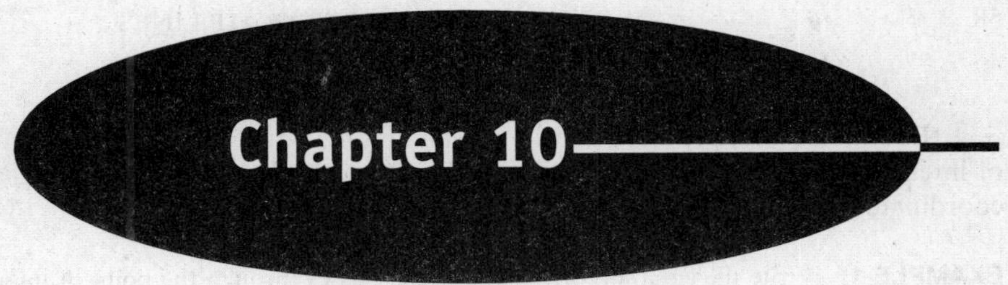

Families of Straight Lines

THE EQUATION $y = 3x + b$ represents the set of all lines having slope 3. The quantity b is a variable over the set of lines, each value of b being associated with one and only one line of the set. To distinguish it from the variables x, y, which vary over the points of each line, b is called a *parameter*. Such sets of lines are also called *one-parameter systems* or *families* of lines. See Fig. 10-1.

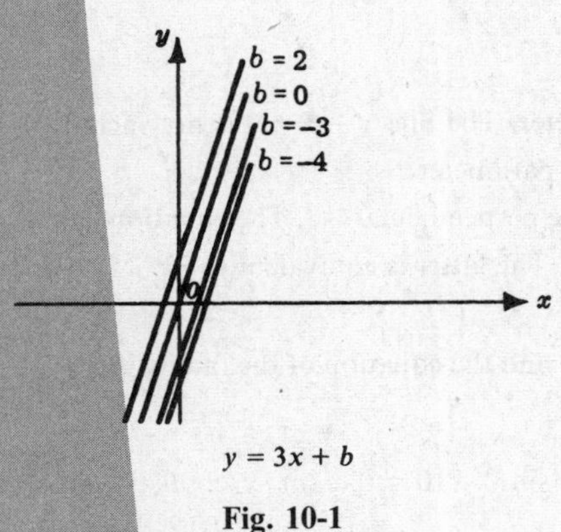

$y = 3x + b$

Fig. 10-1

$y = mx + 5$

Fig. 10-2

Similarly, the equation $y = mx + 5$, in which m is the parameter, represents the one-parameter family of lines having y intercept $= 5$ or passing through the point (0, 5). See Fig. 10-2. It is important to note that one line satisfying the geometric condition is not included since for no value of m does the equation $y = mx + 5$ yield the line $x = 0$. (See Problems 10.1–10.4.)

THE EQUATION

$$A_1x + B_1y + C_1 + k(A_2x + B_2y + C_2) = 0 \tag{10.1}$$

represents the family of all lines, except for l_2, passing through the point of intersection of the lines

$$l_1: \quad A_1x + B_1y + C_1 = 0 \qquad \text{and} \qquad l_2: \quad A_2y + C_2 = 0$$

(See Problem 10.4.)

Equation (*10.1*) provides a means of finding the equation of a line which passes through the point of intersection of two given lines and satisfies one other condition, without having first to compute the coordinate of the point of intersection.

EXAMPLE 1. Write the equation of the line which passes through the point of intersection of the lines

$$l_1: \quad 3x + 5y + 29 = 0 \qquad \text{and} \qquad l_2: \quad 7x - 11y - 13 = 0$$

and through the point (2, 1).

Since the given point is not on l_2, the required line is among those given by

$$l_1 + kl_2: \quad 3x + 5y + 29 + k(7x - 11y - 13) = 0$$

For the line which passes through (2, 1), $3(2) + 5(1) + 29 + k[7(2) - 11(1) - 13] = 0$ and $k = 4$; hence, its equation is

$$l_1 + 4l_2: \quad 3x + 5y + 29 + 4(7x - 11y - 13) = 31x - 39y - 23 = 0$$

(See Problem 10.6.)

Solved Problems

10.1 Write the equation of the family of lines satisfying the given condition. Name the parameter and list any lines satisfying the condition but not obtained by assigning a value to the parameter.

(*a*) Parallel to the x axis.
(*b*) Through the point $(-3, 2)$.
(*c*) At a distance 5 from the origin.
(*d*) Perpendicular to $5x + 3y - 8 = 0$.
(*e*) The sum of whose intercepts is 10.

(*a*) The equation is $y = k$, k being the parameter.

(*b*) The equation is $y - 2 = m(x + 3)$, m being the parameter. The line $x + 3 = 0$ is not included.

(*c*) The equation is $x \cos\omega + y \sin\omega - 5 = 0$, ω being the parameter.

(*d*) The slope of the given lines is $-\frac{5}{3}$; hence, the slope of the perpendicular is $\frac{3}{5}$. The equation of the family is $y = \frac{3}{5}x + c$ or $3x - 5y + 5c = 0$, c being the parameter. The latter is equivalent to $3x - 5y + k = 0$, with k as parameter.

(*e*) Taking the x intercept as $a \neq 0$, the y intercept is $10 - a$ and the equation of the family is $\dfrac{x}{a} + \dfrac{y}{10 - a} = 1$. The parameter is a.

10.2 Describe each family of lines: (*a*) $x = k$, (*b*) $x \cos\omega + y \sin\omega - 10 = 0$, (*c*) $x/\cos\theta + y/\sin\theta = 1$, and (*d*) $kx + \sqrt{1 - k^2}\,y - 10 = 0$.

(*a*) This is the family of all vertical lines.

(*b*) This is the family of all tangents to the circle with center at the origin and radius = 10.

(*c*) This is the family of all lines the sum of the squares of whose intercepts is 1.

(*d*) Since $\sqrt{1 - k^2}$ is to be assumed real, the range of $k = \cos\omega$ is $-1 \leq k \leq 1$, while the range of $\sqrt{1 - k^2} = \sin\omega$ is $0 \leq \sqrt{1 - k^2} \leq 1$. The equation is that of the family of tangents to the upper half circle of (*b*).

10.3 In each of the following write the equation of the family of lines satisfying the first condition and then obtain the equation of the line of the family satisfying the second condition: (*a*) parallel to $3x - 5y + 12 = 0$, through $P(1, -2)$, (*b*) perpendicular to $3x - 5y + 12 = 0$, through $P(1, -2)$, and (*c*) through $P(-3, 2)$, at a distance 3 from the origin.

(*a*) The equation of the family is $3x - 5y + k = 0$. Substituting $x = 1$, $y = -2$ in this equation and solving for k, we find $k = -13$. The required line has equation $3x - 5y - 13 = 0$.

(*b*) The equation of the family is $5x + 3y + k = 0$ [see Problem 10.1(*d*)]. Proceeding as in (*a*) above, we find $k = 1$; the required line has equation $5x + 3y + 1 = 0$.

(*c*) The equation of the family is $y - 2 = k(x + 3)$ or, in normal form, $\dfrac{kx - y + (3k + 2)}{\pm\sqrt{k^2 + 1}} = 0$. Setting the undirected distance of a line of the family from the origin equal to 3, we have $\left|\dfrac{3k + 2}{\pm\sqrt{k^2 + 1}}\right| = 3$. Then $\dfrac{(3k + 2)^2}{k^2 + 1} = 9$ and $k = \frac{5}{12}$. Thus, the required line has equation $y - 2 = \frac{5}{12}(x + 3)$ or $5x - 12y - 12y + 39 = 0$.

Now there is a second line, having equation $x + 3 = 0$, satisfying the conditions, but this line [see Problem 10.1(*b*)] was not included in the equation of the family.

10.4 Write the equation of the line which passes through the point of intersection of the lines l_1: $3x + 5y + 26 = 0$ and l_2: $7x - 11y - 13 = 0$ and satisfies the additional condition: (*a*) Passes through the origin. (*b*) Is perpendicular to the line $7x + 3y - 9 = 0$.

Each of the required lines is a member of the family

$$l_1 + kl_2: \quad 3x + 5y + 26 + k(7x - 11y - 13) = 0 \qquad (A)$$

(*a*) Substitution $x = 0, y = 0$ in (A), we find $k = 2$; the required line has equation

$$l_1 + 2l_2: \quad 3x + 5y + 26 + 2(7x - 11y - 13) = 0 \quad \text{or} \quad x - y = 0$$

(*b*) The slope of the given line is $-\frac{7}{3}$ and the slope of a line of (*A*) is $-\dfrac{3 + 7k}{5 - 11k}$.

Setting one slope equal to the negative reciprocal of the other, we find $-\dfrac{7}{3} = \dfrac{5 - 11k}{3 + 7k}$ and $k = -\frac{9}{4}$. The required line has equation

$$l_1 - \tfrac{9}{4}l_2: \quad 3x + 5y + 26 - \tfrac{9}{4}(7x - 11y - 13) = 0 \quad \text{or} \quad 3x - 7y - 13 = 0$$

Supplementary Problems

10.5 Write the equation of the family of lines satisfying the given condition. List any lines satisfying the condition but not obtained by assigning a value to the parameter.

(*a*) Perpendicular to the x axis.

(*b*) Through the point $(3, -1)$.

(*c*) At the distance 6 units from the origin.

(*d*) Parallel to $2x + 5y - 8 = 0$.

(*e*) The product of whose intercepts on the coordinate axes is 10.

Ans. (*a*) $a = k$ (*c*) $x\cos\omega + y\sin\omega - 6 = 0$ (*e*) $10x + k^2y - 10k = 0$
(*b*) $y = kx - 3k - 1, x = 3$ (*d*) $2x + 5y + k = 0$

10.6 Write the equation of the family of lines satisfying the first condition and then obtain the equation of that line of the family satisfying the second condition.

(*a*) Parallel to $2x - 3y + 8 = 0$; passing through $P(2, -2)$.

(*b*) Perpendicular to $2x - 3y + 8 = 0$; passing through $P(2, -2)$.

(*c*) Sum of the intercepts on the coordinate axes is 2; forms with the coordinate axes a triangle of area 24 square units.

(*d*) At a distance 3 units from the origin; passes through $P(1,3)$.

(*e*) Slope is $\frac{3}{2}$; product of the intercepts on the coordinate axes is -54.

(*f*) The y intercept is 6; makes an angle of 135° with the line $7x - y - 23 = 0$.

(*g*) Slope is $\frac{5}{12}$; 5 units from $(2, -3)$.

Ans. (*a*) $2x - 3y - 10 = 0$

(*b*) $3x + 22y - 2 = 0$

(*c*) $3x - 4y - 24 = 0, \quad 4x - 3y + 24 = 0$

(*d*) $3x + 4y - 15 = 0, \quad y = 3$

(*e*) $3x - 2y \pm 18 = 0$

(*f*) $3x - 4y + 24 = 0$

(*g*) $5x - 12y + 19 = 0, \quad 5x - 12y - 111 = 0$

10.7 Find the equation of the line which passes through the point $P(3, -4)$ and has the additional property:

(*a*) The sum of its intercepts on the coordinate axes is -5.

(*b*) The product of its intercepts on the coordinate axes is the negative reciprocal of its slope.

(*c*) It forms with the coordinate axes a triangle of area 1 square unit.

Ans. (*a*) $2x - y - 10 = 0, \quad 2x + 3y + 6 = 0$

(*b*) $x + y + 1 = 0, \quad 5x + 3y - 3 = 0$

(*c*) $2x + y - 2 = 0, \quad 8x + 9y + 12 = 0$

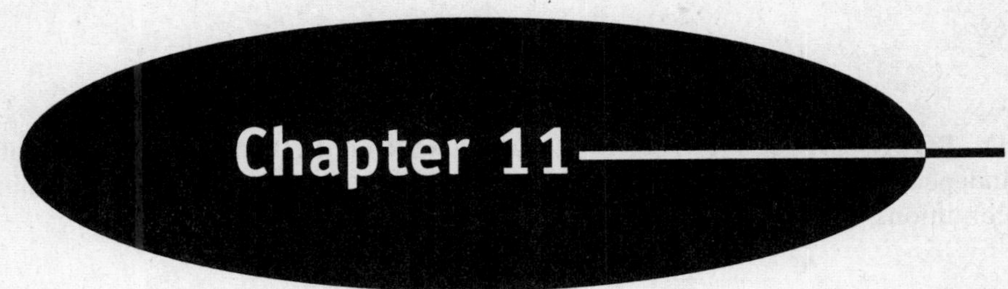

The Circle

A CIRCLE IS THE LOCUS of a point which moves in a plane so that it is always at a constant distance from a fixed point in the plane. The fixed point is called the *center* and the constant distance is the length of the radius of the circle. The circle is one of the "conic sections." That is, the circle also results from taking particular cross sections of the right-circular cone. As you think about slicing a cone with a plane, what additional cross sections are generated?

THE STANDARD FORM of the equation of the circle whose center is at the point $C(h, k)$ and whose radius is the constant r is

$$(x-h)^2 + (y-k)^2 = r^2 \tag{11.1}$$

See Fig. 11-1. (See Problem 11.1.)

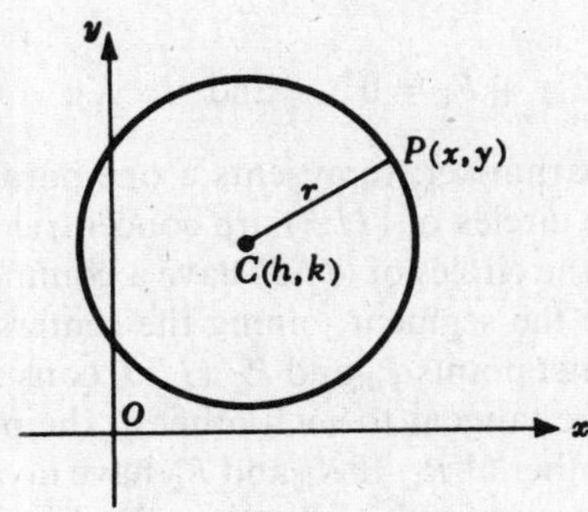

Fig. 11-1

THE GENERAL FORM of the equation of a circle is

$$x^2 + y^2 + 2Dx + 2Ey + F = 0 \tag{11.2}$$

By completing the squares this may be put in the form

$$(x+D)^2 + (x+E)^2 = D^2 + E^2 - F$$

Thus, (*11.2*) represents a circle with center at $C(-D, -E)$ and radius $\sqrt{D^2 + E^2 - F}$ if $(D^2 + E^2 - F) > 0$, a point if $(D^2 + E^2 - F) = 0$, and an imaginary locus if $(D^2 + E^2 - F) = 0$. (See Problems 11.2–11.3.)

IN BOTH THE STANDARD AND GENERAL FORM the equation of a circle contains three independent arbitrary constants. It follows that a circle is uniquely determined by three independent conditions. (See Problems 11.4–11.6.)

THE EQUATION OF A TANGENT to a circle may be found by making use of the fact that a tangent and the radius drawn to the point of tangency are perpendicular. (See Problem 11.7.)

THE LENGTH OF A TANGENT to a circle from an external point P_1 is defined as the distance from the point P_1 to the point of tangency. The two tangents from an external point are of equal length. See Fig. 11-2, where $\overline{P_1T} \cong \overline{P_1T'}$.

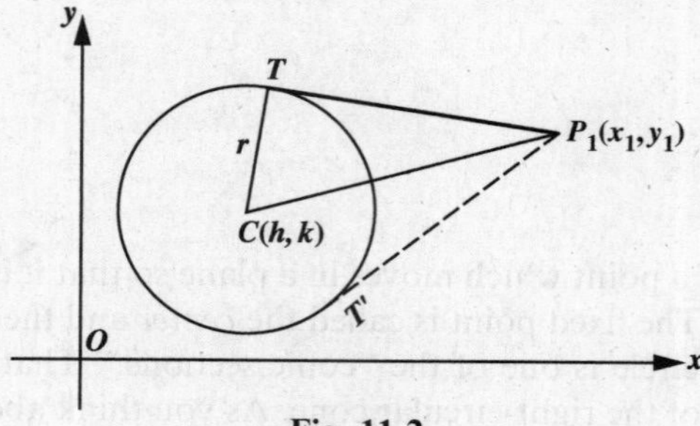

Fig. 11-2

The square of the length of a tangent from the external point $P_1(x_1, y_1)$ to a circle is obtained by substituting the coordinates of the point in the left member of the equation of the circle when written in the form $(x-h)^2 + (y-k)^2 - r^2 = 0$ or $x^2 + y^2 + 2Dx + 2Ey + F = 0$. (See Problems 11.8–11.9.)

THE EQUATION

$$x^2 + y^2 + 2D_1x + 2E_1y + F_1 + k(x^2 + y^2 + 2D_2x + 2E_2y + F_2) = 0 \qquad (11.3)$$

where

$$K_1: \quad x^2 + y^2 + 2D_1x + 2E_1y + F_1 = 0 \qquad \text{and} \qquad K_2: \quad x^2 + y^2 + 2D_2x + 2E_2y + F_2 = 0$$

are distinct circles and $k \neq -1$ is a parameter, represents a one-parameter family of circles.

If K_1 and K_2 are concentric, the circles of (*11.3*) are concentric with them.

If K_1 and K_2 are not concentric, the circles of (*11.3*) have a common line of centers with them and the centers of the circles of (*11.3*) divide the segment joining the centers of K_1 and K_2 in the ratio $k : 1$.

If K_1 and K_2 intersect in two distinct points P_1 and P_2, (*11.3*) consists of all circles except K_2 which pass through these points. If K_1 and K_2 are tangent to each other at the point P_1, (*11.3*) consists of all circles except K_2 which are tangent to each other at P_1. If K_1 and K_2 have no point in common, any two circles of the family (*11.3*) have no point in common with each other. See Fig. 11-3. (See Problems 11.10–11.11.)

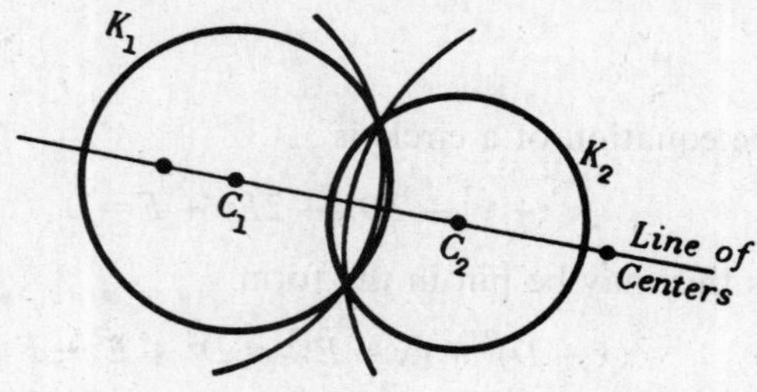

Fig. 11-3

Solved Problems

11.1 Write the equation of the circle satisfying the given conditions.

(*a*) $C(0,\ 0)$, $r = 5$
(*b*) $C(4,\ -2)$, $r = 8$
(*c*) $C(-4,\ -2)$ and passing through $P(1, 3)$
(*d*) $C(-5,\ 6)$ and tangent to the x axis
(*e*) $C(3, 4)$ and tangent to $2x - y + 5 = 0$
(*f*) Center on $y = x$, tangent to both axes, $r = 4$

(*a*) Using $(x-h)^2 + (y-k)^2 = r^2$, the equation is $(x-0)^2 + (y-0)^2 = 25$ or $x^2 + y^2 = 25$.

(*b*) Using $(x-h)^2 + (y-k)^2 = r^2$, the equation is $(x-4)^2 + (y+2)^2 = 64$.

(*c*) Since the center is at $C(-4,\ -2)$, the equation has the form $(x+4)^2 + (y+2)^2 = r^2$. The condition that $P(1, 3)$ lie on this circle is $(1+4)^2 + (3+2)^2 = r^2 = 50$. Hence the required equation is $(x+4)^2 + (y+2)^2 = 50$.

(*d*) The tangent to a circle is perpendicular to the radius drawn to the point of tangency; hence $r = 6$. The equation of the circle is $(x+5)^2 + (y-6)^2 = 36$. See Fig. 11-4(*a*).

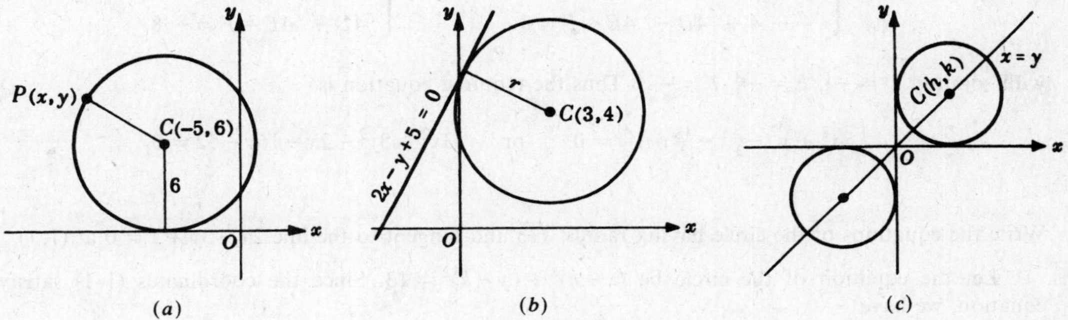

Fig. 11-4

(*e*) The radius is the undirected distance of the point $C(3, 4)$ from the line $2x - y + 5 = 0$; thus $r = \left|\dfrac{2 \cdot 3 - 4 + 5}{-\sqrt{5}}\right| = \dfrac{7}{\sqrt{5}}$. The equation of the circle is $(x-3)^2 + (y-4)^2 = \frac{49}{5}$. See Fig. 11-4(*b*).

(*f*) Since the center $(h,\ k)$ lies on the line $x = y$, $h = k$; since the circle is tangent to both axes, $|h| = |k| = r$. Thus there are two circles satisfying the conditions, one with center $(4,\ 4)$ and equation $(x-4)^2 + (y-4)^2 = 16$, the other with center $(-4,\ -4)$ and equation $(x+4)^2 + (y+4)^2 = 16$. See Fig. 11-4(*c*).

11.2 Describe the locus represented by each of the following equations:

(*a*) $x^2 + y^2 - 10x + 8y + 5 = 0$
(*b*) $x^2 + y^2 - 6x - 8y + 25 = 0$
(*c*) $x^2 + y^2 + 4x - 6y + 24 = 0$
(*d*) $4x^2 + 4y^2 + 80x + 12y + 265 = 0$

(*a*) From the standard form $(x-5)^2 + (y+4)^2 = 36$, the locus is a circle with center at $C(5,\ -4)$ and radius 6.

(*b*) From the standard form $(x-3)^2 + (y-4)^2 = 0$, the locus is a point circle or the point (3, 4).

(*c*) Here we have $(x+2)^2 + (y-3)^2 = -11$; the locus is imaginary.

(*d*) Dividing by 4, we have $x^2 + y^2 + 20x + 3y + \frac{265}{4} = 0$. From $(x+10)^2 + (y+\frac{3}{2})^2 = 36$, the locus is a circle with center at $C(-10,\ -\frac{3}{2})$ and radius 6.

11.3 Show that the circles $x^2+y^2-16x-20y+115=0$ and $x^2+y^2+8x-10y+5=0$ are tangent and find the point of tangency.

The first circle has center $C_1(8,\ 10)$ and radius 7; the second has center $C_2(-4,5)$ and radius 6. The two circles are tangent externally since the distance between their centers $C_1C_2=\sqrt{144+25}=13$ is equal to the *sum* of the radii.

The point of tangency $P(x,y)$ divides the segment C_2C_1 in the ratio 6:7. Then

$$x=\frac{6\cdot 8+7(-4)}{6+7}=\frac{20}{13},\qquad y=\frac{6\cdot 10+7\cdot 5}{6+7}=\frac{95}{13}$$

and the point of tangency has coordinates $(\frac{20}{13},\frac{95}{13})$.

11.4 Find the equation of the circle through the points (5, 1), (4, 6), and (2, −2).

Take the equation in general form $x^2+y^2+2Dx+2Ey+F=0$. Substituting successively the coordinates of the given points, we have

$$\begin{cases}25+1+10D+2E+F=0\\16+36+8D+12E+F=0\\4+4+4D-4E+F=0\end{cases}\qquad\text{or}\qquad\begin{cases}10D+2E+F=-26\\8D+12E+F=-52\\4D-4E+F=-8\end{cases}$$

with solution $D=-\frac{1}{3}$, $E=-\frac{8}{3}$, $F=-\frac{52}{3}$. Thus the required equation is

$$x^2+y^2-\tfrac{2}{3}x-\tfrac{16}{3}y-\tfrac{52}{3}=0\qquad\text{or}\qquad 3x^2+3y^2-2x-16y-52=0$$

11.5 Write the equations of the circle having radius $\sqrt{13}$ and tangent to the line $2x-3y+1=0$ at (1, 1).

Let the equation of the circle be $(x-h)^2+(y-k)^2=13$. Since the coordinates (1, 1) satisfy this equation, we have

$$(1-h)^2+(1-k)^2=13 \tag{1}$$

The undirected distance from the tangent to the center of the circle is equal to the radius; that is,

$$\left|\frac{2h-3k+1}{-\sqrt{3}}\right|=\sqrt{13}\qquad\text{and}\qquad\frac{2h-3k+1}{\sqrt{13}}=\pm\sqrt{13} \tag{2}$$

Finally, the radius through (1, 1) is perpendicular to the tangent there; that is,

$$\text{Slope of radius through }(1,1)=-\frac{1}{\text{slope of given line}}\qquad\text{or}\qquad\frac{k-1}{h-1}=-\frac{3}{2} \tag{3}$$

Since there are only two unknowns, we may solve simultaneously any two of the three equations. Using (2) and (3), noting that there are two equations in (2), we find $h=3,\ k=-2$ and $h=-1, k=4$. The equations of the circles are $(x-3)^2+(y+2)^2=13$ and $(x+1)^2+(y-4)^2=13$.

11.6 Write the equations of the circles satisfying the following sets of conditions:

(*a*) Through (2, 3) and (−1, 6), with center on $2x+5y+1=0$.

(*b*) Tangent to $5x-y-17=0$ at (4, 3) and also tangent to $x-5y-5=0$.

(*c*) Tangent to $x-2y+2=0$ and to $2x-y-17=0$, and passing through (6, −1).

(*d*) With center in the first quadrant and tangent to lines $y=0$, $5x-12y=0$, $12x+5y-39=0$.

Take the equation of the circle in standard form $(x-h)^2+(y-k)^2=r^2$.

(*a*) We obtain the following system of equations:

$$2h+5k+1=0\qquad[\text{center }(h,\ k)\text{ on }2x+5y+1=0] \tag{1}$$

$$(2-h)^2+(3-k)^2=r^2 \quad \text{[point (2, 3) on the circle]} \tag{2}$$

$$(-1-h)^2+(6-k)^2=r^2 \quad \text{[point (−1, 6) on the circle]} \tag{3}$$

The elimination of r between (*2*) and (*3*) yields $h-k+4=0$ and when this is solved simultaneously with (*1*), we obtain $h=-3$, $k=1$. By (*2*), $r^2=(2+3)^2+(3-1)^2=29$; the equation of the circle is $(x+3)^2+(y-1)^2=29$.

(*b*) We obtain the following system of equations:

$$(4-h)^2+(3-k)^2=r^2 \quad \text{[point (4, 3) on the circle]} \tag{1}$$

$$\left(\frac{5h-k-17}{\sqrt{26}}\right)^2=\left(\frac{h-5k-5}{\sqrt{26}}\right)^2 \quad \text{(the square of the directed distance from each tangent to the center is } r^2\text{)} \tag{2}$$

$$\frac{k-3}{h-4}=-\frac{1}{5} \quad \text{[the radius drawn to (4, 3) is perpendicular to the tangent there]} \tag{3}$$

The elimination of h between (*2*) and (*3*) yields

$$\left(\frac{-26k+78}{\sqrt{26}}\right)^2=\left(\frac{-10k+14}{\sqrt{26}}\right)^2$$

or $$9k^2-59k+92=(k-4)(9k-23)=0;\ \text{then } k=4, \tfrac{23}{9}.$$

When $k=4, h=-5k+19=-1$; then $r^2=(4+1)^2+(3-4)^2=26$ and the equation of the circle is $(x+1)^2+(y-4)^2=26$. When $k=\frac{23}{9}, h=\frac{56}{9}$ and $r^2=\frac{416}{81}$; the equation of the circle is $(x-\frac{56}{9})^2+(y-\frac{23}{9})^2=\frac{416}{81}$.

(*c*) Observing the directions indicated in Fig. 11-5(*a*), we obtain for each circle the following system of equations:

$$-\frac{h-2k+2}{-\sqrt{5}}=r \quad (x-2y+2=0 \text{ is tangent to the circle}) \tag{1}$$

$$-\frac{2h-k-17}{\sqrt{5}}=r \quad (2x-y-17=0 \text{ is tangent to the circle}) \tag{2}$$

$$(6-h)^2+(-1-k)^2=r^2 \quad \text{[point (6, −1) is on the circle]} \tag{3}$$

The elimination of r between (*1*) and (*2*) yields

$$h-k-5=0 \tag{4}$$

and the elimination of r between (*1*) and (*3*) yields

$$(6-h)^2+(-1-k)^2=\frac{(h-2k+2)^2}{5} \tag{5}$$

Eliminating h between (*4*) and (*5*), we have $(1-k)^2+(-1-k)^2=(-k+7)^2/5$ or $9k^2+14k-39=(k+3)(9k-13)=0$ and $k=-3, \frac{13}{9}$.

When $k=-3, h=k+5=2, r^2=(6-2)^2+[-1-(-3)]^2=20$ and the equation of the circle is $(x-2)^2+(y+3)^2=20$. When $k=\frac{13}{9}, h=k+5=\frac{58}{9}$, $r^2=(6-\frac{58}{9})^2+(-1-\frac{13}{9})^2=\frac{500}{81}$ and the circle has equation $(x-\frac{58}{9})^2+(y-\frac{13}{9})^2=\frac{500}{81}$.

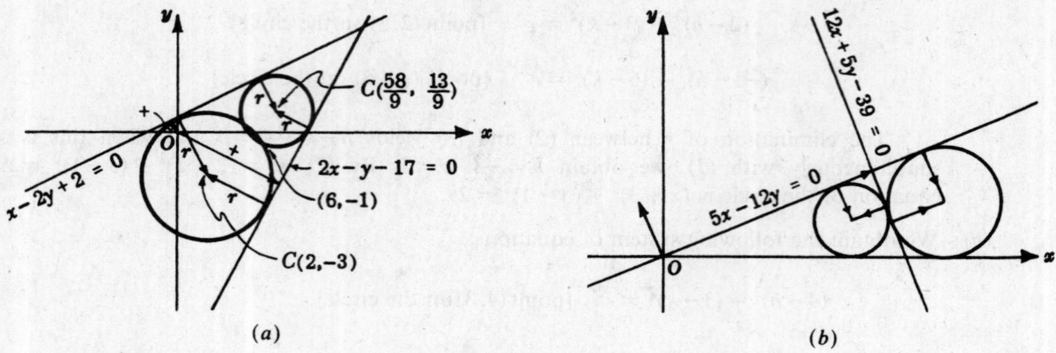

Fig. 11-5

(*d*) Observing the directions indicated in Fig. 11-5(*b*), we have for the inscribed circle the following system of equations:

$$k = r, \quad -\frac{5h-12k}{-13} = r \quad \text{and} \quad -\frac{12h+5k-39}{13} = r$$

Then $h = \frac{5}{2}$, $k = r = \frac{1}{2}$, and the equation of the circle is $(x-\frac{5}{2})^2 + (y-\frac{1}{2})^2 = \frac{1}{4}$.

For the other circle, we have the following system of equations:

$$k = r, \quad -\frac{5h-12k}{-13} = r \quad \text{and} \quad \frac{12h+5k-39}{13} = r$$

Here $h = \frac{15}{4}$, $k = r = \frac{3}{4}$, and the circle has equation $(x-\frac{15}{4})^2 + (y-\frac{3}{4})^2 = \frac{9}{16}$.

11.7 Find the equations of all tangents to the circle $x^2 + y^2 - 2x + 8y - 23 = 0$:

(*a*) at the point $(3, -10)$ on it, (*b*) having slope 3, and (*c*) through the external point $(8, -3)$.

(*a*) The center of the given circle is $C(1, -4)$. Since the tangent at any point *P* on the circle is perpendicular to the radius through *P*, the slope of the tangent is $\frac{1}{3}$. The equation of the tangent is then $y + 10 = \frac{1}{3}(x-3)$ or $x - 3y - 33 = 0$.

(*b*) Each tangent belongs to the family of lines $y = 3x + b$ or, in normal form, $\dfrac{3x - y + b}{\pm\sqrt{10}} = 0$.

The undirected distance from any tangent to the center of the circle is equal to the radius of the circle; thus,

$$\left|\frac{3(1)-(-4)+b}{\pm\sqrt{10}}\right| = 2\sqrt{10} \quad \text{or} \quad \frac{7+b}{\sqrt{10}} = \pm\sqrt{10} \quad \text{and} \quad b = 13, -27$$

There are then two tangents having equations $y = 3x + 13$ and $y = 3x - 27$.

(*c*) The tangents to a circle through the external point $(8, -3)$ belong to the family of lines $y + 3 = m(x-8)$ or $y = mx - 8m - 3$. When this replacement for *y* is made in the equation of the circle, we have

$$x^2 + (mx - 8m - 3)^2 - 2x + 8(mx - 8m - 3) - 23 = (m^2+1)x^2$$
$$+ (-16m^2 + 2m - 2)x + (64m^2 - 16m - 38) = 0$$

Now this equation will have equal roots *x* provided the discriminant is zero, that is, provided $(-16m^2 + 2m - 2)^2 - 4(m^2+1)(64m^2 - 16m - 38) = -4(m-3)(9m+13) = 0$. Then $m = 3, -\frac{13}{9}$ and the equations of the tangents are $y + 3 = 3(x-8)$ or $3x - y - 27 = 0$ and $y + 3 = -\frac{13}{9}(x-8)$ or $13x + 9y - 77 = 0$.

11.8 Find the length of the tangent

(*a*) To the circle $x^2 + y^2 - 2x + 8y - 23 = 0$ from the point $(8, -3)$

(*b*) To the circle $4x^2 + 4y^2 - 2x + 5y - 8 = 0$ from the point $(-4, 4)$

Denote the required length by *t*.

(*a*) Substituting the coordinates (8, −3) in the left member of the equation of the circle, we have

$$t^2 = (8)^2 + (-3)^2 - 2(8) + 8(-3) - 23 = 0 \qquad \text{and} \qquad t = \sqrt{10}$$

(*b*) From the general form of the equation $x^2 + y^2 - \frac{1}{2}x + \frac{5}{4}y - 2 = 0$, we find

$$t^2 = (-4)^2 + (4)^2 - \tfrac{1}{2}(-4) + \tfrac{5}{4}(4) - 2 = 37 \qquad \text{and} \qquad t = \sqrt{37}$$

11.9 For the circle $x^2 + y^2 + 6x - 8y = 0$, find the values of m for which the lines of the family $y = mx - \frac{1}{3}$ (*a*) intersect the circle in two distinct points, (*b*) are tangent to the circle, and (*c*) do not meet the circle.

Eliminating y between the two equations, we have

$$x^2 + \left(mx - \tfrac{1}{3}\right)^2 + 6x - 8\left(mx - \tfrac{1}{3}\right) = (m^2 + 1)x^2 + (6 - \frac{26}{3}m)x + \frac{25}{9} = 0$$

This equation will have real and distinct roots, two equal roots, or two imaginary roots according as its discriminant

$$(6 - \tfrac{26}{3}m)^2 - 4(m^2 + 1)\left(\tfrac{25}{9}\right) = \tfrac{8}{9}(72m^2 - 117m + 28) = \tfrac{8}{9}(3m - 4)(24m - 7) >, =, \text{or} < 0$$

(*a*) The lines will intersect the circle in two distinct points when $m > \frac{4}{3}$ and $m < \frac{7}{24}$.

(*b*) The lines will be tangent to the circle when $m = \frac{4}{3}$ and $m = \frac{7}{24}$.

(*c*) The lines will not meet the circle when $\frac{7}{24} < m < \frac{4}{3}$.

11.10 Write the equation of the family of circles satisfying the given conditions: (*a*) having the common center (−2, 3), (*b*) having radius = 5, and (*c*) with center on the x axis.

(*a*) The equation is $(x + 2)^2 + (y - 3)^2 = r^2$, r being a parameter.

(*b*) The equation is $(x - h)^2 + (y - k)^2 = 25$, h and k being parameters.

(*c*) Let the center have coordinates $(k_1, 0)$ and denote the radius by k_2. The equation of the family is $(x - k_1)^2 + y^2 = k_2^2$, k_1 and k_2 being parameters.

11.11 Write the equation of the family of circles which are tangent to the circles of Problem 11.3 at their common point and determine the circle of the family having the property (*a*) center on $x + 4y + 16 = 0$ (*b*) radius is $\frac{1}{2}$.

The equation of the family is

$$K_1 + kK_2: \quad x^2 + y^2 - 16x - 20y + 115 + k(x^2 + y^2 + 8x - 10y + 5) = 0$$

(*a*) The centers of K_1 and K_2 are (8,10) and (−4, 5), respectively; the equation of their line of centers is $5x - 12y + 80 = 0$. This line meets $x + 4y + 16 = 0$ in the point (−16, 0), the required center.

Now (−16, 0) divides the segment joining (8,10) and (−4, 5) in the ratio $k : 1$; thus,

$$\frac{5k + 10}{k + 1} = 0 \qquad \text{and} \qquad k = -2$$

The equation of the circle is $K_1 - 2K_2 : x^2 + y^2 + 32x - 105 = 0$.

(*b*) Setting the square of the radius of $K_1 + kK_2$ equal to $\left(\frac{1}{2}\right)^2$ and solving for k, we find

$$\left(\frac{4k - 8}{1 + k}\right)^2 + \left(-\frac{10 + 5k}{1 + k}\right)^2 - \frac{115 + 5k}{1 + k} = \frac{1}{4} \qquad \text{and} \qquad k = 1, \tfrac{15}{11}$$

The required circles have equations

$$K_1 + K_2 : x^2 + y^2 - 4x - 15y + 60 = 0 \qquad \text{and} \qquad K_1 + \tfrac{15}{11}K_2 : \quad 13x^2 + 13y^2 - 28x - 185y + 670 = 0$$

Supplementary Problems

11.12 Write the equation of each of the following circles: (*a*) $C(0, 0)$, radius 7 (*b*) $C(-4, 8)$, radius 3 (*c*) $C(5, -4)$, through (0, 0) (*d*) $C(-4, -3)$, through (2, 1) (*e*) $C(-2, 5)$, tangent to x axis (*f*) $C(-2, -5)$, tangent to $2x - y + 3 = 0$ (*g*) tangent to both axes, radius 5 (*h*) circumscribed about the right triangle whose vertices are (3, 4), (−1, −4), (5, −2) (*i*) circumscribed about the triangle of Problem 9.8.

Ans. (*a*) $x^2 + y^2 = 49$

(*b*) $x^2 + y^2 + 8x - 16y + 71 = 0$

(*c*) $x^2 + y^2 - 10x + 8y = 0$

(*d*) $x^2 + y^2 + 8x + 6y - 27 = 0$

(*e*) $x^2 + y^2 + 4x - 10y + 4 = 0$

(*f*) $5x^2 + 5y^2 + 20x + 50y + 129 = 0$

(*g*) $x^2 + y^2 \pm 10x \pm 10y + 25 = 0, x^2 + y^2 \mp 10x \pm 10y + 25 = 0$

(*h*) $x^2 + y^2 - 2x - 19 = 0$

(*i*) $56x^2 + 56y^2 - 260x - y - 5451 = 0$

11.13 Find the center and radius of each of the circles.

(*a*) $x^2 + y^2 - 6x + 8y - 11 = 0$ *Ans.* $C(3, -4)$, $r = 6$

(*b*) $x^2 + y^2 - 4x - 6y - \frac{10}{3} = 0$ *Ans.* $C(2, 3)$, $r = 7\sqrt{3}/3$

(*c*) $7x^2 + 7y^2 + 14x - 56y - 25 = 0$ *Ans.* $C(-1, 4)$, $4 = 12\sqrt{7}/7$

11.14 Explain why any line passing through (4, −1) cannot be tangent to the circle $x^2 + y^2 - 4x + 6y - 12 = 0$.

11.15 (*a*) Show that the circles $x^2 + y^2 + 6x - 2y - 54 = 0$ and $x^2 + y^2 - 22x - 8y + 112 = 0$ do not intersect.

(*b*) Show that the circles $x^2 + y^2 + 2x - 6y + 9 = 0$ and $x^2 + y^2 + 8y - 6y + 9 = 0$ are tangent internally. Do this with and without a graphing calculator.

11.16 The equation of a given circle is $x^2 + y^2 = 36$. Find (*a*) the length of the chord which lies along the line $3x + 4y - 15 = 0$ (*b*) the equation of the chord whose midpoint is (3,2).

Hint: In (*a*) draw the normal to the given line. *Ans.* (*a*) $6\sqrt{3}$ (*b*) $3x + 2y - 13 = 0$

11.17 Find the equation of each circle satisfying the given conditions.

(*a*) Through (6,0) and (−2,−4), tangent to $4x + 3y - 25 = 0$.

(*b*) Tangent to $3x - 4y + 5 = 0$ at (1,2), radius 5.

(*c*) Tangent to $x - 2y - 4 = 0$ and $2x - y - 6 = 0$, passes through (−1, 2).

(*d*) Tangent to $2x - 3y - 7 = 0$ at (2,−1); passes through (4,1).

(*e*) Tangent to $3x + y + 3 = 0$ at (−3, 6), tangent to $x + 3y - 7 = 0$.

Ans. (*a*) $(x-3)^2 + (y+4)^2 = 25, (x - \frac{213}{121})^2 + (y + \frac{184}{121})^2 = 297\,025/14\,641$

(*b*) $(x-4)^2 + (y+2)^2 = 25, (x+2)^2 + (y-6)^2 = 25$

(*c*) $x^2 + y^2 - 2x - 2y - 3 = 0, x^2 + y^2 + 118x - 122y + 357 = 0$

(*d*) $x^2 + y^2 + 4x - 10y - 23 = 0$

(*e*) $x^2 + y^2 - 6x - 16y + 33 = 0, x^2 + y^2 + 9x - 11y + 48 = 0$

11.18 Repeat Problem 11.13 using a graphing calculator.

11.19 Find the equation of the tangent to the given circle at the given point on it.

(*a*) $x^2 + y^2 = 169$, $(5, -12)$ (*b*) $x^2 + y^2 - 4x + 6y - 37 = 0$, $(3, 4)$

Ans. (*a*) $5x - 12y - 169 = 0$ (*b*) $x + 7y - 31 = 0$

11.20 Find the equations of the tangent to each circle through the given external point.

(*a*) $x^2 + y^2 = 25$, $(7, 1)$ (*b*) $x^2 + y^2 - 4x + 2y - 31 = 0$, $(-1, 5)$

Ans. (*a*) $3x + 4y - 25 = 0$, $4x - 3y - 25 = 0$ (*b*) $y - 5 = 0$, $4x - 3y + 19 = 0$

11.21 Show that the circles $x^2 + y^2 + 4x - 6y = 0$ and $x^2 + y^2 + 6x + 4y = 0$ are *orthogonal*, that is, that the tangents to the two circles at a point of intersection are mutually perpendicular. Also, that the square of the distance between the centers of the circles is equal to the sum of the squares of the radii.

11.22 Determine the equation of the circle of the family of Problem 11.11: (*a*) which passes through the point (0, 3) (*b*) which is tangent to the line $4x + 3y - 25 = 0$.

Ans. (*a*) $5x^2 + 5y^2 + 16x - 60y + 135 = 0$

(*b*) $x^2 + y^2 - 40x - 30y + 225 = 0, 8x^2 + 8y^2 - 20x - 115y + 425 = 0$

11.23 Repeat Problem 11.22 using a graphing calculator or a computer software package such as Maple.

PART II

PRE-CALCULUS AND ELEMENTARY CALCULUS

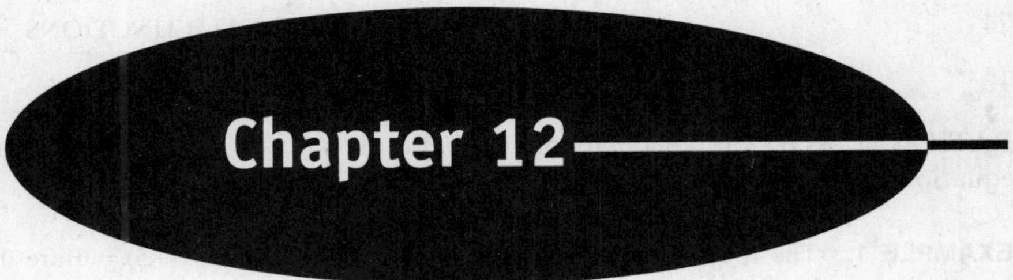

Rational and Polynomial Functions

A POLYNOMIAL EQUATION (or rational integral equation) is obtained when any polynomial in one variable is set equal to zero. We shall work with polynomials having integral coefficients although many of the theorems will be stated for polynomial equations with weaker restrictions on the coefficients.

A polynomial equation is said to be in *standard form* when written as

$$a_0x^n + a_1x^{n-1} + a_2x^{n-2} + \cdots + a_{n-2}x^2 + a_{n-1}x + a_n = 0$$

where the terms are arranged in descending powers of x, a zero has been inserted as coefficient on each missing term, the coefficients have no common factor except ± 1, and $a_0 \neq 0$. (See Problem 12.1)

A NUMBER r IS CALLED A ROOT of $f(x) = 0$ if and only if $f(r) = 0$. It follows that the abscissas of the points of intersection of the graph of $y = f(x)$ and the x axis are roots of $f(x) = 0$.

THE FUNDAMENTAL THEOREM OF ALGEBRA. Every polynomial equation $f(x) = 0$ has at least one root, real or complex.

A polynomial equation of degree n has exactly n roots. These n roots may not all be distinct. If r is one of the roots and occurs just once, it is called a *simple root*; if r occurs exactly $m > 1$ times among the roots, it is called a *root of multiplicity m* or an *m-fold root*. If $m = 2$, r is called a *double root*; if $m = 3$, a *triple root*; and so on. (See Problems 12.2–12.3.)

COMPLEX ROOTS. If the polynomial equation $f(x) = 0$ has real coefficients and if the complex $a + bi$ is a root of $f(x) = 0$, then the *complex conjugate* $a - bi$ is also a root. (For a proof, see Problem 12.11.)

IRRATIONAL ROOTS. Given the polynomial equation $f(x) = 0$, if the irrational number $a + \sqrt{b}$, where a and b are rational, is a root of $f(x) = 0$, then the conjugate irrational $a - \sqrt{b}$ is also a root. (See Problem 12.4.)

LIMITS TO THE REAL ROOTS. A real number L is called an *upper limit* of the real roots of $f(x) = 0$ if no (real) root is greater than L; a real number l is called a *lower limit* if no (real) root is smaller than l.

If $L > 0$ and if, when $f(x)$ is divided by $x - L$ by synthetic division, every number in the third line is nonnegative, then L is an upper limit of the real roots of $f(x) = 0$.

If $l < 0$ and if, when $f(x)$ is divided by $x - l$ by synthetic division, the numbers in the third line alternate in sign, then l is a lower limit of the real roots of $f(x) = 0$.

RATIONAL ROOTS. A polynomial equation has 0 as a root if and only if the constant term of the equation is zero.

EXAMPLE 1. The roots of $x^5 - 2x^4 + 6x^3 - 5x^2 = x^2(x^3 - 2x^2 + 6x - 5) = 0$ are 0, 0, and the three roots of $x^3 - 2x^2 + 6x - 5 = 0$.

If a rational fraction p/q, expressed in lowest terms, is a root of (12.1) in which $a_n \neq 0$, then p is a divisor of the constant term a_n and q is a divisor of the leading coefficient of a_0 of (12.1). (For a proof, see Problem 12.12.)

EXAMPLE 2. The theorem permits us to say that $\frac{2}{3}$ is a *possible* root of the equation $9x^4 - 5x^2 + 8x + 4 = 0$ since the numerator 2 divides the constant term 4 and the denominator 3 divides the leading coefficient 9. It does *not* assure that $\frac{2}{3}$ is a root. However, the theorem does assure that neither $\frac{1}{2}$ nor $-\frac{4}{5}$ is a root. In each case the denominator does not divide the leading coefficient.

If p, an integer, is a root of (12.1) then p is a divisor of its constant term.

EXAMPLE 3. The *possible* rational roots of the equation

$$12x^4 - 40x^3 - 5x^2 + 45x + 18 = 0$$

are all numbers $\pm p/q$ in which the values of p are the positive divisors 1, 2, 3, 6, 9, 18 of the constant term 18 and the values of q are the positive divisors 1, 2, 3, 4, 6, 12 of the leading coefficient 12. Thus the rational roots, if any, of the equation are among the numbers

$$\pm 1, \pm 2, \pm 3, \pm 6, \pm 9, \pm 18, \pm \tfrac{1}{2}, \pm \tfrac{3}{2}, \pm \tfrac{9}{2}, \pm \tfrac{1}{3}, \pm \tfrac{2}{3}, \pm \tfrac{1}{4}, \pm \tfrac{3}{4}, \pm \tfrac{1}{6}, \pm \tfrac{1}{12}$$

THE PRINCIPAL PROBLEM OF THIS CHAPTER is to find the rational roots of a given polynomial equation. The general procedure is this: Test the possible rational roots by synthetic division, accepting as roots all those for which the last number in the third line *is* zero and rejecting all those for which it is not. Certain refinements, which help to shorten the work, are pointed out in the examples and solved problems below.

EXAMPLE 4. Find the rational roots of $x^5 + 2x^4 - 18x^3 - 8x^2 + 41x + 30 = 0$.

Since the leading coefficient is 1, all rational roots p/q are integers. The possible integral roots, the divisors (both positive and negative) of the constant term 30, are

$$\pm 1, \pm 2, \pm 3, \pm 5, \pm 6, \pm 10, \pm 15, \pm 30$$

$$\text{Try 1:}\quad \begin{array}{l} 1 + 2 - 18 - \ \ 8 + 41 + 30 \quad \underline{|1} \\ \underline{\qquad 1 + \ 3 - 15 - 23 + 18} \\ 1 + 3 - 15 - 23 + 18 + 48 \end{array}$$

Then 1 is not a root. This number (+1) should be removed from the list of possible roots lest we forget and try it again later on.

$$\text{Try 2:}\quad \begin{array}{l} 1 + 2 - 18 - \ \ 8 + 41 + 30 \quad \underline{|2} \\ \underline{\qquad 2 + \ 8 - 20 - 56 - 30} \\ 1 + 4 - 10 - 28 + 15 + \ \ 0 \end{array}$$

Then 2 is a root and the remaining rational roots of the given equation are the rational roots of the *depressed equation*

$$x^4 + 4x^3 - 10x^2 - 28x - 15 = 0$$

Now ± 2, ± 6, ± 10, and ± 30 cannot be roots of this equation (they are not divisors of 15) and should be removed from the list of possibilities. We return to the depressed equation.

$$\begin{array}{rl} & 1+4-10-28-15 \quad \underline{|3} \\ \text{Try 3:} & 3+21+33+15 \\ \hline & 1+7+11+\ 5+\ 0 \end{array}$$

Then 3 is a root and the new depressed equation is

$$x^3 + 7x^2 + 11x + 5 = 0$$

Since the coefficients of this equation are nonnegative, it has no positive roots. We now remove +3, +5, +15 from the original list of possible roots and return to the new depressed equation.

$$\begin{array}{rl} & 1+7+11+5 \quad \underline{|-1} \\ \text{Try } -1: & -1-\ 6-5 \\ \hline & 1+6+\ 5+0 \end{array}$$

Then −1 is a root and the depressed equation

$$x^2 + 6x + 5 = (x+1)(x+5) = 0$$

has −1 and −5 as roots.

The necessary computations may be neatly displayed as follows:

$$\begin{array}{ll} 1+2-18-\ 8+41+30 & \underline{|2} \\ 2+\ 8-20-56-\ 30 & \\ 1+4-10-28-\ 15 & \underline{|3} \\ 3+21+33+15 & \\ \hline 1+7+11+\ 5 & \underline{|-1} \\ -1-\ 6-\ 5 & \\ \hline 1+6+\ 5 & \end{array}$$

$$x^2 + 6x + 5 = (x+1)(x+5) = 0 \qquad x = -1, -5$$

The roots are 2, 3, −1, −1, −5.

Note that the roots here are numerically small numbers; that is, 3 is a root but 30 is not, −1 is a root but −15 is not. Hereafter we shall not list integers which are large numerically or fractions with large numerator or denominator among the possible roots. (See Problems 12.5–12.9.)

IF $f(x) = 0$ IS A POLYNOMIAL EQUATION, the equation $f(-x) = 0$ has as roots the negatives of the roots of $f(x) = 0$. When $f(x) = 0$ is written in standard form, the equation whose roots are the negatives of the roots of $f(x) = 0$ may be obtained by changing the signs of alternate terms, beginning with the second.

EXAMPLE 5

(*a*) The roots $x^3 + 3x^2 - 4x - 12 = 0$ are $2, -2, -3$; the roots of $x^3 - 3x^2 - 4x + 12 = 0$ are $-2, 2, 3$.

(*b*) The equation $6x^4 + 13x^3 - 13x - 6 = 0$ has roots $1, -1, -\frac{2}{3}, -\frac{3}{2}$; the equation $6x^4 - 13x^3 + 13x - 6 = 0$ has roots $-1, 1, \frac{2}{3}, \frac{3}{2}$.

(See Problem 12.13.)

VARIATION OF SIGN. If, when a polynomial is arranged in descending powers of the variable, two successive terms differ in sign, the polynomial is said to have a *variation of sign*.

EXAMPLE 6

(a) The polynomial $x^3 - 3x^2 - 4x + 12$ has two variations of sign, one from $+x^3$ to $-3x^2$ and one from $-4x$ to $+12$; the polynomial $x^3 + 3x^2 - 4x - 12$ has one variation of sign.

(b) The polynomial $6x^4 + 13x^3 - 13x - 6$ has one variation of sign; the polynomial $6x^4 - 13x^3 + 13x - 6$ has three. Note that here the term with zero coefficient has not been considered.

DESCARTES' RULE OF SIGNS. The number of positive roots of a polynomial equation $f(x) = 0$, with real coefficients, is equal either to the number of variations of sign in $f(x)$ or to the number diminished by an even number.

The number of negative roots of $f(x) = 0$ is equal to the number of positive roots of $f(-x) = 0$.

EXAMPLE 7. Since $f(x) = x^3 - 3x^2 - 4x + 12$ of Example 6(a) has two variations of sign, $f(x) = 0$ has either two or no positive roots.

Since $x^3 + 3x^2 - 4x - 12$ has one variation of sign, $f(-x) = 0$ has one positive root and $f(x) = 0$ has one negative root. (See Problem 12.14.)

DIMINISHING THE ROOTS OF AN EQUATION. Let

$$f(x) = a_0x^n + a_1x^{n-1} + \cdots + a_{n-1}x + a_n = 0 \tag{12.1}$$

be a polynomial equation of degree n. Let

$$\begin{aligned} f(x) &= (x-h)\cdot q_1(x) + R_1, \\ q_1(x) &= (x-h)\cdot q_2(x) + R_2, \\ &\vdots \\ q_{n-2}(x) &= (x-h)\cdot q_{n-1}(x) + R_{n-1}, \\ q_{n-1}(x) &= (x-h)\cdot q_n(x) + R_n, \end{aligned} \tag{12.2}$$

where each R is a constant and $q_n(x) = a_0$. Then the roots of

$$g(y) = a_0y^n + R_ny^{n-1} + R_{n-1}y^{n-2} + \cdots + R_{2y} + R_1 = 0 \tag{12.3}$$

are the roots of $f(x) = 0$ diminished by h.

We shall show that if r is any root of $f(x) = 0$ then $r - h$ is a root of $g(y) = 0$. Since $f(r) = 0$,

$$\begin{aligned} R_1 &= -(r-h)\cdot q_1(r), \\ R_2 &= q_1(r) - (r-h)\cdot q_2(r), \\ &\vdots \\ R_{n-1} &= q_{n-2}(r) - (r-h)\cdot q_{n-1}(r), \\ R_n &= q_{n-1}(r) - (r-h)a_0. \end{aligned}$$

When these replacements are made in (12.3), we have

$$\begin{aligned} a_0y^n &+ [q_{n-1}(r) - (r-h)a_0]y^{n-1} + [q_{n-2}(r) - (r-h)\cdot q_{n-1}(r)]\,y^{n-2} + \cdots \\ &+ [q_1(r) - (r-h)\cdot q_2(r)]y - (r-h)\cdot q_1(r) \\ &= a_0[y-(r-h)]y^{n-1} + q_{n-1}(r)[y-(r-h)]\,y^{n-2} + \cdots + q_1(r)[\,y-(r-h)] = 0. \end{aligned}$$

It is clear that $r - h$ is a root of (*12.3*) as was to be proved.

EXAMPLE 8. Find the equation each of whose roots is 4 less than the roots of $x^3 + 3x^2 - 4x - 12 = 0$.

$$\begin{array}{llll} 1+ & 3- & 4- & 12 \quad \underline{|4} \\ & 4+ & 28+ & 96 \\ \hline 1+ & 7+ & 24+ & \textcircled{84} \\ & 4+ & 44 & \\ \hline 1+ & 11+ & \textcircled{68} & \\ & 4 & & \\ \hline 1+ & \textcircled{15} & & \end{array} \qquad \begin{array}{llll} 1+ & 15+ & 68+ & 84 \quad \underline{|-2} \\ & -\ 2- & 26- & 84 \\ \hline 1+ & 13+ & 42 & \quad \underline{|-6} \\ & -\ 6- & 42 & \\ \hline 1+ & 7 & & \quad \underline{|-7} \\ & -\ 7 & & \\ \hline 1 & & & \end{array}$$

On the left the successive remainders have been found and circled. The resulting equation is $y^3 + 15y^2 + 68y + 84 = 0$. The given equation has roots $x = 2, -2 - 3$; on the right, it is shown that $2 - 4 = -2, -2 - 4 = -6, -3 - 4 = -7$ are roots of the newly formed equation. (See Problem 12.5.)

APPROXIMATION OF IRRATIONAL ROOTS

LET *f* (*x*) = *0* BE A POLYNOMIAL EQUATION having no rational roots. If the given equation had rational roots, we suppose that they have been found using synthetic division and that $f(x)$ is then the last third line in process.

THE METHOD OF SUCCESSIVE LINEAR APPROXIMATIONS will be explained by means of examples.

EXAMPLE 9. The equation $f(x) = x^3 + x - 4 = 0$ has no rational roots. By Descartes' rule of signs, it has one positive (real) root and two imaginary roots. To approximate the real root, we shall first isolate it as lying between two consecutive intgers. Since $f(1) = -2$ and $f(2) = 6$, the root lies between $x = 1$ and $x = 2$. Figure 12-1 exhibits the portion of the graph of $f(x)$ between $(1, -2)$ and $(2, 6)$.

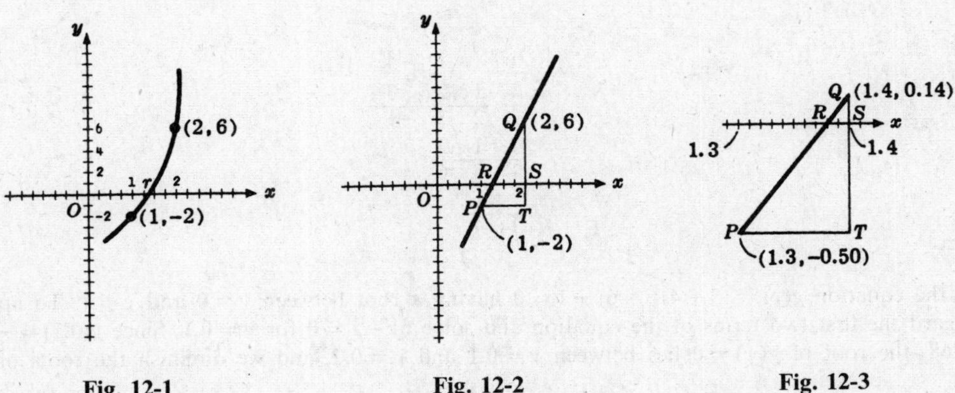

Fig. 12-1 Fig. 12-2 Fig. 12-3

In Fig. 12-2, the curve joining the two points has been replaced by a straight line which meets the x axis at R. We shall take OR, measured to the nearest tenth of a unit, as the first approximation of required root and use it to isolate the root between successive tenths. From the similar triangles RSQ and PTQ,

$$\frac{RS}{PT} = \frac{SQ}{TQ} \quad \text{or} \quad RS = \frac{SQ}{TQ}(PT) = \frac{6}{8}(1) = \frac{3}{4} = 0.7$$

The first approximation of the root is given by $OR = OS - RS = 2 - 0.7 = 1.3$. Since $f(1.3) = -0.50$ and $f(1.4) = 0.14$, the required root lies between $x = 1.3$ and $x = 1.4$.

We now repeat the above process using the points $(1.3, -0.50)$ and $(1.4, 0.14)$ and isolate the root between successive hundredths. From Fig. 12-3,

$$RS = \frac{SQ}{TQ}(PT) = \frac{0.14}{0.64}(0.1) = 0.02 \quad \text{and} \quad OR = OS - RS = 1.4 - 0.02 = 1.38$$

is the next approximation. Since $f(1.38) = 0.008$ (hence, too large) and $f(1.37) = -0.059$, the root lies between $x = 1.37$ and $x = 1.38$.

Using the points $(1.37, -0.059)$ and $(1.38, 0.008)$, we isolate the root between successive thousandths. We find (no diagram needed)

$$RS = \frac{0.008}{0.067}(0.01) = 0.001 \quad \text{and} \quad OR = 1.38 - 0.001 = 1.379$$

Since $f(1.379) = 0.0012$ and $f(1.378) = -0.0054$, the root lies between $x = 1.378$ and $x = 1.379$.

For the next approximation

$$RS = \frac{0.0012}{0.0066}(0.001) = 0.0001 \quad \text{and} \quad OR = 1.379 - 0.0001 = 1.3789$$

The root correct to three decimal places is 1.379. (See Problem 12.16.)

HORNER'S METHOD OF APPROXIMATION. This method will be explained by means of examples.

EXAMPLE 10. The equation $x^3 + x^2 + x - 4 = 0$ has no rational roots. By Descartes' rule of signs, it has one positive root. Since $f(1) = -1$ and $f(2) = 10$, this root is between $x = 1$ and $x = 2$. We first diminish the roots of the given equation by 1.

$$\begin{array}{rrrrl}
1 & 1 & 1 & -4 & \underline{|1} \\
 & 1 & 2 & 3 & \\
\hline
1 & 2 & 3 & -1 & \\
 & 1 & 3 & & \\
\hline
1 & 3 & 6 & & \\
 & 1 & & & \\
\hline
1 & 4 & & &
\end{array}$$

and obtain the equation $g(y) = y^3 + 4y^2 + 6y - 1 = 0$ having a root between $y = 0$ and $y = 1$. To approximate it, we disregard the first two terms of the equation and solve $6y - 1 = 0$ for $y = 0.1$. Since $g(0.1) = -0.359$ and $g(0.2) = 0.368$, the root of $g(y) = 0$ lies between $y = 0.1$ and $y = 0.2$, and we diminish the roots of $g(y) = 0$ by 0.1.

$$\begin{array}{rrrrl}
1 & 4 & 6 & -1 & \underline{|0.1} \\
 & 0.1 & 0.41 & 0.641 & \\
\hline
1 & 4.1 & 6.41 & -0.359 & \\
 & 0.1 & 0.42 & & \\
\hline
1 & 4.2 & 6.83 & & \\
 & 0.1 & & & \\
\hline
1 & 4.3 & & &
\end{array}$$

We obtain the equation $h(z) = z^3 + 4.3z^2 + 6.83z - 0.359 = 0$ having a root between 0 and 0.01. Disregarding the first two terms of this equation and solving $6.83z - 0.359 = 0$, we obtain $z = 0.05$ as an approximation of the root. Since $h(0.05) = -0.007$ and $h(0.06) = 0.07$, the root of $h(z) = 0$ lies between $z = 0.05$ and $z = 0.06$ and we diminish the roots by 0.05.

$$\begin{array}{rrrrl}
1 & 4.3 & 6.83 & -0.359 & \underline{|0.05} \\
 & 0.05 & 0.2175 & 0.352375 & \\
\hline
1 & 4.35 & 7.0475 & -0.006625 & \\
 & 0.05 & 0.2200 & & \\
\hline
1 & 4.40 & 7.2675 & & \\
 & 0.05 & & & \\
\hline
1 & 4.45 & & &
\end{array}$$

and obtain the equation $k(w) = w^3 + 4.45w^2 + 7.2675w - 0.006625 = 0$ having a root between $w = 0$ and $w = 0.001$. An approximation of this root, obtained by solving $7.2675w - 0.006625 = 0$ is $w = 0.0009$.

Without further computation, we are safe in stating the root of the given equation to be

$$x = 1 + 0.1 + 0.05 + 0.0009 = 1.1509$$

The complete solution may be exhibited more compactly, as follows:

$$\begin{array}{rrrrll}
1 & 1 & 1 & -4 & \underline{|1} & \\
 & 1 & 2 & 3 & & \\
\hline
1 & 2 & 3 & -1 & & \\
 & 1 & 3 & & & y = \frac{1}{6} = 0.1 \\
\hline
1 & 3 & 6 & & & \\
 & 1 & & & & \\
\hline
1 & 4 & & & & \\
 & & 6 & -1 & \underline{|0.1} & \\
 & 0.1 & 0.41 & 0.641 & & \\
\hline
1 & 4.1 & 6.41 & -0.359 & & \\
 & 0.1 & 0.42 & & & z = \dfrac{0.359}{6.83} = 0.05 \\
\hline
1 & 4.2 & 6.83 & & & \\
 & 0.1 & & & & \\
\hline
1 & 4.3 & 6.83 & & & \\
 & & & -0.359 & \underline{|0.05} & \\
 & 0.05 & 0.2175 & 0.352375 & & \\
\hline
1 & 4.35 & 7.0475 & -0.006625 & & \\
 & 0.05 & 0.2200 & & & w = \dfrac{0.006625}{7.2675} = 0.0009 \\
\hline
1 & 4.40 & 7.2675 & & & \\
 & 0.05 & & & & \\
\hline
1 & 4.45 & & & & \\
 & & 7.2675 & -0.006625 & &
\end{array}$$

(See Problems 12.17–12.18.)

Solved Problems

12.1 Write each of the following in standard form.

(a) $4x^2 + 2x^3 - 6 + 5x = 0$ *Ans.* $2x^3 + 4x^2 + 5x - 6 = 0$

(b) $-3x^2 + 6x - 4x^2 + 2 = 0$ *Ans.* $3x^3 + 4x^2 - 6x - 2 = 0$

(c) $2x^5 + x^3 + 4 = 0$ *Ans.* $2x^5 + 0 \cdot x^4 + x^3 + 0 \cdot x^2 + 0 \cdot x + 4 = 0$

(d) $x^2 + \frac{1}{2}x^2 - x + 2 = 0$ *Ans.* $2x^3 + x^2 - 2x + 4 = 0$

(e) $4x^4 + 6x^3 - 8x^2 + 12x - 10 = 0$ *Ans.* $2x^4 + 3x^3 - 4x^2 + 6x - 5 = 0$

12.2 (a) Show that -1 and 2 are roots of $x^4 - 9x^2 + 4x + 12 = 0$.

Using synthetic division,

$$\begin{array}{l} 1 + 0 - 9 + \ \ 4 + 12 \quad \lfloor -1 \\ \underline{\quad - 1 + 1 + \ \ 8 - 12} \\ 1 - \ 1 - 8 + 12 + \ 0 \end{array} \qquad \begin{array}{l} 1 + 0 - 9 + \ \ 4 + 12 \quad \lfloor 2 \\ \underline{\quad - 2 + 4 - 10 - 12} \\ 1 + 2 - 5 - \ 6 + \ 0 \end{array}$$

Since $f(-1) = 0$ and $f(2) = 0$, both -1 and 2 are roots.

(b) Show that the equation in (a) has at least two other roots by finding them.

From the synthetic division in (a) and the Factor Theorem,

$$x^4 - 9x^2 + 4x + 12 = (x + 1)(x^3 - x^2 - 8x + 12)$$

Since 2 is also root of the given equation, 2 is a root of $x^3 - x^2 - 8x + 12 = 0$.

Using synthetic division

$$\begin{array}{l} 1 - 1 - 8 + 12 \quad \lfloor 2 \\ \underline{\quad \ \ 2 + 2 - 12} \\ 1 + 1 - 6 + \ 0 \end{array}$$

we obtain $x^3 - x^2 - 8x + 12 = (x - 2)(x^2 + x - 6)$. Then

$$x^4 - 9x^2 + 4x + 12 = (x + 1)(x - 2)(x^2 + x - 6) = (x + 1)(x - 2)(x + 3)(x - 2)$$

Thus, the roots of the given equation are -1, 2, -3, 2.

(NOTE: Since $x - 2$ appears twice among the factors of $f(x)$, 2 appears twice among the roots of $f(x) = 0$ and is a double root of the equation.)

12.3 (a) Find all of the roots of $(x + 1)(x - 2)^3(x + 4)^2 = 0$.

The roots are -1, 2, 2, 2, -4, -4; thus, -1 is a simplest root, 2 is a root of multiplicity three or a triple root, and -4 is a root of multiplicity two or a double root.

(b) Find all the roots of $x^2(x - 2)(x - 5) = 0$.

The roots are 0, 0, 2, 5; 2 and 5 are simple roots and 0 is a double root.

12.4 Form the equation of lowest degree with integral coefficients having the roots $\sqrt{5}$ and $2 - 3i$.

To ensure integral coefficients, the conjugates $-\sqrt{5}$ and $2 + 3i$ must also be roots. Thus the required equation is

$$(x - \sqrt{5})(x + \sqrt{5})[x - (2 - 3i)][x - (2 + 3i)] = x^4 - 4x^3 + 8x^2 + 20x - 65 = 0$$

In Problems 12.5–12.9 find the rational roots; when possible, find all the roots.

12.5 $2x^4 - x^3 - 11x^2 + 4x + 12 = 0$

The possible rational roots are ± 1, ± 2, ± 3, ± 4, $\pm \frac{1}{2}$, $\pm \frac{3}{2}$, ... Discarding all false trials, we find

$$\begin{array}{r|l} 2-1-11+4+12 & \underline{2} \\ 4+6-10-12 & \\ \hline 2+3-5-6 & \underline{-1} \\ -2-1+6 & \\ \hline 2+1-6 & \end{array} \qquad 2x^2 + x - 6 = (2x-3)(x+2) = 0; \qquad x = \tfrac{3}{2}, -2$$

The roots are $2, -1, \frac{3}{2}, -2$.

12.6 $4x^4 - 3x^3 - 4x + 3 = 0$

The possible rational roots are: ± 1, ± 3, $\pm \frac{1}{2}$, $\pm \frac{3}{2}$, $\pm \frac{1}{4}$, $\pm \frac{3}{4}$. By inspection the sum of the coefficients is 0; then $+1$ is a root. Discarding all false trials, we find

$$\begin{array}{r|l} 4-3+0-4+3 & \underline{1} \\ 4+1+1-3 & \\ \hline 4+1+1-3 & \underline{\tfrac{3}{4}} \\ 3+3+3 & \\ \hline 4+4+4 & \text{(Factor out 4.)} \\ 1+1+1 & \end{array} \qquad x^2 + x + 1 = 0; \qquad x = \frac{-1 \pm i\sqrt{3}}{2}$$

The roots are $1, \dfrac{3}{4}, -\dfrac{1}{2} \pm \dfrac{i\sqrt{3}}{2}$.

12.7 $24x^6 - 20x^5 - 6x^4 + 9x^3 - 2x^2 = 0$

Since $24x^6 - 20x^5 - 6x^4 + 9x^3 - 2x^2 = x^2(24x^4 - 20x^3 - 6x^2 + 9x - 2)$, the roots of the given equation are 0, 0 and the roots of $24x^4 - 20x^3 - 6x^2 + 9x - 2 = 0$. Possible rational roots are: ± 1, ± 2, $\pm \frac{1}{2}$, $\pm \frac{1}{3}$, $\pm \frac{2}{3}$, $\pm \frac{1}{4}$, ... Discarding all false trials, we find

$$\begin{array}{r|l} 24-20-6+9-2 & \underline{\tfrac{1}{2}} \\ 12-4-5+2 & \\ \hline 24-8-10+4 & \text{(Factor out 2.)} \\ 12-4-5+2 & \underline{\tfrac{1}{2}} \\ 6+1-2 & \\ \hline 12+2-4 & \text{(Factor out 2.)} \\ 6+1-2 & \end{array} \qquad 6x^2 + x - 2 = (2x-1)(3x+2) = 0; \quad x = \tfrac{1}{2}, -\tfrac{2}{3}$$

The roots are $0, 0, \frac{1}{2}, \frac{1}{2}, \frac{1}{2}, -\frac{2}{3}$.

12.8 $4x^5 - 32x^4 + 93x^3 - 119x^2 + 70x - 25 = 0$

Since the signs of the coefficients alternate, the rational roots (if any) are positive. Possible rational roots are $1, 5, \frac{1}{2}, \frac{5}{2}, \frac{1}{4}, \ldots$. Discarding all false trials, we find

$$\begin{array}{rl}
4 - 32 + 93 - 119 + 70 - 25 & \underline{|\tfrac{5}{2}} \\
10 - 55 + \ 95 - 60 + 25 & \\
\hline
4 - 22 + 38 - \ 24 + 10 & \text{(Factor out 2.)} \\
2 - 11 + 19 - \ 12 + \ 5 & \underline{|\tfrac{5}{2}} \\
5 - 15 + \ 10 - \ 5 & \\
\hline
2 - \ 6 + \ 4 - \ \ 2 & \text{(Factor out 2.)} \\
1 - \ 3 + \ 2 - \ \ 1 &
\end{array}$$

The rational roots are $\frac{5}{2}, \frac{5}{2}$.

The equation $f(x) = x^3 - 3x^2 + 2x - 1 = 0$ has at least one real (irrational) root since $f(0) < 0$ while for sufficiently large $x(x > 3)$, $f(x) > 0$.

The only possible rational root of $x^3 - 3x^2 + 2x - 1 = 0$ is 1; it is not a root.

12.9 $6x^4 + 13x^3 - 11x^2 + 5x + 1 = 0$

The possible rational roots are $\pm 1, \pm\frac{1}{2}, \pm\frac{1}{3}, \pm\frac{1}{6}$. After testing each possibility, we conclude that the equation has no rational roots.

12.10 If $ax^3 + bx^2 + cx + d = 0$, with integral coefficients, has a rational root r and if $c + ar^2 = 0$, then all of the roots are rational.

Using synthetic division to remove the known root, the depressed equation is

$$ax^2 + (b + ar)x + c + br + ar^2 = 0$$

which reduces to $ax^2 + (b + ar)x + br = 0$ when $c + ar^2 = 0$. Since its discriminant is $(b - ar)^2$, a perfect square, its roots are rational. Thus, all the roots of the given equation are rational.

12.11 Prove: If the polynomial $f(x)$ has real coefficients and if the imaginary $a + bi, b \neq 0$, is a root of $f(x) = 0$, then the conjugate imaginary $a - bi$ is also a root.

Since $a + bi$ is a root of $f(x) = 0, x - (a + bi)$ is a factor of $f(x)$. Similarly, if $a - bi$ is to be a root of $f(x) = 0, x - (a - bi)$ must be a factor of $f(x)$. We need to show then that when $a + bi$ is a root of $f(x) = 0$, it follows that

$$[x - (a + bi)][x - (a - bi)] = x^2 - 2ax + a^2 + b^2$$

is a factor of $f(x)$. By division we find

$$\begin{aligned} f(x) &= [x^2 - 2ax + a^2 + b^2] \cdot Q(x) + Mx + N \\ &= [x - (a + bi)][x - (a - bi)] \cdot Q(x) + Mx + N \end{aligned} \tag{1}$$

where $Q(x)$ is a polynomial of degree 2 less than that of $f(x)$ and the remainder $Mx + N$ is of degree at most 1 in x; that is, M and N are constants.

Since $a + bi$ is a root of $f(x) = 0$, we have from *(1)*,

$$f(a + bi) = 0 \cdot Q(a + bi) + M(a + bi) + N = (aM + N) + bMi = 0$$

Then $aM + N = 0$ and $bM = 0$. Now $b \neq 0$; hence, $bM = 0$ requires $M = 0$ and then $aM + N = 0$ requires $N = 0$. Since $M = N = 0$, *(1)* becomes

$$f(x) = (x^2 - 2ax + a^2 + b^2) \cdot Q(x)$$

Then $x^2 - 2ax + a^2 + b^2$ is a factor of $f(x)$ as was to be proved.

12.12 Prove: If a rational fraction p/q, expressed in lowest terms, is a root of the polynomial equation (*12.1*) whose constant terms $a_n \neq 0$, then p is a divisor of the constant term a_n, and q is a divisor of the leading coefficient a_0.

Let the given equation be

$$a_0x^n + a_1x^{n-1} + a_2x^{n-2} + \cdots + a_{n-2}x^2 + a_{n-1}x + a_n = 0, \quad a_0a_n \neq 0$$

If p/q is a root, then

$$a_0(p/q)^n + a_1(p/q)^{n-1} + a_2(p/q)^{n-2} + \cdots + a_{n-2}(p/q)^2 + a_{n-2}(p/q) + a_n = 0$$

Multiplying both members by q^n, this becomes

$$a_0p^n + a_1p^{n-1}q + a_2p^{n-2}q^2 + \cdots + a_{n-2}p^2q^{n-2} + a_{n-1}pq^{n-1} + a_nq^n = 0 \qquad (2)$$

When (*2*) is written as

$$a_0p^n + a_1p^{n-1}q + a_2p^{n-2}q^2 + \cdots + a_{n-2}p^2q^{n-2} + a_{n-1}pq^{n-1} = -a_nq^n$$

it is clear that p, being a factor of every term of the left member of the equality, must divide a_nq^n. Since p/q is expressed in lowest terms, no factor of p will divide q. Hence, p must divide a_n as was to be shown.

Similarly, when (*2*) is written as

$$a_1p^{n-1}q + a_2p^{n-2}q^2 + \cdots + a_{n-2}p^2q^{n-2} + a_{n-1}pq^{n-1} + a_nq^n = -a_0p^n$$

it follows that q must divide a_0.

12.13 For each of the equations $f(x) = 0$, write the equation where roots are the negatives of those of $f(x) = 0$.

(*a*) $x^3 - 8x^2 + x - 1 = 0$ *Ans.* $x^3 + 8x^2 + x + 1 = 0$

(*b*) $x^4 + 3x^2 + 2x + 1 = 0$ *Ans.* $x^4 + 3x^2 - 2x + 1 = 0$

(*c*) $2x^4 - 5x^2 + 8x - 3 = 0$ *Ans.* $2x^4 - 5x^2 - 8x - 3 = 0$

(*d*) $x^5 + x + 2 = 0$ *Ans.* $x^5 + x - 2 = 0$

12.14 Give all the information obtainable from Descartes' rule of signs about the roots of the following equations:

(*a*) $f(x) = x^3 - 8x^2 + x - 1 = 0$ [Problem 12.13(*a*)].

Since there are three variations of sign in $f(x) = 0$ and no variation of sign in $f(-x) = 0$, the given equation has either three positive roots or one positive root or one positive root and two imaginary roots.

(*b*) $f(x) = 2x^4 - 5x^2 + 8x - 3 = 0$ [Problem 12.13(*c*)].

Since there are three variations of sign in $f(x) = 0$ and one variation of sign in $f(-x) = 0$, the given equation has either three positive and one negative root or one positive, one negative, and two imaginary roots.

(*c*) $f(x) = x^5 + x + 2 = 0$ [Problem 12.13(*d*)].

Since there is no variation of sign in $f(x) = 0$ and one in $f(-x) = 0$, the given equation has one negative and four imaginary roots.

12.15 Form the equation whose roots are equal to the roots of the given equation diminished by the indicated number.

(a) $x^3 - 4x^2 + 8x - 5 = 0; 2.$ (b) $2x^3 + 9x^2 - 5x - 8 = 0; -3.$ (c) $x^4 - 8x^3 + 5x^2 + x + 8 = 0; 2.$

(a)

$$\begin{array}{l} 1 - 4 + 8 - 5 \quad \underline{|2} \\ \underline{\phantom{1 -{}} 2 - 4 + 8} \\ 1 - 2 + 4 + 3 \\ \underline{\phantom{1 - {}} 2 + 0} \\ 1 + 0 + 4 \\ \underline{\phantom{1 + {}} 2} \\ 1 + 2 \end{array}$$

The required equation is $y^3 + 2y^2 + 4y + 3 = 0$.

(b)

$$\begin{array}{l} 2 + 9 - 5 - 8 \quad \underline{|-3} \\ \underline{ - 6 - 9 + 42} \\ 2 + 3 - 14 + 34 \\ \underline{ - 6 + 9} \\ 2 - 3 - 5 \\ \underline{ - 6} \\ 2 - 9 \end{array}$$

The required equation is $2y^3 - 9y^2 - 5y + 34 = 0$.

(c)

$$\begin{array}{l} 1 - 8 + 5 + 1 + 8 \quad \underline{|2} \\ \underline{\phantom{1 -{}} 2 - 12 - 14 - 26} \\ 1 - 6 - 7 - 13 - 18 \\ \underline{\phantom{1 - {}} 2 - 8 - 30} \\ 1 - 4 - 15 - 43 \\ \underline{\phantom{1 - {}} 2 - 4} \\ 1 - 2 - 19 \\ \underline{\phantom{1 - {}} 2} \\ 1 + 0 \end{array}$$

The required equation is $y^4 - 19y^2 - 43y - 18 = 0$.

12.16 Use the method of successive linear approximation to approximate the irrational roots of

$$f(x) = x^3 + 3x^2 - 2x - 5 = 0$$

By Descartes' rule of signs the equation has either one positive and two negative roots or one positive and two imaginary roots. By the location principle (see Table 12.1), there are roots between $x = 1$ and $x = 2$, $x = -1$ and $x = -2$, and $x = -3$ and $x = -4$.

Table 12.1

x	2	1	0	−1	−2	−3	−4
$f(x)$	11	−3	−5	−1	3	1	−13

(a) To approximate the positive root, use Fig. 12-4. Then

$$RS = \frac{SQ}{TQ}(PT) = \frac{11}{14}(1) = 0.7 \qquad \text{and} \qquad OR = 2 - 0.7 = 1.3$$

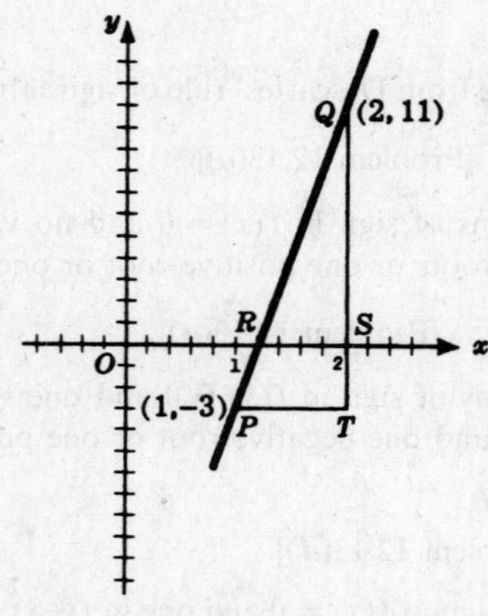

Fig. 12-4

Since $f(1.3) = -0.33$ and $f(1.4) = 0.82$, the root lies between $x = 1.3$ and $x = 1.4$.

For the next approximation,

$$RS = \frac{0.82}{1.15}(0.1) = 0.07 \qquad \text{and} \qquad OR = 1.4 - 0.07 = 1.33$$

Since $f(1.33) = -0.0006$ and $f(1.34) = 0.113$, the root lies between $x = 1.33$ and $x = 1.34$.

For the next approximation,

$$RS = \frac{0.113}{0.1136}(0.01) = 0.009 \qquad \text{and} \qquad OR = 1.34 - 0.009 = 1.331$$

Now $f(1.331) > 0$ so that this approximation is too large; in fact, $f(1.3301) > 0$. Thus, the root to three decimal places is $x = 1.330$.

In approximating a negative root of $f(x) = 0$, it is more convenient to approximate the equally positive root of $f(-x) = 0$.

(*b*) To approximate the root of $f(x) = 0$ between $x = -1$ and $x = -2$, we shall approximate the positive root between $x = 1$ and $x = 2$ of $g(x) = x^3 - 3x^2 - 2x + 5 = 0$. Since $g(1) = 1$ and $g(2) = -3$, we obtain from Fig. 12-5

$$SR = \frac{SP}{TP}(TQ) = \frac{1}{4}(1) = 0.2 \qquad \text{and} \qquad OR = OS + SR = 1.2$$

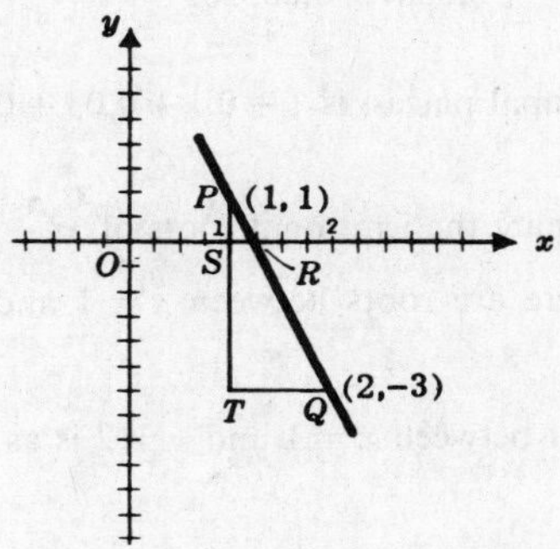

Fig. 12-5

Since $g(1.2) = 0.01$ and $g(1.3) = -0.47$, the root is between $x = 1.2$ and $x = 1.3$.

For the next approximation,

$$SR = \frac{0.01}{0.48}(0.1) = 0.002 \qquad \text{and} \qquad OR = 1.2 + 0.002 = 1.202$$

Since $g(1.202) = -0.0018$ (hence, too large) and $g(1.201) = 0.0031$, the root is between $x = 1.201$ and $x = 1.202$.

For the next approximation,

$$SR = \frac{0.0031}{0.0049}(0.001) = 0.0006 \qquad \text{and} \qquad OR = 1.201 + 0.0006 = 1.2016$$

Since $g(1.2016) = -0.00281$ and $g(1.2015) = 0.00007$, the root of $g(x) = 0$ to three decimal places is $x = 1.202$. The corresponding root of the given equation is $x = -1.202$.

(*c*) The approximation of the root -3.128 between $x = -3$ and $x = -4$ is left as an exercise.

12.17 Use Horner's method to approximate the irrational roots of $x^3 + 2x^2 - 4 = 0$.

By Descartes' rule of signs the equation has either one positive and two negative roots or one positive and two imaginary roots. By the location principle there is one root, between $x = 1$ and $x = 2$.

Arranged in compact form, the computation is as follows:

$$
\begin{array}{llll l}
1+2 & +0 & -4 & \underline{|1} & \\
\quad 1 & +3 & +3 & & \\
\hline
1+3 & +3 & -1 & & \\
\quad 1 & +4 & & & \\
\hline
1+4 & +7 & & & y=\frac{1}{7}=0.1 \\
\quad 1 & & & & \\
\hline
1+5 & +7 & -1 & \underline{|0.1} & \\
\quad 0.1 & +0.51 & +0.751 & & \\
\hline
1+5.1 & +7.51 & -0.249 & & z=\dfrac{0.249}{8.03}=0.03 \\
\quad 0.1 & +0.52 & & & \\
\hline
1+5.2 & +8.03 & & & \\
\quad 0.1 & & & & \\
\hline
1+5.3 & +8.03 & -0.249 & \underline{|0.03} & \\
\quad 0.03 & +0.1599 & +0.245697 & & \\
\hline
1+5.33 & +8.1899 & -0.003303 & & w=\dfrac{0.003303}{8.3507}=0.00039 \\
\quad 0.03 & +0.1608 & & & \\
\hline
1+5.36 & +8.3507 & & & \\
\quad 0.03 & & & & \\
\hline
1+5.39 & +8.3507 & -0.003303 & &
\end{array}
$$

The root, correct to four decimal places, is $1 + 0.1 + 0.03 + 0.00039 = 1.1304$.

12.18 Use Horner's method to approximate the irrational roots of $x^3 + 3x^2 - 2x - 5 = 0$.

By the location principle there are roots between $x = 1$ and $x = 2, x = -1$ and $x = -2, x = -3$ and $x = -4$.

(*a*) The computation for the root between $x = 1$ and $x = 2$ is as follows:

$$
\begin{array}{llll l}
1+3 & -\ 2 & -5 & \underline{|1} & y=\frac{3}{7}=0.4 \\
\quad 1 & +\ 4 & +2 & & \text{but is too large since, when used,} \\
\hline
1+4 & +\ 2 & -3 & & \text{the last number in the third line} \\
\quad 1 & +\ 5 & & & \text{is positive.} \\
\hline
1+5 & +\ 7 & & & \\
\quad 1 & & & & \\
\hline
1+6 & +\ 7 & -3 & \underline{|0.3} & \\
\quad 0.3 & +\ 1.89 & +2.667 & & \\
\hline
1+6.3 & +\ 8.89 & -0.333 & & z=\dfrac{0.333}{10.87}=0.03 \\
\quad 0.3 & +\ 1.98 & & & \\
\hline
1+6.6 & +10.87 & & & \\
\quad 0.3 & & & & \\
\hline
1+6.9 & +10.87 & -0.333 & \underline{|0.03} & \\
\quad 0.03 & +\ 0.2079 & +0.332337 & & \\
\hline
1+6.93 & +11.0779 & -0.000663 & & w=\dfrac{0.000663}{11.2867}=0.000058 \\
\quad 0.03 & +\ 0.2088 & & & \\
\hline
1+6.96 & +11.2867 & & & \\
\quad 0.03 & & & & \\
\hline
1+6.99 & +11.2867 & -0.000663 & &
\end{array}
$$

The root is 1.330.

When approximating a negative root of $f(x) = 0$ using Horner's method, it is more convenient to approximate the equally positive root of $f(-x) = 0$.

(*b*) To approximate the root between $x = -1$ and $x = -2$ of the given equation, we approximate the root between $x = 1$ and $x = 2$ of the equation $x^3 - 3x^2 - 2x + 5 = 0$. The computation is as follows:

$$
\begin{array}{llll}
1 - 3 & -2 & +5 & \underline{|1} \\
\underline{\quad 1} & \underline{-2} & \underline{-4} & \\
1 - 2 & -4 & +1 & \\
\underline{\quad 1} & \underline{-1} & & y = \tfrac{1}{5} = 0.2 \\
1 - 1 & -5 & & \\
\underline{\quad 1} & & & \\
1 + 0 & -5 & +1 & \underline{|0.2} \\
\underline{\quad 0.2} & \underline{+0.04} & \underline{-0.992} & \\
1 + 0.2 & -4.96 & +0.008 & z = \dfrac{0.008}{4.88} = 0.001 \\
\underline{\quad 0.2} & \underline{+0.08} & & \\
1 + 0.4 & -4.88 & & \\
\underline{\quad 0.2} & & & \\
1 + 0.6 & -4.88 & +0.008 & \underline{|0.001} \\
\multicolumn{3}{l}{\underline{\quad 0.001 + 0.000601 - 0.004879399}} & \\
\multicolumn{3}{l}{1 + 0.601 - 4.879399 + 0.003120601} & w = \dfrac{0.003120601}{4.878797} = 0.00063 \\
\multicolumn{3}{l}{\underline{\quad 0.001 + 0.000602}} & \\
\multicolumn{3}{l}{1 + 0.602 - 4.878797} & \\
\multicolumn{3}{l}{\underline{\quad 0.001}} & \\
\multicolumn{3}{l}{1 + 0.603 - 4.878797 + 0.003120601} &
\end{array}
$$

To four decimal places the root is $x = 1.2016$; thus, the root of the given equation is $x = -1.2016$

(*c*) The approximation of the root $x = -3.1284$ between $x = -3$ and $x = -4$ is left as an exercise.

Supplementary Problems

12.19 Find all the roots.

(*a*) $x^5 - 2x^4 - 9x^3 + 22x^2 + 4x - 24 = 0$

(*b*) $18x^4 - 27x^3 + x^2 + 12x - 4 = 0$

(*c*) $12x^4 - 40x^3 - 5x^2 + 45x + 18 = 0$

(*d*) $6x^4 + 5x^3 - 16x^2 - 9x - 10 = 0$

(*e*) $9x^4 - 19x^2 - 6x + 4 = 0$

(*f*) $2x^5 + 3x^4 - 3x^3 - 2x^2 = 0$

Ans. (*a*) $2, 2, 2 - 1, -3$ (*b*) $1, \frac{1}{2}, \frac{2}{3}, -\frac{2}{3}$ (*c*) $-\frac{2}{3}, -\frac{1}{2}, \frac{3}{2}, 2$ (*d*) $-2, \frac{5}{3}, (-1 \pm I\sqrt{7})/4$ (*e*) $-1, \frac{1}{3}, (1 \pm \sqrt{13})/3$ (*f*) $0, 0, 1, -2, -\frac{1}{2}$

12.20 Solve the inequalities.

(*a*) $x^3 - 5x^2 + 2x + 8 > 0$

(*b*) $6x^2 - 17x^2 - 5x + 6 < 0$

(*c*) $x^4 - 3x^3 + x^2 + 4 > 0$

(*d*) $x^5 - x^4 - 2x^3 + 2x^2 + x - 1 > 0$

Ans. (*a*) $x > 4, -1 < x < 2$ (*b*) $x < -\frac{2}{3}, \frac{1}{2} < x < 3$ (*c*) all $x \neq 2$ (*d*) $x > 1$

12.21 Prove: if $r \neq 1$ is a root of $f(x) = 0$, then $r - 1$ divides $f(1)$.

12.22 Use Descartes' rule of signs to show

(*a*) $x^4 + 5x^2 + 24 = 0$ has only complex roots.

(*b*) $x^n - 1 = 0$ has exactly two real roots if n is even and only one real root if n is odd.

(*c*) $x^3 + 3x + 2 = 0$ has exactly one real root.

(*d*) $x^7 - x^5 + 2x^4 + 3x^2 + 5 = 0$ has at least four complex roots.

(*e*) $x^7 - 2x^4 + 3x^3 - 5 = 0$ has at most three real roots.

12.23 Find all the irrational roots of the following equations:

(*a*) $x^3 + x - 3 = 0$ (*c*) $x^2 - 9x + 3 = 0$ (*e*) $x^4 + 4x^3 + 6x^2 - 15x - 40 = 0$

(*b*) $x^3 - 3x + 1 = 0$ (*d*) $x^3 + 6x^2 + 7x - 3 = 0$

Ans. (*a*) 1.2134 (*c*) 0.3376, 2.8169, −3.1546 (*e*) 2.7325, −0.7325

(*b*) 0.3473, 1.5321, −1.8794 (*d*) 0.3301, −2.2016, −4.1284

12.24 Show that

(*a*) The equation $f(x) = x^n p_1 x^{n-1} + \cdots + p_{n-1}x + p_n = 0$, with integral coefficients, has no rational root if $f(0)$ and $f(1)$ are odd integers.

Hint: Suppose r is an integral root; then r is odd and $r - 1$ does not divide $f(1)$.

(*b*) The equation $x^4 - 301x - 1275 = 0$ has no rational roots.

12.25 In a polynomial equation of the form

$$x^n + p_1 x^{n-1} + \cdots + p_{n-1}x + p_n = 0$$

the following relations exist between the coefficients and roots:

(*1*) The sum of the roots is $-p_1$.
(*2*) The sum of the products of the roots taken two at a time is p_2.
(*3*) The sum of the products of the roots taken three at a time is $-p_3$.
⋮
(*n*) The product of the roots is $(-1)^n p_n$.

If a, b, c are the roots of $x^3 - 3x^2 + 4x + 2 = 0$, find

(*a*) $a + b + c$

(*b*) $ab + bc + ca$

(*c*) abc

(*d*) $a^2 + b^2 + c^2 = (a + b + c)^2 - 2(ab + bc + ca)$

(*e*) $a^3 + b^3 + c^3$

(*f*) $\dfrac{1}{a} + \dfrac{1}{b} + \dfrac{1}{c} = \dfrac{ab + bc + ca}{abc}$

(*g*) $\dfrac{1}{ab} + \dfrac{1}{bc} + \dfrac{1}{ca}$

Ans. (*a*) 3, (*b*) 4, (*c*) −2, (*d*) 1, (*e*) −15, (*f*) −2, (*g*) $-\frac{3}{2}$

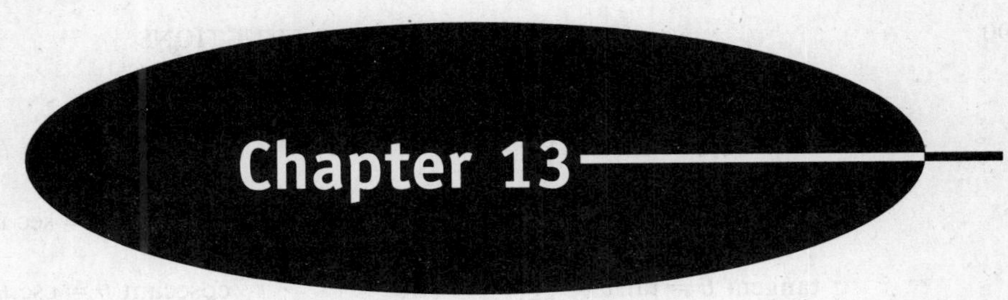

Trigonometric Functions

ANGLES IN STANDARD POSITION. With respect to a rectangular coordinate system, an angle is said to be *in standard position* when its vertex is at the origin and its initial side coincides with the positive x axis.

An angle is said to be a *first quadrant angle* or to be *in the first quadrant* if, when in standard position, its terminal side falls in the quadrant. Similar definitions hold for the other quadrants. For example, the angles $30°, 59°$, and $-330°$ are first quadrant angles; $119°$ is a second quadrant angle; $-119°$ is a third quadrant angle; $-10°$ and $710°$ are fourth quadrant angles. See Figs. 13-1 and 13-2.

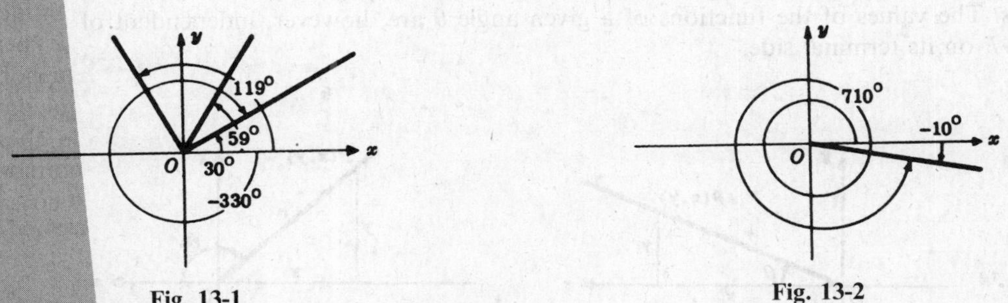

Fig. 13-1

Fig. 13-2

Two angles which, when placed in standard position, have coincident terminal sides are called *coterminal angels*. For example, $30°$ and $-330°$, $-10°$ and $710°$ are pairs of coterminal angles. There are an unlimited number of angles coterminal with a given angle. (See Problem 13.1.)

The angles $0°, 90°, 180°, 270°$, and all angles conterminal with them are called *quadrantal angles*.

TRIGONOMETRIC FUNCTIONS OF A GENERAL ANGLE. Let θ be an angle (not quadrantal) in standard position and let $P(x, y)$ be any point, distinct from the origin, on the terminal side of the angle. The six trigonometric functions of θ are defined, in terms of the abscissa, ordinate, and distance of P from the origin O, as follows:

$$\sin\theta = \sin\theta = \frac{\text{ordinate}}{\text{distance}} = \frac{y}{r} \qquad \text{cotangent}\,\theta = \cot\,\theta = \frac{\text{abscissa}}{\text{ordinate}} = \frac{x}{y}$$

$$\text{cosine}\,\theta = \cos\theta = \frac{\text{abscissa}}{\text{distance}} = \frac{x}{r} \qquad \text{secant}\,\theta = \sec\,\theta = \frac{\text{distance}}{\text{abscissa}} = \frac{r}{x}$$

$$\text{tangent}\,\theta = \tan\theta = \frac{\text{ordinate}}{\text{abscissa}} = \frac{y}{x} \qquad \text{cosecant}\,\theta = \csc\,\theta = \frac{\text{distance}}{\text{ordinate}} = \frac{r}{y}$$

Note that $r = \sqrt{x^2 + y^2}$ (see Fig. 13-3).

As an immediate consequence of these definitions, we have the so-called *reciprocal relations:*

$$\sin\,\theta = \frac{1}{\csc\,\theta} \qquad \tan\,\theta = \frac{1}{\cot\,\theta} \qquad \sec\,\theta = \frac{1}{\cos\theta}$$

$$\cos\theta = \frac{1}{\sec\,\theta} \qquad \cot\,\theta = \frac{1}{\tan\theta} \qquad \csc\,\theta = \frac{1}{\sin\phi}$$

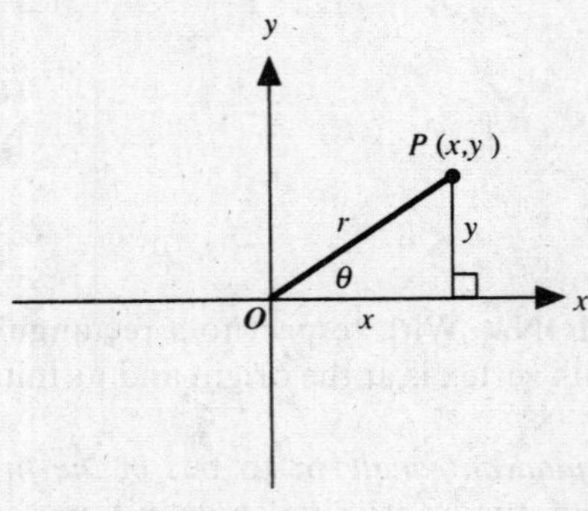

Fig. 13-3

It is evident from Figs. 13-4(*a*)–(*d*) that the values of the trigonometric functions of θ change as θ changes. The values of the functions of a given angle θ are, however, independent of the choice of the point P on its terminal side.

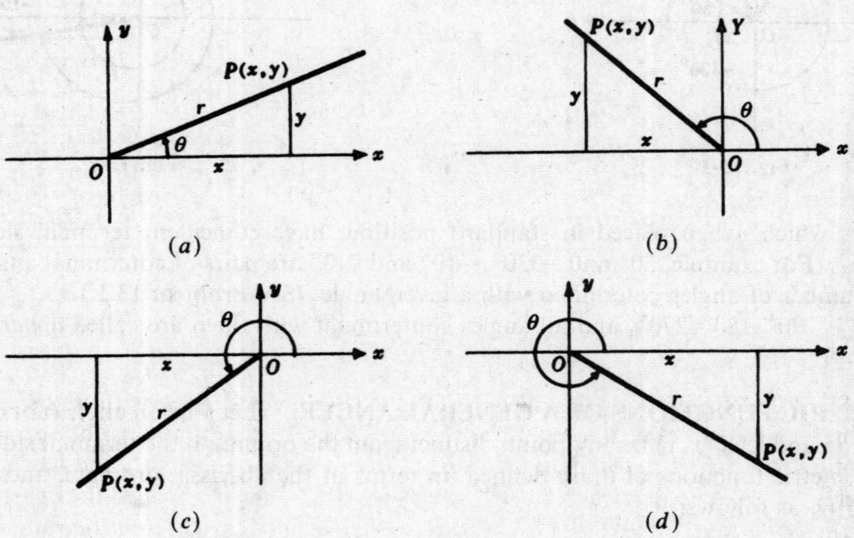

Fig. 13-4

ALGEBRAIC SIGNS OF THE FUNCTIONS. Since r is always positive, the signs of the functions in the various quadrants depend upon the signs of x and y. To determine these signs one may visualize the angle in standard position or use some device as shown in Fig. 13-5 in which only the functions having signs are listed.

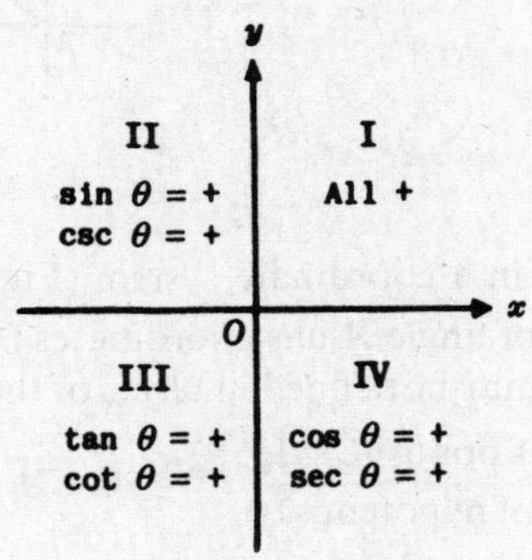

Fig. 13-5

When an angle is given, its trigonometric functions are uniquely determined. When, however, the value of one function of an angle is given, the angle is not uniquely determined. For example, if $\sin\theta = \frac{1}{2}$, then $\theta = 30°, 150°, 390°, 510°, \ldots$. In general, two possible positions of the terminal side are found—for example, the terminal sides of 30° and 150° in Fig. 13-4(*a*), (*b*). The exceptions to this rule occur when the angle is quadrantal. (See Problems 13.2–13.10.)

TRIGONOMETRIC FUNCTIONS OF QUADRANTAL ANGLES. For a quadrantal angle, the terminal side coincides with one of the axes. A point, P, distinct from the origin, on the terminal side has either $x = 0, y \neq 0$, or $x \neq 0, y = 0$. In either case, two of the six functions will not be defined. For example, the terminal side of the angle 0° coincides with the positive x axis and the ordinate of P is 0. Since the ordinate occurs in the denominator of the ratio defining the contangent and cosecant, these functions are not defined. Certain authors indicate this by writing $\cot 0° = \infty$ and others write $\cot 0° = \pm\infty$. The trigonometric functions of the quadrantal angles are given in Table 13.1.

Table 13.1

angle θ	$\sin\theta$	$\cos\theta$	$\tan\theta$	$\cot\theta$	$\sec\theta$	$\csc\theta$
0°	0	1	0		1	undefined
90°	1	0	undefined	0	undefined	1
180°	0	−1	0		−1	undefined
270°	−1	0	undefined	0	undefined	−1

TRIGONOMETRIC FUNCTIONS OF AN ACUTE ANGLE. In dealing with any right triangle, it will be convenient (see Fig. 13-6) to denote the vertices as A, B, C such that C is the vertex of the right triangle; to denote the angles of the triangle as A, B, C, such that $m \measuredangle C = 90°$; and to denote the sides opposite the angles as a, b, c, respectively. With respect to angle A, a will be called the *opposite side* and b will be called the *adjacent side*; with respect to angle B, a will be called the *adjacent side* and b the *opposite side*. Side c will always be called the *hypotenuse*.

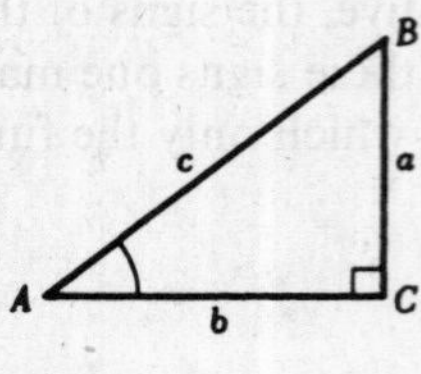

Fig. 13-6

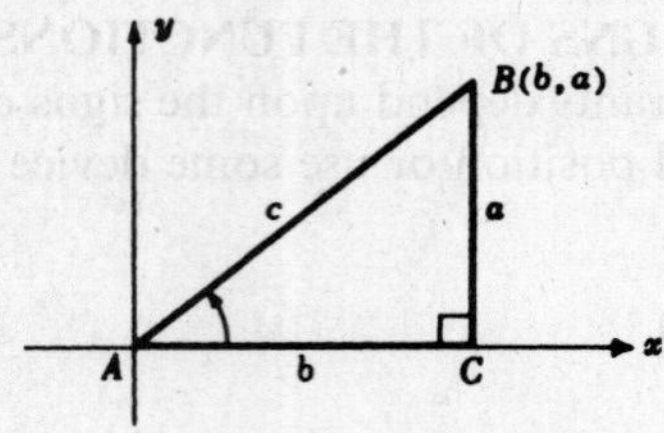

Fig. 13-7

If now the right triangle is placed in a coordinate system (Fig. 13-7) so that angle A is in standard position, point B on the terminal side of angle A has coordinates (b, a) and distance $c = \sqrt{a^2 + b^2}$. Then the trigonometric functions of angle A may be defined in terms of the sides of the right triangle, as follows:

$$\sin A = \frac{a}{c} = \frac{\text{length of opposite side}}{\text{length of hypotenuse}} \qquad \cot A = \frac{b}{a} = \frac{\text{length of adjacent side}}{\text{length of opposite side}}$$

$$\cos A = \frac{b}{c} = \frac{\text{length of adjacent side}}{\text{length of hypotenuse}} \qquad \sec A = \frac{c}{b} = \frac{\text{length of hypotenuse}}{\text{length of adjacent side}}$$

$$\tan A = \frac{a}{b} = \frac{\text{length of opposite side}}{\text{length of adjacent side}} \qquad \csc A = \frac{c}{a} = \frac{\text{length of hypotenuse}}{\text{length of opposite side}}$$

TRIGONOMETRIC FUNCTIONS OF COMPLEMENTARY ANGLES. The acute angles A and B of the right triangle ABC are complementary; that is, $m\measuredangle A + m\measuredangle B = 90°$. From Fig. 13-6, we have

$$\sin B = \frac{b}{c} = \cos A \qquad \cot B = \frac{a}{b} = \tan A$$

$$\cos B = \frac{a}{c} = \sin A \qquad \sec B = \frac{c}{a} = \csc A$$

$$\tan B = \frac{b}{a} = \cot A \qquad \csc B = \frac{c}{b} = \sec A$$

These relations associate the functions in pairs—sine and cosine, tangent and cotangent, secant and cosecant—each function of a pair being called the *cofunction* of the other. Thus, any function of an acute angle is equal to the corresponding cofunction of the complementary angle.

TRIGONOMETRIC FUNCTIONS OF 30°, 45°, 60°. The results in Table 13.2 are obtained in Problems 13.18–13.19.

Table 13.2

Angle θ	$\sin\theta$	$\cos\theta$	$\tan\theta$	$\cot\theta$	$\sec\theta$	$\csc\theta$
30°	$\frac{1}{2}$	$\frac{1}{2}\sqrt{3}$	$\frac{1}{3}\sqrt{3}$	$\sqrt{3}$	$\frac{2}{3}\sqrt{3}$	2
45°	$\frac{1}{2}\sqrt{2}$	$\frac{1}{2}\sqrt{2}$	1	1	$\sqrt{2}$	$\sqrt{2}$
60°	$\frac{1}{2}\sqrt{3}$	$\frac{1}{2}$	$\sqrt{3}$	$\frac{1}{3}\sqrt{3}$	2	$\frac{2}{3}\sqrt{3}$

Problems 13.20–13.26 illustrate a number of simple applications of the trigonometric functions. For this purpose, Table 13.3 will be used.

Table 13.3

Angle θ	$\sin\theta$	$\cos\theta$	$\tan\theta$	$\cot\theta$	$\sec\theta$	$\csc\theta$
15°	0.26	0.97	0.27	3.7	1.0	3.9
20°	0.34	0.94	0.36	2.7	1.1	2.9
30°	0.50	0.87	0.58	1.7	1.2	2.0
40°	0.64	0.77	0.84	1.2	1.3	1.6
45°	0.71	0.71	1.0	1.0	1.4	1.4
50°	0.77	0.64	1.0	0.84	1.6	1.3
60°	0.87	0.50	1.7	0.58	2.0	1.2
70°	0.94	0.34	2.7	0.36	2.9	1.1
75°	0.97	0.26	3.7	0.27	3.9	1.0

COTERMINAL ANGLES. Let θ be any angle; then

$$\sin(\theta + n360°) = \sin\theta \qquad \cot(\theta + n360°) = \cot\theta$$
$$\cos(\theta + n360°) = \cos\theta \qquad \sec(\theta + n360°) = \sec\theta$$
$$\tan(\theta + n360°) = \tan\theta \qquad \csc(\theta + n360°) = \csc\theta$$

where n is any positive or negative integer or zero.

EXAMPLE 1. $\sin 400° = \sin(40° + 360°) = \sin 40°$

$\cos 850° = \cos(130° + 2 \cdot 360°) = \cos 130°$

$\tan(-1000°) = \tan(80° - 3 \cdot 360°) = \tan 80°$

FUNCTIONS OF A NEGATIVE ANGLE. Let θ be an angle; then

$$\sin(-\theta) = -\sin\theta \qquad \cot(-\theta) = -\cot\theta$$
$$\cos(-\theta) = \cos\theta \qquad \sec(-\theta) = \sec\theta$$
$$\tan(-\theta) = -\tan\theta \qquad \csc(-\theta) = -\csc\theta$$

EXAMPLE 2. $\sin(-50°) = -\sin 50°, \cos(-30°) = \cos 30°, \tan(-200°) = -\tan 200°$.

REDUCTION FORMULAS. Let θ be an angle; then

$$\sin(90° - \theta) = \cos\theta \qquad \sin(90° + \theta) = \cos\theta$$
$$\cos(90° - \theta) = \sin\theta \qquad \cos(90° + \theta) = -\sin\theta$$
$$\tan(90° - \theta) = \cot\theta \qquad \tan(90° + \theta) = -\cot\theta$$

$$\cot(90° - \theta) = \tan\theta \qquad \cot(90° + \theta) = -\tan\theta$$
$$\sec(90° - \theta) = \csc\theta \qquad \sec(90° + \theta) = -\csc\theta$$
$$\csc(90° - \theta) = \sec\theta \qquad \csc(90° + \theta) = \sec\theta$$

$$\sin(180° - \theta) = \sin\theta \qquad \sin(180° + \theta) = -\sin\theta$$
$$\cos(180° - \theta) = -\cos\theta \qquad \cos(180° + \theta) = -\cos\theta$$
$$\tan(180° - \theta) = -\tan\theta \qquad \tan(180° + \theta) = \tan\theta$$
$$\cot(180° - \theta) = -\cot\theta \qquad \cot(180° + \theta) = \cot\theta$$
$$\sec(180° - \theta) = -\sec\theta \qquad \sec(180° + \theta) = -\sec\theta$$
$$\csc(180° - \theta) = \csc\theta \qquad \csc(180° + \theta) = -\csc\theta$$

GENERAL REDUCTION FORMULA. Any trigonometric function of $(n \cdot 90° \pm \theta)$, where θ is any angle, is *numerically* equal

(*a*) To the same function of θ if n is an even integer
(*b*) To the corresponding cofunction of θ if n is an odd integer

The algebraic sign in each case is the same as the sign of the given function for that quadrant in which $n \cdot 90° \pm \theta$ lies when θ is a positive acute angle.

EXAMPLE 3.

(*1*) $\sin(180° - \theta) = \sin(2 \cdot 90° - \theta) = \sin\theta$ since 180° is an even multiple of 90° and, when θ is positive acute, the terminal side of $180° - \theta$ lies in quadrant II.

(*2*) $\cos(180° + \theta) = \cos(2 \cdot 90° + \theta) = -\cos\theta$ since 180° is an even multiple of 90° and, when θ is positive acute, the terminal side of $180° + \theta$ lies is quadrant III.

(*3*) $\tan(270° - \theta) = \tan(3 \cdot 90° - \theta) = \cot\theta$ since 270° is an odd multiple of 90° and, when θ is positive acute, the terminal side of $270° - \theta$ lies in quadrant III.

(*4*) $\cos(270° + \theta) = \cos(3 \cdot 90° + \theta) = \sin\theta$ since 270° is an odd multiple of 90° and, when θ is positive acute, the terminal side of $270° + \theta$ lies in quadrant IV.

LINE REPRESENTATIONS OF THE TRIGONOMETRIC FUNCTIONS. Let θ be any given angle in standard position. (See the Figs. 13-8 through 13-11 for θ in each of the quadrants.) With the vertex O as center describe a circle of radius one unit cutting the initial side $\overrightarrow{OX}$ of θ at A, the positive y axis at B, and the terminal side of θ at P. Draw $\overrightarrow{MP}$ perpendicular to $\overrightarrow{OX}$; draw also the tangents to the circle at A and B meeting the terminal side of θ or its extension through O in the points Q and R, respectively.

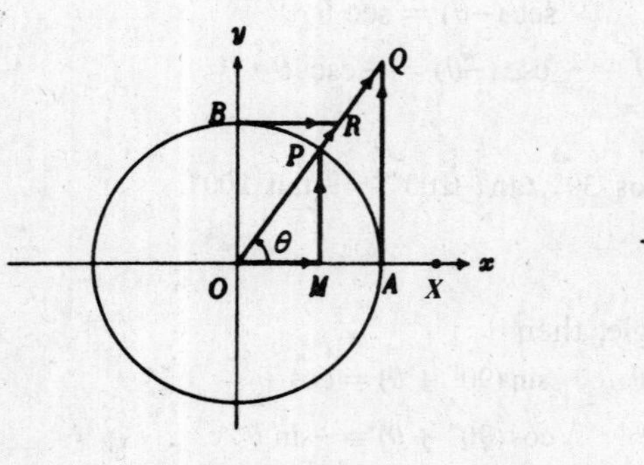

Fig. 13-8

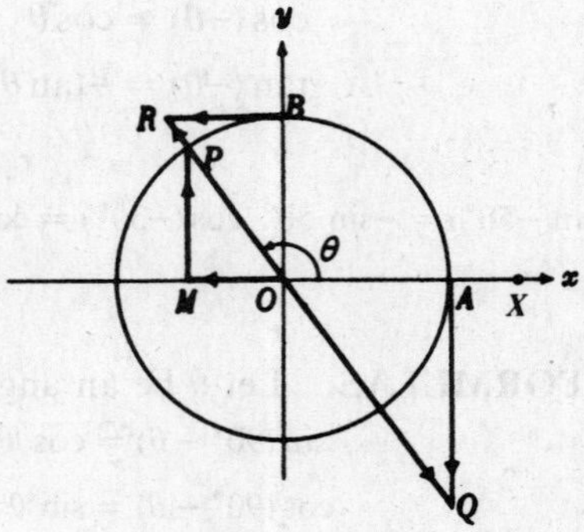

Fig. 13-9

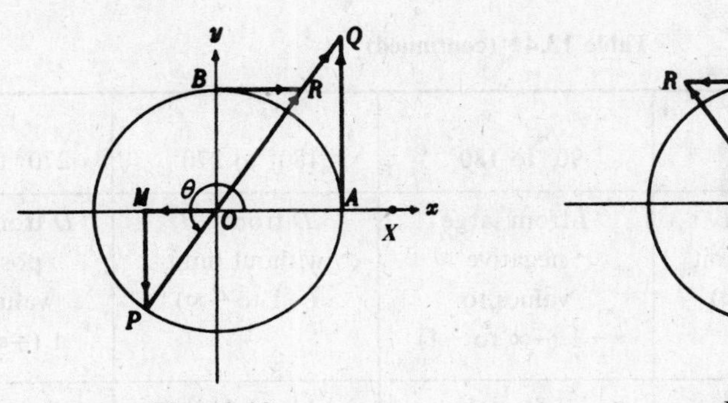

Fig. 13-10

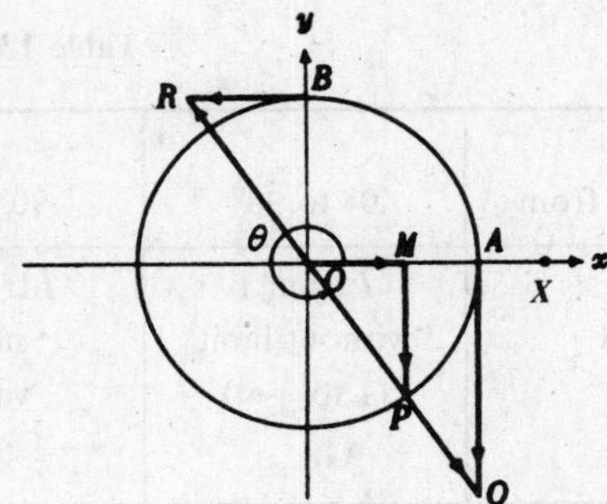

Fig. 13-11

In each of the figures, the right triangles OMP, OAQ, and OBR are similar, and

$$\sin\theta = \frac{MP}{OP} = MP \qquad \cot\theta = \frac{OM}{MP} = \frac{BR}{OB} = BR$$

$$\cos\theta = \frac{OM}{OP} = OM \qquad \sec\theta = \frac{OP}{OM} = \frac{OQ}{OA} = OQ$$

$$\tan\theta = \frac{MP}{OM} = \frac{AQ}{OA} = AQ \qquad \csc\theta = \frac{OP}{MP} = \frac{OR}{OB} = OR$$

The, $\overrightarrow{MP}, \overrightarrow{OM}, \overrightarrow{AQ}$, etc., are directed line segments, the magnitude of a function being given by the length of the corresponding segment and the sign being given by the indicated direction. The directed segments $\overrightarrow{OQ}$ and $\overrightarrow{OR}$ are to be considered positive when measured on the terminal side of the angle and negative when measured on the terminal side extended.

VARIATIONS OF THE TRIGONOMETRIC FUNCTIONS. Let P move counterclockwise about the unit circle, starting at A, so that $m\measuredangle\theta = m\measuredangle XOP$ varies continuously from 0° to 360°. Using Figs. 13-8 through 13-11, Table 13.4 is derived.

Table 13.4

As θ increases from	0° to 90°	90° to 180°	180° to 270°	270° to 360°
$\sin\theta$	*I* from 0 to 1	*D* from 1 to 0	*D* from 0 to −1	*I* from −1 to 0
$\cos\theta$	*D* from 1 to 0	*D* from 0 to −1	*I* from −1 to 0	*I* from 0 to 1
$\tan\theta$	*I* from 0 without limit (0 to +∞)	*I* from large negative values to 0 (−∞ to 0)	*I* from 0 without limit (0 to +∞)	*I* from large negative values to 0 (−∞ to 0)
$\cot\theta$	*D* from large positive values to 0 (+∞ to 0)	*D* from 0 without limit (0 to −∞)	*D* from large positive values to 0 (+∞ to 0)	*D* from 0 without limit (0 to −∞)

Table 13.4 (continued)

As θ increases from	0° to 90°	90° to 180°	180° to 270°	270° to 360°
$\sec\theta$	*I* from 1 without limit (1 to $+\infty$)	*I* from large negative values to -1 ($-\infty$ to -1)	*D* from -1 without limit (-1 to $-\infty$)	*D* from large positive values to 1 ($+\infty$ to 1)
$\csc\theta$	*D* from large positive values to 1 ($+\infty$ to 1)	*I* from 1 without limit (1 to $+\infty$)	*I* from large negative values to -1 ($-\infty$ to -1)	*D* from -1 without limit (-1 to $-\infty$)

I = increases; *D* = decreases.

GRAPHS OF THE TRIGONOMETRIC FUNCTIONS. In Table 13.5, values of the angle x are given in radians.

Table 13.5

x	$y=\sin x$	$y=\cos x$	$y=\tan x$	$y=\cot x$	$y=\sec x$	$y=\csc x$
0	0	1.00	0	$\pm\infty$	1.00	$\pm\infty$
$\pi/6$	0.50	0.87	0.58	1.73	1.15	2.00
$\pi/4$	0.71	0.71	1.00	1.00	1.41	1.41
$\pi/3$	0.87	0.50	1.73	0.58	2.00	1.15
$\pi/2$	1.00	0	$\pm\infty$	0	$\pm\infty$	1.00
$2\pi/3$	0.87	-0.50	-1.73	-0.58	-2.00	1.15
$3\pi/4$	0.71	-0.71	-1.00	-1.00	-1.41	1.41
$5\pi/6$	0.50	-0.87	-0.58	-1.73	-1.15	2.00
π	0	-1.00	0	$+\infty$	-1.00	$+\infty$
$7\pi/6$	-0.50	-0.87	0.58	1.73	-1.15	-2.00
$5\pi/4$	-0.71	-0.71	1.00	1.00	-1.41	-1.41
$4\pi/3$	-0.87	-0.50	1.73	0.58	-2.00	-1.15
$3\pi/2$	-1.00	0	$\pm\infty$	0	$\pm\infty$	-1.00
$5\pi/3$	-0.87	0.50	-1.73	-0.58	2.00	-1.15
$7\pi/4$	-0.71	0.71	-1.00	-1.00	1.41	-1.41
$11\pi/6$	-0.50	0.87	-0.58	-1.73	1.15	-2.00
2π	0	1.00	0	$\pm\infty$	1.00	$\pm\infty$

Note 1. Since $\sin(\frac{\pi}{2}+x)=\cos x$, the graph of $y=\cos x$ may be obtained most easily by shifting the graph of $y=\sin x$ a distance $\frac{\pi}{2}$ to the left. See Fig. 13-12.

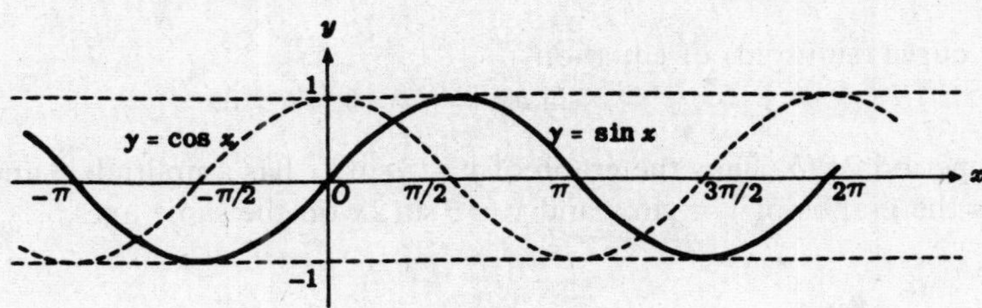

Fig. 13-12

Note 2. Since $\csc\left(\frac{\pi}{2}+x\right)=\sec x$, the graph of $y=\csc x$ may be obtained by shifting the graph of $y=\sec x$ a distant $\frac{\pi}{2}$ to the right. Notice, too, the relationship between the graphs for $\tan x$ and $\cot x$. See Figs. 13-13 through 13-16.

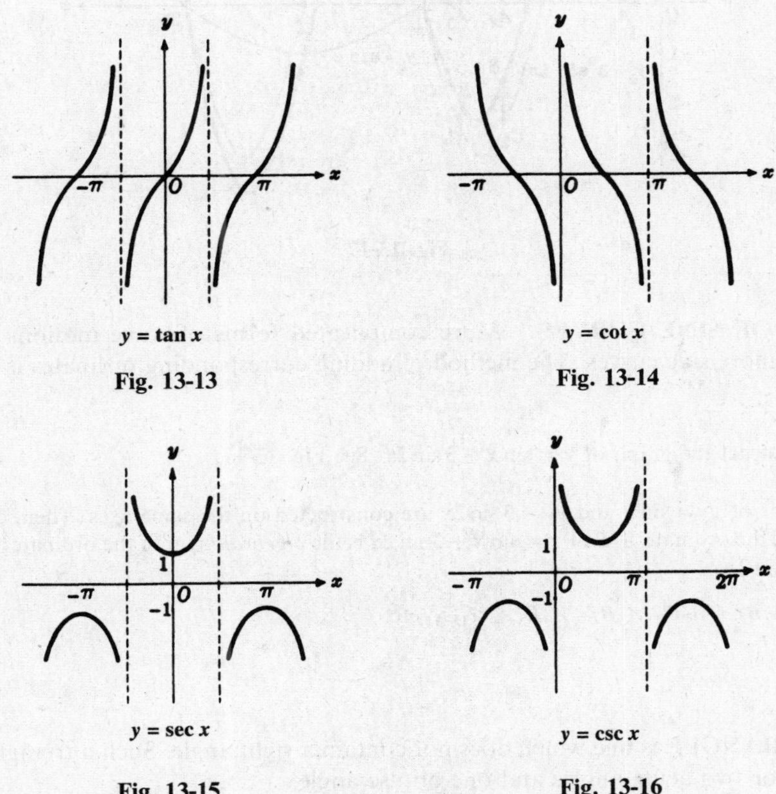

$y=\tan x$

Fig. 13-13

$y=\cot x$

Fig. 13-14

$y=\sec x$

Fig. 13-15

$y=\csc x$

Fig. 13-16

PERIODIC FUNCTIONS. Any function of a variable x, $f(x)$, which repeats its values in definite cycles, is called *periodic*. The smallest range of values of x which corresponds to a complete cycle of values of the function is called the period of the function. It is evident from the graphs of the trigonometric functions that the sine, cosine, secant, and cosecant are of period 2π while the tangent and cotangent are of period π.

THE GENERAL SINE CURVE. The *amplitude* (*maximum ordinate*) *and period* (*wavelength*) *of* $y=\sin x$ are, respectively, 1 and 2π. For a given value of x, the value of $y=a\sin x, a>0$, is a times the value of $y=\sin x$. Thus, the amplitude of $y=a\sin x$ is a and the period is 2π. Since when $bx=2\pi, x=2\pi/b$, the amplitude of $y=\sin bx$, $b>0$, is 1 and the period is $2\pi/b$.

The general sine curve (sinusoid) of equation

$$y = a \sin bx, \qquad a > 0, \quad b > 0,$$

has amplitude a and period $2\pi/b$. Thus the graph of $y = 3\sin 2x$ has amplitude 3 and period $2\pi/2 = \pi$. Figure 13-17 exhibits the graphs of $y = \sin x$ and $y = 3\sin 2x$ on the same axes.

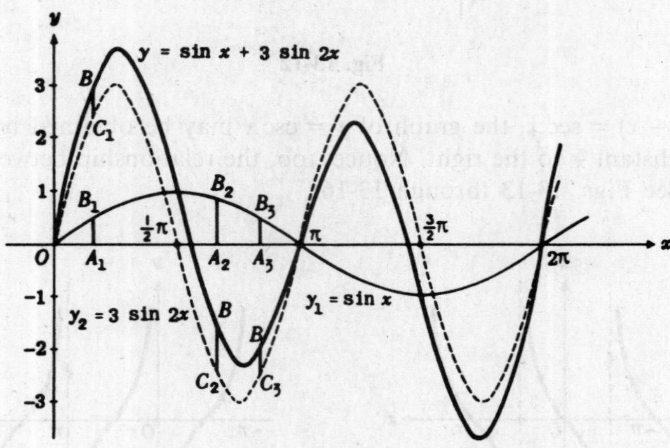

Fig. 13-17

COMPOSITION OF SINE CURVES. More complicated forms of wave motions are obtained by combining two or more sine curves. The method of adding corresponding ordinates is illustrated in the following example.

EXAMPLE 4. Construct the graph of $y = \sin x + 3\sin 2x$. See Fig. 13-17.

First the graphs of $y_1 = \sin x$ and $y_2 = 3\sin 2x$ are constructed on the same axes. Then, corresponding to a given value $x = OA_1$, the ordinate A_1B of $y = \sin x + 3\sin 2x$ is the *algebraic* sum of the ordinates A_1B_1 of $y_1 = \sin x$ and A_1C_1 of $y_2 = 3\sin 2x$.

Also, $A_2B = A_2B_2 + A_2C_2, A_3B = A_3B_3 + A_3C_3$, etc.

AN OBLIQUE TRIANGLE is one which does not contain a right angle. Such a triangle contains either three acute angles or two acute angles and one obtuse angle.

The convention of denoting the measures of the angles by A, B, C and the lengths of the corresponding opposite sides by a, b, c will be used here. See Figs. 13-18 and 13-19.

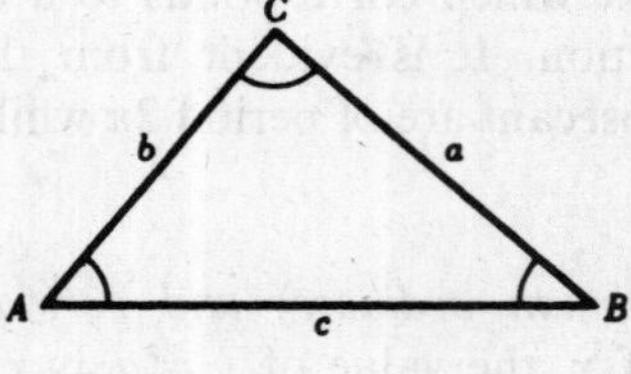

Fig. 13-18

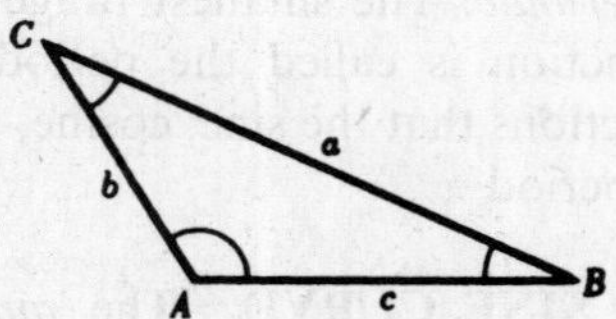

Fig. 13-19

When three parts, not all angles, are known, the triangle is uniquely determined, except in one case to be noted below. The four cases of oblique triangles are

Case I. Given one side and two angles

Case II. Given two sides and the angle opposite one of them

Case III. Given two sides and the included angle

Case IV. Given the three sides

THE LAW OF SINES. In any triangle, the sides are proportional to the sines of the opposite angles, i.e.,

$$\frac{a}{\sin A} = \frac{b}{\sin B} = \frac{c}{\sin C}$$

The following relations follow readily:

$$\frac{a}{b} = \frac{\sin A}{\sin B}, \quad \frac{b}{c} = \frac{\sin B}{\sin C}, \quad \frac{c}{a} = \frac{\sin C}{\sin A}.$$

(For a proof of the law of sines, see Problem 13.39.)

PROJECTION FORMULAS. In any triangle ABC,

$$a = b \cos C + c \cos B$$
$$b = c \cos A + a \cos C$$
$$c = a \cos B + b \cos A$$

(For the derivation of these formulas, see Problem 13.40.)

CASE I. Given one side and two angles

EXAMPLE 5. Suppose a, B, and C are given.

To find A, use $A = 180° - (B + C)$.

To find b, use $\frac{b}{a} = \frac{\sin B}{\sin A}$ whence $b = \frac{a \sin B}{\sin A}$.

To find c, use $\frac{c}{a} = \frac{\sin C}{\sin A}$ whence $c = \frac{a \sin C}{\sin A}$.

(See Problems 13.42–13.44.)

CASE II. Given two sides and the angle opposite one of them

EXAMPLE 6. Suppose b, c, and B are given.

From $\frac{\sin C}{\sin B} = \frac{c}{b}$, $\quad \sin C = \frac{c \sin B}{b}$.

If $\sin C > 1$, no angle C is determined.
If $\sin C = 1, C = 90°$ and a right triangle is determined.
If $\sin C < 1$, two angles are determined: an acute angle C and an obtuse angle $C' = 180° - C$. Thus, there may be one or two triangles determined. This is known as the "ambiguous case."

This case is discussed geometrically in Problem 13.44. The results obtained may be summarized as follows: When the given angle is *acute*, there will be

(*a*) *One* solution if the side opposite the given angle is equal to or greater than the other given side.
(*b*) *No* solution, *one* solution (right triangle), or *two* solutions if the side opposite the given angle is less than the other given side.

When the given angle is *obtuse*, there will be

(*c*) *No* solution when the side opposite the given angle is less than or equal to the other given side.
(*d*) *One* solution if the side opposite the given angle is greater than the other given side.

EXAMPLE 7.

(*1*) When $b = 30$, $c = 20$, and $B = 40°$, there is one solution since B is acute and $b > c$.

(*2*) When $b = 20$, $c = 30$, and $B = 40°$, there is either no solution, one solution, or two solutions. The particular subcase is determined after computing $\sin C = \dfrac{c \sin B}{b}$.

(*3*) When $b = 30$, $c = 20$, and $B = 140°$, there is one solution.
(*4*) When $b = 20$, $c = 30$, and $B = 140°$, there is no solution.

This, the so-called ambiguous case, is solved by the law of sines and may be checked by the projection formulas. (See Problems 13.46–13.48.)

THE LAW OF COSINES. In any triangle *ABC*, the square of any side is equal to the sum of the squares of the other two sides diminished by twice the product of these sides and the cosine of their included angle; i.e.,

$$a^2 = b^2 + c^2 - 2bc \cos A$$
$$b^2 = c^2 + a^2 - 2ca \cos B$$
$$c^2 = a^2 + b^2 - 2ab \cos C$$

(For the derivation of these formulas, see Problem 31.48.)

CASE III. Given two sides and the included angle

EXAMPLE 8. Suppose a, b, and C are given.

To find c, use $c^2 = a^2 + b^2 - 2ab \cos C$.

To find A, use $\sin A = \dfrac{a \sin C}{c}$. To find B, use $\sin B = \dfrac{b \sin C}{c}$.

To check, use $A + B + C = 180°$.
(See Problems 31.50–13.52.)

CASE IV. Given three sides

EXAMPLE 9. With a, b, and c given, solve the law of cosines for each of the angles.

To find the angles, use $\cos A = \dfrac{b^2 + c^2 - a^2}{2bc}$, $\cos B = \dfrac{c^2 + a^2 - b^2}{2ca}$, $\cos C = \dfrac{a^2 + b^2 - c^2}{2ab}$.

To check, use $A + B + C = 180°$.
(See Problems 13.52–13.53.)

INVERSE FUNCTIONS. The equation

$$y = 2x + 3$$

defines a unique value of y for each value of x. Similarly, the equation

$$y = \frac{x}{2} - 3$$

does the same; however, these two equations have an interesting, even provocative, relationship: If

$$f(x) = 2x + 3 \quad \text{and} \quad g(x) = \frac{x}{2} - 3,$$

then

$$f(g(x)) = g(f(x)) = x$$

That is, f and g "undo" each other. We call the function g the inverse of f and we call f the inverse of g.

NOTATION. This relationship is written as follows:

$$f = g^{-1}$$

and

$$g = f^{-1}$$

DEFINITION. If f and g are functions and if

$$f(g(x)) = g(f(x)) = x$$

for all values of x for which these composites are defined, then we say that f and g are each other's inverses.

To determine the equation of an inverse function for $y = f(x)$, simply solve $y = f(x)$ for x and then interchange the roles of the two variables.

EXAMPLE 10. If $f(x) = y = 3x - 5$,

then $3x = y + 5$ (interchange x and y)

and $x = \dfrac{y+5}{3}$ (solve for x)

Thus, $f^{-1}(x) = y = \dfrac{x+5}{3}$ (interchange x and y)

Note that in Fig. 13-20 $f(x)$ and $f^{-1}(x)$ are mirror images of each other in the line $y = x$:

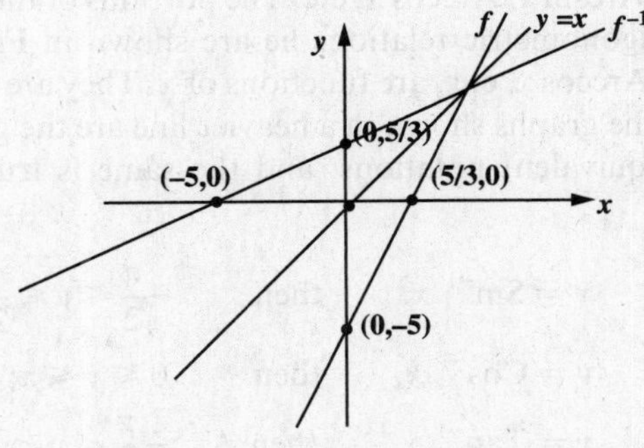

Fig. 13-20

The reader should verify that for $$g(x) = x^3, \quad g^{-1}(x) = x^{1/3}.$$

For
$$h(x) = x^2, \quad x > 0, \quad h^{-1}(x) = \sqrt{x} \tag{13.1}$$

Note also that if $f(x)$ is *not* one-to-one, then $f^{-1}(x)$ is *not* a function. Thus, functions that are not one-to-one do not have inverse functions. For $f(x) = x^2$, unless we restrict the domain (as we did above in (*13.1*)), $f(x)$ does not possess an inverse function.

One particularly important class of inverse functions is the class of inverse trigonometric functions.

INVERSE TRIGONOMETRIC RELATIONS. The equation
$$x = \sin y \tag{13.2}$$

defines a unique value of x for each given angle y. But when x is given, the equation may have no solution or many solutions. For example: if $x = 2$, there is no solution, since the sine of an angle never exceeds 1; if $x = \frac{1}{2}$, there are many solutions: $y = 30°, 150°, 390°, 510°, -210°, -330°, \ldots$.

To express y in terms of x, we will write
$$y = \arcsin x \tag{13.3}$$

In spite of the use of the word *arc*, (*13.3*) is to be interpreted as stating that "y is an angle whose sine is x." Similarly, we shall write $y = \arccos x$ if $x = \cos y$, $y = \arctan x$ if $x = \tan y$, etc. An alternate notation for $y = \arcsin x$ is $y = \sin^{-1} x$ (and similarly, $y = \cos^{-1} x$, etc., for the other functions). Note that $y = \arcsin x$, $\arccos x$, etc., are all relations but not functions of x.

GRAPHS OF THE INVERSE TRIGONOMETRIC RELATIONS. The graph of $y = \arcsin x$ is the graph of $x = \sin y$ and differs from the graph $y = \sin x$ in that the roles of x and y are interchanged. Thus, the graph of $y = \arcsin x$ is a sine curve drawn on the y axis instead of the x axis.

Similarly, the graphs of the remaining inverse trigonometric relations are those of the corresponding trigonometric functions except that the roles of x and y are interchanged.

INVERSE TRIGONOMETRIC FUNCTIONS. It is at times necessary to consider the inverse trigonometric relations as single-valued (i.e., one value of y corresponding to each admissible value of x). To do this, we agree to select one out of the many angles corresponding to the given value of x. For example, when $x = \frac{1}{2}$, we shall agree to select the value $y = 30°$, and when $x = -\frac{1}{2}$, we shall agree to select the value $y = -30°$. This selected value is called the *principal value* of $\arcsin x$. When only the principal value is called for, we shall write $\text{Arcsin}\, x$, $\text{Arccos}\, x$, etc. The portions of the graphs on which the principal values of each of the inverse trigonometric relations lie are shown in Figs. 13-21(*a*) through (*f*) by a heavier line. Note that $\text{Arcsin}\, x$, $\text{Arccos}\, x$, etc., are functions of x. They are called the *inverse trigonometric functions*. Thus, the portions of the graphs shown in a heavier line are the graphs of these functions. Note that $\text{Sin}^{-1} x$ and $\text{Arcsin}\, x$ are equivalent notations, and the same is true for the other trigonometric functions.

If $$y = \text{Sin}^{-1} x, \quad \text{then} \quad -\frac{\pi}{2} \leq y \leq \frac{\pi}{2}.$$

If $$y = \text{Cos}^{-1} x, \quad \text{then} \quad 0 \leq y \leq \pi;$$

and if $$y = \text{Tan}^{-1} x, \quad \text{then} \quad -\frac{\pi}{2} < y < \frac{\pi}{2}.$$

Similarly,

if $y = \text{Sec}^{-1}\, x$ then $0 \leqslant y \leqslant \pi, \quad y \neq \dfrac{\pi}{2}$;

if $y = \text{Cot}^{-1}\, x$ then $0 < y < \pi$;

and if $y = \text{Csc}^{-1}\, x$, then $-\dfrac{\pi}{2} \leqslant y \leqslant \dfrac{\pi}{2}, \quad y \neq 0$.

For example, $\text{Sin}^{-1}\dfrac{\sqrt{3}}{2} = \dfrac{\pi}{3}$, $\text{Arctan}\ 1 = \dfrac{\pi}{4}$, $\text{Sin}^{-1}\dfrac{-\sqrt{3}}{2} = \dfrac{-\pi}{3}$, $\text{Arccos}\left(\dfrac{-1}{2}\right) = \dfrac{2\pi}{3}$, $\text{Sec}^{-1}\left(\dfrac{-2}{\sqrt{3}}\right) = \dfrac{-5\pi}{6}$ and $\text{Arccsc}\,(-\sqrt{2}) + \dfrac{-\pi}{4}$.

Note that the inverse trigonometric functions are inverses of the trigonometric functions. For example $\sin(\text{Sin}^{-1}\, x) = y$.

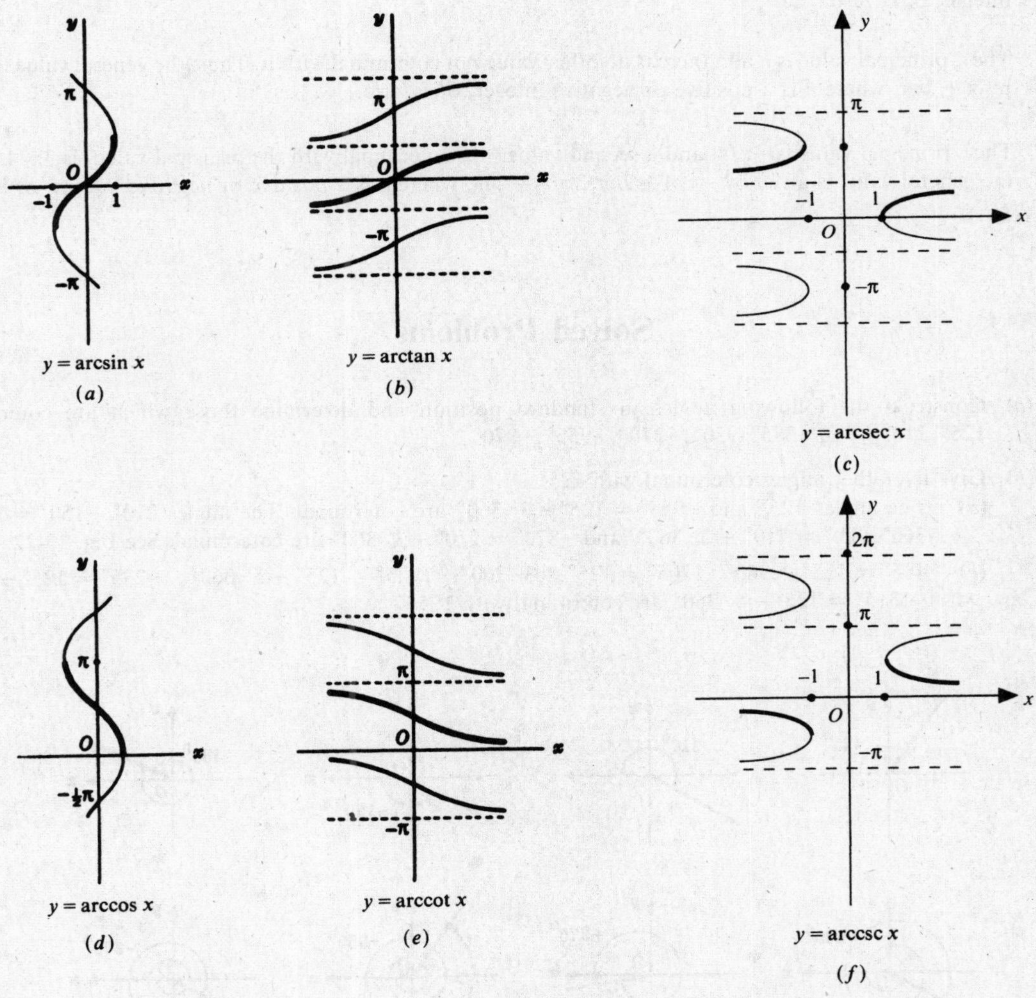

$y = \arcsin x$ (a) $y = \arctan x$ (b) $y = \text{arcsec}\ x$ (c)

$y = \arccos x$ (d) $y = \text{arccot}\ x$ (e) $y = \text{arccsc}\ x$ (f)

Fig. 13-21

GENERAL VALUES OF THE INVERSE TRIGONOMETRIC RELATIONS. Let y be an inverse trigonometric relation of x. Since the value of a trigonometric function of y is known, there are determined in general two positions for the terminal side of the angle y (see Fig. 13-4.). Let y_1 and y_2 respectively be angles determined by the two positions of the terminal side. Then the totality of values of y consist of the angles y_1 and y_2, together with all angles coterminal with them, that is,

$$y_1 + 2n\pi \quad \text{and} \quad y_2 + 2n\pi$$

where n is any positive or negative integer, or is zero.

One of the values y_1 or y_2 may always be taken as the principal value of the inverse trigonometric relation with the domains properly restricted.

EXAMPLE 11. Write expressions for the general value of (*a*) $\arcsin \frac{1}{2}$, (*b*) $\arccos(-1)$, (*c*) $\arctan(-1)$.

(*a*) The principal value of $\arcsin \frac{1}{2}$ is $\pi/6$, and a second value (not coterminal with the principal value) is $5\pi/6$. The general value of $\arcsin \frac{1}{2}$ is given by $\pi/6 + 2n\pi$, $5\pi/6 + 2n\pi$, where n is any positive or negative integer, or is zero.

(*b*) The principal value is π and there is no other value not coterminal with it. Thus, the general value is given by $\pi + 2n\pi$, where n is a positive or negative integer, or is zero.

(*c*) The principal value is $-\pi/4$, and a second value (not coterminal with the principal value) is $3\pi/4$. Thus, the general value is given by $-\pi/4 + 2n\pi$, $3\pi/4 + 2n\pi$, where n is a positive or negative integer, or is zero.

Solved Problems

13.1 (*a*) Construct the following angles in standard position and determine those which are coterminal: $125°, 210°, -150°, 385°, 930°, -370°, -955°, -870°$.

(*b*) Give five other angles coterminal with $125°$.

(*a*) The angles $125°$ and $-955° = 125° - 3 \cdot 360°$ are coterminal. The angles $210°, -150° = 210° - 360°, 930° = 210° + 2 \cdot 360°$, and $-870° = 210° - 3 \cdot 360°$ are coterminal. See Fig. 13-22.

(*b*) $485° = 125° + 360°$, $1205° = 125° + 3 \cdot 360°$, $1925° = 125° + 5 \cdot 360°$, $-235° = 125° - 360°$, $-1315° = 125° - 4 \cdot 360°$ are coterminal with $125°$.

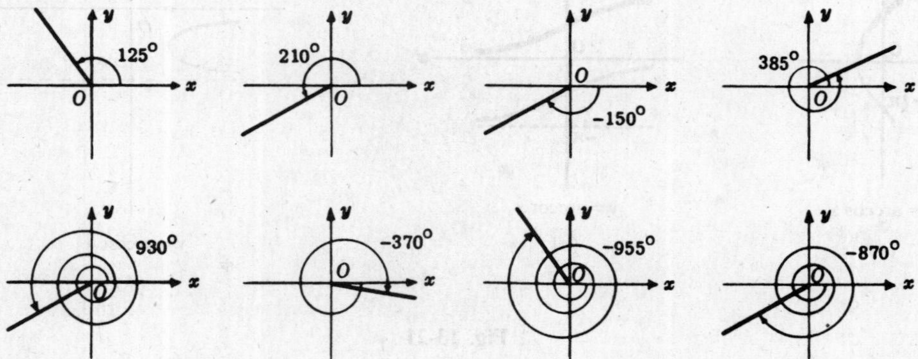

Fig. 13-22

13.2 Determine the values of the trigonometric functions of angle θ (smallest positive angle in standard position) if P is a point on the terminal side of θ and the coordinates of P are (*a*) $P\ (3,\ 4)$, (*b*) $P(-3, 4)$, (*c*) $P(-1, -3)$. See Fig. 13-23.

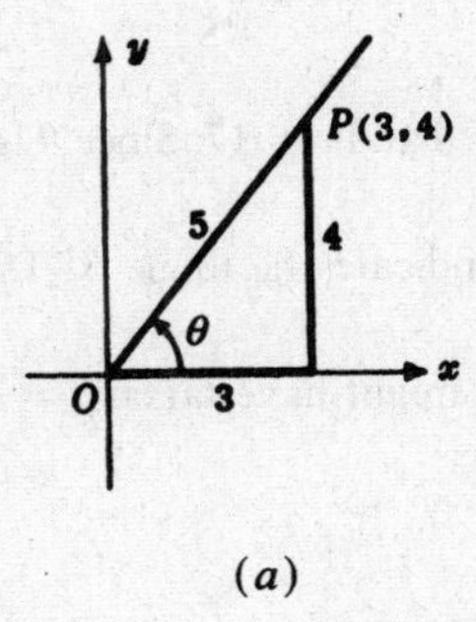

(*a*)

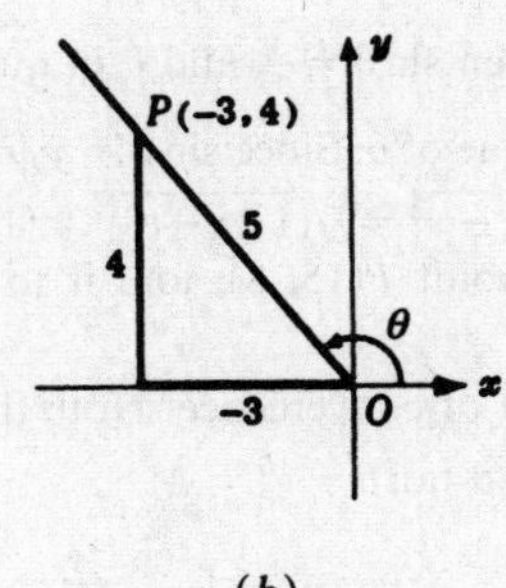

(*b*)

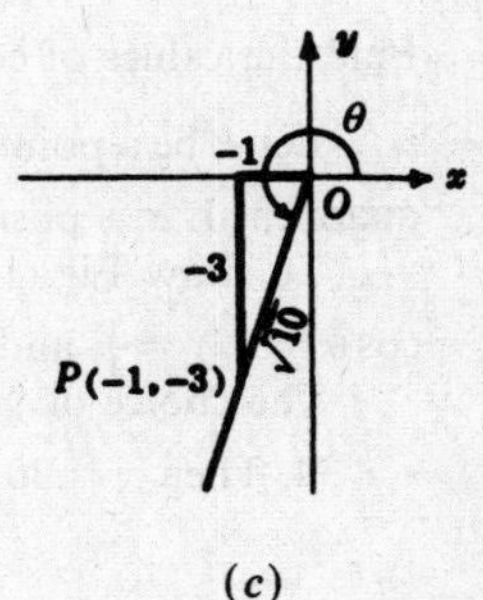

(*c*)

Fig. 13-23

(*a*) $r = \sqrt{3^2 + 4^2} = 5$ (*b*) $r = \sqrt{(-3)^2 + 4^2} = 5$ (*c*) $r = \sqrt{(-1)^2 + (-3)^2} = \sqrt{10}$

(*a*)

$\sin\theta = y/r = \frac{4}{5}$
$\cos\theta = x/r = \frac{3}{5}$
$\tan\theta = y/x = \frac{4}{3}$
$\cot\theta = x/y = \frac{3}{4}$
$\sec\theta = r/x = \frac{5}{3}$
$\csc\theta = r/y = \frac{5}{4}$

(*b*)

$\sin\theta = \frac{4}{5}$
$\cos\theta = -\frac{3}{5}$
$\tan\theta = \frac{4}{-3} = -\frac{4}{3}$
$\cot\theta = -\frac{3}{4}$
$\sec\theta = \frac{5}{-3} = -\frac{5}{3}$
$\csc\theta = \frac{5}{4}$

(*c*)

$\sin\theta = -3\sqrt{10} = -3\sqrt{10}/10$
$\cos\theta = -1/\sqrt{10} = -\sqrt{10}/10$
$\tan\theta = -\frac{3}{1} = 3$
$\cot\theta = -\frac{1}{3} = \frac{1}{3}$
$\sec\theta\sqrt{10}/(-1) = -\sqrt{10}$
$\csc\theta\sqrt{10}(-3) = -\sqrt{10}/3$

Note the reciprocal relationships. For example, in (*b*), $\sin\theta = 1/\csc\theta = \frac{4}{5}, \cos\theta = 1/\sec\theta = -\frac{3}{5}$, $\tan\theta = 1/\cot\theta = -\frac{4}{3}$, etc.

13.3 In what quadrant will θ terminate, if

(*a*) $\sin\theta$ and $\cos\theta$ are both negative?
(*b*) $\sin\theta$ and $\tan\theta$ are both positive?
(*c*) $\sin\theta$ is positive and secant θ is negative?
(*d*) $\sec\theta$ is negative and $\tan\theta$ is negative?

(*a*) Since $\sin\theta = y/r$ and $\cos\theta = x/r$, both x and y are negative. (Recall that r is always positive.) Thus, θ is a third quadrant angle.

(*b*) Since $\sin\theta$ is positive, y is positive; since $\tan\theta = y/x$ is positive, x is also positive. Thus, θ is a first quadrant angle.

(*c*) Since $\sin\theta$ is positive, y is positive; since $\sec\theta$ is negative, x is negative. Thus, θ is a second quadrant angle.

(*d*) Since $\sec\theta$ is negative, x is negative; since $\tan\theta$ is negative, y is then positive. Thus, θ is a second quadrant angle.

13.4 In what quadrants may θ terminate, if (*a*) $\sin\theta$ is positive? (*b*) $\cos\theta$ is negative? (*c*) $\tan\theta$ is negative? (*d*) $\sin\theta$ is positive?

(*a*) Since $\sin\theta$ is positive, y is positive. Then x may be positive or negative and θ is a first or second quadrant angle.

(*b*) Since $\cos\theta$ is negative, x is negative. Then y may be positive or negative and θ is a second or third quadrant angle.

(*c*) Since $\tan\theta$ is negative, either y is positive and x is negative or y is negative and x is positive. Thus, θ may be a second or fourth quadrant angle.

(*d*) Since $\sec\theta$ is positive, x is positive. Thus, θ may be a first or fourth quadrant angle.

13.5 Find the values of $\cos\theta$ and $\tan\theta$, given $\sin\theta = \frac{8}{17}$ and θ in quadrant I.

Let P be a point on the terminal line of θ. Since $\sin\theta = y/r = \frac{8}{17}$, we take $y = 8$ and $r = 17$. Since θ is in quadrant I, x is positive; thus $x = \sqrt{r^2 - y^2} = \sqrt{(17)^2 - (8)^2} = 15$.

To draw Fig. 13-24, locate the point $P(15, 8)$, join it to the origin, and indicate the angle θ. Then $\cos\theta = x/y = \frac{15}{17}$ and $\tan\theta = y/x = \frac{8}{15}$.

The choice of $y = 8, r = 17$ is one of convenience. Note that $\frac{8}{17} = \frac{16}{34}$ and we might have taken $y = 16$, $r = 34$. Then $x = 30, \cos\theta = \frac{30}{34} = \frac{15}{17}$ and $\tan\theta = \frac{16}{30} = \frac{8}{15}$.

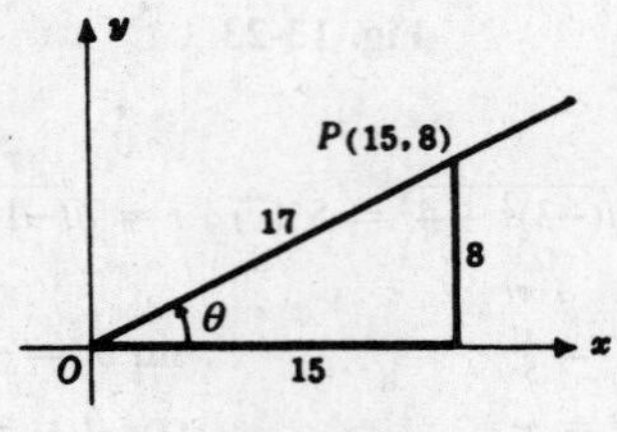

Fig. 13-24

13.6 Find the possible values of $\sin\theta$ and $\tan\theta$, given $\cos\theta = \frac{5}{6}$.

Since $\cos\theta$ is positive, θ is in quadrant I or IV. Since $\cos\theta = x/r = \frac{5}{6}$, we take $x = 5, r = 6$; $y = \pm\sqrt{(6)^2 - (5)^2} = \pm\sqrt{11}$.

(*a*) For θ in quadrant I [Fig. 13-25(*a*)] we have $x = 5, y = \sqrt{11}, r = 6$; then $\sin\theta = y/r = \sqrt{11}/6$ and $\tan\theta = y/x = \sqrt{11}/5$.

(*b*) For θ in quadrant IV [Fig. 13-25(*b*)] we have $x = 5, y = -\sqrt{11}, r = 6$; then $\sin\theta = y/r = -\sqrt{11}/6$ and $\tan\theta = y/x = -\sqrt{11}/5$.

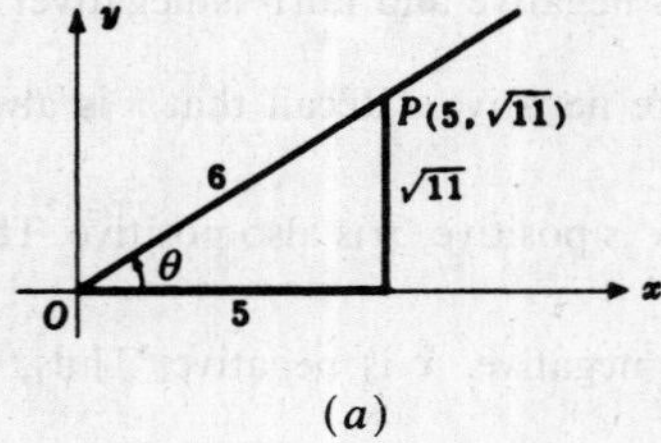

(*a*)

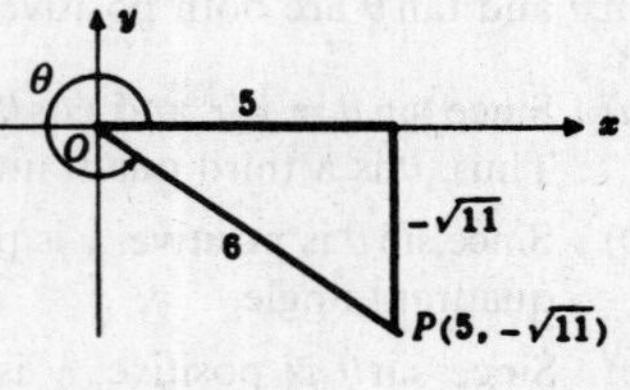

(*b*)

Fig. 13-25

13.7 Find the possible values of $\sin\theta$ and $\cos\theta$, given $\tan\theta = -\frac{3}{4}$.

Since $\tan\theta = y/x$ is negative, θ is in quadrant II (take $x = -4$, $y = 3$) or in quadrant IV (take $x = 4, y = -3$). In either case $r = \sqrt{16 + 9} = 5$.

(*a*) For θ in quadrant II [Fig. 13-26(*a*)], $\sin\theta = y/r = \frac{3}{5}$ and $\cos\theta = x/r = -\frac{4}{5}$.

(*b*) For θ in quadrant IV [Fig. 13-26(*b*)], $\sin\theta = y/r = -\frac{3}{5}$ and $\cos\theta = x/r = \frac{4}{5}$.

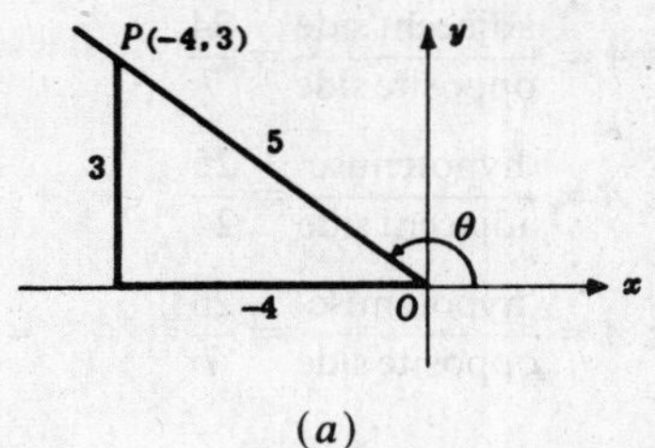

(a)

(b)

Fig. 13-26

13.8 Find the values of the remaining functions of θ, given $\sin\theta = \sqrt{3}/2$ and $\theta = -\frac{1}{2}$.

Since $\sin\theta = y/r$ is positive, y is positive. Since $\cos\theta = x/r$ is negative, x is negative. Thus, θ is in quadrant II.

Taking $x = -1, y = \sqrt{3}, r = \sqrt{(-1)^2 + (\sqrt{3})^2} = 2$ (Fig. 13-27), we have

$$\tan\theta = y/x = \sqrt{3}/-1 = -\sqrt{3} \qquad \cot\theta = 1/\tan\theta = -1\sqrt{3} = -\sqrt{3}/3$$
$$\sec\theta = 1/\cos\theta = -2 \qquad \csc\theta = 1/\sin\theta = 2/\sqrt{3} = 2\sqrt{3}/3$$

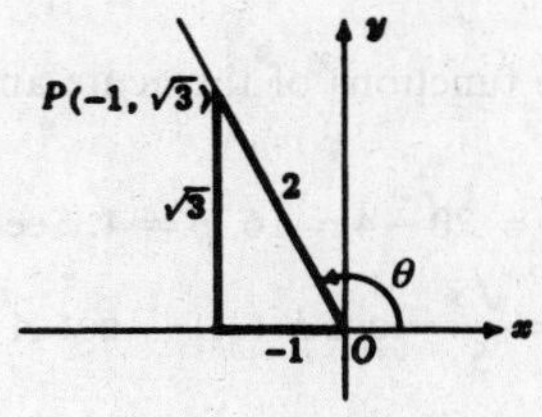

Fig. 13-27

13.9 Determine the possible values of $\cos\theta$ and $\tan\theta$ if $\sin\theta = m/n$, a negative fraction.

Since $\sin\theta$ is negative, θ is in quadrant III or IV.

(a) In quadrant III: Take $y = m, r = n, x = -\sqrt{n^2 - m^2}$; then $\cos\theta = x/r = -\sqrt{n^2 - m^2}/n$ and $\tan\theta = y/x = -m/\sqrt{n^2 - m^2}$.

(b) In quadrant IV: Take $y = m, r = n, x = +\sqrt{n^2 - m^2}$; then $\cos\theta = x/r = \sqrt{n^2 - m^2}/n$ and $\tan\theta = y/x = m\sqrt{n^2 - m^2}$.

13.10 Evaluate:

(a) $\sin 0° + 2\cos 0° + 3\sin 90° + 4\cos 90° + 5\sec 0° + 6\csc 90°$

(b) $\sin 180° + 2\cos 180° + 3\sin 270° + 4\cos 270° - 5\sec 180° - 6\csc 270°$

(a) $0 + 2(1) + 3(1) + 4(0) + 5(1) + 6(1) = 16$

(b) $0 + 2(-1) + 3(-1) + 4(0) - 5(-1) - 6(-1) = 6$

[NOTATION: We will write AB (or c) to denote the length of AB, and $\overline{AB}$ to denote "the segment AB." $\overleftrightarrow{AB}$ denotes "the line AB."]

13.11 Find the values of the trigonometric functions of the acute angles of the right triangle ABC given $b = 24$ and $c = 25$.

Since $a^2 = c^2 - b^2 = (25)^2 - (24)^2 = 49$, $a = 7$. See Fig. 13-28. Then

$$\sin A = \frac{\text{opposite side}}{\text{hypotenuse}} = \frac{7}{25} \qquad \cos A = \frac{\text{adjacent side}}{\text{opposite side}} = \frac{24}{7}$$

$$\cos A = \frac{\text{adjacent side}}{\text{hypotenuse}} = \frac{24}{25} \qquad \sec A = \frac{\text{hypotenuse}}{\text{adjacent side}} = \frac{25}{24}$$

$$\tan A = \frac{\text{opposite side}}{\text{adjacent side}} = \frac{7}{24} \qquad \csc A = \frac{\text{hypotenuse}}{\text{opposite side}} = \frac{25}{7}$$

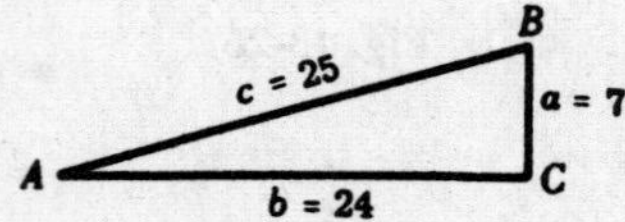

Fig. 13-28

and

$$\sin B = \tfrac{24}{25} \qquad \cot B = \tfrac{7}{24}$$

$$\cos B = \tfrac{7}{25} \qquad \sec B = \tfrac{25}{7}$$

$$\tan B = \tfrac{24}{7} \qquad \csc B = \tfrac{25}{24}$$

13.12 Find the values of the trigonometric functions of the acute angles of the right triangle ABC, given $a = 2$, $c = 2\sqrt{5}$.

Since $b^2 = c^2 - a^2 = (2\sqrt{5})^2 - 2^2 = 20 - 4 = 16$, $b = 4$. See Fig. 13-29. Then

$$\sin A = \frac{2}{2\sqrt{5}} = \frac{\sqrt{5}}{5} = \cos B \qquad \cot A = \tfrac{4}{2} = 2 = \tan B$$

$$\cos A = \frac{4}{2\sqrt{5}} = \frac{2\sqrt{5}}{5} = \sin B \qquad \sec A = \frac{2\sqrt{5}}{4} = \frac{\sqrt{5}}{2} = \csc B$$

$$\tan A = \tfrac{2}{4} = \tfrac{1}{2} = \cot B \qquad \csc A = \frac{2\sqrt{5}}{2} = \sqrt{5} = \sec B$$

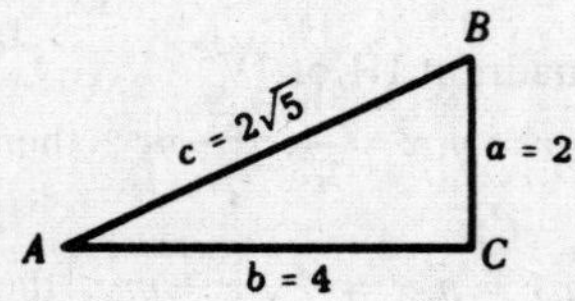

Fig. 13-29

13.13 Find the values of the trigonometric functions of the acute angle A, given $\sin A = \frac{3}{7}$.

Construct the right triangle ABC having $a = 3$, $c = 7$, and $b = \sqrt{7^2 - 3^2} = 2\sqrt{10}$ units. See Fig. 13-30. Then

$$\sin A = \tfrac{3}{7} \qquad \cot A = \frac{2\sqrt{10}}{3}$$

$$\cos A = \frac{2\sqrt{10}}{7} \qquad \sec A = \frac{7}{2\sqrt{10}} = \frac{7\sqrt{10}}{20}$$

$$\tan A = \frac{3}{2\sqrt{10}} = \frac{3\sqrt{10}}{20} \qquad \csc A = \tfrac{7}{3}$$

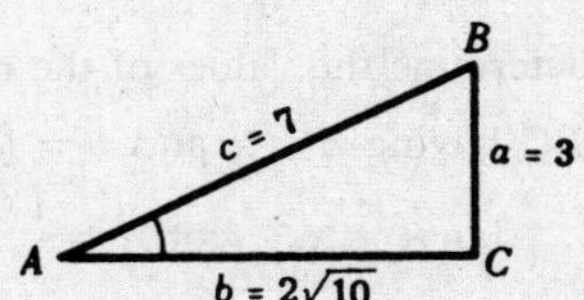

Fig. 13-30

13.14 Find the values of the trigonometric functions of the acute angle B, given $\tan B = 1.5$.

Refer to Fig. 13-31. Construct the right triangle ABC having $b = 15$ and $a = 10$ units. (Note that $1.5 = \frac{3}{2}$ and a right triangle with $b = 3$, $a = 2$ will serve equally well.)

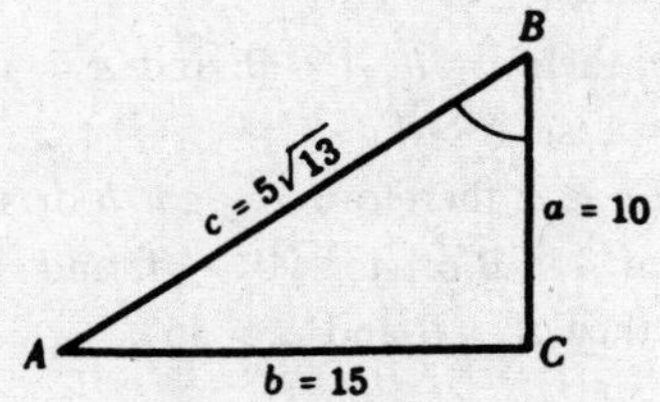

Fig. 13-31

Then $c = \sqrt{a^2 + b^2} = \sqrt{10^2 + 15^2} = 5\sqrt{13}$ and

$$\sin B = \frac{15}{5\sqrt{13}} = \frac{3\sqrt{13}}{13} \qquad \cot B = \tfrac{2}{3}$$

$$\cos B = \frac{10}{5\sqrt{13}} = \frac{2\sqrt{13}}{13} \qquad \sec B = \frac{5\sqrt{13}}{10} = \frac{\sqrt{13}}{2}$$

$$\tan B = \tfrac{15}{10} = \tfrac{3}{2} \qquad \csc B = \frac{5\sqrt{13}}{15} = \frac{\sqrt{13}}{3}$$

13.15 If A is acute and $\sin A = 2x/3$, determine the values of the remaining functions.

Construct the right triangle ABC having $a = 2x < 3$ and $c = 3$, as in Fig. 13-32.
Then $b = \sqrt{c^2 - a^2} = \sqrt{9 - 4x^2}$ and

$$\sin A = \frac{2x}{3}, \qquad \cos A = \frac{\sqrt{9-4x^2}}{3}, \qquad \tan A = \frac{2x}{\sqrt{9-4x^2}}, \qquad \cot A = \frac{\sqrt{9-4x^2}}{2x},$$

$$\sec A = \frac{3}{\sqrt{9-4x^2}}, \qquad \csc A = \frac{3}{2x}.$$

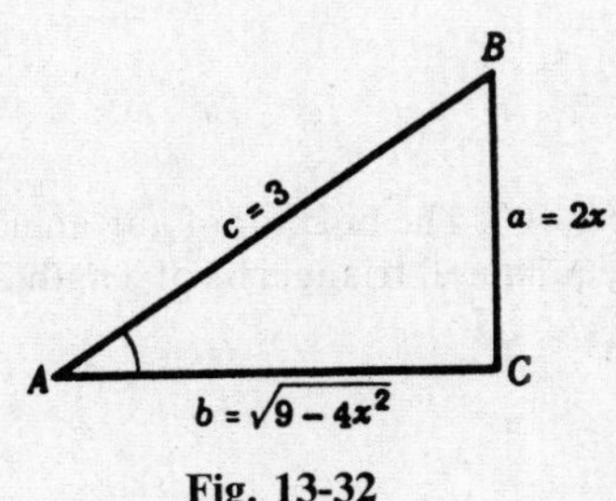

Fig. 13-32

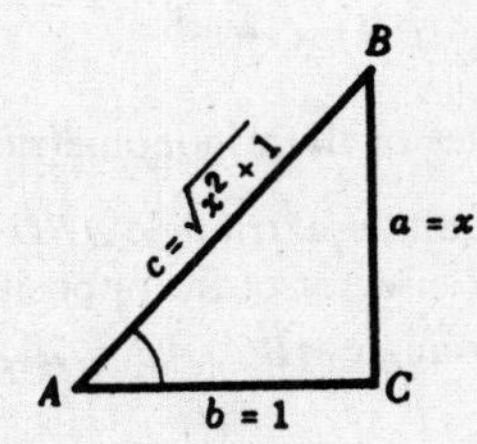

Fig. 13-33

13.16 If A is acute and $\tan A = x = x/1$, determine the values of the remaining functions.

Construct the right triangle ABC having $a = x$ and $b = 1$, as in Fig. 13-33. Then $c = \sqrt{x^2+1}$ and

$$\sin A = \frac{x}{\sqrt{x^2+1}}, \quad \cos A = \frac{1}{\sqrt{x^2+1}}, \quad \tan A = x, \quad \cot A = \frac{1}{x}, \quad \sec A = \sqrt{x^2+1}, \quad \csc A = \frac{\sqrt{x^2+1}}{x}.$$

13.17 If A is an acute angle:

(a) Why is $\sin A < 1$?
(b) When is $\sin A = \cos A$?
(c) Why is $\sin A < \csc A$?
(d) Why is $\sin A < \tan A$?
(e) When is $\sin A < \cos A$?
(f) When is $\tan A > 1$?

In any right triangle ABC:

(a) Side $a <$ side c; therefore $\sin A = a/c < 1$.
(b) $\sin A = \cos A$ when $a/c = b/c$; then $a = b$, $A = B$, and $A = 45°$.
(c) $\sin A < 1$ (above) and $\csc A = 1/\sin A > 1$.
(d) $\sin A = a/c$, $\tan A = a/b$, and $b < c$; therefore $a/c < a/b$ or $\sin A < \tan A$.
(e) $\sin A < \cos A$ when $a < b$; then $A < B$ or $A < 90° - A$, and $A < 45°$.
(f) $\tan A = a/b > 1$ when $a > b$; then $A > B$ and $A > 45°$.

13.18 Find the value of the trigonometric functions of 45°.

In any isosceles right triangle ABC, $A = B = 45°$ and $a = b$. See Fig. 13-34. Let $a = b = 1$; then $c = \sqrt{1+1} = \sqrt{2}$ and

$$\sin 45° = \frac{1}{\sqrt{2}} = \frac{1}{2}\sqrt{2} \qquad \cot 45° = 1$$

$$\cos 45° = \frac{1}{\sqrt{2}} = \frac{1}{2}\sqrt{2} \qquad \sec 45° = \sqrt{2}$$

$$\tan 45° = \tfrac{1}{1} = 1 \qquad \csc 45° = \sqrt{2}$$

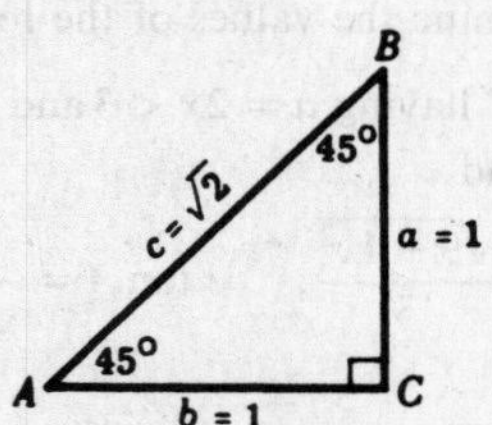

Fig. 13-34

13.19 Find the values of the trigonometric functions of 30° and 60°.

In any equilateral triangle ABD (see Fig. 13-35), each angle is 60°. The bisector of any angle, as B, is the perpendicular bisector of the opposite side. Let the sides of the equilateral triangle be of length 2 units. Then in the right triangle ABC, $AB = 2$, $AC = 1$, and $BC = \sqrt{2^2 - 1^2} = \sqrt{3}$.

$$\sin 30° = \tfrac{1}{2} = \cos 60° \qquad \cot 30° = \sqrt{3} = \tan 60°$$

$$\cos 30° = \frac{\sqrt{3}}{2} = \sin 60° \qquad \sec 30° = \frac{2}{\sqrt{3}} = \frac{2\sqrt{3}}{3} = \csc 60°$$

$$\tan 30° = \frac{1}{\sqrt{3}} = \frac{\sqrt{3}}{3} = \cot 60° \qquad \csc 30° = 2 = \sec 60°$$

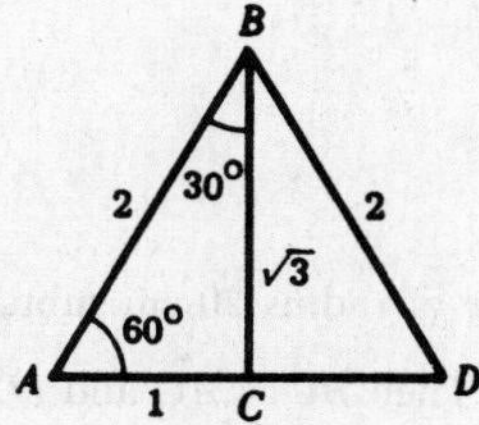

Fig. 13-35

13.20 When the sun is 20° above the horizon, how long is the shadow cast by a building 150 ft high?

In Fig. 13-36, $A = 20°$ and $CB = 150$. Then $\cot A = AC/CB$ and $AC = CB \cot A = 150 \cot 20° = 150(2.7) = 405$ ft.

13.21 A tree 100 ft tall casts a shadow 120 ft long. Find the measure of the angle of elevation of the sun.

In Fig. 13-37, $CB = 100$ and $AC = 120$. Then $\tan A = CB/AC = \frac{100}{120} = 0.83$ and m $\measuredangle$ $A = 40°$.

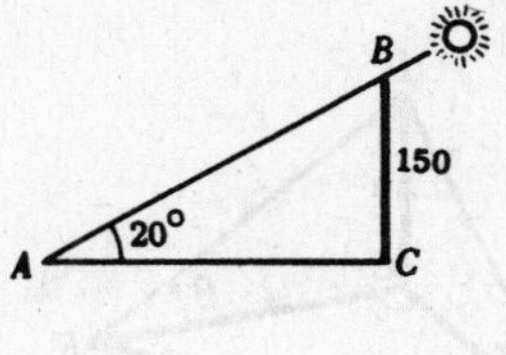

Fig. 13-36

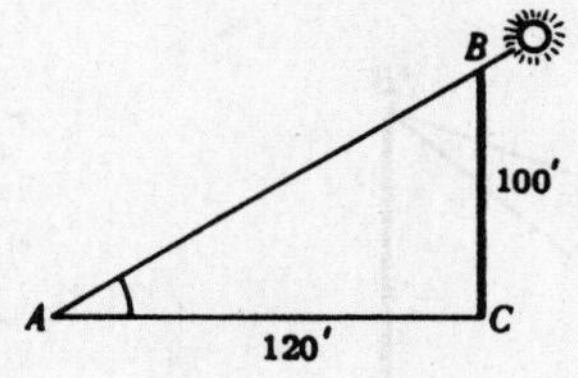

Fig. 13-37

Fig. 13-38

13.22 A ladder leans against the side of a building with its foot 12 m from the building. How far from the ground is the top of the ladder and how long is the ladder if it makes an angle of 70° with the ground?

From Fig. 13-38, $\tan A = CB/AC$; then $CB = AC \tan A = 12 \tan 70° = 12(2.7) = 32.4$. The top of ladder is 32 m above the ground.

Sec $A = AB/AC$; then $AB = AC \sec A = 12 \sec 70° = 12(2.9) = 34.8$. The ladder is 35 m long.

13.23 From the top of a lighthouse, 120 ft above the sea, the angle of depression of a boat is 15°. How far is the boat from the lighthouse?

In Fig. 13-39, the right triangle ABC has A measuring 15° and $CB = 120$; then $\cot A = AC/CB$ and $AC = CB \cot A = 120 \cot 15° = 120(3.7) = 440$ ft.

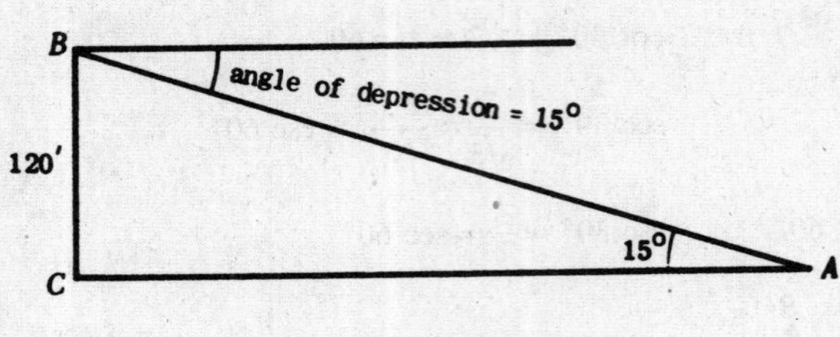

Fig. 13-39

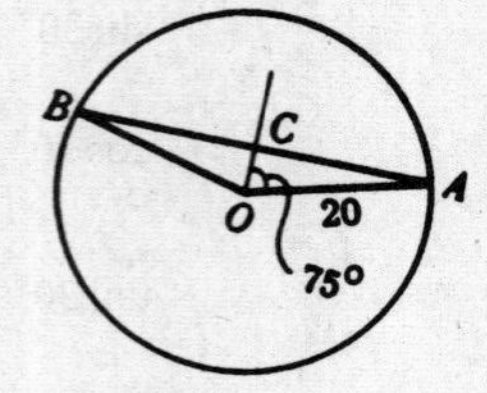

Fig. 13-40

13.24 Find the length of the chord of a circle of radius 20 cm subtended by a central angle of 150°.

In Fig. 13-40, OC bisects $\angle AOB$. Then $BC = AC$ and OAC is a right triangle. In $\triangle OAC$,

$$\sin\angle COA = \frac{AC}{OA} \quad \text{and } AC = OA \sin\angle COA = 20 \sin 75^\circ = 20(0.97) = 19.4$$

Then $BA = 38.8$ and the length of the chord is 39 cm.

13.25 Find the height of a tree if the angle of elevation of its top changes from 20° to 40° as the observer advances 75 ft toward its base. See Fig. 13-41.

In the right triangle ABC, $\cot A = AC/CB$; then $AC = CB \cot A$ or $DC + 75 = CB \cot 20^\circ$.
In the right triangle DBC, $\cot D = DC/CB$; then $DC = CB \cot 40^\circ$.

Then $\quad DC = CB \cot 20^\circ - 75 = CB \cot 40^\circ, \qquad CB(\cot 20^\circ - \cot 40^\circ) = 75,$

$$CB(2.7 - 1.2) = 75, \quad \text{and} \quad CB = \frac{75}{1.5} = 50 \text{ ft.}$$

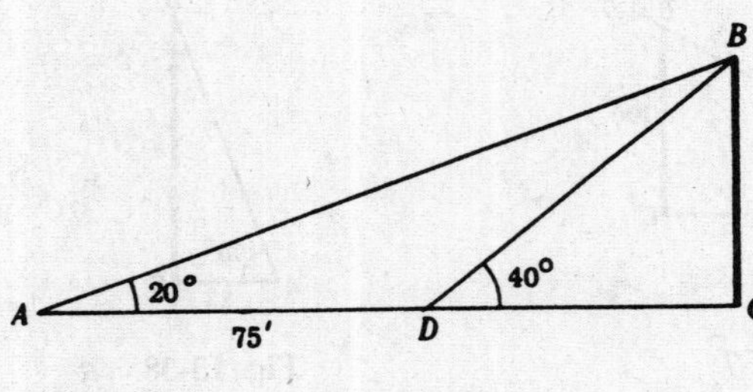

Fig. 13-41

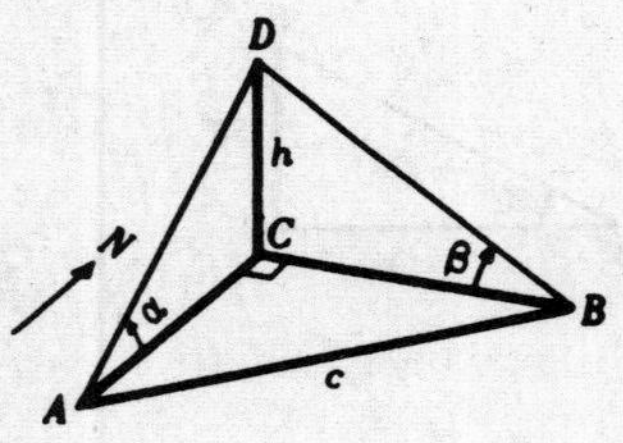

Fig. 13-42

13.26 A tower standing on level ground is due north of point A and due west of point B, a distance c ft from A. If the angles of elevation of the top of the tower as measured from A and B are α and β, respectively, find the height h of the tower.

In the right triangle ACD of Fig. 13-42, $\cot\alpha = AC/h$; in the right triangle BCD, $\cot\beta = BC/h$. Then $AC = h \cot\alpha$ and $BC = h \cot\beta$.

Since ABC is a right triangle, $(AC)^2 + (BC)^2 = c^2 = h^2(\cot\alpha)^2 + h^2(\cot\beta)^2$ and

$$h = \frac{c}{\sqrt{(\cot\alpha)^2 + (\cot\beta)^2}}$$

13.27 If holes are to be spaced regularly on a circle, show that the distance d between the centers of two successive holes is given by $d = 2r\sin(180°/n)$, where r = radius of the circle and n = number of holes. Find d when $r = 20$ in. and $n = 4$.

Let A and B be the centers of two consecutive holes on the circle of radius r and center O. See Fig. 13-43. Let the bisector of the angle O of the triangle AOB meet AB at C. In right triangle AOC, $\sin\angle AOC = AC/r = \frac{1}{2}d/r = d/2r$. Then

$$d = 2r\sin\angle AOC = 2r\sin\tfrac{1}{2}\angle AOB = 2r\sin\frac{1}{2}\left(\frac{360°}{n}\right) = 2r\sin\frac{180°}{n}$$

When $r = 20$ and $n = 4$, $d = 2\cdot 20\sin 45° = 2\cdot 20\cdot\frac{\sqrt{2}}{2} = 20\sqrt{2}$ in.

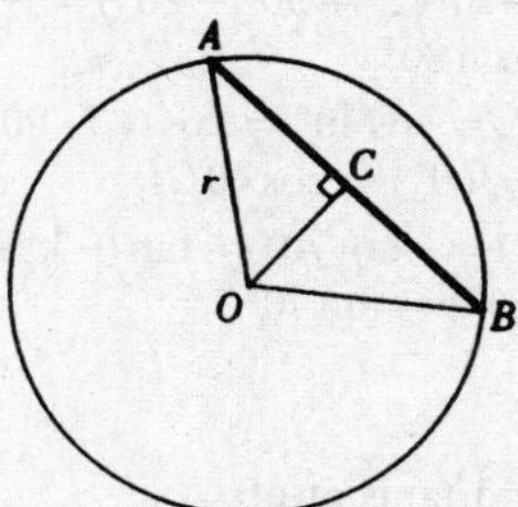

Fig. 13-43

13.28 Express each of the following in terms of a function of θ:

(a) $\sin(\theta - 90°)$ (d) $\cos(-180° + \theta)$ (g) $\sin(540° + \theta)$ (j) $\cos(-450° - \theta)$

(b) $\cos(\theta - 90°)$ (e) $\sin(-270° - \theta)$ (h) $\tan(720° - \theta)$ (k) $\csc(-900° + \theta)$

(c) $\sec(-\theta - 90°)$ (f) $\tan(\theta - 360°)$ (i) $\tan(720° + \theta)$ (l) $\sin(-540° - \theta)$

(a) $\sin(\theta - 90°) = \sin(-90° + \theta) = \sin(-1\cdot 90° + \theta) = -\cos\theta$, the sign being negative since, when θ is positive acute, the terminal side of $\theta - 90°$ lies in quadrant IV.

(b) $\cos(\theta - 90°) = \cos(-90° + \theta) = \cos(-1\cdot 90° + \theta) = \sin\theta$.

(c) $\sec(-\theta - 90°) = \sec(-90° - \theta) = \sec(-1\cdot 90° - \theta) = -\csc\theta$, the sign being negative since, when θ is positive acute, the terminal side of $-\theta - 90°$ lies in quadrant III.

(d) $\cos(-180° + \theta) = \cos(-2\cdot 90° + \theta) = -\cos\theta$. (quadrant III)

(e) $\sin(-270° - \theta) = \sin(-3\cdot 90° - \theta) = \cos\theta$. (quadrant I)

(f) $\tan(\theta - 360°) = \sin(-4\cdot 90° + \theta) = \tan\theta$. (quadrant I)

(g) $\sin(540° + \theta) = \sin(6\cdot 90° + \theta) = -\sin\theta$. (quadrant III)

(h) $\tan(720° - \theta) = \tan(8\cdot 90° - \theta) = -\tan\theta = \tan(2\cdot 360° - \theta) = \tan(-\theta) = -\tan\theta$.

(i) $\tan(720° + \theta) = \tan(8\cdot 90° + \theta) = -\tan\theta = \tan(2\cdot 360° + \theta) = \tan\theta$.

(j) $\cos(-450° - \theta) = \cos(-5\cdot 90° - \theta) = -\sin\theta$.

(k) $\csc(-900° + \theta) = \csc(-10\cdot 90° + \theta) = -\csc\theta$.

(l) $\sin(-540° - \theta) = \sin(-6\cdot 90° - \theta) = \sin\theta$.

13.29 Express each of the following in terms of functions of a positive acute angle in two ways:

(a) $\sin 130°$ (c) $\sin 200°$ (e) $\tan 165°$ (g) $\sin 670°$ (i) $\csc 865°$ (k) $\cos(-680°)$

(b) $\tan 325°$ (d) $\cos 310°$ (f) $\sec 250°$ (h) $\cot 930°$ (j) $\sin(-100°)$ (l) $\tan(-290°)$

(a) $\sin 130° = \sin(2\cdot 90° - 50°) = \sin 50° = \sin(1\cdot 90° + 40°) = \cos 40°$.

(b) $\tan 325° = \tan(4\cdot 90° - 35°) = -\tan 35° = \tan(3\cdot 90° + 55°) = -\cot 55°$.

(*c*) $\sin 200° = \sin(2 \cdot 90° + 20°) = -\sin 20° = \sin(3 \cdot 90° - 70°) = -\cos 70°$.
(*d*) $\cos 310° = \cos(4 \cdot 90° - 50°) = \cos 50° = \cos(3 \cdot 90° + 40°) = \sin 40°$.
(*e*) $\tan 165° = \tan(2 \cdot 90° - 15°) = -\tan 15° = \tan(1 \cdot 90° + 75°) = -\cot 75°$.
(*f*) $\sec 250° = \sec(2 \cdot 90° + 70°) = -\sec 70° = \sec(3 \cdot 90° - 20°) = -\csc 20°$.
(*g*) $\sin 670° = \sin(8 \cdot 90° - 50°) = -\sin 50° = \sin(7 \cdot 90° + 40°) = -\cos 40°$
or $\sin 670° = \sin(310° + 360°) = \sin 310° = \sin(4 \cdot 90° - 50°) = -\sin 50°$.
(*h*) $\cot 930° = \cot(10 \cdot 90° + 30°) = \cot 30° = \cot(11 \cdot 90° - 60°) = \tan 60°$
or $\cot 930° = \cot(210° + 2 \cdot 360°) = \cot 210° = \cot(2 \cdot 90° + 30°) = \cot 30°$.
(*i*) $\csc 865° = \csc(10 \cdot 90° - 35°) = \csc 35° = \csc(9 \cdot 90° + 55°) = \sec 55°$
or $\csc 865° = \csc(145° + 2 \cdot 360°) = \csc 145° = \csc(2 \cdot 90° - 35°) = \csc 35°$.
(*j*) $\sin(-100°) = \sin(-2 \cdot 90° + 80°) = -\sin 80° = \sin(-1 \cdot 90° - 10°) = -\cos 10°$
or $\sin(-100°) = -\sin 100° = -\sin(2° + 90° - 80°) = -\sin 80°$ or $\sin(-100°) = \sin(-100° + 360°) = \sin 260° = \sin(2 \cdot 90° + 80°) = -\sin 80°$.
(*k*) $\cos(-680°) = \cos(-8 \cdot 90° + 40°) = \cos 40° = \cos(-7 \cdot 90° - 50°) = \sin 50°$
or $\cos(-680°) = \cos(-680° + 2 \cdot 360°) = \cos 40°$.
(*l*) $\tan(-290°) = \tan(-4 \cdot 90° + 70°) = -\tan 70° = \tan(-3 \cdot 90° - 20°) = -\cot 20°$
or $\tan(-290°) = \tan(-290° + 360°) = \tan 70°$.

13.30 Find the values of the sine, cosine, and tangent of

(*a*) 120° (*b*) 210° (*c*) 315° (*d*) −135° (*d*) −240° (*f*) −330°

Call θ, always positive acute, the *related angle* of ϕ when $\phi = 180° - \theta, 180° + \theta$, or $360° - \theta$. Then any function of ϕ is numerically equal to the same function of θ. The algebraic sign in each case is that of the function in the quadrant in which the terminal side of ϕ lies.

(*a*) $120° = 180° - 60°$. The related angle is 60°; 120° is in quadrant II; $\sin 210° = \sin 60° = \sqrt{3}/2$, $\cos 120° = -\cos 60° = -\frac{1}{2}$, $\tan 120° = -\tan 60° = -\sqrt{3}$.

(*b*) $210° = 180° + 30°$. The related angle is 30°; 120° is in quadrant III; $\sin 210° = -\sin 30° = -\frac{1}{2}$, $\cos 210° = -\cos 30° = -\sqrt{3}/2$, $\tan 210° = \tan 30° = \sqrt{3}/3$.

(*c*) $315° = 360° - 45°$. The related angle 45°; 315° is in quadrant IV; $\sin 315° = -\sin 45° = -\sqrt{2}/2$, $\cos 315° = \cos 45° = \sqrt{2}/2$, $\tan 315° = -\tan 45° = -1$.

(*d*) Any function of −135° is the same function of $-135° + 360° = 225° = \phi$; $225° = 180° + 45°$. The related angle is 45°; 225° is in quadrant III. $\sin(-135°) = -\sin 45° = -\sqrt{2}/2$, $\cos(-135°) = -\cos 45° = -\sqrt{-2}/2$, $\tan(-135°) = 1$.

(*e*) Any function of −240° is the same function of $-240° + 360° = 120°$; $120° = 180° - 60°$. The related angle is 60°; 120° is in quadrant II; $\sin(-240°) = \sin 60° = \sqrt{3}/2$, $\cos(-240°) = -\cos 60° = -\frac{1}{2}$, $\tan(-240°) = -\tan 60° = -\sqrt{3}$.

(*f*) Any function of −330° is the same function of $-330° + 360° = 30°$; $\sin(-330°) = \sin 30° = \frac{1}{2}$, $\cos(-330°) = \cos 30° = \sqrt{3}/2$, $\tan(-330°) = \tan(-30°) = \sqrt{3}/3$.

13.31 Using a calculator, verify that

(*a*) $\sin 125°14' = \sin(180° - 54°46') = \sin 54°46' = 0.8168$
(*b*) $\cos 169°40' = \cos(180° - 10°20') = -\cos 10°20' = -0.9838$
(*c*) $\tan 200°23' = \tan(180° + 20°23') = \tan 20°23' = 0.3716$
(*d*) $\cot 250°44' = \cot(180° + 70°44') = \cot 70°44' = 0.3495$
(*e*) $\cos 313°18' = \cos(360° - 46°42') = \cos 46°42' = 0.6858$
(*f*) $\sin 341°25' = \sin(360° - 18°8') = -\sin 18°8' = -0.3112$

13.32 If $\tan 25° = a$, find

(a) $\dfrac{\tan 155° - \tan 115°}{1 + \tan 155° \tan 115°} = \dfrac{-\tan 25° - (-\cot 25°)}{1 + (-\tan 25°)(-\cot 25°)} = \dfrac{-a + 1/a}{1 + a(1/a)} = \dfrac{-a^2 + 1}{a + a} = \dfrac{1 - a^2}{2a}$.

(b) $\dfrac{\tan 205° - \tan 115°}{\tan 245° - \tan 335°} = \dfrac{\tan 25° - (-\cot 25°)}{\cot 25° + (-\tan 25°)} = \dfrac{a + 1/a}{1/a - a} = \dfrac{a^2 + 1}{1 - a^2}$.

13.33 If $m\measuredangle A + m\measuredangle B + m\measuredangle C = 180°$, then show

(a) $\sin (B + C) = \sin A$
(b) $\sin \frac{1}{2}(B + C) = \cos \frac{A}{2}$

(a) $\sin (B + C) = \sin (180° - A) = A$
(b) $\sin \frac{1}{2}(B + C) = \sin \frac{1}{2}(180° - A) = \sin (90° - \frac{A}{2}) = \cos \frac{A}{2}$

13.34 Show that $\sin \theta$ and $\tan \frac{\theta}{2}$ have the same sign.

(a) Suppose $m\measuredangle\theta = n \cdot 180°$ If n is even (including zero), say $2m$, then $\sin(2m \cdot 180°) = \tan (m \cdot 180°) = 0$. The case when n is odd is excluded since then $\tan \frac{\theta}{2}$ is not defined.

(b) Suppose $m\measuredangle\theta = n \cdot 180° + \phi$, where $0 < m\measuredangle\phi < 180°$. If n is even, including zero, θ is in quadrant I or quadrant II and $\sin \theta$ is positive while $\frac{\theta}{2}$ is in quadrant I or quadrant III and $\tan \frac{\theta}{2}$ is positive. If n is odd, θ is in quadrant III or IV and $\sin \theta$ is negative while $\frac{\theta}{2}$ is in quadrant II or IV and $\tan \frac{\theta}{2}$ is negative.

13.35 Find all positive values of θ less than 360° for which $\sin \theta = -\frac{1}{2}$.

There will be two angles, one in the third quadrant and one in the fourth quadrant. The related angle of each has its sine equal to $+\frac{1}{2}$ and is 30°. Thus, the required angles are θ with measure $180° + 30° = 210°$ and θ with measure $360° - 30° = 330°$.

(NOTE: To obtain *all* values of θ for which $\sin \theta = -\frac{1}{2}$, and $n \cdot 360°$ to each of the above solutions; thus $\theta = 210° + n \cdot 360°$ and $\theta = 330° + n \cdot 360°$, where n is any integer.)

13.36 Find all positive values of θ less than 360° which satisfy $\sin 2\theta = \cos \frac{\theta}{2}$.

Since $\cos \frac{\theta}{2} = \sin (90° - \frac{\theta}{2}) = \sin 2\theta$, $2\theta = 90° - \frac{\theta}{2}$, $450° - \frac{\theta}{2}$, $810° - \frac{\theta}{2}, 1170° - \frac{\theta}{2}, \ldots$. Then $\frac{5}{2}\theta = 90°, 450°, 810°, 1170°, \ldots$ and $m\measuredangle\theta = 36°, 180°, 324°, 468°, \ldots$.

Since $\cos \frac{1}{2}\theta = \sin (90° + \frac{\theta}{2}) = \sin 2\theta$, $2\theta = 90° + \frac{\theta}{2}$, $450° + \frac{\theta}{2}$, $810° + \frac{\theta}{2}, \ldots$. Then $\frac{\theta}{2} = 90°, 450°, 810°, \ldots$ and $\theta = 60°, 300°, 540°, \ldots$.

The required solutions have measures $36°, 180°, 324°; 60°, 300°$.

13.37 Sketch the graphs of the following for one period:

(a) $y = 4 \sin x$ (c) $y = 3 \sin \frac{x}{2}$ (e) $y = 3 \cos \frac{x}{2} = 3 \sin (\frac{x}{2} + \frac{\pi}{2})$
(b) $y = \sin 3x$ (d) $y = 2 \cos x = 2 \sin (x + \frac{\pi}{2})$

In each case we use the same curve and then put in the y axis and choose the units on each axis to satisfy the requirements of amplitude and period of each curve.

(a) $y = 4 \sin x$ has amplitude $= 4$ and period $= 2\pi$. See Fig. 13-44(a).

(b) $y = \sin 3x$ has amplitude $= 1$ and period $= 2\pi/3$. See Fig. 13-44(b).

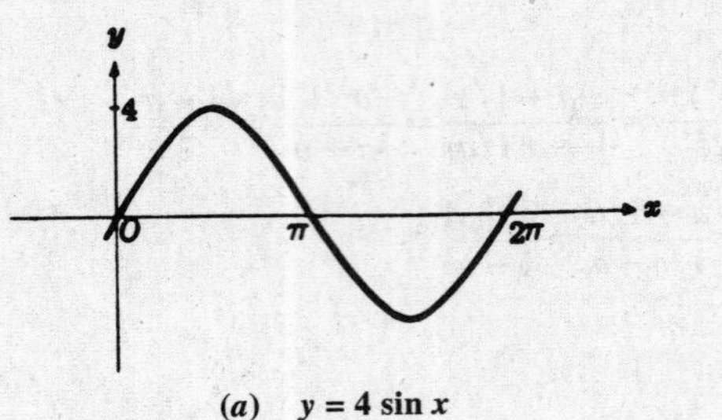
(a) y = 4 sin x

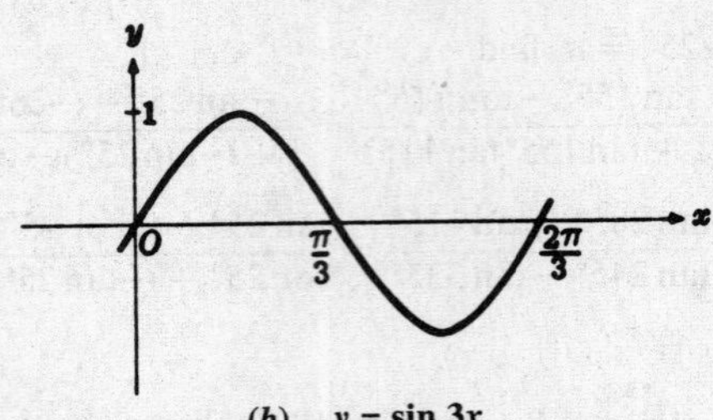
(b) y = sin 3x

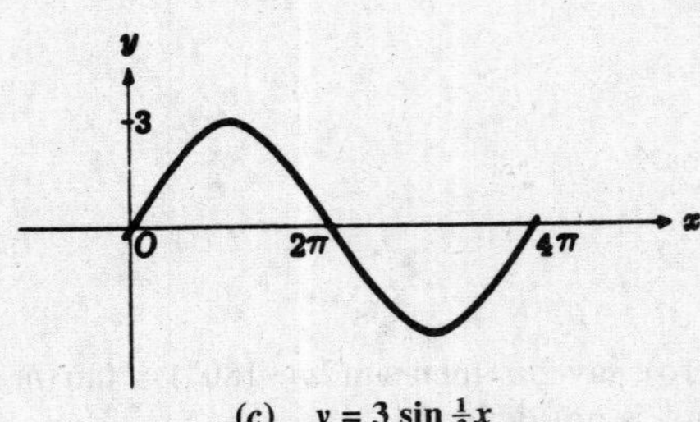
(c) y = 3 sin ½x

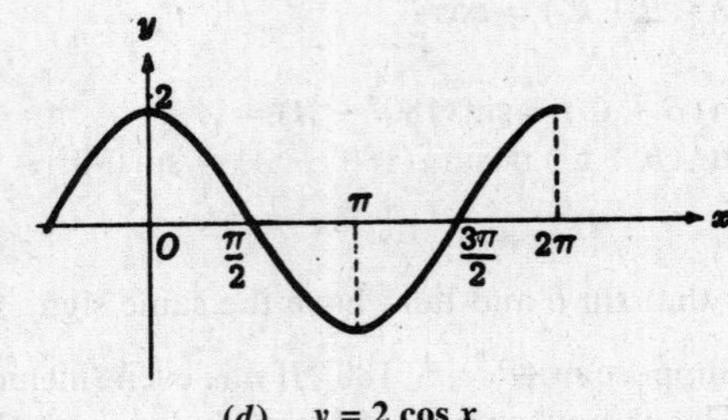
(d) y = 2 cos x

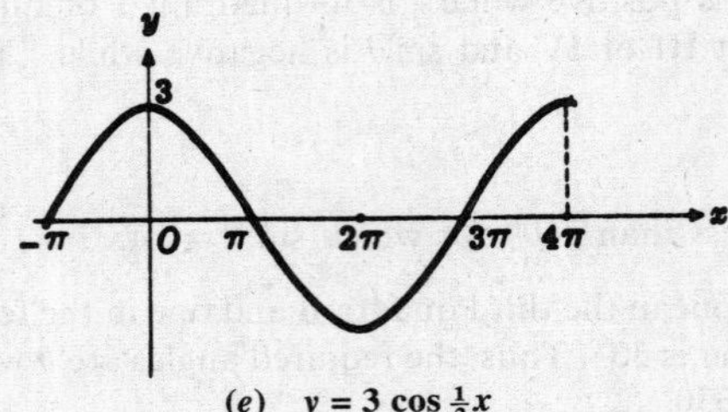
(e) y = 3 cos ½x

Fig. 13-44

(c) $y = 3 \sin \frac{x}{2}$ has amplitude $= 3$ and period $= 2\pi/\frac{1}{2} = 4\pi$. See Fig. 13-44(c).

(d) $y = 2 \cos x$ has amplitude $= 2$ and period $= 2\pi$. Note the position of the y axis. See Fig. 13-44(d).

(e) $y = 3 \cos \frac{x}{2}$ has amplitude $= 3$ and period $= 4\pi$. See Fig. 13-44(e).

13.38 Construct the graph of each of the following:

(a) $y = \sin x + \cos x$

(b) $y = \sin 2x - \cos 3x$

(c) $y = \sin 2x - \cos 3x$

(d) $y = 3 \sin 2x + 2 \cos 3x$

See Fig. 13-73(a)–(d).

13.39 Derive the law of sines.

Let ABC be any oblique triangle. In Fig. 13-45(a), angles A and B are acute while in Fig. 13-45(b), angle B is obtuse. Draw $\overline{CD}$ perpendicular to $\overline{AB}$ or $\overleftrightarrow{AB}$ extended and denote its length by h.

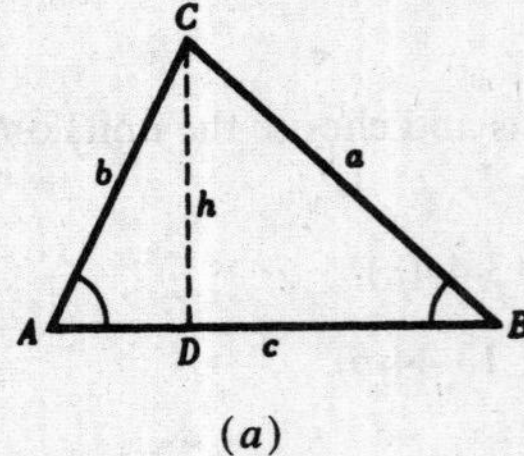
(a)

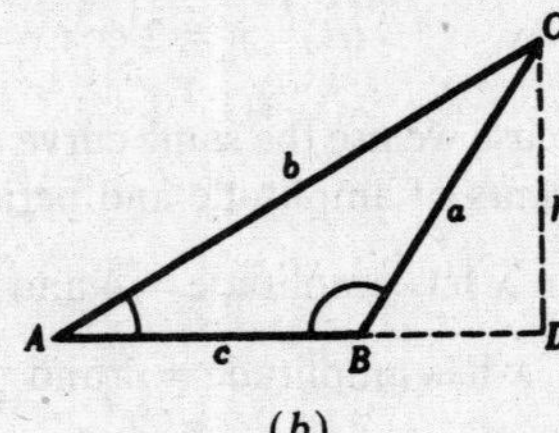
(b)

Fig. 13-45

In the right triangle ACD of either figure, $h = b \sin A$ while in the right triangle BCD, $h = a \sin B$ since in Fig. 13-45(b), $h = a \sin \angle DBC = a \sin (180° - B) = a \sin B$. Thus,

$$a \sin B = b \sin A \quad \text{or} \quad \frac{a}{\sin A} = \frac{b}{\sin B}.$$

In a similar manner (by drawing a perpendicular from B to $\overline{AC}$ or a perpendicular from A to $\overline{BC}$), we obtain

$$\frac{a}{\sin A} = \frac{c}{\sin C} \quad \text{or} \quad \frac{b}{\sin B} = \frac{c}{\sin C}.$$

Thus, finally,

$$\frac{a}{\sin A} = \frac{b}{\sin B} = \frac{c}{\sin C}.$$

13.40 Derive one of the projection formulas.

Refer to Fig. 13-45. In the right triangle ACD of either figure, $AD = b \cos A$. In the right triangle BCD of Fig. 13-45(a), $DB = a \cos B$. Thus, in Fig. 13-45(a)

$$c = AB = AD + DB = b \cos A + a \cos B = a \cos B + b \cos A$$

In the right triangle BCD of Fig. 31-45(b), $BD = a \cos \angle DBC = a \cos (180° - B) = -a \cos B$. Thus, in Fig. 13-45($b$),

$$c = AB = AD - BD = b \cos A - (-a \cos B) = a \cos B + b \cos A$$

13.41 Solve the triangle ABC, given $c = 25, A = 35°$, and $B = 68°$. See Fig. 13-46.

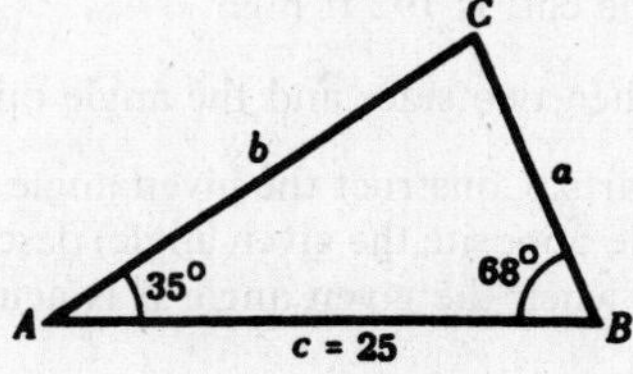

Fig. 13-46

To find C: $\quad C = 180° - (A + B) = 180° - 103° = 77°$

To find a: $\quad a = \dfrac{c \sin A}{\sin C} = \dfrac{25 \sin 35°}{\sin 77°} = \dfrac{25(0.5736)}{0.9744} = 15$

To find b: $\quad b = \dfrac{c \sin B}{\sin C} = \dfrac{25 \sin 68°}{\sin 77°} = \dfrac{25(0.9272)}{0.9744} = 24$

To check by projection formula:

$$c = a \cos B + b \cos A = 15 \cos 68° + 24 \cos 35° = 15(0.3746) + 24(0.8192) = 25.3$$

The required parts are $a = 15, b = 24$, and $C = 77°$.

13.42 A and B are two points on opposite banks of a river. From A a line $\overline{AC} = 275$ ft is laid off and the angles $CAB = 125°40'$ and $ACB = 48°50'$ are measured. Find the length of $\overline{AB}$. See Fig 13-47.

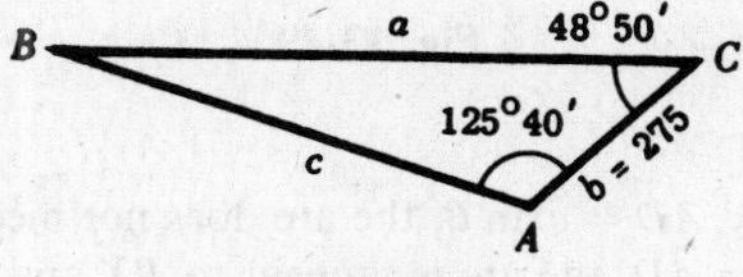

Fig. 13-47

In the triangle ABC, $B = 180° - (C + A) = 5°30'$ and

$$AB = c = \frac{b \sin C}{\sin B} = \frac{275 \sin 48°50'}{\sin 5°30'} = \frac{275(0.7528)}{0.0958} = 2160 \text{ ft}$$

13.43 A tower 125 ft high is on a cliff on the bank of a river. From the top of the tower the angle of depression of a point on the opposite shore is 28°40′ and from the base of the tower the angle of depression of the same point is 18°20′. Find the width of the river and the height of the cliff.

In Fig. 13-48 BC represents the tower, $\overline{DB}$ represents the cliff, and A is the point on the opposite shore.

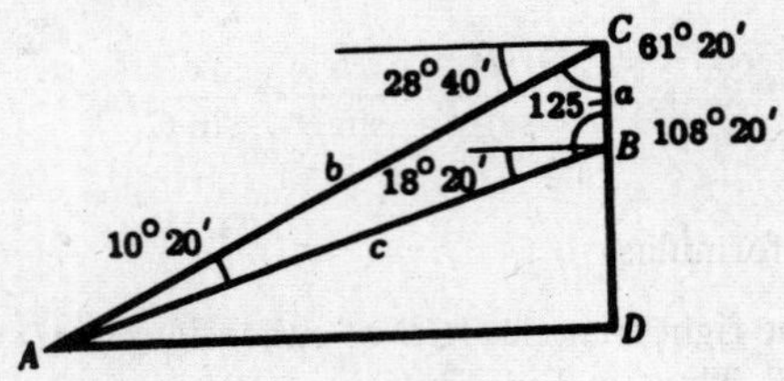

Fig. 13-48

In triangle ABC, $C = 90° - 28°40' = 61°20'$, $B = 90° + 18°20' = 108°20'$, $A = 180° - (B + C) = 10°20'$.

$$c = \frac{a \sin C}{\sin A} = \frac{125 \sin 61°20'}{\sin 10°20'} = \frac{125(0.8774)}{0.1794} = 611$$

In the right triangle ABD, $DB = c \sin 18°20' = 611(0.3145) = 192$, $AD = c \cos 18°20' = 611(0.9492) = 580$.

The river is 580 ft wide and the cliff is 192 ft high.

13.44 Discuss the several special cases when two sides and the angle opposite one of them are given.

Let b, c, and B be the given parts. Construct the given angle B and lay off the side $BA = c$. With A as center and radius equal to b (the side opposite the given angle) describe an arc. Figures 13-49(a)–(e) illustrate the special cases which may occur when the given angle B is acute, while Figs. 13-49(f)–(g) illustrate the cases when B is obtuse.

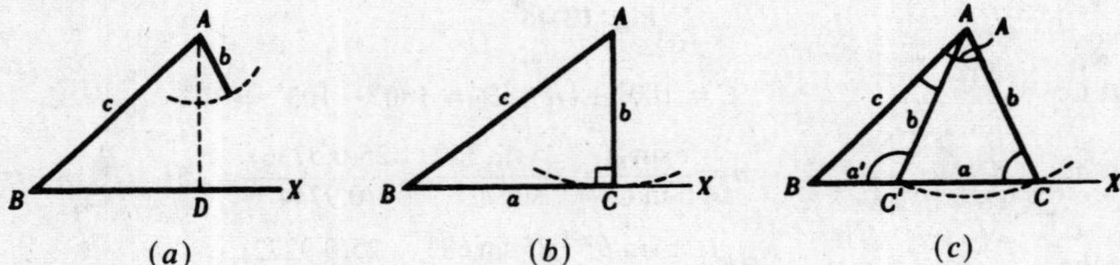

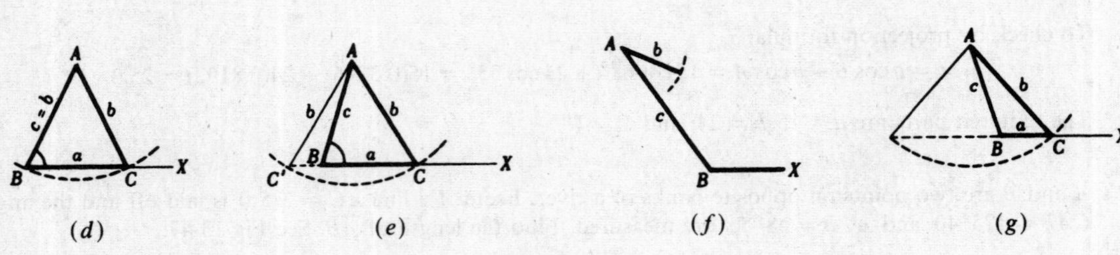

Fig. 13-49

The given angle B is acute.

Fig. 13-49(a). When $b < AD = c \sin B$, the arc does not meet BX and no triangle is determined.

Fig. 13-49(b). When $b = AD$, the arc is tangent to BX and one triangle—a right triangle with the right angle at C—is determined.

Fig. 13-49(c). When $b > AD$ and $b < c$, the arc meets BX in two points C and C' on the same side of B. Two triangles ABC, in which C is acute, and ABC', in which $C' = 180° - C$ is obtuse, are determined.

Fig. 13-49(d). When $b > AD$ and $b = c$, the arc meets $\overrightarrow{BX}$ in C and B. One triangle (isosceles) is determined.

Fig. 13-49(e). When $b > c$, the arc meets BX in C and $\overrightarrow{BX}$ extended in C'. Since the triangle ABC' does not contain the given angle B, only one triangle ABC is determined.

The given angle is obtuse.

Fig. 13-49(f). When $b < c$ or $b = c$, no triangle is formed.

Fig. 13-49(g). When $b > c$, only one triangle is formed as in Fig 13.49(e).

13.45 Solve the triangle ABC, given $c = 628$, $b = 480$, and $C = 55°10'$. Refer to Fig. 13-50.

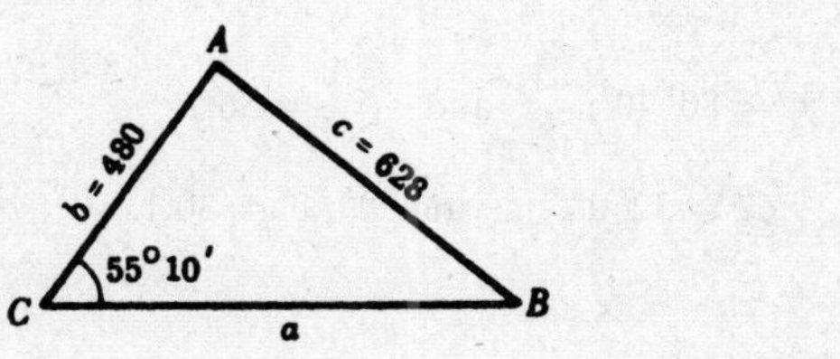

Fig. 13-50

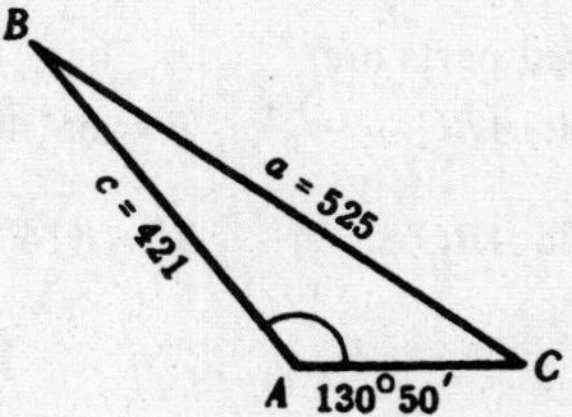

Fig. 13-51

Since C is acute and $c > b$, there is only one solution.

For B: $\sin B = \dfrac{b \sin C}{c} = \dfrac{480 \sin 55°10'}{628} = \dfrac{480(0.8208)}{628} = 0.6274$ and $B = 38°50'$

For A: $A = 180° - (B + C) = 86°0'$

For a: $a = \dfrac{b \sin A}{\sin B} = \dfrac{480 \sin 86°0'}{\sin 38°50'} = \dfrac{480(0.9976)}{0.6271} = 764$

The required parts are $B = 38°50'$, $A = 86°0'$, and $a = 764$.

13.46 Solve the triangle ABC, given $a = 525$, $c = 421$, and $A = 130°50'$. Refer to Fig. 13-51.

Since A is obtuse and $a > c$, there is one solution.

For C: $\sin C = \dfrac{c \sin A}{a} = \dfrac{421 \sin 130°50'}{525} = \dfrac{421(0.7566)}{525} = 0.6067$ and $C = 37°20'$

For B: $B = 180° - (C + A) = 11°50'$

For b: $b = \dfrac{a \sin B}{\sin A} = \dfrac{525 \sin 11°50'}{\sin 130°50'} = \dfrac{525(0.2051)}{0.7566} = 142$

The required parts are $C = 37°20'$, $B = 11°50'$, and $b = 142$.

13.47 Solve the triangle ABC, given $a = 31.5$, $b = 51.8$, and $A = 33°40'$. Refer to Fig. 13-52.

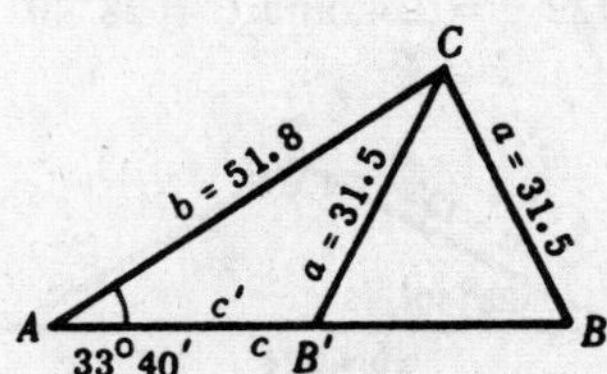

Fig. 13-52

Since A is acute and $a < b$, there is the possibility of two solutions.

For B: $\sin B = \dfrac{b \sin A}{a} = \dfrac{51.8 \sin 33°40'}{31.5} = \dfrac{51.8(0.5544)}{31.5} = 0.9117$

There are two solutions, $B = 65°40'$ and $B' = 180° - 65°40' = 114°20'$.

For C: $C = 180° - (A + B) = 80°40'$

For C': $C' = 180° - (A + B') = 32°0'$

For c: $c = \dfrac{a \sin C}{\sin A} = \dfrac{31.5 \sin 80°40'}{\sin 33°40'} = \dfrac{31.5(0.9868)}{0.5544} = 56.1$

For c': $c' = \dfrac{a \sin C'}{\sin A} = \dfrac{31.5 \sin 32°0'}{\sin 33°40'} = \dfrac{31.5(0.5299)}{0.5544} = 30.1$

The required parts are

For triangle ABC: $B = 65°40'$, $C = 80°40'$, and $c = 56.1$.

For triangle ABC': $B' = 114°20'$, $C' = 32°0'$, and $c' = 30.1$.

13.48 Derive the law of cosines.

In the right triangle ABC of either figure, $b^2 = h^2 + (AD)^2$.

In the right triangle BCD of Fig. 13-53(a), $h = a \sin B$ and $DB = a \cos B$. Then $AD = AB - DB = c - a \cos$ and

$$b^2 = h^2 + (AD)^2 = a^2 \sin^2 B + c^2 - 2ca \cos B + a^2 \cos^2 B$$
$$= a^2(\sin^2 B + \cos^2 B) + c^2 - 2ca \cos B = c^2 + a^2 - 2ca \cos B$$

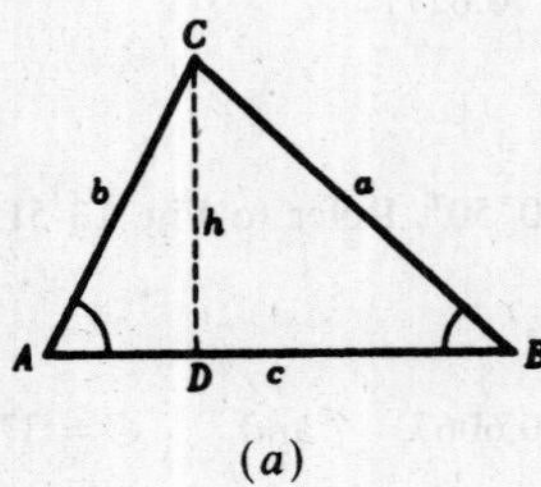

(a)

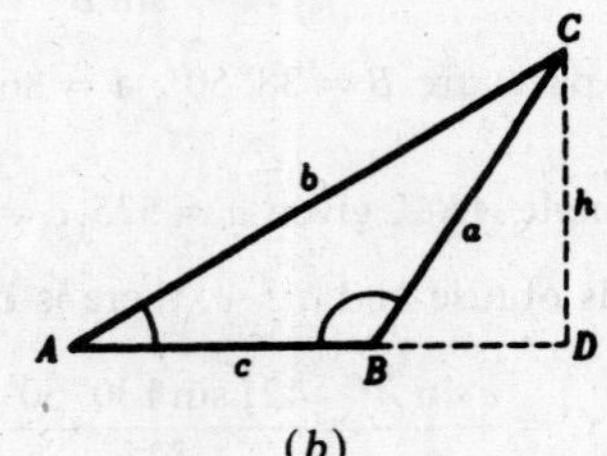

(b)

Fig. 13-53

In the right triangle BCD of Fig. 13-54 (b), $h = a \sin \angle CBD = a \sin(180° - B) = a \sin B$ and $BD = a \cos \angle CBD = a \cos(180° - B) = -a \cos B$. Then $AD = AB + BD = c - a \cos B$ and $b^2 = c^2 + a^2 - 2ca \cos B$.

The remaining equations may be obtained by cyclic changes of the letters.

13.49 Solve the triangle ABC, given $a = 132, b = 224$, and $C = 28°40'$. See Fig. 13-54.

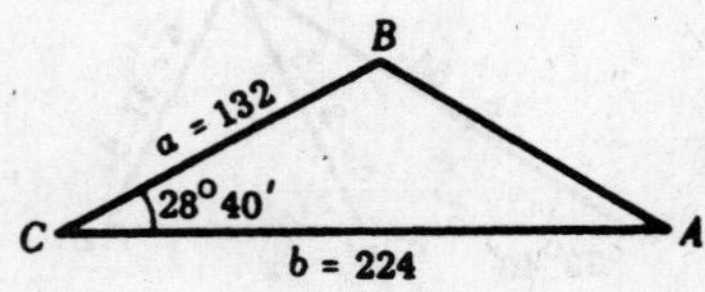

Fig. 13-54

For c:

$$\begin{aligned} c^2 &= a^2 + b^2 - 2ab\cos C \\ &= (132)^2 + (224)^2 - 2(132)(224)\cos 28°40' \\ &= (132)^2 + (224)^2 - 2(132)(224)(0.8774) \\ &= 15\,714 \quad \text{and} \quad c = 125 \end{aligned}$$

For A: $$\sin A = \frac{a\sin C}{c} = \frac{132\sin 28°40'}{125} = \frac{132(0.4797)}{125} = 0.5066 \quad \text{and} \quad A = 30°30'$$

For B: $$\sin B = \frac{b\sin C}{c} = \frac{224\sin 28°40'}{125} = \frac{224(0.4797)}{125} = 0.8596 \quad \text{and} \quad B = 120°40'$$

(Since $b > a$, A is acute; since $A + C < 90°$, $B > 90°$.)
Check: $A + B + C = 179°50'$. The required parts are $A = 30°30'$, $B = 120°40'$, $c = 125$.

13.50 Two forces of 17.5 N and 22.5 N act on a body. If their directions make an angle of 50° 10′ with each other, find the magnitude of their resultant and the angle which it makes with the larger force.

In the parallelogram $ABCD$ (see Fig. 13-55), $A + B = C + D = 180°$ and $B = 180° - 50°10' = 129°50'$. In the triangle ABC,

$$\begin{aligned} b^2 &= c^2 + a^2 - 2ca\cos B \qquad [\cos 129°50' = -\cos(180° - 129°50') = -\cos 50°10'] \\ &= (22.5)^2 + (17.5)^2 - 2(22.5)(17.5)(-0.6406) = 1317 \quad \text{and} \quad b = 36.3 \end{aligned}$$

$$\sin A = \frac{a\sin B}{b} = \frac{17.5\sin 129°50'}{36.3} = \frac{17.5(0.7679)}{36.3} = 0.3702 \quad \text{and} \quad A = 21°40'$$

The resultant is a force of 36.3 N; the required angle is 21° 40′.

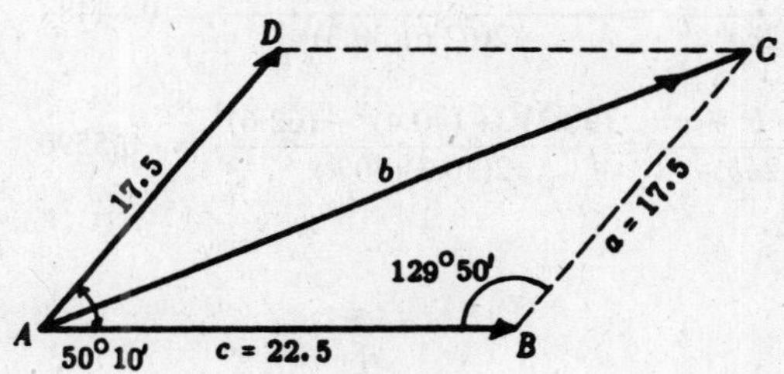

Fig. 13-55

13.51 From A a pilot flies 125 mi in the direction N 38° 20′ W and turns back. Through an error, he then flies 125 mi in the direction S 51° 40′E. How far and in what direction must he now fly to reach his intended destination A?

Denote the turn back point as B and his final position as C. In the triangle ABC (see Fig. 13-56),

$$\begin{aligned} b^2 &= c^2 + a^2 - 2ca\cos B = (125)^2 + (125)^2 - 2(125)(125)\cos 13°20' \\ &= 2(125)^2(1 - 0.9730) = 843.7 \quad \text{and} \quad b = 29.0 \end{aligned}$$

$$\sin A = \frac{a\sin B}{b} = \frac{125\sin 13°20'}{29.0} = \frac{125(0.2306)}{29.0} = 0.9940 \quad \text{and} \quad A = 83°40'$$

Since $\angle CAN_1 = A - \angle N_1AB = 45°20'$, the pilot must fly a course S 45° 20′ W for 29.0 mi in going from C to A.

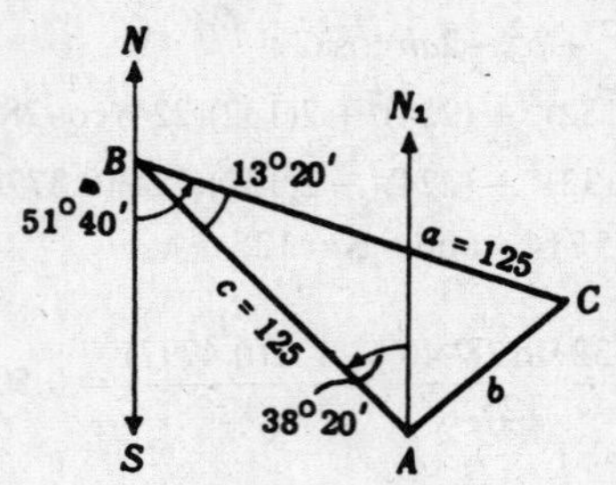

Fig. 13-56

13.52 Solve the triangle ABC, given $a = 30.3$, $b = 40.4$, and $c = 62.6$. Refer to Fig. 13-57.

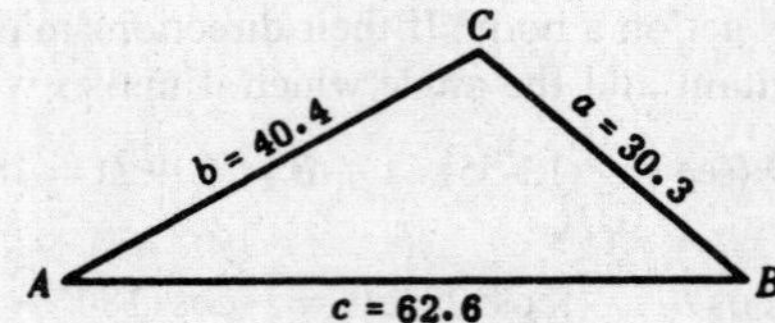

Fig. 13-57

For A: $\cos A = \dfrac{b^2 + c^2 - a^2}{2bc} = \dfrac{(40.4)^2 + (62.6)^2 - (30.3)^2}{2(40.4)(62.6)} = 0.9159$ and $A = 23°40'$

For B: $\cos B = \dfrac{c^2 + a^2 - b^2}{2ca} = \dfrac{(62.6)^2 + (30.3)^2 + (40.4)^2}{2(62.6)(30.3)} = 0.8448$ and $B = 32°20'$

For C: $\cos C = \dfrac{a^2 + b^2 - c^2}{2ab} = \dfrac{(30.3)^2 + (40.4)^2 - (62.6)^2}{2(30.3)(40.4)} = -0.5590$ and $C = 124°0'$

Check: $A + B + C = 180°$.

13.53 The distances of a point C from two points A and B, which cannot be measured directly, are required. The line CA is continued through A for a distance 175 m to D, the line $\overleftrightarrow{CB}$ is continued through B for 225 m to E, and the distances $AB = 300$ m, $DB = 326$ m, and $DE = 488$ m are measured. Find AC and BC. Refer to Fig. 13-58.

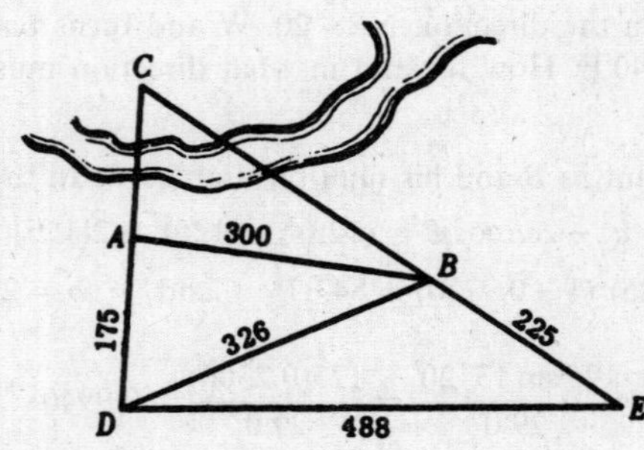

Fig. 13-58

Triangle ABC may be solved for the required parts after the angles $\angle BAC$ and $\angle ABC$ have been found. The first angle is the supplement of $\angle BAD$ and the second is the supplement of the sum of $\angle ABD$ and $\angle DBE$.

In the triangle ABD whose sides are known,

$$\cos\angle BAD = \frac{(175)^2 + (300)^2 - (326)^2}{2(175)(300)} = 0.1367 \quad \text{and} \quad \angle BAD = 82°10'$$

$$\cos\angle ABD = \frac{(300)^2 + (326)^2 - (175)^2}{2(300)(326)} = 0.8469 \quad \text{and} \quad \angle ABD = 32°10'$$

In the triangle BDE whose sides are known,

$$\cos\angle DBE = \frac{(225)^2 + (326)^2 - (488)^2}{2(225)(326)} = -0.5538 \quad \text{and} \quad \angle DBE = 123°40'$$

In the triangle ABC: $AB = 300$,

$$\angle BAC = 180° - \angle BAD = 97°50'$$
$$\angle ABC = 180° - (\angle ABD + \angle DBE) = 24°10'$$
$$\angle ACB = 180° - (\angle BAC + \angle ABC) = 58°0'$$

Then $$AC = \frac{AB \sin\angle ABC}{\sin\angle ACB} = \frac{300 \sin 24°10'}{\sin 58°0'} = \frac{300(0.4094)}{0.8480} = 145$$

and $$BC = \frac{AB \sin\angle BAC}{\sin\angle ACB} = \frac{300 \sin 97°50'}{\sin 58°0'} = \frac{300(0.9907)}{0.8480} = 350$$

The required distances are $AC = 145$ m and $BC = 350$ m.

13.54 Verify each of the following:

(*a*) $\text{Arcsin } 0 = 0$
(*b*) $\text{Arccos } (-1) = \pi$
(*c*) $\text{Arctan } \sqrt{3} = \pi/3$
(*d*) $\text{Arccot } \sqrt{3} = \pi/6$
(*e*) $\text{Arcsec } 2 = \pi/3$
(*f*) $\text{Arccsc } (-\sqrt{2}) = -3\pi/4$
(*g*) $\text{Arccos } \theta = \pi/2$
(*h*) $\text{Arcsin } (-1) = -\pi/2$
(*i*) $\text{Arctan } (-1) = -\pi/4$
(*j*) $\text{Arccot } 0 = \pi/2$
(*k*) $\text{Arcsec } (-\sqrt{2}) = -3\pi/4$
(*l*) $\text{Arccsc } (-2) = -5\pi/6$

13.55 Verify each of the following:

(*a*) $\text{Arcsin } 0.3333 = 19°28'$
(*b*) $\text{Arccos } 0.4000 = 66°25'$
(*c*) $\text{Arctan } 1.5000 = 56°19'$
(*d*) $\text{Arccot } 1.1875 = 40°6'$
(*e*) $\text{Arcsec } 1.0324 = 14°24'$
(*f*) $\text{Arccsc } 1.5082 = 41°32'$
(*g*) $\text{Arcsin } (-0.6439) = -40°5'$
(*h*) $\text{Arccos } (-0.4519) = 116°52'$
(*i*) $\text{Arctan } (-1.4400) = -55°13'$
(*j*) $\text{Arccot } (-0.7340) = 126°17'$
(*k*) $\text{Arcsec } (-1.2067) = -145°58'$
(*l*) $\text{Arccsc } (-4.1923) = -166°12'$

13.56 Verify each of the following:

(*a*) $\sin(\text{Arcsin } \frac{1}{2}) = \sin \pi/6 = \frac{1}{2}$
(*b*) $\cos[\text{Arccos}(-\frac{1}{2})] = \cos 2\pi/3 = -\frac{1}{2}$
(*c*) $\cos[\text{Arcsin}(-\sqrt{2}/2)] = \cos(-\pi/4) = \sqrt{2}/2$
(*d*) $\text{Arcsin}(\sin \pi/3) = \text{Arcsin } \sqrt{3}/2 = \pi/3$
(*e*) $\text{Arccos } [\cos(-\pi/4)] = \text{Arccos } \sqrt{2}/2 = \pi/4$
(*f*) $\text{Arcsin } (\tan 3\pi/4) = \text{Arcsin}(-1) = -\pi/2$
(*g*) $\text{Arccos } [\tan(-5\pi/4)] = \text{Arccos}(-1) = \pi$

13.57 Verify each of the following:

(*a*) $\text{Arcsin } \sqrt{2}/2 - \text{Arcsin } \frac{1}{2} = \pi/4 - \pi/6 = \pi/12$
(*b*) $\text{Arccos } 0 + \text{Arctan}(-1) = \pi/2 + (-\pi/4) = \pi/4 = \text{Arctan } 1$

13.58 Evaluate each of the following:

(*a*) $\cos(\text{Arcsin } \frac{3}{5})$, (*b*) $\sin[\text{Arccos}(-\frac{2}{3})]$, (*c*) $\tan[\text{Arcsin}(-\frac{3}{4})]$.

(*a*) Let $\theta = \text{Arcsin } \frac{3}{5}$; then $\sin\theta = \frac{3}{5}$, θ being a first quadrant angle. From Fig. 13-59(*a*), $\cos(\text{Arcsin } \frac{3}{5}) = \cos\theta = \frac{4}{5}$.

(*b*) Let $\theta = \text{Arccos}(-\frac{2}{3})$; then $\cos\theta = -\frac{2}{3}$, θ being a second quadrant angle. From Fig. 13-59(*b*), $\sin[\text{Arccos}(-\frac{2}{3})] = \sin\theta = \sqrt{5}/3$.

(*c*) Let $\theta = \text{Arcsin}(-\frac{3}{4})$; then $\sin\theta = -\frac{3}{4}$, θ being a fourth quadrant angle. From Fig. 13-59(*c*), $\tan[\text{Arcsin}(-\frac{3}{4})] = \tan\theta = -3/\sqrt{7} = -3\sqrt{7}/7$.

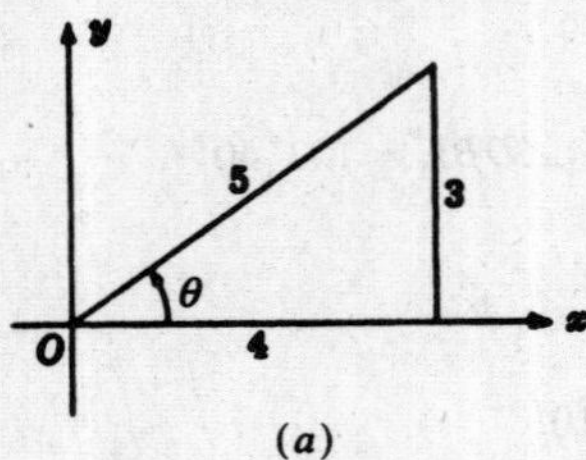

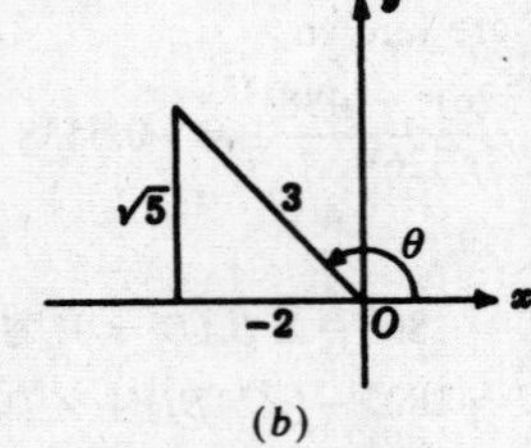

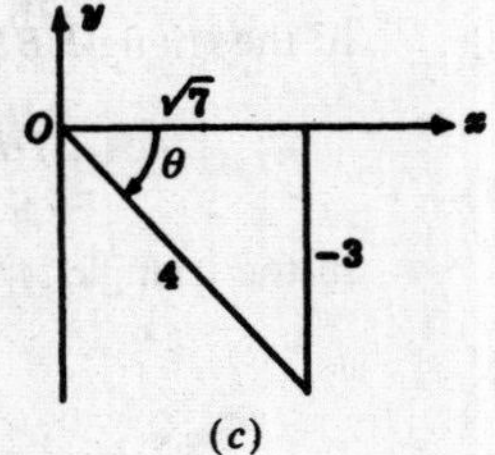

Fig. 13-59

13.59 Evaluate $\sin(\text{Arcsin } \frac{12}{13} + \text{Arcsin } \frac{4}{5})$.

Let $\theta = \text{Arcsin } \frac{12}{13}$ and $\phi = \text{Arcsin } \frac{4}{5}$. Then $\sin\theta = \frac{12}{13}$ and $\sin\phi = \frac{4}{5}$, θ and ϕ being first quadrant angles. From Fig. 13-60 and Fig. 13-61,

$$\sin(\text{Arcsin } \tfrac{12}{13} + \text{Arcsin } \tfrac{4}{5} = \sin(\theta + \phi) = \sin\theta\cos\phi + \cos\theta\sin\phi$$
$$= \tfrac{12}{13}\cdot\tfrac{3}{5} + \tfrac{5}{13}\cdot\tfrac{4}{5} = \tfrac{56}{65}$$

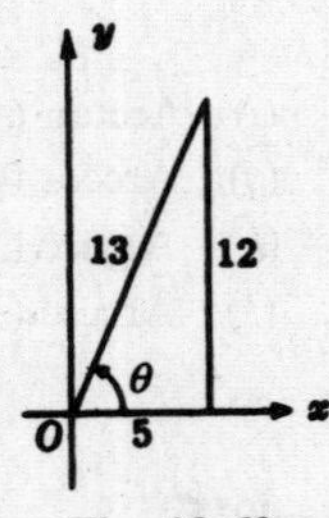

Fig. 13-60

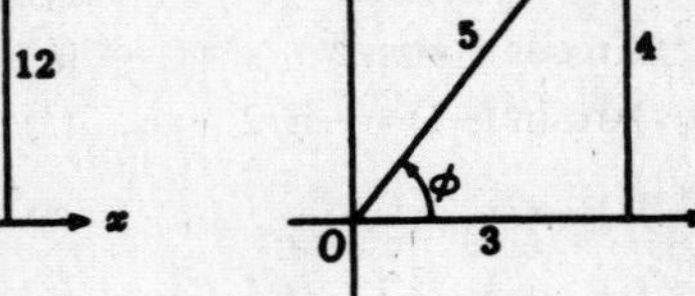

Fig. 13-61

13.60 Evaluate $\cos(\text{Arctan } \frac{15}{8} - \text{Arcsin } \frac{7}{25})$.

Let $\theta = \text{Arctan } \frac{15}{8}$ and $\phi = \text{Arcsin } \frac{7}{25}$. Then $\tan\theta = \frac{15}{8}$ and $\sin\phi = \frac{7}{25}$, θ and ϕ being first quadrant angles. From Fig. 13-62 and Fig. 13-63,

$$\cos(\text{Arctan } \tfrac{15}{8} - \text{Arcsin } \tfrac{7}{25}) = \cos(\theta - \phi) = \cos\theta\cos\phi + \sin\theta\sin\phi$$
$$= \tfrac{8}{17}\cdot\tfrac{24}{25} + \tfrac{15}{17}\cdot\tfrac{7}{25} = \tfrac{297}{425}$$

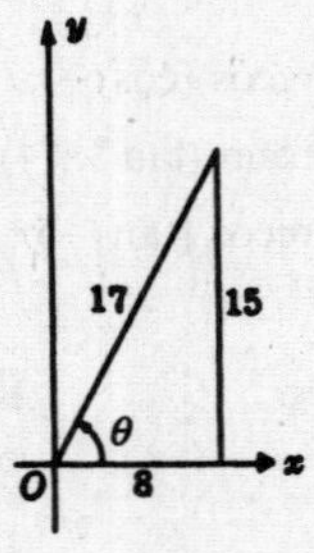

Fig. 13-62

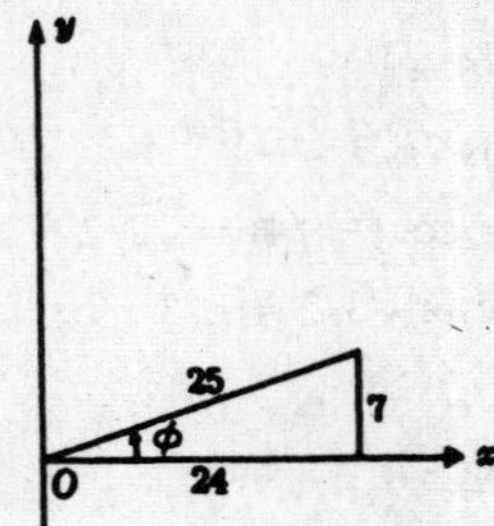

Fig. 13-63

13.61 Evaluate sin (2 Arctan 3).

Let $\theta = \text{Arctan } 3$; then $\tan\theta = 3$, θ being a first quadrant angle. From Fig. 13-64,

$$\sin(2 \text{ Arctan } 3) = \sin 2\theta = 2\sin\theta\cos\theta = 2\left(\frac{3}{\sqrt{10}}\right)\left(\frac{1}{\sqrt{10}}\right) = \frac{3}{5}$$

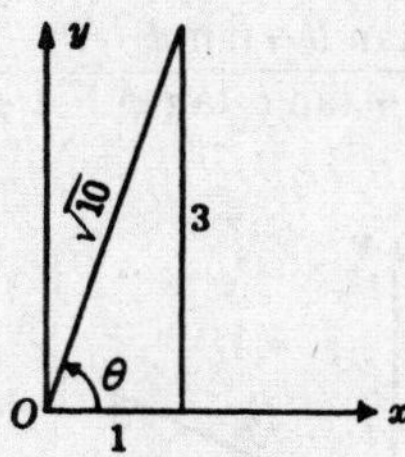

Fig. 13-64

13.62 Show that $\text{Arcsin } 1/\sqrt{5} + \text{Arcsin } 2/\sqrt{5} = \pi/2$.

Let $\theta = \text{Arcsin } 1/\sqrt{5}$ and $\phi = \text{Arcsin } 2/\sqrt{5}$; then $\sin\theta = 1/\sqrt{5}$ and $\sin\phi = 2/\sqrt{5}$, each angle terminating in the first quadrant. We are to show that $\theta + \phi = \pi/2$ or, taking the sines of both members, that $\sin(\theta + \phi) = \sin \pi/2$.

From Fig. 13-65 and Fig. 13-66,

$$\sin(\theta + \phi) = \sin\theta\cos\phi + \cos\theta\sin\phi = \frac{1}{\sqrt{5}}\cdot\frac{1}{\sqrt{5}} + \frac{2}{\sqrt{5}}\cdot\frac{2}{\sqrt{5}} = 1 = \sin\frac{\pi}{2}$$

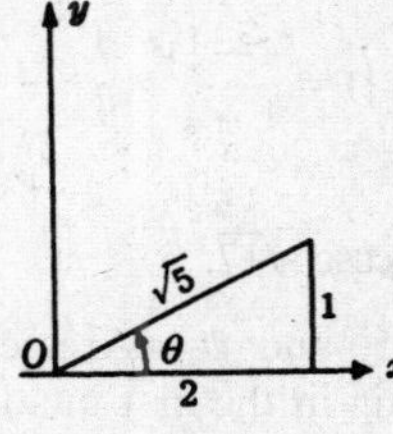

Fig. 13-65

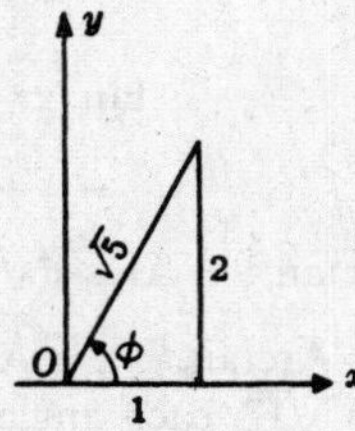

Fig. 13-66

13.63 Show that $2 \text{ Arctan } \frac{1}{2} = \text{Arctan } \frac{4}{3}$.

Let $\theta = \text{Arctan } \frac{1}{2}$ and $\phi = \text{Arctan } \frac{4}{3}$; then $\tan\theta = \frac{1}{2}$ and $\tan\phi = \frac{4}{3}$. We are to show that $2\theta = \phi$ or, taking the tangents of both members, that $\tan 2\theta = \tan\phi$. Now

$$\tan 2\theta = \frac{2\tan\theta}{1 - \tan^2\theta} = \frac{2(\frac{1}{2})}{1 - (\frac{1}{2})^2} = \frac{4}{3} = \tan\phi$$

13.64 Show that $\text{Arcsin } \frac{77}{85} - \text{Arcsin } \frac{3}{5} = \text{Arccos } \frac{15}{17}$.

Let $\theta = \text{Arcsin } \frac{77}{85}$, $\phi = \text{Arcsin } \frac{3}{5}$, and $\psi = \text{Arccos } \frac{15}{17}$ then $\sin\theta = \frac{77}{85}$, $\sin\phi = \frac{3}{5}$, and $\cos\psi = \frac{15}{17}$, each angle terminating in the first quadrant. Taking the sine of both members of the given relation, we are to show that $\sin(\theta - \phi) = \sin\psi$. From Figs. 13-67, 13-68, and 13-69,

$$\sin(\theta - \phi) = \sin\theta\cos\phi - \cos\theta\sin\phi = \tfrac{77}{85}\cdot\tfrac{4}{3} - \tfrac{36}{85}\cdot\tfrac{3}{5} = \tfrac{8}{17} = \sin\psi$$

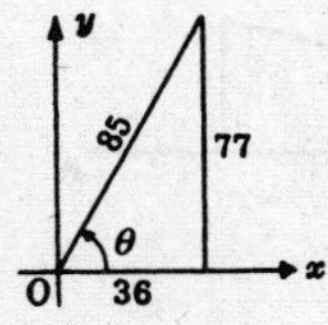

Fig. 13-67

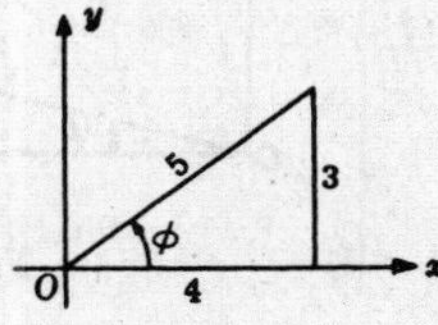

Fig. 13-68

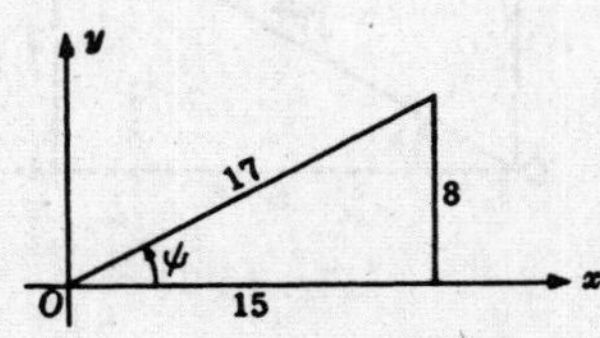

Fig. 13-69

13.65 Show that Arccot $\frac{43}{32}$ − Arccot $\frac{1}{4}$ = Arccos $\frac{12}{13}$.

Let $\theta = \text{Arccot}\ \frac{43}{32}$, $\phi = \text{Arctan}\ \frac{1}{4}$, and $\psi = \text{Arccos}\ \frac{12}{13}$ (see Fig. 32-70); then $\cot\theta = \frac{43}{32}$, $\tan\phi = \frac{1}{4}$, and $\cos\psi = \frac{12}{13}$, each angle terminating in the first quadrant. Taking the tangent of both members of the given relation, we are to show that $\tan(\theta - \phi) = \tan\psi$.

$$\tan(\theta-\phi) = \frac{\tan\theta - \tan\phi}{1+\tan\theta\tan\phi} = \frac{\frac{32}{43}-\frac{1}{4}}{1+(\frac{32}{43})(\frac{1}{4})} = \frac{5}{12} = \tan\psi$$

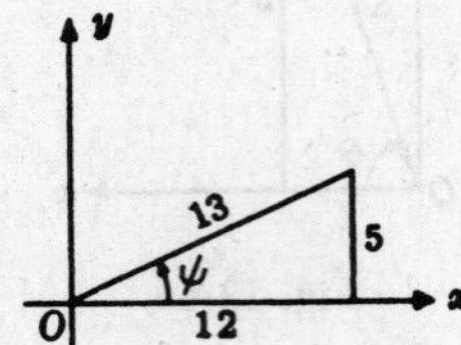

Fig. 13-70

13.66 Show that Arctan $\frac{1}{2}$ + Arctan $\frac{1}{5}$ + Arctan $\frac{1}{8}$ = $\pi/4$.

We shall show that Arctan $\frac{1}{2}$ + Arctan $\frac{1}{5}$ = $\pi/4$ − Arctan $\frac{1}{8}$.

$$\tan(\text{Arctan}\ \tfrac{1}{2} + \text{Arctan}\ \tfrac{1}{5}) = \frac{\frac{1}{2}+\frac{1}{5}}{1-(\frac{1}{2})(\frac{1}{5})} = \frac{7}{9}$$

and

$$\tan(\pi/4 - \text{Arctan}\ \tfrac{1}{8}) = \frac{1-\frac{1}{8}}{1+\frac{1}{8}} = \frac{7}{9}$$

13.67 Show that 2 Arctan $\frac{1}{3}$ + Arctan $\frac{1}{7}$ = Arcsec $\sqrt{34}/5$ + Arccsc $\sqrt{17}$.

Let $\theta = \text{Arctan}\ \frac{1}{3}$, $\phi = \text{Arctan}\ \frac{1}{7}$, $\lambda = \text{Arcsec}\sqrt{34}/5$, and $\psi = \text{Arccsc}\ \sqrt{17}$; then $\tan\theta = \frac{1}{3}$, $\tan\phi = \frac{1}{7}$, $\sec\lambda = \sqrt{34}/5$, and $\csc\psi = \sqrt{17}$, each angle terminating in the first quadrant.

Taking the tangent of both members of the given relation, we are to show that $\tan(2\theta + \phi) = \tan(\lambda + \psi)$. Now

$$\tan 2\theta = \frac{2\tan\theta}{1-\tan^2\theta} = \frac{2(\frac{1}{3})}{1-(\frac{1}{3})^2} = \frac{3}{4}$$

$$\tan(2\theta+\phi) = \frac{\tan 2\theta + \tan\phi}{1-\tan 2\theta\tan\phi} = \frac{\frac{3}{4}+\frac{1}{7}}{1-(\frac{3}{4})(\frac{1}{7})} = 1$$

and, using Fig. 13-71 and Fig. 13-72,

$$\tan(\lambda+\psi) = \frac{\frac{3}{5}+\frac{1}{4}}{1-(\frac{3}{5})(\frac{1}{4})} = 1$$

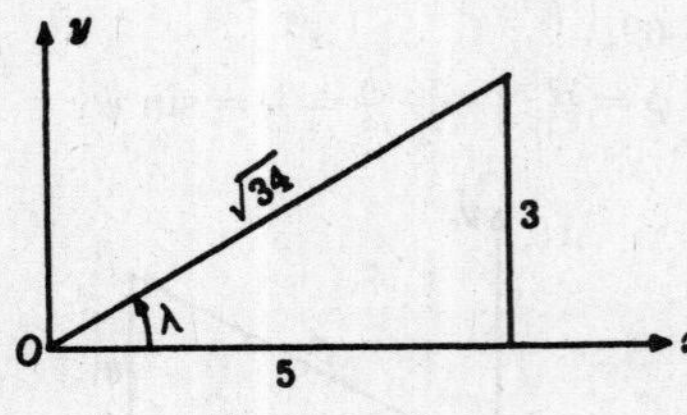

Fig. 13-71

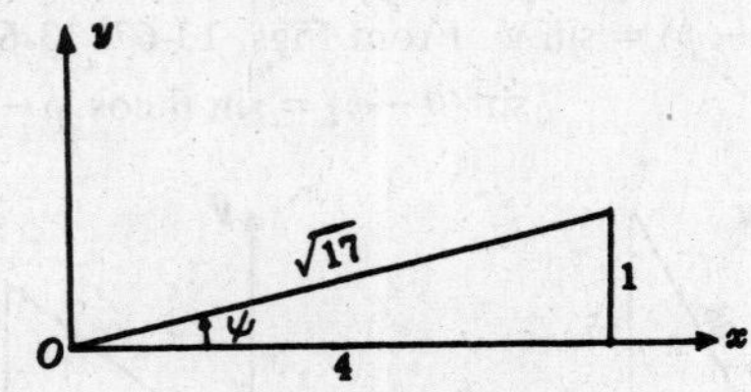

Fig. 13-72

13.68 Find the general value of each of the following:

(a) $\arcsin \sqrt{2}/2 = \pi/4 + 2n\pi, 3\pi/4 + 2n\pi$

(b) $\arccos \frac{1}{2} = \pi/3 + 2n\pi, 5\pi/3 + 2n\pi$

(c) $\arctan 0 = 2n\pi, (2n+1)\pi$

(d) $\arcsin(-1) = -\pi/2 + 2n\pi$

(e) $\arccos 0 = \pi/2 + 2n\pi, 3\pi/2 + 2n\pi$

(f) $\arctan(-\sqrt{3}) = -\pi/3 + 2n\pi, 2\pi/3 + 2n\pi$

where n is a positive or negative integer, or is zero.

13.69 Show that the general value of

(a) $\arcsin\ x = n\pi + (-1)^n \text{Arcsin}\, x$,

(b) $\arccos\ x = 2n\pi \pm \text{Arccos}\, x$,

(c) $\arctan\ x = n\pi + \text{Arctan}\, x$,

where n is any positive or negative integer, or is zero.

(a) Let $\theta = \text{Arcsin}\, x$. Then since $\sin(\pi - \theta) = \sin\theta$, all values of $\arcsin x$ are given by

$$(1)\quad \theta + 2m\pi \qquad \text{and} \qquad (2)\quad \pi - \theta + 2m\pi = (2m+1)\pi - \theta$$

Now, when $n = 2m$, that is, n is an even integer, (1) may be written as $n\pi + \theta = n\pi + (-1)^n\theta$; and when $n = 2m + 1$, that is, n is an odd integer, (2) may be written as $n\pi - \theta = n\pi + (-1)^n\theta$. Thus, $\arcsin x = n\pi + (-1)^n \text{ Arcsin}\, x$, where n is any positive or negative integer, or is zero.

(b) Let $\theta = \text{Arccos}\, x$. Then since $\cos(-\theta) = \cos\theta$, all values of $\arccos x$ are given by $\theta + 2n\pi$ and $-\theta + 2n\pi$ or $2n\pi \pm \theta = 2n\pi \pm \text{Arccos}\, x$, where n is any positive or negative integer, or is zero.

(c) Let $\theta = \text{Arctan}\, x$. Then since $\tan(\pi + \theta) = \tan\theta$, all values of $\arctan x$ are given by $\theta + 2m\pi$ and $(\pi + \theta) + 2m\pi = \theta + (2m+1)\pi$ or, as in (a), by $n\pi + \text{Arctan}\, x$, where n is any positive or negative integer, or is zero.

13.70 Express the general value of each of the functions of Problem 13.68, using the form of Problem 13.69.

(a) $\arcsin \sqrt{2}/2 = n\pi + (-1)^n \pi/4$

(b) $\arccos \frac{1}{2} = 2n\pi \pm \pi/3$

(c) $\arctan 0 = n\pi$

(d) $\arcsin(-1) = n\pi + (-1)^n(-\pi/2)$

(e) $\arccos 0 = 2n\pi \pm \pi/2$

(f) $\arctan(-\sqrt{3}) = n\pi - \pi/3$

where n is any positive or negative integer, or is zero.

Supplementary Problems

13.71 State the quadrant in which each angle terminates and the signs of the sine, cosine, and tangent of each angle.

(a) 125° (b) 75° (c) 320° (d) 212° (e) 460° (f) 750° (g) −250° (h) −1000°

Ans. (a) II; +, −, − (b) I; +, +, + (c) IV; −, +, − (d) III; −, −, + (e) II (f) I (g) II (h) I

13.72 In what quadrant will θ terminate if

(a) $\sin\theta$ and $\cos\theta$ are both positive?

(b) $\cos\theta$ and $\tan\theta$ are both positive?

(c) $\sin\theta$ and $\sec\theta$ are both negative?

(d) $\cos\theta$ and $\cot\theta$ are both negative?

(e) $\tan\theta$ is positive and $\sec\theta$ is negative?

(f) $\tan\theta$ is negative and $\sec\theta$ is positive?

(g) $\sin\theta$ is positive and $\cos\theta$ is negative?

(h) $\sec\theta$ is positive and $\csc\theta$ is negative?

Ans. (a) I (b) I (c) III (d) II (e) III (f) IV (g) II (h) IV

13.73 Denote by θ the smallest positive angle whose terminal side passes through the given point and find the trigonometric functions of θ:

(*a*) $P(-5, 12)$ (*b*) $P(7, -24)$ (*c*) $P(2, 3)$ (*d*) $P(-3, -5)$

Ans. (*a*) $\frac{12}{13}, -\frac{5}{13}, -\frac{12}{5}, -\frac{5}{12}, -\frac{13}{5}, \frac{13}{12}$

(*b*) $-\frac{24}{25}, \frac{7}{25}, -\frac{24}{7}, -\frac{7}{24}, \frac{25}{7}, -\frac{25}{24}$

(*c*) $3/\sqrt{13}, 2/\sqrt{13}, \frac{3}{2}, \frac{2}{3}, \sqrt{13}/2, \sqrt{13}/3$

(*d*) $-5/\sqrt{34}, -3/\sqrt{34}, \frac{5}{3}, \frac{3}{5}, -\sqrt{34}/3, -\sqrt{34}/5$

13.74 Find the possible values of the trigonometric functions of θ, given

(*a*) $\sin\theta = \frac{7}{25}$
(*b*) $\cos\theta = -\frac{4}{5}$
(*c*) $\tan\theta = -\frac{5}{12}$
(*d*) $\cot\theta = \frac{24}{7}$
(*e*) $\sin\theta = -\frac{2}{3}$
(*f*) $\cos\theta = \frac{5}{6}$
(*g*) $\tan\theta = \frac{3}{5}$
(*h*) $\cot\theta = \sqrt{6}/2$
(*i*) $\sec\theta = -\sqrt{5}$
(*j*) $\csc\theta = -2/\sqrt{3}$

Ans. (*a*) I: $\frac{7}{25}, \frac{24}{25}, \frac{7}{24}, \frac{24}{7}, \frac{25}{24}, \frac{25}{7}$; II: $\frac{7}{25}, -\frac{24}{25}, -\frac{7}{24}, -\frac{24}{7}, -\frac{25}{24}, \frac{25}{7}$

(*b*) II: $\frac{3}{5}, -\frac{4}{5}, -\frac{3}{4}, -\frac{4}{3}, -\frac{5}{4}, \frac{5}{3}$; III: $-\frac{3}{5}, -\frac{4}{5}, \frac{3}{4}, \frac{4}{3}, -\frac{5}{4}, -\frac{5}{3}$

(*c*) II: $\frac{5}{13}, -\frac{12}{13}, -\frac{5}{12}, -\frac{12}{5}, -\frac{13}{12}, \frac{13}{5}$; IV: $-\frac{5}{13}, \frac{12}{13}, -\frac{5}{12}, -\frac{12}{5}, \frac{13}{12}, -\frac{13}{5}$

(*d*) I: $\frac{7}{25}, \frac{24}{25}, \frac{7}{24}, \frac{24}{7}, \frac{25}{24}, \frac{25}{7}$; III: $-\frac{7}{25}, -\frac{24}{25}, \frac{7}{24}, \frac{24}{7} - \frac{25}{24}, -\frac{25}{7}$

(*e*) III: $-\frac{2}{3}, -\sqrt{5}/3, 2/\sqrt{5}, \sqrt{5}/2, -3/\sqrt{5}, -\frac{3}{2}$; IV: $-\frac{2}{3}, \sqrt{5}/3, -2\sqrt{5}, -\sqrt{5}/2, 3/\sqrt{5}, -\frac{3}{2}$

(*f*) I: $\sqrt{11}/6, \frac{5}{6}, \sqrt{11}/5, 5/\sqrt{11}, \frac{6}{5}, 6/\sqrt{11}$; IV: $-\sqrt{11}/6, \frac{5}{6}, -\sqrt{11}/5, -5/\sqrt{11}, \frac{6}{5}, -6\sqrt{/11}$

(*g*) I: $3/\sqrt{34}, 5\sqrt{34}, \frac{3}{5}, \frac{5}{3}, \sqrt{34}/5, \sqrt{34}/3$; III: $-3/\sqrt{34}, -5/\sqrt{34}, \frac{3}{5}, \frac{5}{3}, -\sqrt{34}/5, -\sqrt{34}/3$

(*h*) I: $2/\sqrt{10}, \sqrt{3}/\sqrt{5}, 2/\sqrt{6}, \sqrt{6}/2, \sqrt{5}/\sqrt{3}, \sqrt{10}/2$; III: $-2/\sqrt{10}, -\sqrt{3}/\sqrt{5}, 2/\sqrt{6}, \sqrt{6}/2, -\sqrt{5}/\sqrt{3}, -\sqrt{10}/2$

(*i*) II: $2/\sqrt{5}, -1\sqrt{5}, -2, -\frac{1}{2}, -\sqrt{5}, \sqrt{5}/2$; III: $-2/\sqrt{5}, -1\sqrt{5}, 2, \frac{1}{2}, -\sqrt{5}, -\sqrt{5}/2$

(*j*) III: $-\sqrt{3}/2, -\frac{1}{2}, \sqrt{3}, 1/\sqrt{3}, -2, -2/\sqrt{3}$; IV: $-\sqrt{3}/2, \frac{1}{2}, -\sqrt{3}, -1/\sqrt{3}, 2, -2\sqrt{3}$

13.75 Evaluate each of the following:

(*a*) $\tan 180° - 2\cos 180° + 3\csc 270° + \sin 90°$

(*b*) $\sin 0° + 3\cot 90° + 5\sec 180° - 4\cos 270°$

Ans. (*a*) 0
(*b*) -5

13.76 Find the values of the trigonometric functions of the acute angles of the right triangle *ABC*, given (*a*) $a = 3, b = 1$ (*b*) $a = 2, c = 5$ (*c*) $b = \sqrt{7}, c = 4$

Ans. (*a*) *A:* $3/\sqrt{10}, 1/\sqrt{10}, 3, \frac{1}{3}, \sqrt{10}, \sqrt{10}/3$; *B:* $1/\sqrt{10}, 3/\sqrt{10}, \frac{1}{3}, 3, \sqrt{10}/3, \sqrt{10}$

(*b*) *A:* $\frac{2}{5}, \sqrt{21}/5, 2/\sqrt{21}, \sqrt{21}/2, 5/\sqrt{21}, \frac{5}{2}$; *B:* $\sqrt{21}/5, \frac{2}{5}, \sqrt{21}/2, 2/\sqrt{21}, \frac{5}{2}, 5/\sqrt{21}$

(*c*) *A:* $\frac{3}{4}, \sqrt{7}/4, 3/\sqrt{7}, \sqrt{7}/3, 4/\sqrt{7}, \frac{4}{3}$; *B:* $\sqrt{7}/4, \frac{3}{4}, \sqrt{7}/3, 3/\sqrt{7}, \frac{4}{3}, 4/\sqrt{7}$

13.77 Which is the greater and why:

(*a*) $\sin 55°$ or $\cos 55°$? (*c*) $\tan 15°$ or $\cot 15°$?
(*b*) $\sin 40°$ or $\cos 40°$? (*d*) $\sec 55°$ or $\csc 55°$?

Hint: Consider a right triangle having as acute angle the given angle.
Ans. (*a*) $\sin 55°$ (*b*) $\cos 40°$ (*c*) $\cot 15°$ (*d*) $\sec 55°$

13.78 Find the value of each of the following:

(*a*) $\sin 30° + \tan 45°$

(*b*) $\cot 45° + \cos 60°$

(*c*) $\sin 30° \cos 60° + \cos 30° \sin 60°$

(*d*) $\cos 30° \cos 60° - \sin 30° \sin 60°$

(*e*) $\dfrac{\tan 60° - \tan 30°}{1 + \tan 60° \tan 30°}$

(*f*) $\dfrac{\csc 30° + \csc 60° + \csc 90°}{\sec 0° + \sec 30° + \sec 60°}$

Ans. (*a*) $\frac{3}{2}$ (*b*) $\frac{3}{2}$ (*c*) 1 (*d*) 0 (*e*) $1/\sqrt{3}$ (*f*) 1

13.79 A man drives 500 m along a road which is inclined 20° to the horizontal. How high above his starting point is he?

Ans. 170 m

13.80 A tree broken over by the wind forms a right triangle with the ground. If the broken part makes an angle of 50° with the ground and if the top of the tree is now 20 m from its base, how tall was the tree?

Ans. 56 m

13.81 Two straight roads intersect to form an angle of 75°. Find the shortest distance from one road to a gas station on the other road 1000 m from the junction.

Ans. 970 m

13.82 Two buildings with flat roofs are 60 ft apart. From the roof of the shorter building, 40 ft in height, the angle of elevation to the edge of the roof of the taller building is 40°. How high is the taller building?

Ans. 90 ft

13.83 A ladder, with its foot in the street, makes an angle of 30° with the street when its top rests on a building on one side of the street and makes an angle of 40° with the street when its top rests on a building on the other side of the street. If the ladder is 50 m long, how wide is the street?

Ans. 82 m

13.84 Find the perimeter of an isosceles triangle whose base is 40 cm and whose base angle is 70°.

Ans. 156 cm

13.85 Express each of the following in terms of functions of a positive acute angle:

(*a*) $\sin 145°$ (*d*) $\cot 155°$ (*g*) $\sin(-200°)$ (*j*) $\cot 610°$

(*b*) $\cos 215°$ (*e*) $\sec 325°$ (*h*) $\cos(-760°)$ (*k*) $\sec 455°$

(*c*) $\tan 440°$ (*f*) $\csc 190°$ (*i*) $\tan(-1385°)$ (*l*) $\csc 825°$

Ans.

(*a*) $\sin 35°$ or $\cos 55°$ (*g*) $\sin 20°$ or $\cos 70°$

(*b*) $-\cos 35°$ or $-\sin 55°$ (*h*) $\cos 40°$ or $\sin 50°$

(*c*) $\tan 80°$ or $\cot 10°$ (*i*) $\tan 55°$ or $\cot 35°$

(*d*) $-\cot 25°$ or $-\tan 65°$ (*j*) $\cot 70°$ or $\tan 20°$

(*e*) $\sec 35°$ or $\csc 55°$ (*k*) $-\sec 85°$ or $-\csc 5°$

(*f*) $-\csc 10°$ or $\sec 80°$ (*l*) $\csc 75°$ or $\sec 15°$

13.86 Find the exact values of the sine, cosine, and tangent of

(*a*) 150° (*b*) 225° (*c*) 300° (*d*) −120° (*e*) −120° (*f*) −315°

Ans. (*a*) $\frac{1}{2}, -\sqrt{3}/2, 1/\sqrt{3}$ (*d*) $-\sqrt{3}/2, -\frac{1}{2}, \sqrt{3}$

(*b*) $-\sqrt{2}/2, -\sqrt{2}/2, 1$ (*e*) $\frac{1}{2}, \sqrt{3}/2, -1\sqrt{3}$

(*c*) $-\sqrt{3}/2, \frac{1}{2}, -\sqrt{3}$ (*f*) $\sqrt{2}/2, \sqrt{2}/2, 1$

13.87 Using a calculator, verify that

(*a*) $\sin 155°13' = 0.4192$

(*b*) $\cos 104°38' = -0.2526$

(*c*) $\tan 305°24' = -1.4071$

(*d*) $\sin 114°18' = 0.9114$

(*e*) $\cos 166°51' = -0.9738$

13.88 Find all angles, $0 \leqslant \theta < 360°$, for which

(*a*) $\sin\theta = \sqrt{2}/2$ (*b*) $\cos\theta = -1$

Ans. (*a*) 45°, 135°

(*b*) 180°

13.89 When θ is a second quadrant angle for which $\tan\theta = -\frac{2}{3}$, show that

(*a*) $\dfrac{\sin(90° - \theta) - \cos(180° - \theta)}{\tan(270° + \theta) + \cot(360° - \theta)} = -\dfrac{2}{\sqrt{3}}$ (*b*) $\dfrac{\tan(90° - \theta) + \cos(180° - \theta)}{\sin(270° + \theta) - \cot(-\theta)} = \dfrac{2 + \sqrt{13}}{2 - \sqrt{13}}$

13.90 Sketch the graph of each of the following for one period:

(*a*) $y = 3\sin x$ (*b*) $y = \sin 2x$ (*c*) $y = 4\sin(x/2)$ (*d*) $y = 4\cos x$

(*e*) $y = 2\cos(x/3)$

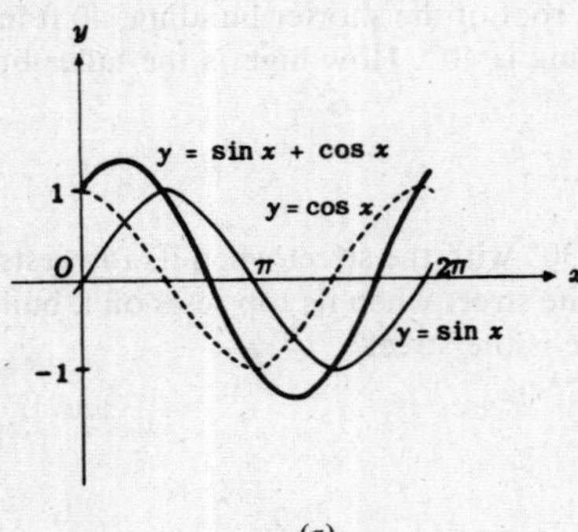

(*a*)

y = sin 2x + cos 3x
y = sin 2x
y = cos 3x

(*b*)

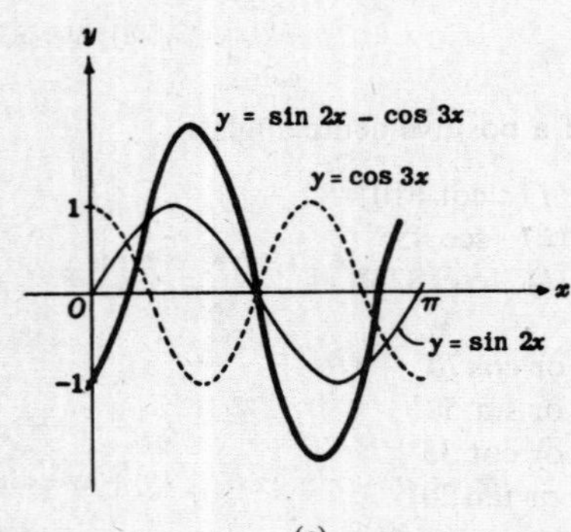

(*c*)

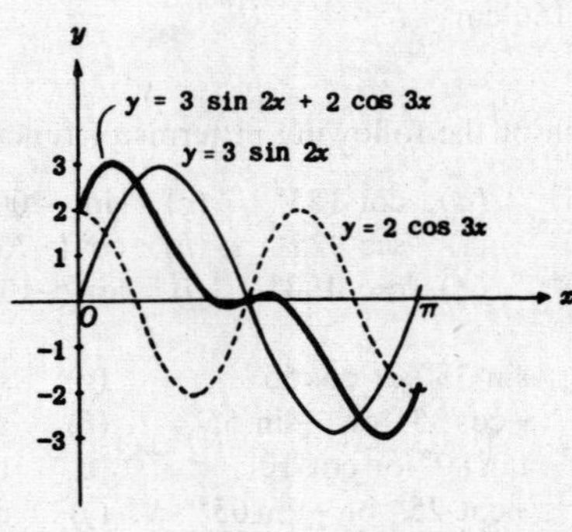

(*d*)

Fig. 13-73

13.91 Construct the graph of each of the following for one period:

(*a*) $y = \sin x + 2\cos x$ (*d*) $y = \sin 2x + \sin 3x$

(*b*) $y = \sin 3x + \cos 2x$ (*e*) $y = \sin 3x - \cos 2x$

(*c*) $y = \sin x + \sin 2x$ (*f*) $y = 2\sin 3x + 3\cos 2x$

Solve each of the following oblique triangles ABC, given:

13.92 $a = 125,\ A = 54°40',\ B = 65°10'$. Ans. $b = 139,\ c = 133,\ C = 60°10'$

13.93 $b = 321,\ A = 75°20', C = 38°30'$. Ans. $a = 339,\ c = 218,\ B = 66°10'$

13.94 $b = 215,\ c = 150,\ B = 42°40'$. Ans. $a = 300,\ A = 109°10',\ C = 28°10'$

13.95 $a = 512,\ b = 426,\ A = 48°50'$. Ans. $c = 680,\ B = 38°50',\ C = 92°20'$

13.96 $b = 50.4,\ c = 33.3,\ B = 118°30'$. Ans. $a = 25.1,\ A = 26°0',\ C = 35°30'$

13.97 $b = 40.2,\ a = 31.5,\ B = 112°20'$. Ans. $c = 15.7,\ A = 46°30',\ C = 21°10'$

13.98 $b = 51.5,\ a = 62.5,\ B = 40°40'$. Ans. $c = 78.9,\ A = 52°20',\ C = 87°0',\ c' = 16.0,\ A' = 127°40',$ $C' = 11°40'$

13.99 $a = 320,\ c = 475,\ B = 35°20'$. Ans. $b = 552,\ B = 85°30',\ C' = 59°10',\ b' = 224,\ B' = 23°50',$ $C' = 120°50'$

13.100 $b = 120,\ c = 270,\ A = 118°40'$. Ans. $a = 344,\ B = 17°50',\ C = 43°30'$

13.101 $a = 24.5,\ b = 18.6,\ c = 26.4$. Ans. $A = 63°10',\ B = 42°40',\ C = 74°10'$

13.102 $a = 6.34,\ b = 7.30,\ c = 9.98$. Ans. $A = 39°20',\ B = 46°50',\ C = 93°50'$

13.103 Two ships have radio equipment with a range of 200 mi. One is 155 mi N 42°40′ E and the other is 165 mi N 45°10′ W of a shore station. Can the two ships communicate directly?

Ans. No; they are 222 mi apart.

13.104 A ship sails 15.0 mi on a course S 40°10′ W and then 21.0 mi on a course N 28°20′ W. Find the distance and direction of the last position from the first.

Ans. 20.9 mi, N 70°40′ W

13.105 A lighthouse is 10 mi northwest of a dock. A ship leaves the dock at 9 A.M. and steams west at 12 mi per hr. At what time will it be 8 mi from the lighthouse?

Ans. 9:16 A.M. and 9:54 A.M.

13.106 Two forces of 115 and 215 acting on an object have a resultant of magnitude 275. Find the angle between the directions in which the given forces act.

Ans. 70°50′

13.107 A tower 150 m high is situated at the top of a hill. At a point 650 m down the hill the angle between the surface of the hill and the line of sight to the top of the tower is $12°30'$. Find the inclination of the hill to a horizontal plane.

Ans. $7°50'$

13.108 Write the following in inverse function notation:

(*a*) $\sin\theta = \frac{3}{4}$, (*b*) $\cos\alpha = -1$, (*c*) $\tan x = -2$, (*d*) $\cot\beta = \frac{1}{2}$.

Ans. (*a*) $\theta = \arcsin\frac{3}{4}$ (*b*) $\alpha = \arccos(-1)$ (*c*) $x = \arctan(-2)$ (*d*) $\beta = \operatorname{arccos}\frac{1}{2}$

13.109 Find the value of each of the following:

(*a*) $\text{Arcsin}\,\sqrt{3}/2$ (*d*) $\text{Arccot}\,1$ (*g*) $\text{Arctan}\,(-\sqrt{3})$ (*j*) $\text{Arccsc}\,(-1)$
(*b*) $\text{Arccos}\,(-\sqrt{2}/2)$ (*e*) $\text{Arcsin}\,(-\frac{1}{2})$ (*h*) $\text{Arccot}\,0$
(*c*) $\text{Arctan}\,1/\sqrt{3}$ (*f*) $\text{Arccos}\,(-\frac{1}{2})$ (*i*) $\text{Arcsec}\,(-\sqrt{2})$

Ans. (*a*) $\pi/3$ (*b*) $3\pi/4$ (*c*) $\pi/6$ (*d*) $\pi/4$ (*e*) $-\pi/6$ (*f*) $2\pi/3$
(*g*) $-\pi/3$ (*h*) $\pi/2$ (*i*) $-3\pi/4$ (*j*) $-\pi/2$

13.110 Evaluate each the following:

(*a*) $\sin[\text{Arcsin}\,(-\frac{1}{2})]$ (*f*) $\sin(\text{Arccos}\,\frac{4}{5})$ (*k*) $\text{Arctan}\,(\cot 230°)$
(*b*) $\cos(\text{Arccos}\,\sqrt{3}/2)$ (*g*) $\cos[\text{Arcsin}\,(-\frac{12}{13})]$ (*l*) $\text{Arccot}\,(\tan 100°)$
(*c*) $\tan[\text{Arctan}\,(-1)]$ (*h*) $\sin(\text{Arctan}\,2)$ (*m*) $\sin(2\,\text{Arcsin}\,\frac{2}{3})$
(*d*) $\sin[\text{Arccos}\,(-\sqrt{3}/2)]$ (*i*) $\text{Arccos}\,(\sin 220°)$ (*n*) $\cos(2\,\text{Arcsin}\,\frac{3}{5})$
(*e*) $\tan(\text{Arcsin}\,0)$ (*j*) $\text{Arcsin}\,[\cos(-105°)]$ (*o*) $\sin(\frac{1}{2}\,\text{Arccos}\,\frac{4}{5})$

Ans. (*a*) $-\frac{1}{2}$ (*f*) $\frac{3}{5}$ (*k*) $40°$
(*b*) $\sqrt{3}/2$ (*g*) $\frac{5}{13}$ (*l*) $170°$
(*c*) -1 (*h*) $2/\sqrt{5}$ (*m*) $4\sqrt{5}/9$
(*d*) $\frac{1}{2}$ (*i*) $130°$ (*n*) $\frac{7}{25}$
(*e*) 0 (*j*) $-15°$ (*o*) $1/\sqrt{10}$

13.111 Show that

(*a*) $\sin(\text{Arcsin}\,\frac{5}{13} + \text{Arcsin}\,\frac{4}{5}) = \frac{63}{65}$

(*b*) $\cos(\text{Arccos}\,\frac{15}{17} - \text{Arccos}\,\frac{7}{25}) = \frac{297}{425}$

(*c*) $\sin\left(\text{Arcsin}\,\frac{1}{2} - \text{Arccos}\,\frac{1}{3}\right) = \frac{1-2\sqrt{6}}{6}$

(*d*) $\tan(\text{Arctan}\,\frac{3}{4} + \text{Arccot}\,\frac{15}{3}) = \frac{77}{36}$

(*e*) $\cos\left(\text{Arctan}\,\frac{-4}{3} + \text{Arcsin}\,\frac{12}{13}\right) = \frac{63}{65}$

(*f*) $\tan\left(\text{Arcsin}\,\frac{-3}{5} - \text{Arccos}\,\frac{5}{13}\right) = \frac{63}{16}$

(*g*) $\tan\left(2\,\text{Arcsin}\,\frac{4}{5} + \text{Arccos}\,\frac{12}{13}\right) = -\frac{253}{204}$

(*h*) $\sin(2\,\text{Arcsin}\,\frac{4}{5} - \text{Arccos}\,\frac{12}{13}) = \frac{323}{325}$

13.112 Show that

(*a*) $\text{Arctan}\,\frac{1}{2} + \text{Arctan}\,\frac{1}{3} = \frac{\pi}{4}$

(*b*) $\text{Arcsin}\,\frac{4}{5} + \text{Arctan}\,\frac{3}{4} = \frac{\pi}{4}$

(*c*) $\text{Arctan}\,\frac{4}{3} - \text{Arctan}\,\frac{1}{7} = \frac{\pi}{4}$

(*d*) $2\ \text{Arctan}\,\frac{1}{3} + \text{Arctan}\,\frac{1}{7} = \frac{\pi}{4}$

(*e*) $\text{Arccos}\,\frac{12}{13} + \text{Arctan}\,\frac{1}{4} = \text{Arccot}\,\frac{43}{32}$

(*f*) $\text{Arcsin}\,\frac{3}{5} + \text{Arcsin}\,\frac{15}{17} = \text{Arccos}\,\frac{-13}{85}$

(*g*) $\text{Arctan}\,\alpha + \text{Arctan}\,\frac{1}{\alpha} = \frac{\pi}{2} \quad (a > 0)$

13.113 Prove: The area of the segment cut from a circle of radius r by a chord at a distance d from the center is given by $K = r^2\,\text{Arccos}\,\frac{d}{r} - d\sqrt{r^2 - d^2}$.

13.114 Determine whether the following functions possess an inverse function:

(*a*) $y = 5x - 3$ (*b*) $y = \sqrt{x}$ (*c*) $y = x^4$ (*d*) $y = x^5 - 6$

Ans. (*a*) yes (*b*) yes (*c*) no (*d*) yes

Chapter 14

Exponential and Logarithmic Functions

THE LOGARITHM OF A POSITIVE NUMBER N to a given base b (written $\log_b N$) is the exponent of the power to which b must be raised to produce N. It will be understood throughout this chapter that b is positive and different from 1.

EXAMPLE 1.

(*a*) Since $9 = 3^2$, $\log_3 9 = 2$.
(*b*) Since $64 = 4^3, \log_4 64 = 3$.
(*c*) Since $64 = 2^6, \log_2 64 = 6$.
(*d*) Since $1000 = 10^3, \log_{10} 1000 = 3$.
(*e*) Since $0.01 = 10^{-2}, \log_{10} 0.01 = -2$.

(See Problems 14.1–14.3.) Note that if $f(x) = b^x$ and $g(x) = \log_b x$ (where $b > 0, b \neq 1$), Then $f(g(x)) = b\log_b x = x$, and $g(f(x)) = \log_b(b^x) = x$. Thus in, f and g are inverse functions.

FUNDAMENTAL LAWS OF LOGARITHMS

(*1*) The logarithm of the product of two or more positive numbers is equal to the sum of the logarithms of the several numbers. For example,

$$\log_b(P \cdot Q \cdot R \cdot) = \log_b P + \log_b Q + \log_b R$$

(*2*) The logarithm of the quotient of two positive numbers is equal to the logarithm of the dividend minus the logarithm of the divisor. For example,

$$\log_b \frac{P}{Q} = \log_b P - \log_b Q$$

(*3*) The logarithm of a power of a positive number is equal to the logarithm of the number, multiplied by the exponent of the power. For example,

$$\log_b P^n = n \log_b P$$

(*4*) The logarithm of a root of a positive number is equal to the logarithm of the number, divided by the index of the root. For example,

$$\log_b \sqrt[n]{P} = \frac{1}{n}\log_b P$$

(See Problems 14.4–14.7.)

IN NUMERICAL COMPUTATIONS a widely used base for a system of logarithms is 10. Such logarithms are called *common logarithms*. The common logarithm of a positive number $N/1$ is written $\log N$. These interact well with the decimal number system; for example, $\log 1000 = \log 10^3 = 3$.

AN EXPONENTIAL EQUATION is an equation involving one or more unknowns in an exponent. For example, $2^x = 7$ and $(1.03)^{-x} = 2.5$ are exponential equations. Such equations are solved by means of logarithms.

EXAMPLE 2. Solve the exponential equation $2^x = 7$.
Take logarithms of both sides: $x \log 2 = \log 7$
Solve for $x: x = \dfrac{\log 7}{\log 2} = \dfrac{0.8451}{0.3010} = 2.808$, approximately.
(See Problem 14.16.)

IN THE CALCULUS the most useful logarithmic function is the *natural logarithm* in which the base is a certain irrational number $e = 2.71828$, approximately.

The natural logarithm of N, $\ln N$, and the common logarithm of $N, \log N$, are related by the formula

$$\ln N \cong 2.3026 \log N$$

THE CALCULATOR can be used to do logarithmic calculation with extreme ease.

EXAMPLE 3. Evaluate log 82,734 rounded to six decimal places.

Press: [8] [2] [7] [3] [4] [log]. On screen: 4.917684.

EXAMPLE 4. Solve for x to four significant digits: $\ln x = -0.3916$. Press: .3916 [±] [e^x]. On screen: 0.6759745. Then $x \approx 0.6760$.
Note that we use the inverse function of $\ln x$, e^x, to find the number corresponding to a given $\ln x$. For $\log x$, we use its inverse, 10^x.

POWER FUNCTIONS in x are of the form x^n. If $n > 0$, the graph of $y = x^n$ is said to be of the *parabolic* type (the curve is a parabola for $n = 2$). If $n < 0$, the graph of $y = x^n$ is said to be of the *hyperbolic* type (the curve is a hyperbola for $n = -1$).

EXAMPLE 5 Sketch the graphs of (*a*) $y = x^{3/2}$, (*b*) $y = -x^{-3/2}$.

Table 14.1 has been computed for selected values of x. We shall assume that the points corresponding to intermediate values of x lie on a smooth curve joining the points given in the table. See Figs. 14-1 and 14-2. (See Problems 14.12–14.14.)

Table 14.1

x	$y = x^{3/2}$	$y = -x^{-3/2}$
9	27	$-\frac{1}{27}$
4	8	$-\frac{1}{8}$
1	1	-1
$\frac{1}{4}$	$\frac{1}{8}$	-8
$\frac{1}{9}$	$\frac{1}{27}$	-27
0	0	—

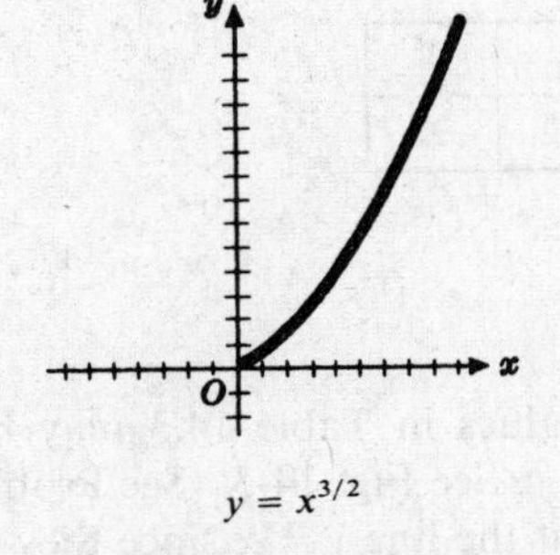

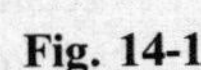

Fig. 14-1

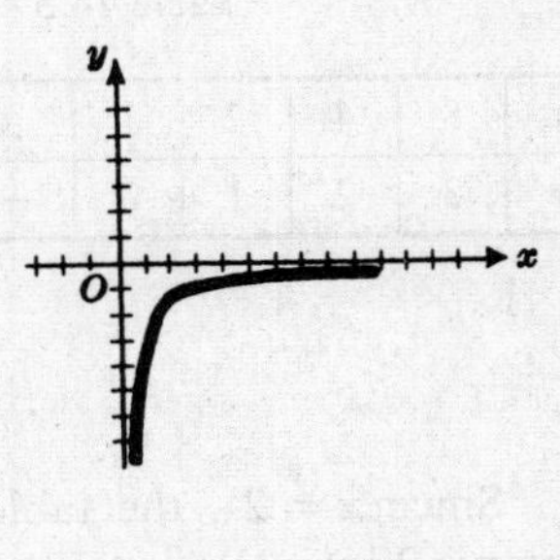

Fig. 14-2

EXPONENTIAL FUNCTIONS in x are of the form b^x where b is a constant. The discussion will be limited here to the case $b > 1$.

The curve whose equation is $y = b^x$ is called an *exponential curve*. The general properties of such curves are

(*a*) The curve passes through the point (0, 1).
(*b*) The curve lies above the x axis and has that axis as an asymptote.

EXAMPLE 6. Sketch the graphs of (*a*) $y = 2^x$, (*b*) $y = 3^x$.

Table 14.2

x	$y = 2^x$	$y = 3^x$
3	8	27
2	4	9
1	2	3
0	1	1
−1	$\frac{1}{2}$	$\frac{1}{3}$
−2	$\frac{1}{4}$	$\frac{1}{9}$
−3	$\frac{1}{8}$	$\frac{1}{27}$

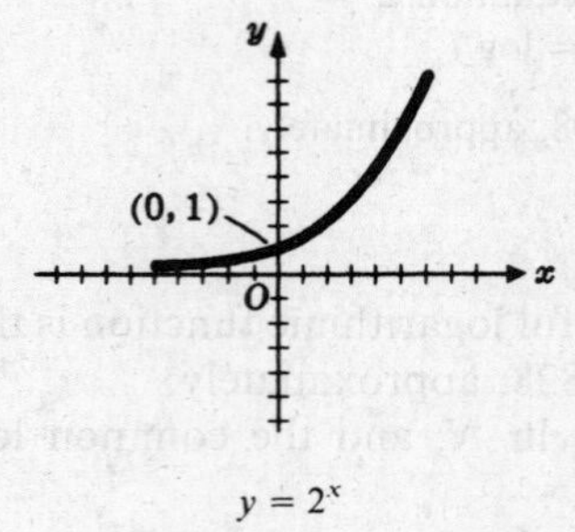

Fig. 14-3

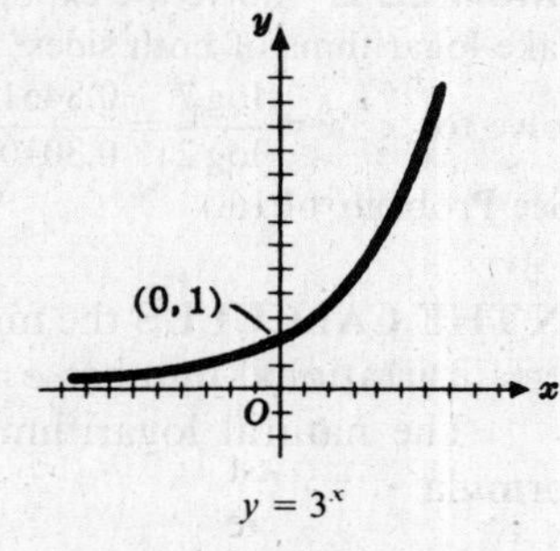

Fig. 14-4

The exponential equation appears frequently in the form $y = c\,e^{kx}$ where c and k are nonzero constants and $e = 2.71828\ldots$ is the natural logarithmic base. See Table 14.2 and Figs. 14-3 and 14-4. (See Problems 14.16 and 14.17.)

THE CURVE WHOSE EQUATION IS $y = \log_b x, b > 1$, is called a *logarithmic curve*. The general properties are

(*a*) The curve passes through the point (1, 0).
(*b*) The curve lies to the right of the y axis and has that axis as an asymptote.

EXAMPLE 7. Sketch the graph of $y = \log_2 x$.

Table 14.3

x	8	4	2	1	$\frac{1}{2}$	$\frac{1}{4}$	$\frac{1}{8}$
y	3	2	1	0	−1	−2	−3

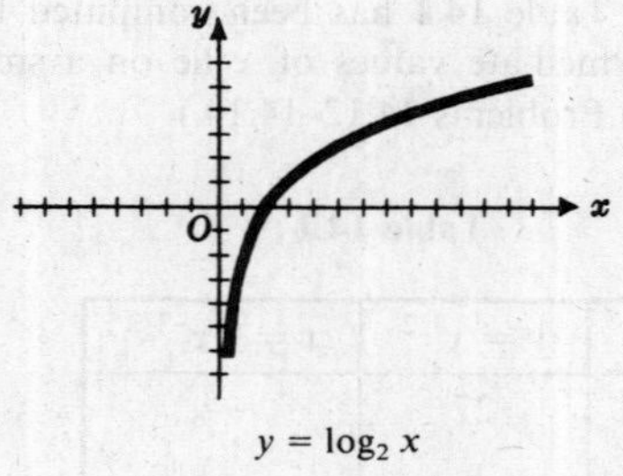

Fig. 14-5

Since $x = 2^y$, the table of values in Table 14.3 may be obtained from the table for $y = 2^x$ of Example 2 by interchanging x and y. See Fig. 14-5. (See Problem 14.18.) Note that the graphs of $y = 2^x$ and $y = \log_2 x$ are symmetric about the line $y = x$ since they are graphs of inverse functions. The same will be true for $y = a^x$ and $y = \log_a x$ for all $a > 0, a \neq 1$.

Exponential functions arise naturally in situations where the rate of change of something is proportional to the amount of that "something". For example, the more money you have in an interest-bearing account, the faster the growth of your savings will be. When the interest accumulates on a continuous basis, the **exponential growth** is of the general shape shown in Fig. 14-6 (the speed with which the curve rises reflects a very favorable interest rate).

The change in a physical quantity may either be in a positive or a negative direction, corresponding, respectively, to exponential growth and **exponential decay.** The previous example was one of growth.

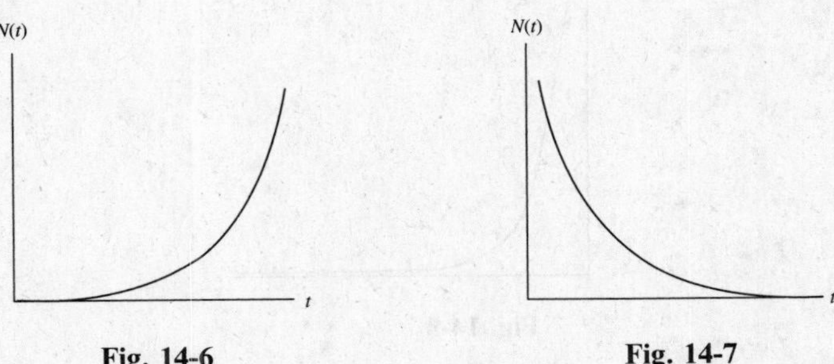

Fig. 14-6 **Fig. 14-7**

In contrast, Fig. 14-7 shows a pattern of exponential decay—a pattern in which something decreases relatively rapidly and then more slowly. Consider, e.g., the change in the speed of a book as it slides horizontally across a floor. The friction between the book and the floor will slow its velocity. The faster the book is going, the more molecules per second it will encounter that will slow its motion. Hence, in this situation, a fast moving object will slow down by a greater number of meters per second than the same object moving at a slower speed on the same surface. Qualitatively, the behavior is shown in Fig. 14-7. The speed decreases quickly—hence the steep negative slope of the curve while the object is moving quickly. Later, at slower speeds, there are fewer collisions per second with opposing molecules, and so the opposing force—and, hence, the decrease in velocity—is less.

Continuing with this example, Fig. 14-8 shows the slowing down of two identical books, "launched" with different initial velocities (all other conditions, such as the roughness of the surface, are assumed identical). Clearly, the faster-moving book loses more speed in the same time interval than the slower-moving book. Using calculus, one can derive the mathematical form of this curve. (As happens throughout physics, the mathematical model that we are presenting here is a simplified but useful approximation to the actual force, which, for the frictional force, is a complex interaction between the molecules on two surfaces depending upon many factors and ultimately requiring a quantum mechanical treatment.)

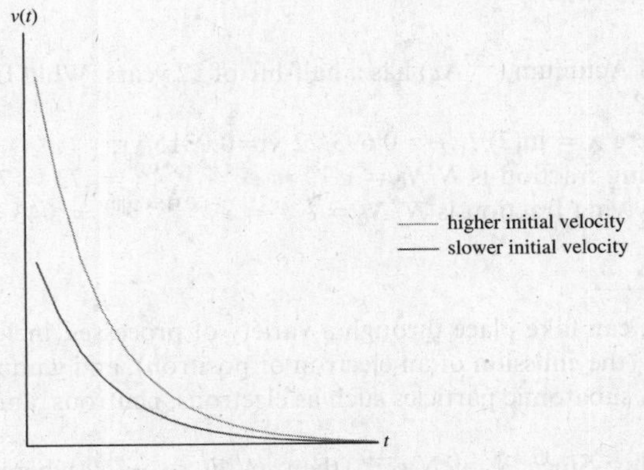

Fig. 14-8

RADIOACTIVE DECAY The decay of unstable elements (otherwise known as radioactive elements), provides an example of exponential decay. Atoms with the same number of protons but different neutrons are called isotopes and can decay into different isotopes of the same atom or into different atoms*.

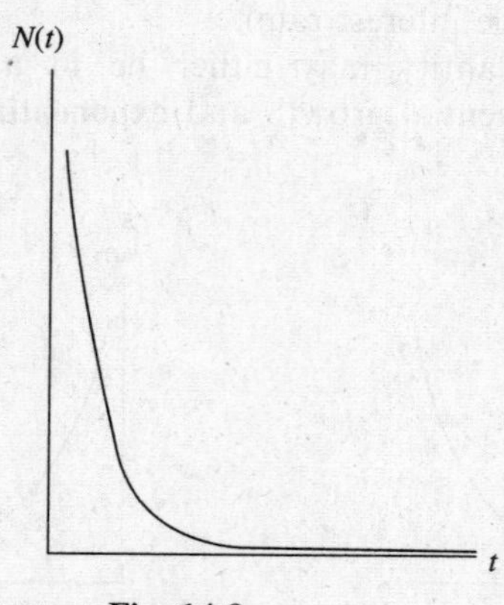

Fig. 14-9

According to the law of radioactive decay, the number of surviving atoms is given by

$$N = N_0 e^{-\lambda t}$$

where N_0 is the initial number of atoms, λ is the **decay constant** (in s^{-1}) and t is the elapsed time.

The original sample of atoms is reduced over time as they transform in such a manner that after a characteristic time (called the **half-life**), only one half of the original sample remains. After two half-lives, only one quarter (½ of ½) of the original sample will remain. After three half-lives, only 1/8 (½ of ½ of ½) of the original sample will survive, and so on. For example, if the half-life is 3 min, then 50% of the initial sample will remain after 3 min, 25% of the sample remains after 6 min, 12.5% remains after 9 min, and so on. These results are plotted in the graph.

The **activity** or **decay rate** R is the number of decays per second, which is the product of the decay constant λ and the number of radioactive nuclei N present at a particular time:

$$R = N\lambda$$

Recall that, in the beginning of this chapter, we described exponential behavior as describing a situation in which the rate of change of something at a particular time depends upon the quantity of that "something" present at that same time. In this case, the decay rate (or activity) depends upon the number of radioactive atoms at a particular time. As that number declines, the activity itself declines according to an exponential decay.†

EXAMPLE 8. A 1 g sample of Actinium (^{227}Ac) has a half-life of 22 years. What fraction of it will survive 10 years from now? 100 years from now?
We know that $N = N_0 e^{-\lambda t}$, where $\lambda = \ln(2)/t_{1/2} = 0.693/22$ yr=0.0315/yr;
so, for after 10 years the surviving fraction is $N/N_0 = e^{-\lambda t} = e^{-(.0315*10)} = .73 = 73\%$;
and for after 100 years, the surviving fraction is $N/N_0 = e^{-\lambda t} = e^{-(.0315*100)} = .043 = 4.3\%$.

* This decay, or transformation, can take place through a variety of processes, including alpha decay (the emission of a helium nucleus), beta decay (the emission of an electron or positron), and gamma decay (the emission of a very energetic photon). In both cases, subatomic particles such as electrons, neutrons, and photons may be emitted in the process.

† For those familiar with calculus: Since $N = N_0 e^{-\lambda t}$, then $dN/dt = -\lambda N$, where N refers to the number in the sample at any time t. Hence, the decay rate is directly proportional to the number in the sample.

CARBON DATING—AN APPLICATION OF RADIOACTIVE DECAY. Carbon occurs in two different isotopes. Its typical form is Carbon 12, which has a total of 12 protons and neutrons in its nucleus, but a radioactive form, Carbon 14 has 14 protons and neutrons. By measuring the relative amount of C^{14} and C^{12} in an animal or plant remain, we can date the specimen. The process relies upon the following facts:

(*1*) that bombardment of the atmosphere by high energy cosmic rays produces the C^{14} isotope, and that the amount of this isotope—and hence, its percentage of total carbon—has remained approximately constant over many thousands of years.
(*2*) that living things—animals and plants maintain this same ratio of C^{14}/C^{12} during their lives as they incorporate atmospheric carbon into their tissues.
(*3*) that once death occurs, the C^{14}/C^{12} ratio steadily decreases as C^{14} decays with a half-life of 5730 years

Hence, animal or plant remains can be dated according to their C^{14}/C^{12} ratio (or, equivalently, as discussed above, according to their activity). According to the law of radioactive decay,

$$N = N_0 e^{-\lambda t} \tag{14.1}$$

Since we know that, in the case of C^{14}, $t_{1/2} = 5730\ yr$,

$$\lambda = \ln(2)/t_{1/2} = .693/5730yr = 1.21 \times 10^{-4}/yr$$

EXAMPLE 9. A fossilized piece of wood has a C^{14}/C^{12} ratio that is 10% of that found in a living tree. Approximately how old is it?
In this case, we know the remaining fraction N/N_0 and need to find the elapsed time t. Rearranging (14.1) gives $e^{-\lambda t} = N/N_0$ so that $t = [\ln(N/N_0)]/\lambda$
Substituting $N/N_0 = 10\%$ and the computed value of λ in the above expression then yields:
$t = 19045$ yr. Hence, the sample is approximately 19000 years old.

POWER LAW BEHAVIOR. Exponential growth is a special example of the power law $y = Ax^n$.

Nuclear fission provides an important example of power law behavior. Under certain circumstances, a neutron can "split" an atomic nucleus, causing it to fragment into smaller nuclei, releasing energy as well as additional neutrons. These additional neutrons can split other nuclei, releasing more energy and neutrons and continuing the process of what is called a "chain reaction".

Suppose, e.g., during a particular time interval Δt, a neutron splits a nucleus, resulting in the liberation of energy E_0 and 2 additional neutrons. During the following time interval Δt, each of these 2 neutrons will each liberate 2 more neutrons and a total energy $2E_0$. The process is schematically indicated in Fig. 14-10. During each successive time interval, the number of neutrons and energy released will double over the previous one. While we are glossing over many important details in order to simplify the mathematics (and the physics), the essential idea is that—in this idealized model—the rate at which nuclei are split will follow a power law. In this particular example, the energy increases with time approximately according to $E = k \cdot 2^t$ with time, as indicated in Fig. 14-11 below*.

* This process does not continue indefinitely without limit because the density of neutrons required to continue the chain reaction is ultimately limited by either the explosive effects of energy production (as in nuclear weapons) or through deliberate intervention (as in the use of control rods in fission reactors).

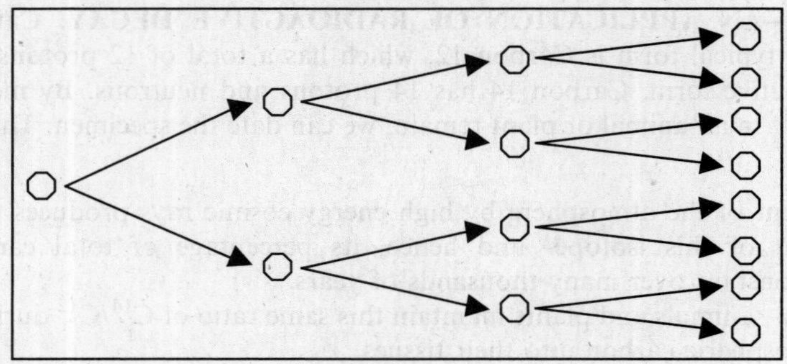

Fig. 14-10

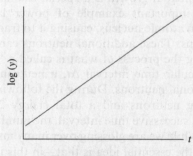

Fig. 14-11

One can show using the power law and the properties of logarithms that, for a power law $y = k \cdot N^t$ a plot of the logarithm of y as a function of t yields a linear relationship (Fig. 14-12).

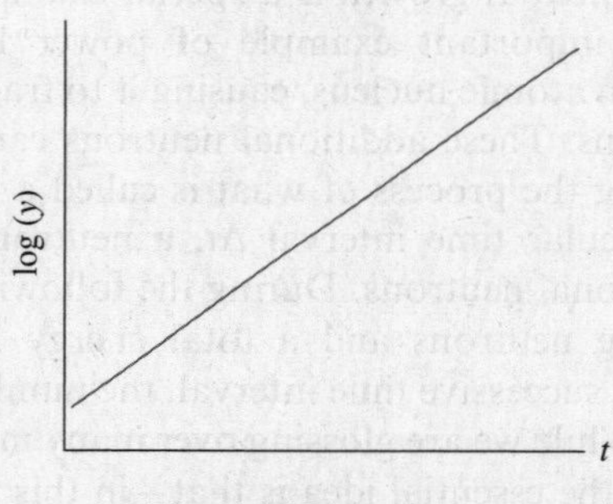

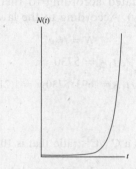

Fig. 14-12

Solved Problems

14.1 Change the following from exponential to logarithmic form:

(*a*) $7^2 = 49$, (*b*) $6^{-1} = \frac{1}{6}$, (*c*) $10^0 = 1$, (*d*) $4^0 = 1$, (*e*) $\sqrt[3]{8} = 2$.

Ans. (*a*) $\log_7 49 = 2$, (*b*) $log_6 \frac{1}{6} = -1$, (*c*) $\log_{10} 1 = 0$, (*d*) $log_4 = 0$, (*e*) $\log_8 2 = \frac{1}{3}$.

14.2 Change the following from logarithmic to exponential form:

(a) $\log_3 81 = 4$, (b) $\log_5 \frac{1}{625} = -4$, (c) $\log_{10} 10 = 1$, (d) $\log_9 27 = \frac{3}{2}$.

Ans. (a) $3^4 = 81$, (b) $5^{-4} = \frac{1}{625}$, (c) $10^1 = 10$, (d) $9^{3/2} = 27$.

14.3 Evaluate x, given:

(a) $x = \log_5 125$ (d) $x = \log_2 \frac{1}{16}$ (g) $\log_x \frac{1}{16} = -2$
(b) $x = \log_{10} 0.001$ (e) $x = \log_{1/2} 32$ (h) $\log_6 x = 2$
(c) $x = \log_8 2$ (f) $\log_x 243 = 5$ (i) $\log_a x = 0$

Ans. (a) 3, since $5^3 = 125$ (d) -4, since $2^{-4} = \frac{1}{16}$ (g) 4, since $4^{-2} = \frac{1}{16}$
(b) -3, since $10^{-3} = 0.001$ (e) -5, since $(\frac{1}{2})^{-5} = 32$ (h) 36, since $6^2 = 36$
(c) $\frac{1}{3}$, since $8^{1/3} = 2$ (f) 3, since $3^5 = 243$ (i) 1, since $a^0 = 1$

14.4 Prove the four laws of logarithms.

Let $P = b^p$ and $Q = b^q$; then $\log_b P = p$ and $\log_b Q = q$.

(1) Since $P \cdot Q = b^p \cdot b^q = b^{p+q}$, $\log_b PQ = p + q = \log_b P + \log_b Q$; that is, the logarithm of the product of two positive numbers is equal to the sum of the logarithms of the numbers.

(2) Since $P/Q = b^p/b^q = b^{p-q}$, $\log_b(P/Q) = p - q = \log_b P - \log_b Q$; that is, the logarithm of the quotient of two positive numbers is the logarithm of the numerator minus the logarithm of the denominator.

(3) Since $P^n = (b^P)^n = b^{np}$, $\log_b P^n = np = n \log_b P$; that is, the logarithm of a power of a positive number is equal to the product of the exponent and the logarithm of the number.

(4) Since $\sqrt[n]{P} = P^{1/n} = b^{p/n}$, $\log_b \sqrt[n]{P} = \dfrac{p}{n} = \dfrac{1}{n}\log_b P$; that is, the logarithm of a root of a positive number is equal to the logarithm of the number divided by the index of the root.

14.5 Express the logarithms of the given expressions in terms of the logarithms of the individual letters involved.

(a) $\log_b \dfrac{P \cdot Q}{R} = \log_b(P \cdot Q) - \log_b R = \log_b P + \log_b Q - \log_b R$

(b) $\log_b \dfrac{P}{Q \cdot R} = \log_b P - \log_b(Q \cdot R) = \log_b P - (\log_b Q + \log_b R) = \log_b P - \log_b Q - \log_b R$

(c) $\log_b P^2 \cdot \sqrt[3]{Q} = \log_b P^2 + \log_b \sqrt[3]{Q} = 2\log_b P + \frac{1}{3}\log_b Q$

(d) $$\log_b \sqrt{\frac{P \cdot Q^3}{R^{1/2} \cdot S}} = \frac{1}{2}\log_b \frac{P \cdot Q^3}{R^{1/2} \cdot S} = \tfrac{1}{2}[\log_b(P \cdot Q^3) - \log_b(R^{1/2} \cdot S)]$$
$$= \tfrac{1}{2}(\log_b P + 3\log_b Q - \tfrac{1}{2}\log_b R - \log_b S)$$

14.6 Express each of the following as a single logarithm:

(a) $\log_b x - 2\log_b y + \log_b z = (\log_b x + \log_b z) - 2\log_b y = \log_b xz - \log_b y^2 = \log_b \dfrac{xz}{y^2}$

(b) $$\log_b 2 + \log_b \pi + \tfrac{1}{2}\log_b l - \tfrac{1}{2}\log_b g = (\log_b 2 + \log_b \pi) + \tfrac{1}{2}(\log_b l - \log_b g)$$
$$= \log_b(2\pi) + \tfrac{1}{2}\log_b \frac{l}{g} = \log_b\left(2\pi\sqrt{\frac{l}{g}}\right)$$

14.7 Show that $b^{3\log_b x} = x^3$.

Let $3\log_b x = t$. Then $\log_b x^3 = t$ and $x^3 = b^t = b^{3\log_b x}$. Alternatively, $b^{3\log_b x} = b^{\log_b x^3} = x^3$ (using the inverse function property).

14.8 Verify that

(a) $\log 3860 = 3.5866$
(b) $\log 52.6 = 1.7210$
(c) $\log 7.84 = 0.8943$
(d) $\log 728\,000 = 5.8621$
(e) $\log 5.463 = 0.7374$
(f) $\log 77.62 = 1.8900$
(g) $\log 2.866 = 0.4573$

14.9 Find (*a*) $\log 2.864^3$, (*b*) $\log \sqrt{2.864}$.

Since $\log 2.864 = 0.4570$,

(*a*) $\log 2.864^3 = 3\log 2.864 = 3(0.4570) = 1.3710$

(*b*) $\log \sqrt{2.864} = \frac{1}{2}\log 2.864 = \frac{1}{2}(0.4570) = 0.2285$

14.10 Find the number whose log is 1.4232. Using a calculator, press 1.4232 $\boxed{10^x}$. The result on the screen is 26.50 (rounded to two places).

14.11 Solve.

(*a*) $(1.06)^x = 3$.
Taking logarithms, $x \log 1.06 = \log 3$.

$$x = \frac{\log 3}{\log 1.06} = \frac{0.4771}{0.0253} \qquad x = 18.86$$

(*b*) $12^{2x+5} = 55(7^{3x})$.
Taking logarithms, $(2x+5)\log 12 = \log 55 + 3x \log 7$

$$2x \log 12 - 3x \log 7 = \log 55 - 5\log 12$$

$$x = \frac{\log 55 - 5\log 12}{2\log 12 - 3\log 7} = \frac{1.7404 - 5(1.0792)}{2(1.0792) - 3(0.8451)} = \frac{3.6556}{0.3769} = 9.700$$

(*c*) $41.2^x = 12.6^{x-1}$.
Taking logarithms, $x \log 41.2 = (x-1)\log 12.6$.

$$x \log 41.2 - x \log 12.6 = -\log 12.6 \quad \text{or} \quad x = \frac{-\log 12.6}{\log 41.2 - \log 12.6}$$

$$y = -x = \frac{\log 12.6}{\log 41.2 - \log 12.6} = \frac{1.1004}{0.5145} = 2.138$$

$$x = -2.138$$

14.12 Sketch the graph of the *semicubic parabola* $y^2 = x^3$.

Since the given equation is equivalent to $y = \pm x^{3/2}$, the graph consists of the curve of Example 1(*a*) together with its reflection in the x axis. See Fig. 14-13.

14.13 Sketch the graph of $y^3 = x^2$.

Refer to Fig. 14-14 and Table 14.4.

Table 14.4

x	± 3	± 2	± 1	0
y	2.1	1.6	1	0

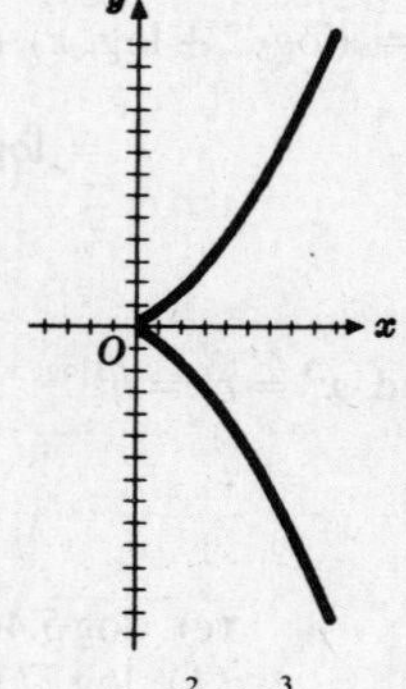

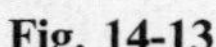

$y^2 = x^3$

Fig. 14-13

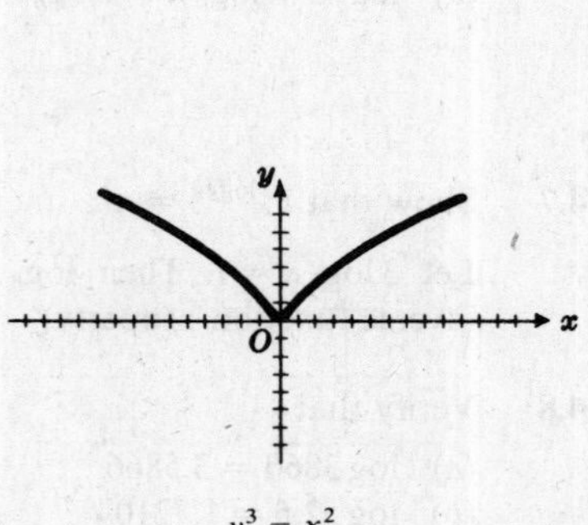

$y^3 = x^2$

Fig. 14-14

14.14 Sketch the graph of $y = x^{-2}$.

See Fig 14-15 and Table 14.5.

Table 14.5

x	± 3	± 2	± 1	$\pm \frac{1}{2}$	$\pm \frac{1}{4}$	0
y	$\frac{1}{9}$	$\frac{1}{4}$	1	4	16	—

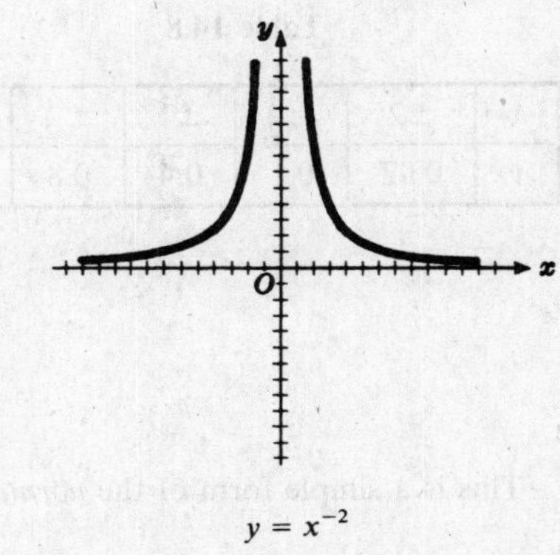

Fig. 14-15

14.15 Sketch the graph of $y = 3^{-x}$.

See Fig. 14-16 and Table 14.6.
Note that the graph of $y = b^{-x}$ is a reflection in the y axis of the graph of $y = b^x$.

Table 14.6

x	3	2	1	0	−1	−2	−3
y	$\frac{1}{27}$	$\frac{1}{9}$	$\frac{1}{3}$	1	3	9	27

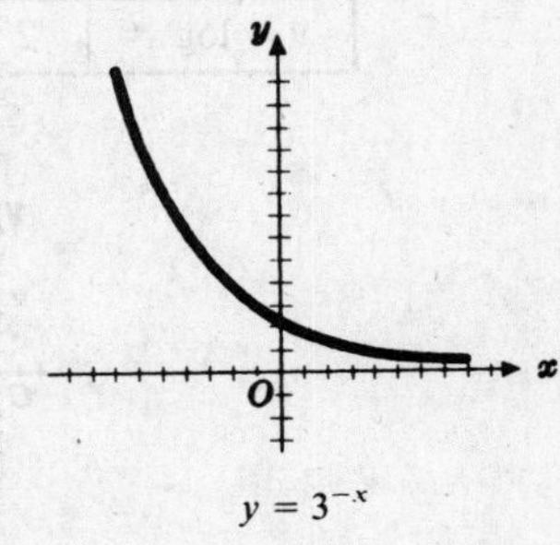

Fig. 14-16

14.16 Sketch the graph of $y = e^{2x}$.

See Fig. 14-17 and Table 14.7.

Table 14.7

x	2	1	$\frac{1}{2}$	0	$-\frac{1}{2}$	−1	−2
$y = e^{2x}$	54.6	7.4	2.7	1	0.4	0.14	0.02

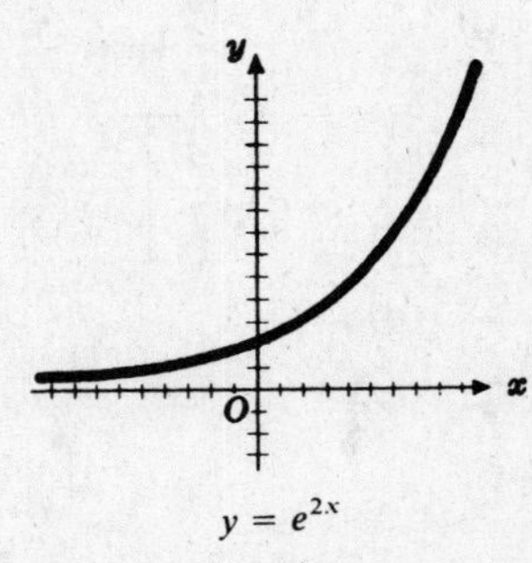

Fig. 14-17

14.17 Sketch the graph of $y = e^{-x^2}$.

Refer to Fig. 14-18 and Table 14.8.

Table 14.8

x	± 2	$\pm\frac{3}{2}$	± 1	$\pm\frac{1}{2}$	0
y	0.02	0.1	0.4	0.8	1

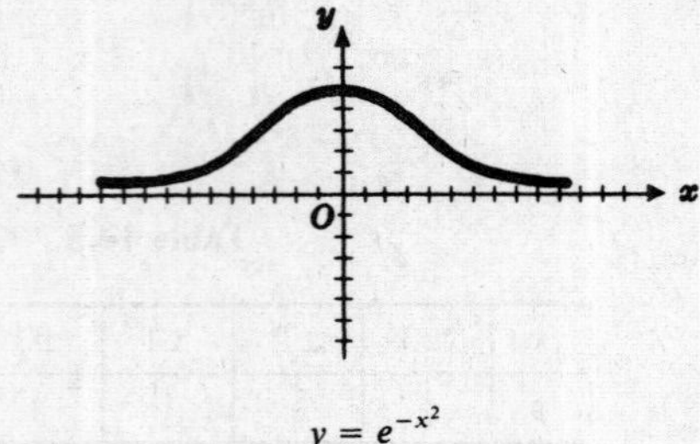

$y = e^{-x^2}$

Fig. 14-18

This is a simple form of the *normal probability curve* used in statistics.

14.18 Sketch the graphs of (*a*) $y = \log x$, (*b*) $y = \log x^2 = 2\log x$.

See Table 14.9 and Figs. 14-19 and 14-20.

Table 14.9

x	10	5	4	3	2	1	0.5	0.25	0.1	0.01
$y = \log x$	1	0.7	0.6	0.5	0.3	0	−0.3	−0.6	−1	−2
$y = \log x^2$	2	1.4	1.2	1	0.6	0	−0.6	−1.2	−2	−4

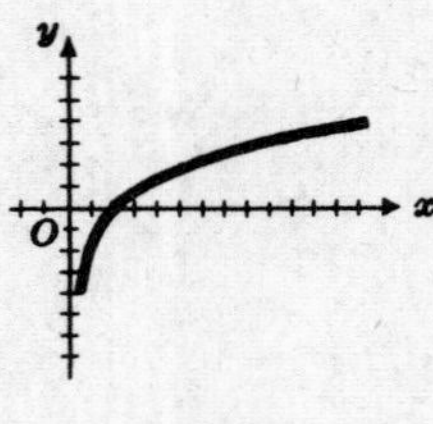

$y = \log x$

Fig. 14-19

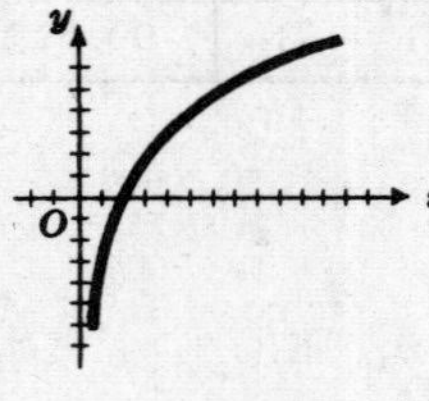

$y = \log x^2$

Fig. 14-20

14.19 Polonium has a half-life of approximately 3 min. Plot the fraction of original atoms that survives as a function of time.

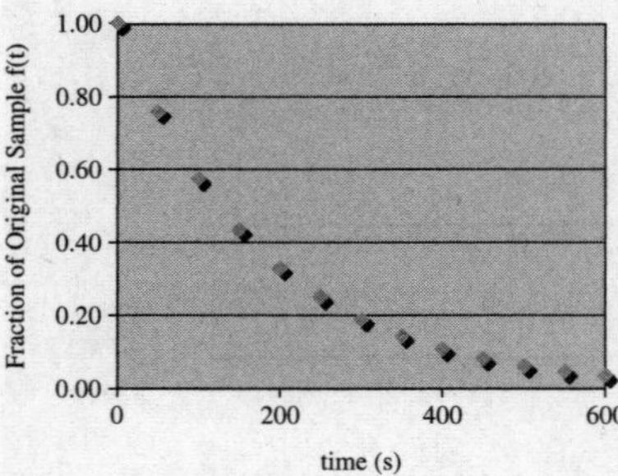

Fig. 14-21

14.20 A capacitor in a circuit charges according to the law:

$$V(t) = V_0(1 - e^{-t/RC})$$

What fraction of the limiting voltage V_0 across the capacitor is reached after an elapsed time equal to:
(*a*) one time constant?
(*b*) two time constants?
(*c*) five time constants?

(*a*) $V(t) = V_0(1 - e^{-t/RC}) = V_0(1 - e^{-RC/RC}) = V_0(1 - e^{-1}) = V_0(1 - .367) = .632V_0$
(*b*) $V(t) = V_0(1 - e^{-t/RC}) = V_0(1 - e^{-2\,RC/RC}) = V_0(1 - e^{-2}) = V_0(1 - .135) = .865V_0$
(*c*) $V(t) = V_0(1 - e^{-t/RC}) = V_0(1 - e^{-5\,RC/RC}) = V_0(1 - e^{-5}) = V_0(1 - .00673) = .993V_0$

14.21 Find the activity of a sample of 10^{21} atoms of Cobalt-60 (Co^{60}) (half-life=5.26 yr).

Recall that the activity R is the number of decays per second, which is the product of the decay rate λ and the number of radioactive nuclei N:

$$R = N\lambda$$

From the conversion between the half-life and the decay rate (see previous problem):

$$\lambda = \ln(2)/t_{1/2}$$

It is convenient to express the half-life in seconds:

$$t_{1/2} = \frac{5.26y}{} \times \frac{365.25d}{y} \times \frac{24h}{d} \times \frac{3600\text{ s}}{h} = 1.65 \times 10^8\text{ s}$$

[a useful mnemonic in situations such as these is to remember that $1y \approx \pi \times 10^7\text{ s} = 3.14 \times 10^7\text{ s}$]

$$\lambda = \frac{.693}{1.65 \times 10^8\text{ s}} = 4.2 \times 10^{-9}\text{ s}^{-1}$$

Therefore, the activity is:

$$R = (4.2 \times 10^{-9}\text{ s}^{-1})(10^{21}\,decays) = 4.2 \times 10^{12}\text{ s}^{-1}$$

14.22 The current flowing through an electrical circuit is given by the equation $I = I_0(1 - e^{-t/\tau})$, where I_0 is a constant (measured in A, the SI unit of current), t is the time in s (after a switch is thrown), and τ is a constant characteristic of the circuit. τ has the dimensions of seconds and is called a "time constant".

(*a*) Plot the growth in current as a function of time (where time is expressed in units of the time constant τ).
(*b*) What is the limiting value of the current?
(*c*) After what length of time will the current reach 90% of its final value?
(*d*) After what length of time will the current reach 99% of its final value?

(*a*) [In units of I_0 for the vertical axis, we have:]

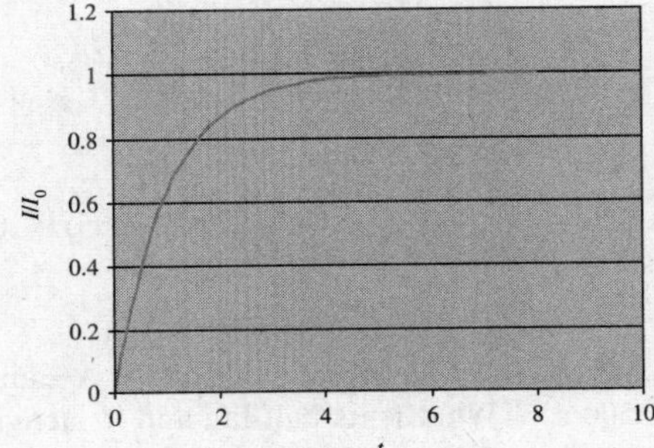

Fig. 14-22

(*b*) As t increases without limit, the value of exp($-t/\tau$) approaches 0. Therefore, the limiting value is $I=I_0$

(*c*) We need to compute the elapsed time as a function of the fraction of the initial current remaining (I/I_0)

$$I = I_0(1 - e^{-t/\tau})$$
$$\frac{I}{I_0} = 1 - e^{-t/\tau}$$
$$e^{-t/\tau} = 1 - \frac{I}{I_0}$$
$$-t/\tau = \ln\left(1 - \frac{I}{I_0}\right)$$
$$t = -\tau \ln\left(1 - \frac{I}{I_0}\right)$$

When I/I_0 is 90%,

$$t = -\tau \ln(1-0.9) = -\tau \ln(0.1) = 2.30\,\tau$$

(*d*) When $I/I_0$is 99%,

$$t = -\tau \ln(1 - 0.99) = -\tau \ln(0.01) = 4.61\,\tau$$

14.23 A flashbulb in an RC circuit discharges with a time constant equal to 5E-03 s. If the resistance of the circuit is 300 Ω, what is the circuit capacitance?

We know that $t = RC$, and so $C = t/R = \dfrac{5 \times 10^{-3}\text{ s}}{300\,\Omega} = 1.67 \times 10^{-5}\text{ F} = 16.7\,\mu\text{F}$

14.24 A mass on a spring oscillates according to the following displacement:

$$y = y_0 e^{-bt} \sin(\omega t)$$

where $y_0 = 25$ cm, $\omega = .1\text{ s}^{-1}$ and $b = .05\text{ s}^{-1}$

(*a*) What is the displacement after 10 s?

(*b*) What is the ratio between the displacement at 10 s and that at 20 s?

(*a*) $y_{10} = y_0 e^{-bt}\sin(\omega t) = (.25\text{ m})e^{-(.05\text{ s}^{-1}(10\text{ s}))}\sin(.1\text{ s}^{-1} \times 10\text{ s}) = .128$

(*b*) $y_{20} = y_0 e^{-bt}\sin(\omega t) = (.25\text{ m})e^{-(.05\text{ s}^{-1}(20\text{ s}))}\sin(.1\text{ s}^{-1} \times 20\text{ s}) = .0836$

Therefore, $\dfrac{y_{20}}{y_{10}} = \dfrac{.128}{.0836} = 1.53$

14.25 A charging capacitor in an RC reaches 95% of its steady-state value in 10 s. The resistor has a value of 500 Ω. What is the value of the capacitor?

We know that, for a charging capacitor, $V = V_0\,(1 - e^{-t/RC})$

Therefore, $\frac{V}{V_0} = 1 - e^{-t/RC}$ and $e^{-t/RC} = 1 - \frac{V}{V_0}$

Rearranging terms, we get:

$$\frac{-t}{RC} = \ln\left(1 - \frac{V}{V_0}\right)$$

Isolating C, we get

$$C = \frac{-t}{R}\frac{1}{\ln\left(1 - \frac{V}{V_0}\right)} = \left(\frac{-10}{500}\right)\ln(1. - .95) = .060\text{ F}$$

14.26 *[Note: this problem requires knowledge of elementary calculus].* A sample of 10^{20} atoms of a radioactive substance has a decay rate of 500 s^{-1}. What is its half-life and its activity?

For radioactive decay, we assume that each atom decays in a random fashion.

The total number of atoms decaying per second, dN/dt, is proportional to the number of atoms in the sample. Hence, $dN/dt = -\lambda N$

Therefore $dN/N = -\lambda t$

$$\ln(N) = -\lambda t + c$$

$$N = e^{-\lambda t + c} = e^c e^{-\lambda t}$$

At t = 0, we know that $N = N_0$, so that $N_0 = e^c$. Therefore:

$$N = N_0 e^{-\lambda t}$$

We would like to convert this expression—involving the decay rate λ—to one involving the half-life $t_{1/2}$. To do so, we note that the half-life is the time at which ½ of the sample remains:

$$N_0/2 = N_0 e^{-\lambda t}$$

$$1/2 = e^{-\lambda t_{1/2}}$$

$$\ln(1/2) = -\lambda t_{1/2}$$

$$-\ln(2) == -\lambda t_{1/2}$$

Therefore, the half-life is related to the decay rate through:

$$t_{1/2} = \ln(2)/\lambda = 0.693/\lambda = 0.1386\,\text{s}^{-1}$$

Its activity $R = \lambda N_0 = 10^{20}(.139\,\text{s}^{-1}) = 1.39 \times 10^{19}\,\text{s}^{-1}7$

Supplementary Problems

14.27 Solve for x using a calculator.

(*a*) $3^x = 30$ (*b*) $1.07^x = 3$ (*c*) $5.72^x = 8.469$ (*d*) $38.5^x = 6.5^{x-2}$

Ans. (*a*) 3.096 (*b*) 16.23 (*c*) 1.225 (*d*) −2.104

14.28 Show that $10^{n \log(a+b)} = (a+b)^n$.

14.29 Find (*a*) $\log 3.64$, (*b*) $\log 36.4$, (*c*) $\log 364$.

14.30 Find the number whose natural logarithm is (*a*) 10, (*b*) 130, (*c*) 407.1.

14.31 Sketch the graphs of (*a*) $y^2 = x^{-3}$, (*b*) $y^3 = x^{-2}$, (*c*) $y^2 = \frac{1}{x}$, (*d*) the cubical parabola $y = x^3$.

14.32 Sketch the graphs of

(*a*) $y = (2.5)^x$ (*c*) $y = 2^{-1/x}$ (*e*) $y = e^{x/2}$ (*g*) $y = e^{x+2}$

(*b*) $y = 2^{x+1}$ (*d*) $y = \frac{1}{2}e^x$ (*f*) $y = e^{-x/2}$ (*h*) $y = xe^{-x}$

14.33 Sketch the graphs of (*a*) $y = \frac{1}{2}\log x$, (*b*) $y = \log(3x+2)$, (*c*) $y = \log(x^2+1)$.

14.34 Show that the curve $y^q = x^p$, where p and q are positive integers, lies entirely in

(*a*) Quadrants I and III if p and q are both odd

(*b*) Quadrants I and IV if p is odd and q is even

(*c*) Quadrants I and II if p is even and q is odd

14.35 Show that the curve $y^q = x^{-p}$, where p and q are positive integers, lies entirely in

(*a*) Quadrants I and III if p and q are both odd

(*b*) Quadrants I and II if p is even and q is odd

(*c*) Quadrants I and IV if p is odd and q is even

14.36 The half-life of radium-228 is 5.75 years. Approximately how many years will elapse until a 1 g sample of radium decays to 1% of its original mass?

Ans. 11.5 s

14.37 Gamma rays – high energy electromagnetic radiation – is absorbed in matter according to the following law:

$$I = I_0 e^{-\mu x}$$

where I_0 is the initial beam intensity and μ is a constant that depends upon the material. What thickness of aluminum is required to reduce an initial gamma ray beam intensity to ¼ of its original value? ($\mu = .075/\text{cm}$ in aluminum).

Ans. 18.5 cm

14.38 The entropy S is related to the number of accessible states in a system through the relation $S = \text{k}\ \ln(W)$, where W is the number of accessible states and k is Boltzmann's constant ($=1.38 \times 10^{-23}$ J/K). Find the number of states for a system with an entropy of 5.3×10^{-22} J/K.

Ans. 1.02×10^6

14.39 If a 10 μF capacitor carrying a potential difference of 25 V is discharged through a 1 KΩ resistor at time t = 0, how long will it take for the potential difference across the capacitor to drop to 1 V?

Ans. 0.032 s

14.40 For distances greater than about 1 Fermi (1 fm), the nuclear force can be described by the Yukawa potential

$$U(r) = -g^2 \frac{e^{-r/R}}{r}$$

where R, the range, is equal to about 2 fm. How much stronger is the potential at 1 fm than at 10 fm?

Ans. 11 times stronger

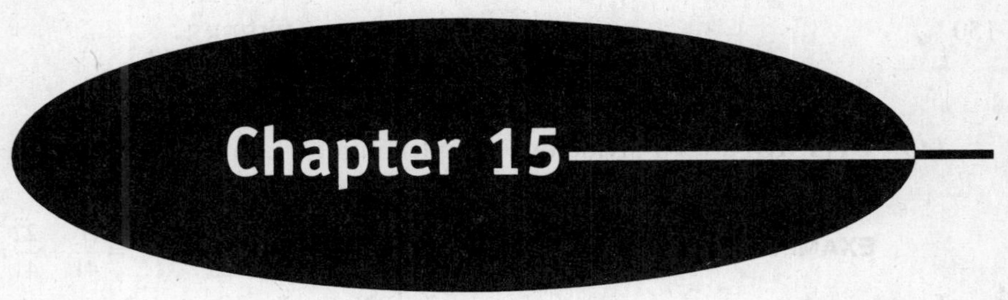

Complex Numbers

PURE IMAGINARY NUMBERS. The square root of a negative number (i.e., $\sqrt{-1}, \sqrt{-5}, \sqrt{-9}$) is called a *pure imaginary number*. Since by definition $\sqrt{-5} = \sqrt{5} \cdot \sqrt{-1}$ and $\sqrt{-9} = \sqrt{9} \cdot \sqrt{-1} = 3\sqrt{-1}$, it is convenient to introduce the symbol $i = \sqrt{-1}$ and to adopt $\sqrt{-5} = i\sqrt{5}$ and $\sqrt{-9} = 3i$ as the standard form for these numbers.

The symbol i has the property $i^2 = -1$, and for higher integral powers we have $i^3 = i^2 \cdot i = (-1)i = -i$, $i^4 = (i^2)^2 = (-1)^2 = 1$, $i^5 = i^4 \cdot i = i$, etc.

The use of the standard form simplifies the operation on pure imaginaries and eliminates the possibility of certain common errors. Thus, $\sqrt{-9} \cdot \sqrt{4} = \sqrt{-36} = 6i$ since $\sqrt{-9} \cdot \sqrt{4} = 3i(2) = 6i$ but $\sqrt{-9} \cdot \sqrt{-4} \neq \sqrt{36}$ since $\sqrt{-9} \cdot \sqrt{-4} = (3i)(2i) = 6i^2 = -6$.

Notice the cyclic nature of the powers of i. i^m equals $i, -1, -i, 1$ for every natural number, $m \cdot i^6 = -1, i^7 = -i, i^8 = 1$, etc.

COMPLEX NUMBERS. A number $a + bi$, where a and b are real numbers, is called a *complex number*. The first term a is called the *real part* of the complex number and the second term bi is called the *pure imaginary part*.

Complex numbers may be thought of as including all real numbers and all pure imaginary numbers. For example, $5 = 5 + 0i$ and $3i = 0 + 3i$.

Two complex numbers $a + bi$ and $c + di$ are said to be *equal* if and only if $a = c$ and $b = d$.

The *conjugate* of a complex number $a + bi$ is the complex number $a - bi$. Thus, $2 + 3i$ and $2 - 3i$, $-3 + 4i$ and $-3 - 4i$ are pairs of conjugate complex numbers.

ALGEBRAIC OPERATIONS

(*1*) **ADDITION**. To add complex numbers, add the real parts and add the pure imaginary parts.

EXAMPLE 1. $(2 + 3i) + (4 - 5i) = (2 + 4) + (3 - 5)i = 6 - 2i$.

(*2*) **SUBTRACTION**. To subtract two complex numbers, subtract the real parts and subtract the pure imaginary parts.

EXAMPLE 2. $(2 + 3i) - (4 - 5i) = (2 - 4) + [3 - (-5)]i = -2 + 8i$.

(*3*) **MULTIPLICATION**. To multiply two complex numbers, carry out the multiplication as if the numbers were ordinary binomials and replace i^2 by -1.

EXAMPLE 3. $(2 + 3i)(4 - 5i) = 8 + 2i - 15i^2 = 8 + 2i - 15(-1) = 23 + 2i$.

(*4*) **DIVISION.** To divide two complex numbers, multiply both numerator and denominator of the fraction by the conjugate of the denominator.

EXAMPLE 4. $\dfrac{2+3i}{4-5i}=\dfrac{(2+3i)(4+5i)}{(4-5i)(4+5i)}=\dfrac{(8-15)+(10+12)i}{16+25}=\dfrac{7}{41}+\dfrac{22}{41}i.$

$\left[\text{Note the form of the result; it is neither } \dfrac{-7+22i}{41} \text{ nor } \dfrac{1}{41}(-7+22i).\right]$

(See Problems 15.1–15.9.)

GRAPHIC REPRESENTATION OF COMPLEX NUMBERS. The complex number $x+yi$ may be represented graphically by the point P (see Fig. 15-1) whose rectangular coordinates are (x, y).

The point O, having coordinates (0, 0), represents the complex number $0+0i=0$. All points on the x axis have coordinates of the form $(x, 0)$ and correspond to real numbers $x+0i=x$. For this reason, the x axis is called the *axis of reals*. All points on the y axis have coordinates of the form $(0, y)$ and correspond to pure imaginary numbers $0+yi=yi$. The y axis is called the *axis of imaginaries*. The plane on which the complex numbers are represented is called the *complex plane*. See Fig. 15-1.

In addition to representing a complex number by a point P in the complex plane, the number may be represented by the directed line segment or vector OP. See Fig. 15-2. The vector OP is sometimes denoted by $\overrightarrow{OP}$ and is the directed line segment beginning at O and terminating at P.

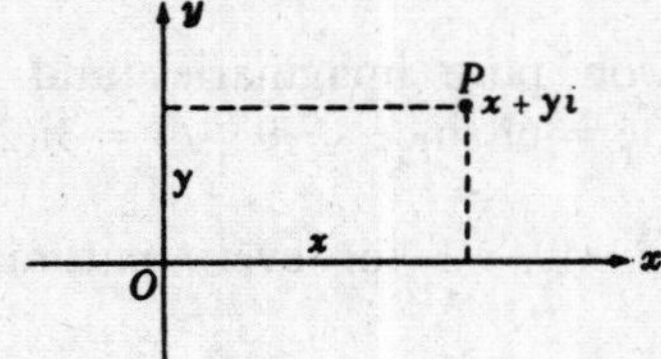

Fig. 15-1

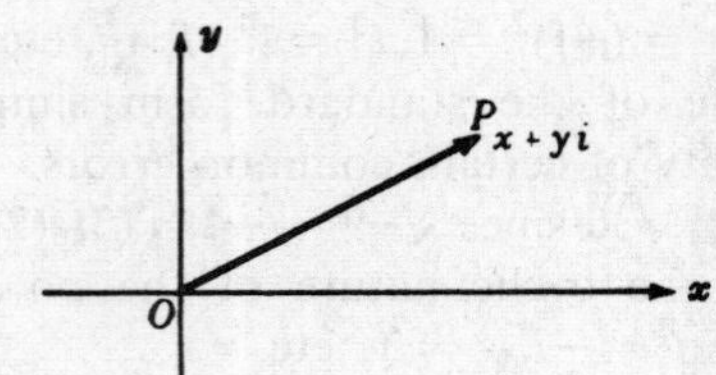

Fig. 15-2

GRAPHIC REPRESENTATION OF ADDITION AND SUBTRACTION. Let $z_1=x_1+iy_1$ and $z_2=x_2+iy_2$ be two complex numbers. The vector representation of these numbers suggests the illustrated parallelogram law for determining graphically the sum $z_1+z_2=(x_1+iy_1)+(x_2+iy_2)$, since the coordinates of the endpoint of the vector z_1+z_2 must be, for each of the x coordinates and the y coordinates, the sum of the corresponding x or y values. See Fig. 15-3.

Since $z_1-z_2=(x_1+iy_1)-(x_2+iy_2)=(x_1+iy_1)+(-x_2-iy_2)$, the difference z_1-z_2 of the two complex numbers may be obtained graphically by applying the parallelogram law to x_1+iy_1 and $-x_2-iy_2$. (See Fig. 15-4.)

In Fig. 15-5 both the sum $OR=z_1+z_2$ and the difference $OS=z_1-z_2$ are shown. Note that the segments $\overline{OS}$ and $\overline{P_2P_1}$ (the other diagonal of OP_2RP_1) are congruent. (See Problem 15.11.)

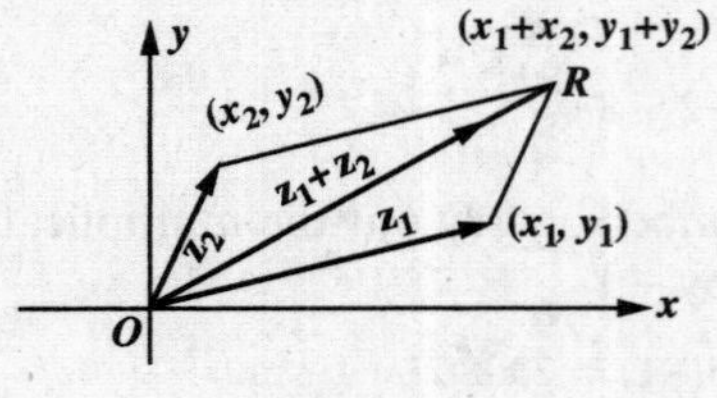

Fig. 15-3

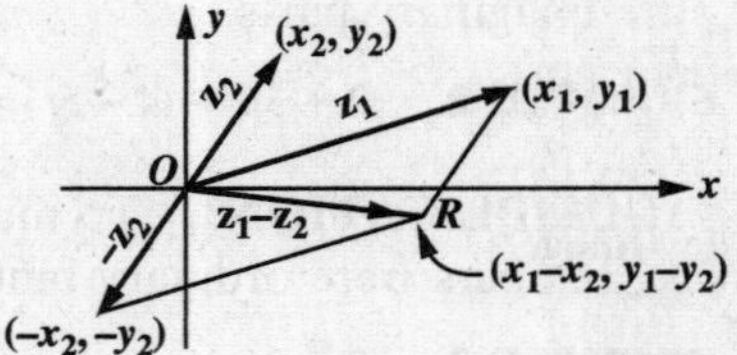

Fig. 15-4

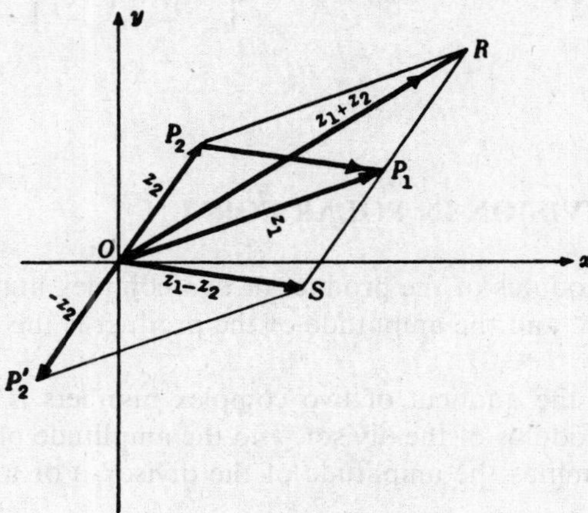

Fig. 15-5

POLAR OR TRIGONOMETRIC FORM OF COMPLEX NUMBERS. Let the complex number $x+yi$ be represented (Fig. 15-6) by the vector OP. This vector (and hence the complex number) may be described in terms of the length r of the vector and *any* positive angle θ which the vector makes with the positive x axis (axis of positive reals). The number $r=\sqrt{x^2+y^2}$ is called the *modulus* or *absolute value* of the complex number. The angle θ, called the *amplitude* of the complex number, is usually chosen as the smallest, positive angle for which $\tan\theta = y/x$ but at times it will be found more convenient to choose some other angle coterminal with it.

From Fig. 15-6, $x = r\cos\theta$; and $y = r\sin\theta$; then $z = x+yi = r\cos\theta + ir\sin\theta = r(\cos\theta + i\sin\theta)$. We call $z = r(\cos\theta + i\sin\theta)$ the *polar* or *trigonometric form* and $z = x+yi$ the *rectangular form* of the complex number z.

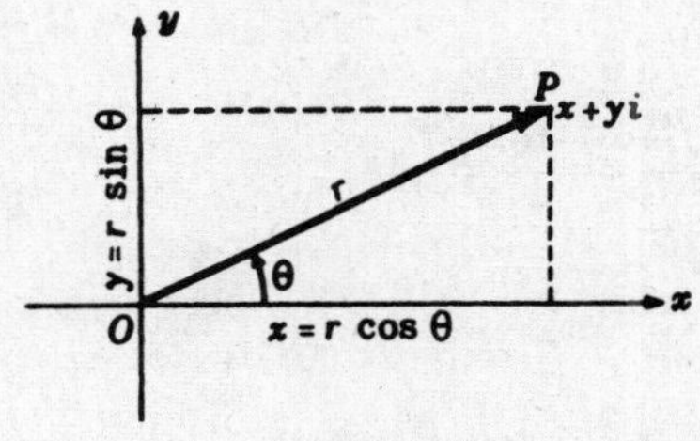

Fig. 15-6

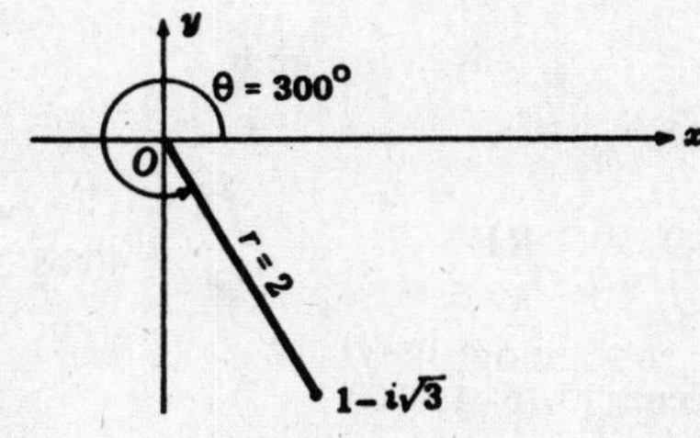

Fig. 15-7

EXAMPLE 5. Express $z = 1 - i\sqrt{3}$ in polar form. (See Fig. 15-7.)

The modulus is $r = \sqrt{(1)^2 + (-\sqrt{3})^2} = 2$. Since $\tan\theta = y/x = -\sqrt{3}/1 = -\sqrt{3}$, the amplitude θ is either 120° or 300°. Now we know that P lies in quadrant IV; hence, $\theta = 300°$ and the required polar form is

$$z = r(\cos\theta + i\sin\theta) = 2(\cos 300° + i\sin 300°)$$

Note that z may also be represented in polar form by

$$z = 2[\cos(300° + n360°) + i\sin(300° + n360°)]$$

where n is any integer.

EXAMPLE 6. Express the complex number $z = 8(\cos 210° + i\sin 210°)$ in rectangular form.

Since $\cos 210° = -\sqrt{3}/2$ and $\sin 210° = -\frac{1}{2}$,

$$z = 8(\cos 210° + i \sin 210°) = 8\left[-\frac{\sqrt{3}}{2} + i\left(-\frac{1}{2}\right)\right] = -4\sqrt{3} - 4i$$

is the required rectangular form.
(See Problems 15.12–15.13.)

MULTIPLICATION AND DIVISION IN POLAR FORM

MULTIPLICATION. The modulus of the product of two complex numbers is the product of their moduli, and the amplitude of the product is the sum of their amplitudes.

DIVISION. The modulus of the quotient of two complex numbers is the modulus of the dividend divided by the modulus of the divisor, and the amplitude of the quotient is the amplitude of the dividend minus the amplitude of the divisor. For a proof of these theorems, see Problem 15.14.

EXAMPLE 7. Find (*a*) the product z_1z_2, (*b*) the quotient and z_1/z_2, and (*c*) the quotient z_2/z_1 where $z_1 = 2(\cos 300° + i \sin 300°)$ and $z_2 = 8(\cos 210° + i \sin 210°)$.

(*a*) The modulus of the product is $2(8) = 16$. The amplitude is $300° + 210° = 510°$, but following the convention, we shall use the smallest positive coterminal angle $510° - 360° = 150°$. Thus, $z_1z_2 = 16(\cos 150° + i \sin 150°)$.

(*b*) The modulus of the quotient z_1/z_2 is $\frac{2}{8} = \frac{1}{4}$ and the amplitude is $300° - 210° = 90°$. Thus, $z_1/z_2 = \frac{1}{4}(\cos 90° + i \sin 90°)$.

(*c*) The modulus of the quotient z_2/z_1 is $\frac{8}{2} = 4$. The amplitude is $210° - 300° = -90°$ but we shall use the smallest positive coterminal angle $-90° + 360° = 270°$. Thus $z_2/z_1 = 4(\cos 270° + i \sin 270°)$.

[NOTE: From Examples 5 and 6 the numbers are $z_1 = 1 - i\sqrt{3}$ and $z_2 = -4\sqrt{3} - 4i$ in rectangular form. Then

$$z_1z_2 = (1 - i\sqrt{3})(-4\sqrt{3} - 4i) = -8\sqrt{3} + 8i = 16(\cos 150° + i \sin 150°)$$

as in (*a*), and

$$\frac{z_2}{z_1} = \frac{-4\sqrt{3} - 4i}{1 - i\sqrt{3}} = \frac{(-4\sqrt{3} - 4i)(1 + i\sqrt{3})}{(1 - i\sqrt{3})(1 + i\sqrt{3})} = \frac{-16i}{4} = -4i$$
$$= 4(\cos 270° + i \sin 270°)$$

as in (*c*).
(See Problems 15.15–15.16.)

DE MOIVRE'S THEOREM. If n is any rational number,

$$[r(\cos \theta + i \sin \theta)]^n = r^n(\cos n\theta + i \sin n\theta)$$

A proof of this theorem is beyond the scope of this book; a verification for $n = 2$ and $n = 3$ is given is Problem 15.17.

EXAMPLE 8.

$$\begin{aligned}(\sqrt{3} - i)^{10} &= [2(\cos 330° + i \sin 330°)]^{10} \\ &= 2^{10}(\cos 10 \cdot 330° + i \sin 10 \cdot 330°) \\ &= 1024(\cos 60° + i \sin 60°) = 1024\left(\frac{1}{2} + \frac{i\sqrt{3}}{2}\right) \\ &= 512 + 512i\sqrt{3}\end{aligned}$$

(See Problem 15.18.)

ROOTS OF COMPLEX NUMBERS. We state, without proof, the theorem:

A complex number $a + bi = r(\cos\theta + i\sin\theta)$ has exactly n distinct nth roots.

The procedure for determining these roots is given in Example 9.

EXAMPLE 9. Find all fifth roots of $4 - 4i$.

The usual polar form of $4 - 4i$ is $4\sqrt{2}(\cos 315^\circ + i\sin 315^\circ)$, but we shall need the more general form $4\sqrt{2}[\cos(315^\circ + k360^\circ) + i\sin(315^\circ + k360^\circ)]$, where k is any integer, including zero.

Using De Moiver's theorem, a fifth root of $4 - 4i$ is given by

$$\{4\sqrt{2}[\cos(315^\circ + k360^\circ) + i\sin(315^\circ + k360^\circ)]\}^{1/5} = (4\sqrt{2})^{1/5}\left(\cos 315^\circ + \frac{k360^\circ}{5} + i\sin\frac{315^\circ + k360^\circ}{5}\right)$$
$$= \sqrt{2}[\cos(63^\circ + k72^\circ) + i\sin(63^\circ + k72^\circ)]$$

Assigning in turn the values $k = 0, 1, \ldots,$ we find

$$k = 0: \quad \sqrt{2}(\cos 63^\circ + i\sin 63^\circ) = R_1$$
$$k = 1: \quad \sqrt{2}(\cos 135^\circ + i\sin 135^\circ) = R_2$$
$$k = 2: \quad \sqrt{2}(\cos 207^\circ + i\sin 207^\circ) = R_3$$
$$k = 3: \quad \sqrt{2}(\cos 279^\circ + i\sin 279^\circ) = R_4$$
$$k = 4: \quad \sqrt{2}(\cos 351^\circ + i\sin 351^\circ) = R_5$$
$$k = 5: \quad \sqrt{2}(\cos 423^\circ + i\sin 423^\circ) = \sqrt{2}(\cos 63^\circ + i\sin 63^\circ) = R_1, \quad \text{etc.}$$

Thus, the five fifth roots are obtained by assigning the values $0, 1, 2, 3, 4$ (i.e., $0, 1, 2, 3, \ldots, n - 1$) to k. (See also Problem 15.19.)

The modulus of each of the roots is $\sqrt{2}$; hence these roots lie on a circle of radius $\sqrt{2}$ with center at the origin. The difference in amplitude of two consecutive roots is 72°; hence the roots are equally spaced on this circle, as shown in Fig. 15-8.

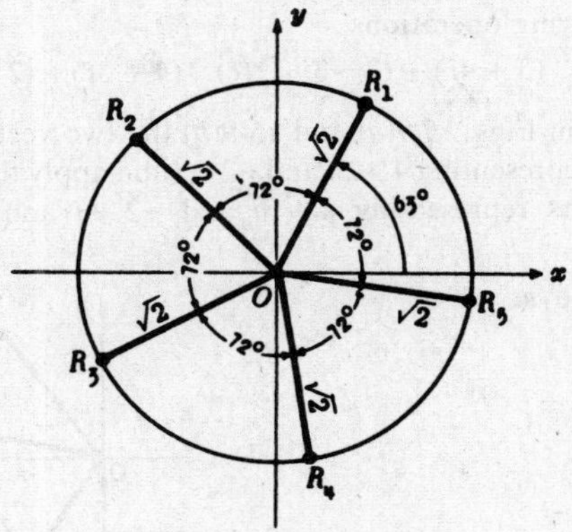

Fig. 15-8

Solved Problems

In Problems 15.1–15.6, perform the indicated operations, simplify, and write the results in the form $a + bi$.

15.1 $(3 - 4i) + (-5 + 7i) = (3 - 5) + (-4 + 7)i = -2 + 3i$

15.2 $(4 + 2i) - (-1 + 3i) = [4 - (-1)] + (2 - 3)i = 5 - i$

15.3 $(2 + i)(3 - 2i) = (6 + 2) + (-4 + 3)i = 8 - i$

15.4 $(3+4i)(3-4i) = 9+16 = 25$

15.5 $\dfrac{1+3i}{2+i} = \dfrac{(1+3i)(2-i)}{(2+i)(2-i)} = \dfrac{(2+3)+(-1+6)i}{4+1} = 1+i$

15.6 $\dfrac{3-2i}{2-3i} = \dfrac{(3-2i)(2+3i)}{(2-3i)(2+3i)} = \dfrac{(6+6)+(9-4)i}{4+9} = \dfrac{12}{13} + \dfrac{5}{13}i$

15.7 Find x and y such that $2x - yi = 4 + 3i$.

Here $2x = 4$ and $-y = 3$; then $x = 2$ and $y = -3$.

15.8 Show that the conjugate complex numbers $2+i$ and $2-i$ are roots of the equation $x^2 - 4x + 5 = 0$.

For $x = 2+i$: $(2+i)^2 - 4(2+i) + 5 = 4 + 4i + i^2 - 8 - 4i + 5 = 0$.

For $x = 2-i$: $(2-i)^2 - 4(2-i) + 5 = 4 - 4i + i^2 - 8 + 4i + 5 = 0$.

Since each number satisfies the equation, it is a root of the equation.

15.9 Show that the conjugate of the sum of two complex numbers is equal to the sum of their conjugates.

Let the complex numbers be $a+bi$ and $c+di$. Their sum is $(a+c)+(b+d)i$ and the conjugate of the sum is $(a+c)-(b+d)i$.

The conjugates of the two given numbers are $a-bi$ and $c-di$, and their sum is $(a+c)+(-b-d)i = (a+c)-(b+d)i$.

15.10 Represent graphically (as a vector) the following complex numbers: (*a*) $3+2i$ (*b*) $2-i$ (*c*) $-2+i$ (*d*) $-1-3i$.

We locate, in turn, the points whose coordinates are $(3,2), (2,-1), (-2,1), (-1,-3)$ and join each to the origin O.

15.11 Perform graphically the following operations:
(*a*) $(3+4i)+(2+5i)$, (*b*) $(3+4i)+(2-3i)$ (*c*) $(4+3i)-(2+i)$, (*d*) $(4+3i)-(2-i)$.

For (*a*) and (*b*), draw as in Figs. 15-9(*a*) and 15-9(*b*) the two vectors and apply the parallelogram law. For (*c*) draw the vectors representing $4+3i$ and $-2-i$ and apply the parallelogram law as in Fig. 15-9(*c*). For (*d*) draw the vectors representing $4+3i$ and $-2+i$ and apply the parallelogram law as in Fig. 15-9(*d*).

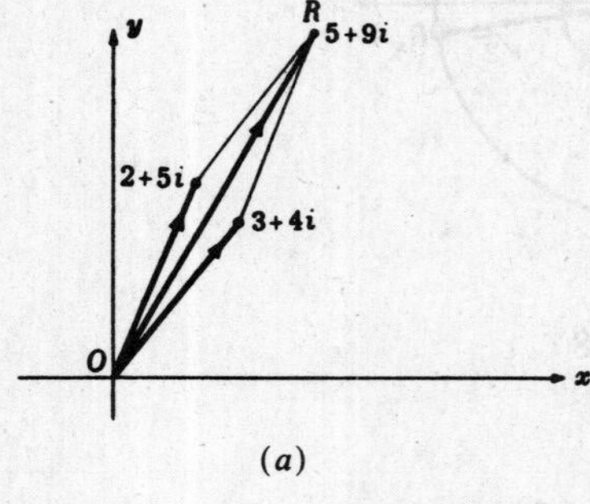

(*a*)

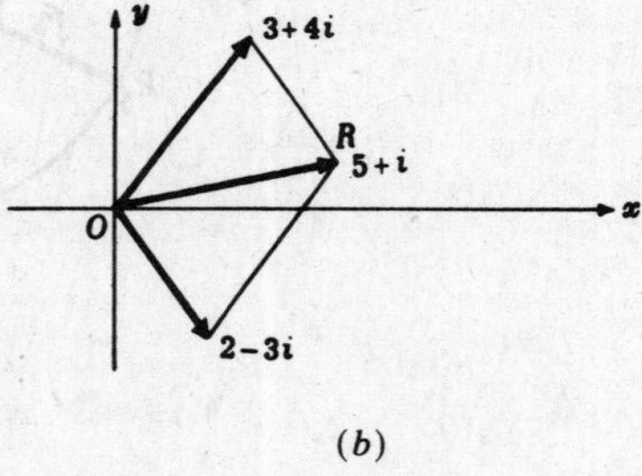

(*b*)

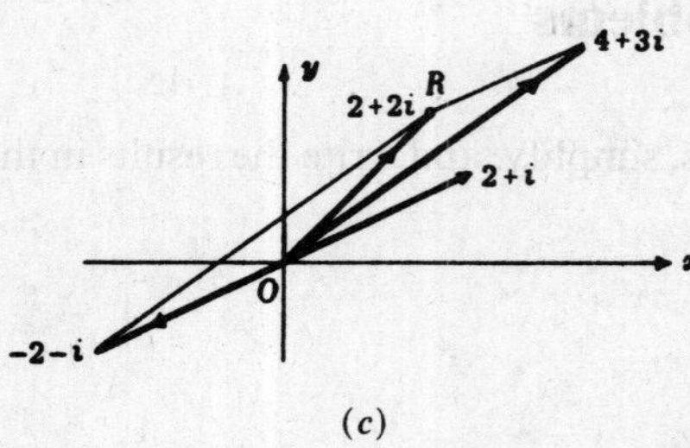

(*c*)

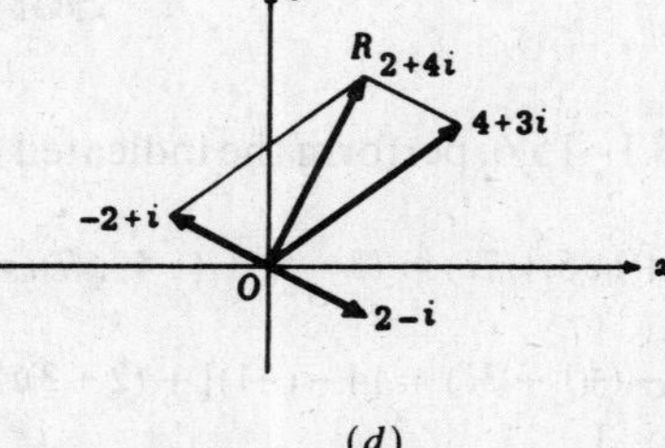

(*d*)

Fig. 15-9

15.12 Express each of the following complex numbers z in polar form:

(a) $-1+i\sqrt{3}$, (b) $6\sqrt{3}+6i$, (c) $2-2i$, (d) $-3=-3+0i$, (e) $4i=0+4i$, (f) $-3-4i$.

(a) P lies in the second quadrant; $r=\sqrt{(-1)^2+(\sqrt{3})^2}=2$; $\tan\theta=\sqrt{3}/(-1)=-\sqrt{3}$ and $\theta=120°$. Thus, $z=2(\cos 120°+i\sin 120°)$.

(b) P lies in the first quadrant; $r=\sqrt{(6\sqrt{3})^2+6^2}=12$; $\tan-\theta=6/6\sqrt{3}=1/\sqrt{3}$ and $\theta=30°$. Thus, $z=12(\cos 30°+i\sin 30°)$.

(c) P lies in the fourth quadrant; $r=\sqrt{2^2+(-2)^2}=2\sqrt{2}$; $\tan\theta=-\frac{2}{2}=-1$ and $\theta=315°$. Thus, $z=2\sqrt{2}(\cos 315°+i\sin 315°)$.

(d) P lies on the negative x axis and $\theta=180°$; $r=\sqrt{(-3)^2+0^2}=3$. Thus, $z=3(\cos 180°+i\sin 180°)$.

(e) P lies on the positive y axis and $\theta=90°$; $r=\sqrt{0^2+4^2}=4$. Thus, $z=4(\cos 90°+i\sin 90°)$.

(f) P lies in the third quadrant; $r=\sqrt{(-3)^2+(-4)^2}=5$; $\tan\theta=-4/(-3)=1.3333$, $\theta=233°8'$. Thus, $z=5(\cos 233°8'+i\sin 233°8')$.

15.13 Express each of the following complex numbers z in rectangular form:

(a) $4(\cos 240°+i\sin 240°)$ (c) $3(\cos 90°+i\sin 90°)$
(b) $2(\cos 315°+i\sin 315°)$ (d) $5(\cos 128°+i\sin 128°)$

(a) $4(\cos 240°+i\sin 240°)=4[-\frac{1}{2}+i(-\sqrt{3}/2)]=-2-2i\sqrt{3}$

(b) $2(\cos 315°+i\sin 315°)=2[1/\sqrt{2}+i(-1/\sqrt{2})]=\sqrt{2}-i\sqrt{2}$

(c) $3(\cos 90°+i\sin 90°)=3[0+i(1)]=3i$

(d) $5(\cos 128°+i\sin 128°)=5[-0.6157+i(0.7880)]=-3.0785+3.9400i$

15.14 Prove:

(a) The modulus of the product of two complex numbers is the product of their moduli, and the amplitude of the product is the sum of their amplitudes.

(b) The modulus of the quotient of two complex numbers is the modulus of the dividend divided by the modulus of the divisor, and the amplitude of the quotient is the amplitude of the dividend minus the amplitude of the divisor.

Let $z_1=r_1(\cos\theta_1+i\sin\theta_1)$ and $z_2=r_2(\cos\theta_2+i\sin\theta_2)$.

(a)
$$\begin{aligned}z_1z_2&=r_1(\cos\theta_1+i\sin\theta_1)\cdot r_2(\cos\theta_2+i\sin\theta_2)\\&=r_1r_2[(\cos\theta_1\cos\theta_2-\sin\theta_1\sin\theta_2)+i(\sin\theta_1\cos\theta_2+\cos\theta_1\sin\theta_2)]\\&=r_1r_2[\cos(\theta_1+\theta_2)+i\sin(\theta_1+\theta_2)]\end{aligned}$$

(b)
$$\begin{aligned}\frac{r_1(\cos\theta_1+i\sin\theta_1)}{r_2(\cos\theta_2+i\sin\theta_2)}&=\frac{r_1(\cos\theta_1+i\sin\theta_1)(\cos\theta_2-i\sin\theta_2)}{r_2(\cos\theta_2+i\sin\theta_2)(\cos\theta_2-i\sin\theta_2)}\\&=\frac{r_1}{r_2}\cdot\frac{\cos\theta_1\cos\theta_2+\sin\theta_1\sin\theta_2)+i(\sin\theta_1\cos\theta_2-\cos\theta_1\sin\theta_2)}{\cos^2\theta_2+\sin^2\theta_2)}\\&=\frac{r_1}{r_2}[\cos(\theta_1-\theta_2)+i\sin(\theta_1-\theta_2)]\end{aligned}$$

15.15 Perform the indicated operations, giving the result in both polar and rectangular form.

(a) $5(\cos 170°+i\sin 170°)\cdot(\cos 55°+i\sin 55°)$
(b) $2(\cos 50°+i\sin 50°)\cdot 3(\cos 40°+i\sin 40°)$
(c) $6(\cos 110°+i\sin 110°)\cdot\frac{1}{2}(\cos 212°+i\sin 212°)$
(d) $10(\cos 305°+i\sin 305°)\div 2(\cos 65°+i\sin 65°)$

(*e*) $4(\cos 220° + i \sin 220°) \div 2(\cos 40° + i \sin 40°)$
(*f*) $6(\cos 230° + i \sin 230°) \div 3(\cos 75° + i \sin 75°)$

(*a*) The modulus of the product is $5(1) = 5$ and the amplitude is $170° + 55° = 225°$.
In polar form the product is $5(\cos 225° + i \sin 225°)$ and in rectangular form the product is $5(-\sqrt{2}/2 - i\sqrt{2}/2) = -5\sqrt{2}/2 - 5i\sqrt{2}/2$.

(*b*) The modulus of the product is $2(3) = 6$ and the amplitude is $50° + 40° = 90°$.
In polar form the product is $6(\cos 90° + i \sin 90°)$ and in rectangular form it is $6(0 + i) = 6i$.

(*c*) The modulus of the product is $6(\frac{1}{2}) = 3$ and the amplitude is $110° + 212° = 322°$.
In polar form the product is $3(\cos 322° + i \sin 322°)$ and in rectangular form it is $3(0.7880 - 0.6157i) = 2.3640 - 1.8471i$.

(*d*) The modulus of the quotient is $\frac{10}{2} = 5$ and the amplitude is $305° - 65° = 240°$.
In polar form the product is $5(\cos 240° + i \sin 240°)$ and in rectangular form it is $(5 - \frac{1}{2} - i\sqrt{3}/2) = -\frac{5}{2} - 5i\sqrt{3}/2$.

(*e*) The modulus of the quotient is $\frac{4}{2} = 2$ and the amplitude is $220° - 40° = 180°$.
In polar form the quotient is $2(\cos 180° + i \sin 180°)$ and in rectangular form it is $2(-1 + 0i) = -2$.

(*f*) The modulus of the quotient is $\frac{6}{3} = 2$ and the amplitude is $230° - 75° = 155°$.
In polar form the quotient is $2(\cos 155° + i \sin 155°)$ and in rectangular form it is $2(-0.9063 + 0.4226i) = -1.8126 + 0.8452i$.

15.16 Express each of the numbers in polar form, perform the indicated operation, and give the result in rectangular form

(*a*) $(-1 + i\sqrt{3})(\sqrt{3} + i)$ (*d*) $-2 \div (-\sqrt{3} + i)$ (*g*) $(3 + 2i)(2 + i)$
(*b*) $(3 - 3i\sqrt{3})(-2 - 2i\sqrt{3})$ (*e*) $6i \div (-3 - 3i)$ (*h*) $(2 + 3i) \div (2 - 3i)$
(*c*) $(4 - 4i\sqrt{3}) \div (-2\sqrt{3} + 2i)$ (*f*) $(1 + i\sqrt{3})(1 + i\sqrt{3})$

(*a*)
$$\begin{aligned}(-1 + i\sqrt{3})(\sqrt{3} + i) &= 2(\cos 120° + i \sin 120°) \cdot 2(\cos 30° + i \sin 30°)\\ &= 4(\cos 150° + i \sin 150°) = 4(-\sqrt{3}/2 + \tfrac{1}{2}i) = -2\sqrt{3} + 2i\end{aligned}$$

(*b*)
$$\begin{aligned}(3 - 3i\sqrt{3})(-2 - 2i\sqrt{3}) &= 6(\cos 300° + i \sin 300°) \cdot 4(\cos 240° + i \sin 240°)\\ &= 24(\cos 540° + i \sin 540°) = 24(-1 + 0i) = -24\end{aligned}$$

(*c*)
$$\begin{aligned}(4 - 4i\sqrt{3}) \div (-2\sqrt{3} + 2i) &= 8(\cos 300° + i \sin 300°) + 4(\cos 150° + i \sin 150°)\\ &= 2(\cos 150° + i \sin 150°) = 2(-\sqrt{3}/2 + \tfrac{1}{2}i) = -\sqrt{3} + i\end{aligned}$$

(*d*)
$$\begin{aligned}-2 + (-\sqrt{3} + i) &= 2(\cos 180° + i \sin 180°) \div 2(\cos 150° + i \sin 150°)\\ &= \cos 30° + i \sin 30° = \tfrac{1}{2}\sqrt{3} + \tfrac{1}{2}i\end{aligned}$$

(*e*)
$$\begin{aligned}6i \div (-3 - 3i) &= 6(\cos 90° + i \sin 90°) \div 3\sqrt{2}(\cos 225° + i \sin 225°)\\ &= \sqrt{2}(\cos 225° + i \sin 225°) = -1 - i\end{aligned}$$

(*f*)
$$\begin{aligned}(1 + i\sqrt{3})(1 + i\sqrt{3}) &= 2(\cos 60° + i \sin 60°) \cdot 2(\cos 60° + i \sin 60°)\\ &= 4(\cos 120° + i \sin 120°) = 4(-\tfrac{1}{2} + \tfrac{1}{2}i\sqrt{3}) = -2 + 2i\sqrt{3}\end{aligned}$$

(*g*)
$$\begin{aligned}(3 + 2i)(2 + i) &= \sqrt{13}(\cos 33°41' + i \sin 33°41') \cdot \sqrt{5}(\cos 26°34' + i \sin 26°34')\\ &= \sqrt{65}(\cos 60°15' + i \sin 60°15')\\ &= \sqrt{65}(0.4962 + 0.8682i) = 4.001 + 7.000i = 4 + 7i\end{aligned}$$

(*h*)
$$\begin{aligned}\frac{2 + 3i}{2 - 3i} &= \frac{\sqrt{13}(\cos 56°19' + i \sin 56°19')}{\sqrt{13}(\cos 303°41' + i \sin 303°41')} = \frac{\cos 416°19' + i \sin 416°19'}{\cos 303°41'' + i \sin 303°41'}\\ &= \cos 112°38' + i \sin 112°38' = -0.3849 + 0.9230i\end{aligned}$$

15.17 Verify De Moivre's theorem for $n = 2$ and $n = 3$.

Let $z = r(\cos\theta + i\sin\theta)$.

For $n = 2$:

$$z^2 = [r(\cos\theta + i\sin\theta)][r(\cos\theta + i\sin\theta)]$$
$$= r^2[(\cos^2\theta - \sin^2\theta) + i(2\sin\theta\cos\theta)] = r^2(\cos 2\theta + i\sin 2\theta)$$

For $n = 3$:

$$z^3 = z^2\cdot z = [r^2(\cos 2\theta + i\sin 2\theta)][r(\cos\theta + i\sin\theta)]$$
$$= r^3[(\cos 2\theta\cos\theta - \sin 2\theta\sin\theta) + i(\sin 2\theta\cos\theta + \cos 2\theta\sin\theta)]$$
$$= r^3(\cos 3\theta + i\sin 3\theta).$$

The theorem may be established for n a positive integer by mathematical induction.

15.18 Evaluate each of the following using De Moivre's theorem and express each result in rectangular form:

(*a*) $(1 + i\sqrt{3})^4$, (*b*) $(\sqrt{3} - i)^5$, (*c*) $(-1 + i)^{10}$, (*d*) $(2 + 3i)^4$.

(*a*)
$$(1 + i\sqrt{3})^4 = [2(\cos 60^\circ + i\sin 60^\circ)]^4 = 2^4(\cos 4\cdot 60^\circ + i\sin 4\cdot 60^\circ)$$
$$= 2^4(\cos 240^\circ + i\sin 240^\circ) = -8 - 8i\sqrt{3}$$

(*b*)
$$(\sqrt{3} - i)^5 = [2(\cos 330^\circ + i\sin 330^\circ)]^5 = 32(\cos 1650^\circ + i\sin 1650^\circ)$$
$$= 32(\cos 210^\circ + i\sin 210^\circ) = -16\sqrt{3} - 16i$$

(*c*)
$$(-1 + i)^{10} = [\sqrt{2}(\cos 135^\circ + i\sin 135^\circ)]^{10} = 32(\cos 270^\circ + i\sin 270^\circ) = -32i$$

(*d*)
$$(2 + 3i)^4 = [\sqrt{13}(\cos 56^\circ 19' + i\sin 56^\circ 19')]^4 = 13^2(\cos 225^\circ 16' + i\sin 225^\circ 16')$$
$$= 169(-0.7038 - 7104i) = -118.9 - 120.1i$$

15.19 Find the indicated roots in rectangular form, except when this would necessitate the use of tables.

(*a*) Square roots of $2 - 2i\sqrt{3}$
(*b*) Fourth roots of $-8 - 8i\sqrt{3}$
(*c*) Cube roots of $-4\sqrt{2} + 4i\sqrt{2}$
(*d*) Cube roots of 1
(*e*) Fourth roots of i
(*f*) Sixth roots of -1
(*g*) Fourth roots of $-16i$
(*h*) Fifth roots of $1 + 3i$

(*a*)
$$2 - 2i\sqrt{3} = 4[\cos(300^\circ + k360^\circ) + i\sin(300^\circ + k360^\circ)]$$
and
$$(2 - 2i\sqrt{3})^{1/2} = 2[\cos(150^\circ + k180^\circ) + i\sin(150^\circ + k180^\circ)]$$

Putting $k = 0$ and 1, the required roots are

$$R_1 = 2(\cos 150^\circ + i\sin 150^\circ) = 2(-\tfrac{1}{2}\sqrt{3} + \tfrac{1}{2}i) = -\sqrt{3} + i$$
$$R_2 = 2(\cos 330^\circ + i\sin 330^\circ) = 2(\tfrac{1}{2}\sqrt{3} - \tfrac{1}{2}i) = \sqrt{3} - i$$

(*b*)
$$-8 - 8i\sqrt{3} = 16[\cos(240^\circ + k360^\circ) + i\sin(240^\circ + k360^\circ)]$$
and
$$(-8 - 8i\sqrt{3})^{1/4} = 2[\cos(60^\circ + k90^\circ) + i\sin(60^\circ + k90^\circ)]$$

Putting $k = 0, 1, 2, 3$, the required roots are

$$R_1 = 2(\cos 60^\circ + i\sin 60^\circ) = 2(\tfrac{1}{2} + i\tfrac{1}{2}\sqrt{3}) = 1 + i\sqrt{3}$$
$$R_2 = 2(\cos 150^\circ + i\sin 150^\circ) = 2(-\tfrac{1}{2}\sqrt{3} + \tfrac{1}{2}i) = -\sqrt{3} + i$$
$$R_3 = 2(\cos 240^\circ + i\sin 240^\circ) = 2(-\tfrac{1}{2} - i\tfrac{1}{2}\sqrt{3}) = -1 - i\sqrt{3}$$
$$R_4 = 2(\cos 330^\circ + i\sin 330^\circ) = 2(\tfrac{1}{2}\sqrt{3} - \tfrac{1}{2}i) = \sqrt{3} - i$$

(*c*)
$$-4\sqrt{2} + 4i\sqrt{2} = 8[\cos(135^\circ + k360^\circ) + i\sin(135^\circ + k360^\circ)]$$
and
$$(-4\sqrt{2} + 4i\sqrt{2})^{1/3} = 2[\cos(45^\circ + k120^\circ) + i\sin(45^\circ + k120^\circ)]$$

Putting $k = 0, 1, 2$, the required roots are

$$R_1 = 2(\cos 45^\circ + i \sin 45^\circ) = 2(1/\sqrt{2} + i/\sqrt{2}) = \sqrt{2} + i\sqrt{2}$$
$$R_2 = 2(\cos 165^\circ + i \sin 165^\circ)$$
$$R_3 = 2(\cos 285^\circ + i \sin 285^\circ)$$

(*d*) $1 = \cos(0^\circ + k360^\circ) + i \sin(0^\circ + k360^\circ)$ and $1^{1/3} = \cos(k120^\circ) + i \sin(k120^\circ)$. Putting $k = 0, 1, 2$, the required roots are

$$R_1 = \cos 0^\circ + i \sin 0^\circ = 1$$
$$R_2 = \cos 120^\circ + i \sin 120^\circ = -\tfrac{1}{2} + i\tfrac{1}{2}\sqrt{3}$$
$$R_3 = \cos 240^\circ + i \sin 240^\circ = -\tfrac{1}{2} - i\tfrac{1}{2}\sqrt{3}$$

Note that $R_2^2 = \cos 2(120^\circ) + i \sin 2(120^\circ) = R_3$,

$$R_3^2 = \cos 2(240^\circ) + i \sin 2(240^\circ) = R_2,$$

and $R_2R_3 = (\cos 120^\circ + i \sin 120^\circ)(\cos 240^\circ + i \sin 240^\circ)$

$$= \cos 0^\circ + i \sin 0^\circ = R_1.$$

(*e*) $i = \cos(90^\circ + k360^\circ) + i \sin(90^\circ + k360^\circ)$ and $i^{1/4} = \cos(22\tfrac{1}{2}^\circ + k90^\circ) + i \sin(22\tfrac{1}{2}^\circ + k90^\circ)$. Thus, the required roots are

$$R_1 = \cos 22\tfrac{1}{2}^\circ + i \sin 22\tfrac{1}{2}^\circ \qquad R_3 = \cos 202\tfrac{1}{2}^\circ + i \sin 202\tfrac{1}{2}^\circ$$
$$R_2 = \cos 112\tfrac{1}{2}^\circ + i \sin 112\tfrac{1}{2}^\circ \qquad R_4 = \cos 292\tfrac{1}{2}^\circ + i \sin 292\tfrac{1}{2}^\circ$$

(*f*) $-1 = \cos(180^\circ + k360^\circ) + i \sin(180^\circ + k360^\circ)$ and $(-1)^{1/6} = \cos(30^\circ + k60^\circ) + i \sin(30^\circ + k60^\circ)$. Thus, the required roots are

$$R_1 = \cos 30^\circ + i \sin 30^\circ = \tfrac{1}{2}\sqrt{3} + \tfrac{1}{2}i$$
$$R_2 = \cos 90^\circ + i \sin 90^\circ = i$$
$$R_3 = \cos 150^\circ + i \sin 150^\circ = -\tfrac{1}{2}\sqrt{3} + \tfrac{1}{2}i$$
$$R_4 = \cos 210^\circ + i \sin 210^\circ = -\tfrac{1}{2}\sqrt{3} - \tfrac{1}{2}i$$
$$R_5 = \cos 270^\circ + i \sin 270^\circ = -i$$
$$R_6 = \cos 330^\circ + i \sin 330^\circ = \tfrac{1}{2}\sqrt{3} - \tfrac{1}{2}i$$

Note that $R_2^2 = \cos R_5^2 = \cos 180^\circ + i \sin 180^\circ$ and thus R_2 and R_5 are the square roots of -1; that $R_1^3 = R_3^3 = R_5^3 = \cos 90^\circ + i \sin 90^\circ = i$ and thus R_1, R_3, R_5 are the cube roots of i; and that $R_2^3 = R_4^3 = R_6^3 = \cos 270^\circ + i \sin 270^\circ = -i$ and thus R_2, R_4, R_6 are the cube roots of $-i$.

(*g*) $-16i = 16[\cos(270^\circ + K360^\circ) + i \sin(270^\circ + k360^\circ)]$ and $(-16i.)^{1/4} = 2[\cos(67\tfrac{1}{2}^\circ + k90^\circ) + i \sin(67\tfrac{1}{2}^\circ + k90^\circ)]$. Thus, the required roots are

$$R_1 = 2(\cos 67\tfrac{1}{2}^\circ + i \sin 67\tfrac{1}{2}^\circ) \qquad R_3 = 2(\cos 247\tfrac{1}{2}^\circ + i \sin 247\tfrac{1}{2}^\circ)$$
$$R_2 = 2(\cos 157\tfrac{1}{2}^\circ + i \sin 157\tfrac{1}{2}^\circ) \qquad R_4 = 2(\cos 337\tfrac{1}{2}^\circ + i \sin 337\tfrac{1}{2}^\circ)$$

(*h*) $1 + 3i = \sqrt{10}[\cos(71^\circ 34' + k360^\circ) + i \sin(71^\circ 34' + k360^\circ)]$ and $(1 + 3i)^{1/5} = \sqrt[10]{10}[\cos(14^\circ 19' + k72^\circ) + i \sin(14^\circ 19' + k72^\circ)]$. The required roots are

$$R_1 = \sqrt[10]{10}(\cos 14^\circ 19' + i \sin 14^\circ 19')$$
$$R_2 = \sqrt[10]{10}(\cos 86^\circ 19' + i \sin 86^\circ 19')$$
$$R_3 = \sqrt[10]{10}(\cos 158^\circ 19' + i \sin 158^\circ 19')$$
$$R_4 = \sqrt[10]{10}(\cos 230^\circ 19' + i \sin 230^\circ 19')$$
$$R_5 = \sqrt[10]{10}(\cos 302^\circ 19' + i \sin 302^\circ 19').$$

Supplementary Problems

15.20 Perform the indicated operations, writing the results in the form $a + bi$.

(a) $(6-2i)+(2+3i)=8+i$

(b) $(6-2i)-(2+3i)=4-5i$

(c) $(3+2i)+(-4-3i)=-1-i$

(d) $(3-2i)-(4-3i)=-1+i$

(e) $3(2-i)=6-3i$

(f) $2i(3+4i)=-8+6i$

(g) $(2+3i)(1+2i)=-4+7i$

(h) $(2-3i)(5+2i)=16-11i$

(i) $(3-2i)(-4+i)=-10+11i$

(j) $(2+3i)(3+2i)=13i$

(k) $(2+\sqrt{-5})(3-2\sqrt{-4})=(6+4\sqrt{5})+(3\sqrt{5}-8)i$

(l) $(1+2\sqrt{-3})(2-\sqrt{-3})=8+3\sqrt{3}i$

(m) $(2-i)^2=3-4i$

(n) $(4+2i)^2=12+16i$

(o) $(1+i)^2(2+3i)=-6+4i$

(p) $\dfrac{2+3i}{1+i}=\dfrac{5}{2}+\dfrac{1}{2}i$

(q) $\dfrac{3-2i}{3-4i}=\dfrac{17}{25}+\dfrac{6}{25}i$

(r) $\dfrac{3-2i}{2+3i}=-i$

15.21 Show that $3 + 2i$ and $3 - 2i$ are roots of $x^2 - 6x + 13 = 0$.

15.22 Perform graphically the following operations:

(a) $(2+3i)+(1+4i)$

(b) $(4-2i)+(2+3i)$

(c) $(2+3i)-(1+4i)$

(d) $(4-2i)-(2+3i)$

15.23 Express each of the following complex numbers in polar form:

(a) $3+3i=32(\cos 45^\circ + i\sin 45^\circ)$

(b) $1+\sqrt{3}i=2(\cos 60^\circ + i\sin 60^\circ)$

(c) $-2\sqrt{3}-2i=4(\cos 210^\circ + i\sin 210^\circ)$

(d) $\sqrt{2}-i\sqrt{2}=2(\cos 315^\circ + i\sin 315^\circ)$

(e) $-8=8(\cos 180^\circ + i\sin 180^\circ)$

(f) $-2i=2(\cos 270^\circ + i\sin 270^\circ)$

(g) $-12+5i=13(\cos 157^\circ 23' + i\sin 157^\circ 23')$

(h) $-4-3i=5(\cos 216^\circ 52' + i\sin 216^\circ 52')$

15.24 Perform the indicated operation and express the results in the form $a + bi$.

(a) $3(\cos 25^\circ + i\sin 25^\circ)8(\cos 200^\circ + i\sin 200^\circ)=-12\sqrt{2}-12\sqrt{2}i$

(b) $4(\cos 50^\circ + i\sin 50^\circ)2(\cos 100^\circ + i\sin 100^\circ)=-4\sqrt{3}+4i$

(c) $\dfrac{4(\cos 190^\circ + i\sin 190^\circ)}{2(\cos 70^\circ + i\sin 70^\circ)}=-1+i\sqrt{3}$

(d) $\dfrac{12(\cos 200^\circ + i\sin 200^\circ)}{3(\cos 350^\circ + i\sin 350^\circ)}=-2\sqrt{3}-2i$

15.25 Use the polar form in finding each of the following products and quotients, and express each result in the form $a + bi$:

(a) $(1+i)(\sqrt{2}-i\sqrt{2})=2\sqrt{2}$

(b) $(-1-i\sqrt{3})(-4\sqrt{3}+4i)=8\sqrt{3}+8i$

(c) $\dfrac{1-i}{1+i}=-i$

(d) $\dfrac{4+4\sqrt{3}i}{\sqrt{3}+i}=2\sqrt{3}+2i$

(e) $\dfrac{-1+i\sqrt{3}}{\sqrt{2}+i\sqrt{2}}=0.2588+0.9659i$

(f) $\dfrac{3+i}{2+i}=1.4-0.2i$

15.26 Use De Moivre's theorem to evaluate each of the following and express each result in the form $a + bi$:

(a) $[2(\cos 6° + i \sin 6°)]^5 = 16\sqrt{3} + 16i$
(b) $[\sqrt{2}(\cos 75° + i \sin 75°)]^4 = 2 - 2\sqrt{3}i$
(c) $(1 + i)^8 = 16$
(d) $(1 - i)^6 = 8i$
(e) $(\frac{1}{2} - i\sqrt{3}/2)^{20} = -\frac{1}{2} - i\sqrt{3}/2$
(f) $(\sqrt{3}/2 + i/2)^9 = -i$
(g) $(3 + 4i)^4 = -526.9 - 336.1i$
(h) $\dfrac{(1 - i\sqrt{3})^3}{(-2 + 2i)} = \dfrac{1}{8}$
(i) $\dfrac{(1 + i)(\sqrt{3} + i)^3}{(1 - i\sqrt{3})^3} = 1 - i$

15.27 Find all the indicated roots, expressing the results in the form $a + bi$ unless tables would be needed to do so.

(a) The square roots of i. *Ans.* $\sqrt{2}/2 + i\sqrt{2}/2, -\sqrt{2}/2 - i\sqrt{2}/2$

(b) The square roots of $1 + i\sqrt{3}$. *Ans.* $\sqrt{6}/2 + i\sqrt{2}/2, -\sqrt{6}/2 - i\sqrt{2}/2$

(c) The cube roots of -8. *Ans.* $1 + i\sqrt{3}, -2, 1 - i\sqrt{3}$

(d) The cube roots of $27i$. *Ans.* $3\sqrt{3}/2 + 3i/2, -3\sqrt{3}/2 + 3i/2, -3i$

(e) The cube roots of $-4\sqrt{3} + 4i$.
Ans. $2(\cos 50° + i \sin 50°)$, $2(\cos 170° + i \sin 170°)$, $2(\cos 290° + i \sin 290°)$

(f) The fifth roots of $1 + i$.
Ans. $\sqrt[10]{2}(\cos 9° + i \sin 9°)$, $\sqrt[10]{2}(\cos 81° + i \sin 81°)$, etc.

(g) The sixth roots of $-\sqrt{3} + i$.
Ans. $\sqrt[6]{2}(\cos 25° + i \sin 25°)$, $\sqrt[6]{2}(\cos 85° + i \sin 85°)$, *etc.*

15.28 Find the tenth roots of 1 and show that the product of any two of them is again one of the tenth roots of 1.

15.29 Show that the reciprocal of any one of the tenth roots of 1 is again a tenth root of 1.

15.30 Denote either of the complex cube roots of 1 [Problem 15.19(*d*)] by ω_1 and the other by ω_2. Show that $\omega_1^2\omega_2 = \omega_1$ and $\omega_1\omega_2^2 = \omega_2$.

15.31 Show that $(\cos \theta + i \sin \theta)^{-n} = \cos n\theta - i \sin n\theta$.

15.32 Use the fact that the segments OS and P_2P_1 in Fig. 15-5 are equal to devise a second procedure for constructing the differences $OS = z_1 - z_2$ of two complex numbers z_1 and z_2.

Chapter 16

The Calculus of Single-Variable Functions: A Mathematics Approach

OVERVIEW: In this chapter, the notions of limit, derivative, and integral will be introduced using a family standard mathematics approach.

It will be understood that number refers always to a real number, that the range of any variable (such as x) is a set of real numbers, and that a function of one variable [such as $f(x)$] is a single-valued function.

In this chapter a procedure is given by which from a given function $y = f(x)$ another function, denoted by y' or $f'(x)$ and called the derivative of y or of $f(x)$ with respect to x, is obtained. Depending upon the quantities denoted by x and $y = f(x)$, the derivative may be interpreted as the slope of a tangent line to a curve, as velocity, as acceleration, etc.

LIMIT OF A FUNCTION. A given function $f(x)$ is said to have a *limit M* as x approaches c [in symbols, $\lim_{x \to c} f(x) = M$] if $f(x)$ can be made as close to M as we please for all values of $x \neq c$ but sufficiently near to c, by having x get sufficiently close to c (approaching both from the left and right).

EXAMPLE 1. Consider $f(x) = x^2 - 2$ for x near 3.

If x is near to 3, say $2.99 < x < 3.01$, then $(2.99)^2 - 2 < f(x) < (3.01)^2 - 2$ or $6.9401 < f(x) < 7.0601$.

If x is nearer to 3, say $2.999 < x < 3.001$, then $(2.999)^2 - 2 < f(x) < (3.001)^2 - 2$ or $6.994001 < f(x) < 7.006001$.

If x is still nearer to 3, say $2.9999 < x < 3.0001$, then $(2.9999)^2 - 2 < f(x) < (3.0001)^2 - 2$ or $6.99940001 < f(x) < 7.00060001$.

It appears reasonable to conclude that as x is taken in a smaller and smaller interval about 3, the corresponding $f(x)$ will lie a smaller and smaller interval about 7. Conversely, it seems reasonable to conclude that if we demand that $f(x)$ have values lying in smaller and smaller intervals about 7, we need only to choose x in sufficiently smaller and smaller intervals about 3. Thus we conclude

$$\lim_{x \to 3} (x^2 - 2) = 7$$

EXAMPLE 2. Consider $f(x) = \dfrac{x^2 - x - 6}{x - 3}$ for x near 3.

When $x \neq 3$, $f(x) = \dfrac{x^2 - x - 6}{x - 3} = x + 2$. Thus, for x near 3, $x + 2$ is near to 5 and

$$\lim_{x \to 3} \frac{x^2 - x - 6}{x - 3} = 5$$

ONE-SIDED LIMITS. We say that the limit of $f(x)$ as x approaches a from the left is $L[\lim_{x \to a^-} f(x) = L]$ when $f(x)$ gets arbitrarily close to L as x approaches a from the left-hand side of a. [Similarly, $\lim_{x \to b^+} f(x) = M$ is the right-hand limit.] Clearly, if $\lim_{x \to a^-} f(x) = \lim_{x \to a^+} f(x) = L$, then $\lim_{x \to a} f(x) = L$.

THEOREMS ON LIMITS. If $\lim_{x \to c} f(x) = M$ and $\lim_{x \to c} g(x) = N$, then

I. $\lim_{x \to c} [f(x) \pm g(x)] = \lim_{x \to c} f(x) \pm \lim_{x \to c} g(x) = M \pm N.$

II. $\lim_{x \to c} [kf(x)] = k \lim_{x \to c} f(x) = kM$, where k is a constant.

III. $\lim_{x \to c} [f(x) \cdot g(x)] = \lim_{x \to c} f(x) \cdot \lim_{x \to c} g(x) = MN.$

IV. $\lim_{x \to c} \dfrac{f(x)}{g(x)} = \dfrac{\lim_{x \to c} f(x)}{\lim_{x \to c} g(x)} = \dfrac{M}{N}$, provided $N \neq 0$.

Note that in these four statements, the assumption that M and N exist is essential. (See Problem 16.2.)

CONTINUOUS FUNCTIONS. A function $f(x)$ is called continuous at $x = c$, provided

(1) $f(c)$ is defined,

(2) $\lim_{x \to c} f(x)$ exists,

(3) $\lim_{x \to c} f(x) = f(c)$.

EXAMPLE 3.

(a) The function $f(x) = x^2 - 2$ of Example 1 is continuous at $x = 3$ since *(1)* $f(3) = 7$, *(2)* $\lim_{x \to 3} (x^2 - 2) = 7$, *(3)* $\lim_{x \to 3} (x^2 - 2) = f(3)$.

(b) The function $f(x) = \dfrac{x^2 - x - 6}{x - 3}$ of Example 2 is not continuous at $x = 3$, since $f(3)$ is not defined. (See Problem 16.2.)

A function $f(x)$ is said to be *continuous* on the interval (a, b) if it is continuous for every value of x of the interval. A polynomial in x is continuous since it continuous for all values of x. A rational function in x, $f(x) = P(x)/Q(x)$, where $P(x)$ and $Q(x)$ are polynomials, is continuous for all values of x except those for which $Q(x) = 0$. Thus, $f(x) = \dfrac{x^2 + x + 1}{(x - 1)(x^2 + 2)}$ is continuous for all values of x, except $x = 1$.

CONTINUITY ON A CLOSED INTERVAL. If a function $y = g(x)$ is continuous for all values of x in $[a, b]$, then it is continuous on (a, b) and also at a and b.

However, in this case, $g(x)$ is continuous at a means $\lim_{x \to a^+} g(x) = g(a)$; similarly, $\lim_{x \to b^-} g(x) = g(b)$. (See Problem 16.4.)

INCREMENTS. Let x_0 and x_1 be two distinct values of x. It is customary to denote their difference $x_1 - x_0$ by Δx (read, delta x) and to write $x_0 + \Delta x$ for x_1.

Now if $y = f(x)$ and if x changes in value from $x = x_0$ to $x = x_0 + \Delta x$, y will change in value from $y_0 = f(x_0)$ to $y_0 + \Delta y = f(x_0 + \Delta x)$. The change in y due to a change in x from $x = x_0$ to $x = x_0 + \Delta x$ is $\Delta y = f(x_0 + \Delta x) - f(x_0)$.

EXAMPLE 4. Compute the change in $y = f(x) = x^2 - 2x + 5$ when x changes in value from *(a)* $x = 3$ to $x = 3.2$, *(b)* $x = 3$ to $x = 2.9$.

(a) Take $x_0 = 3$ and $\Delta x = 0.2$. Then $y_0 = f(x_0) = f(3) = 8$, $y_0 + \Delta y = f(x_0 + \Delta x) = f(3.2) = 8.84$, and $\Delta y = 8.84 - 8 = 0.84$.

(b) Take $x_0 = 3$ and $\Delta x = -0.1$. Then $y_0 = f(3) = 8$, $y_0 + \Delta y = f(2.9) = 7.61$, and $\Delta y = 7.61 - 8 = -0.39$.

(See Problems 16.5–16.6.)

THE DERIVATIVE. The *derivative* of $y = f(x)$ at $x = x_0$ is

$$\lim_{\Delta x \to 0} \frac{\Delta y}{\Delta x} = \lim_{\Delta x \to 0} \frac{f(x_0 + \Delta x) - f(x_0)}{\Delta x}$$

provided the limit exists. $\dfrac{\Delta y}{\Delta x}$ is called the *difference quotient*.

To find derivatives, we shall use the following five-step rule:

(1) Write $y_0 = f(x_0)$.
(2) Write $y_0 + \Delta y = f(x_0 + \Delta x)$.
(3) Obtain $\Delta y = f(x_0 + \Delta x) - f(x_0)$.
(4) Obtain $\Delta y / \Delta x$.
(5) Evaluate $\lim\limits_{\Delta x \to 0} \dfrac{\Delta x}{\Delta y}$. The result is the derivative of $y = f(x)$ at $x = x_0$.

EXAMPLE 5. Find the derivative of $y = f(x) = 2x^2 - 3x + 5$ at $x = x_0$.

(1) $$y_0 = f(x_0) = 2x_0^2 - 3x_0 + 5$$

(2) $$\begin{aligned} y_0 + \Delta y = f(x_0 + \Delta x) &= 2(x_0 + \Delta x)^2 - 3(x_0 + \Delta x) + 5 \\ &= 2x_0^2 + 4x_0 \cdot \Delta x + 2(\Delta x)^2 - 3x_0 - 3 \cdot \Delta x + 5 \end{aligned}$$

(3) $$\Delta y = f(x_0 + \Delta x) - f(x_0) = 4x_0 \cdot \Delta x - 3 \cdot \Delta x + 2(\Delta x)^2$$

(4) $$\frac{\Delta y}{\Delta x} = 4x_0 - 3 + 2 \cdot \Delta x$$

(5) $$\lim_{\Delta x \to 0} \frac{\Delta y}{\Delta x} = \lim_{\Delta x \to 0} (4x_0 - 3 + 2 \cdot \Delta x) = 4x_0 - 3.$$

If in the example above the subscript 0 is deleted, the five-step rule yields a function of x (here, $4x - 3$) called the derivative with respect to x of the given function. The derivative with respect to x of the function $y = f(x)$ is denoted by one of the symbols y', $\dfrac{dy}{dx}$, $f'(x)$, or $D_x y$.

Provided it exists, the value of the derivative for any given value of x, say x_0, will be denoted by $y'\Big|_{x=x_0}$, $\dfrac{dy}{dx}\Big|_{x=x_0}$, or $f'(x_0)$. (See Problems 16.8–16.14.)

HIGHER-ORDER DERIVATIVES. The process of finding the derivative of a given function is called *differentiation*.

By differentiation, we obtain from a given function $y = f(x)$ another function $y' = f'(x)$ which will now be called the *first derivative* of y or of $f(x)$ with respect to x. If, in turn, $y' = f'(x)$ is differentiated with respect to x, another function $y'' = f''(x)$, called the *second derivative* of y or of $f(x)$ is obtained. Similarly, a third derivative may be found, and so on.

EXAMPLE 6. Let $y = f(x) = x^4 - 3x^2 + 8x + 6$. Then $y' = f'(x) = 4x^3 - 6x + 8$, $y'' = f''(x) = 12x^2 - 6$, and $y''' = f'''(x) = 24x$. (See Problem 16.15.)

DIFFERENTIATION FORMULAS

I. If $y = f(x) = kx^n$, where k and n are constants, then $y' = f'(x) = knx^{n-1}$. (See for example, Problem 16.16.)

II. If $y = f(x) \pm g(x)$, then $y' = f'(x) \pm g'(x)$ provided $f'(x)$ and $g'(x)$ exist.

III. If $y = k \cdot u^n$, where k and n are constants and u is a function of x, then $y' = knu^{n-1} \cdot u'$, provided u' exists. This is a form of the chain rule.

EXAMPLE 7. Find y', given (*a*) $y = 8x^{5/4}$, (*b*) $y = (x^2 + 4x - 1)^{3/2}$.

(*a*) Here $k = 8$, $n = \frac{5}{4}$. Then $y' = knx^{n-1} = 8 \cdot \frac{5}{4}x^{5/4-1} = 10x^{1/4}$.
(*b*) Let $u = x^2 + 4x - 1$ so that $y = u^{3/2}$. Then differentiating with respect to x, $u' = 2x + 4$ and

$$y' = \tfrac{3}{2}u^{1/2} \cdot u' = \tfrac{3}{2}\sqrt{x^2 + 4x - 1}(2x + 4) = 3(x + 2)\sqrt{x^2 + 4x - 1}$$

(See Problem 16.18.)

IV. If $y = f(x) \cdot g(x)$, then $y' = f(x) \cdot g'(x) + g(x) \cdot f'(x)$, provided $f'(x)$ and $g'(x)$ exist. (For the derivation, see Problem 16.19.)

EXAMPLE 8. Find y' when $y = (x^3 + 3x^2 + 1)(x^2 + 2)$.
Take $f(x) = x^3 + 3x^2 + 1$ and $g(x) = x^2 + 2$. Then $f'(x) = 3x^2 + 6x$, $g'(x) = 2x$, and

$$\begin{aligned} y' &= f(x) \cdot g'(x) + g(x) \cdot f'(x) \\ &= (x^3 + 3x^2 + 1)(2x) + (x^2 + 2)(3x^2 + 6x) \\ &= 5x^4 + 12x^3 + 6x^2 + 14x \end{aligned}$$

V. If $y = \dfrac{f(x)}{g(x)}$, then $y' = \dfrac{g(x) \cdot f'(x) - f(x) \cdot g'(x)}{[g(x)]^2}$, when $f'(x)$ and $g'(x)$ exist and $g(x) \neq 0$. (For a derivation, see Problem 16.21.)

EXAMPLE 9. Find y', given $y = \dfrac{x + 1}{x^2 + 1}$.

Take $f(x) = x + 1$ and $g(x) = x^2 + 1$. Then

$$y' = \frac{g(x) \cdot f'(x) - f(x) \cdot g'(x)}{[g(x)]^2} = \frac{(x^2 + 1)(1) - (x + 1)(2x)}{(x^2 + 1)^2} = \frac{1 - 2x - x^2}{(x^2 + 1)^2}.$$

INCREASING AND DECREASING FUNCTIONS. A function $y = f(x)$ is said to be an *increasing function* if y *increases* as x increases, and a *decreasing function* if y *decreases* as x *increases*.

Let the graph of $y = f(x)$ be as shown in Fig. 16-1. Clearly $y = f(x)$ is an increasing function from A to B and from C to D, and is a decreasing function from B to C and from D to E. At any point of the curve between A and B (also, between C and D), the inclination θ of the tangent line to the curve is acute;

hence, $f'(x) = \tan\theta > 0$. At any point of the curve between B and C (also, between D and E), the inclination θ of the tangent line is obtuse; hence, $f'(x) = \tan\theta < 0$.

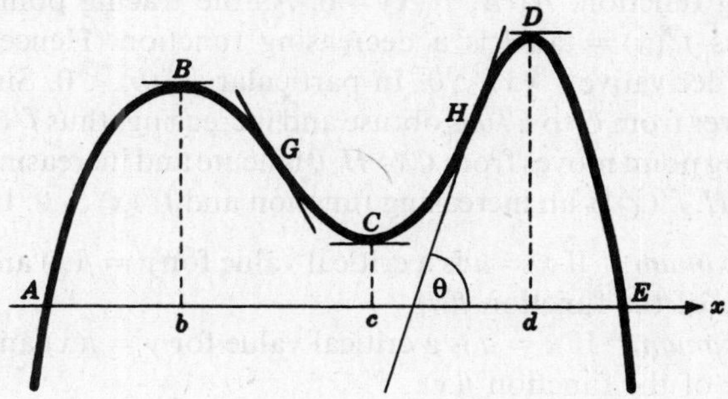

Fig. 16-1

Thus, for values of x for which $f'(x) > 0$, the function $f(x)$ is an increasing function; for values of x for which $f'(x) < 0$, the function is a decreasing function.

When $x = b, x = c$, and $x = d$, the function is neither increasing nor decreasing since $f'(x) = 0$. Such values of x are called *critical values* for the function $f(x)$.

EXAMPLE 10. For the function $f(x) = x^2 - 6x + 8, f'(x) = 2x - 6$.

Setting $f'(x) = 0$, we find the critical value $x = 3$. Now $f'(x) < 0$ when $x < 3$, and $f'(x) > 0$ when $x > 3$. Thus, $f(x) = x^2 - 6x + 8$ is a decreasing function when $x < 3$ and an increasing function when $x > 3$. (See Problem 16.24)

RELATIVE MAXIMUM AND MINIMUM VALUES. Let the curve of Fig. 16-1 be traced from left to right. Leaving A, the tracing point rises to B and then begins to fall. At B the ordinate $f(b)$ is greater than at any point of the curve near to B. We say that the point $B(b, f(b))$ is a *relative maximum point* of the curve or that the function $f(x)$ has a *relative maximum* $[= f(b)]$ when $x = b$. By the same argument $D(d, f(d))$ is also a relative maximum point of the curve or $f(x)$ has a relative maximum $[= f(d)]$ when $x = d$.

Leaving B, the tracing point falls to C and then begins to rise. At C the ordinate $f(c)$ is smaller than at any point of the curve near to C. We say that the point $C(c, f(c))$ is a *relative minimum point* of the curve or that $f(x)$ has a *relative minimum* $[= f(c)]$ when $x = c$.

Note that the relative maximum and minimum of this function occur at the critical values. While not true for all functions, the above statement is true for all of the functions considered in this chapter.

Test for relative maximum. If $x = a$ is a critical value for $y = f(x)$ and if $f'(x) > 0$ for all values of x less than but near to $x = a$ while $f'(x) < 0$ for all values of x greater than but near to $x = a$, then $f(a)$ is a relative maximum value of the function.

Test for relative minimum. If $x = a$ is a critical value for $y = f(x)$ and if $f'(x) < 0$ for all values of x less than but near to $x = a$ while $f'(x) > 0$ for all values of x greater than but near to $x = a$, then $f(a)$ is a relative minimum value of the function.

If as x increase, in value through a critical value, $x = a, f'(x)$ does not change sign, then $f(a)$ is neither a relative maximum nor a relative minimum value of the function.

EXAMPLE 11. For the function of Example 1, the critical value is $x = 3$.

Since $f'(x) = 2(x - 3) < 0$ for $x < 3$ and $f'(x) > 0$ for $x > 3$, the given function has a relative minimum value $f(3) = -1$.

In geometric terms, the point $(3, -1)$ is a relative minimum point of the curve $y = x^2 - 6x + 8$. (See Problems 16.24 and 16.25.)

ANOTHER TEST FOR MAXIMUM AND MINIMUM VALUES. At A on the curve of Fig. 16-1, the inclination θ of the tangent line is acute. As the tracing point moves from A to B, θ decreases; thus $f'(x) = \tan\theta$ is a decreasing function. At B, $f'(x) = 0$. As the tracing point moves from B to G, θ is obtuse and decreasing; thus $f'(x) = \tan\theta$ is a decreasing function. Hence, from A to G, $f'(x)$ is a decreasing function and its derivative $f''(x) < 0$. In particular, $f''(b) < 0$. Similarly, $f''(d) < 0$.

As the tracing point moves from G to C, θ is obtuse and increasing; thus $f'(x)$ is an increasing function. At C, $f'(x) = 0$. As the tracing point moves from C to H, θ is acute and increasing; thus $f'(x)$ is an increasing function. Hence, from G to H, $f'(x)$ is an increasing function and $f''(x) > 0$. In particular, $f''(c) > 0$.

Test for relative maximum. If $x = a$ is a critical value for $y = f(x)$ and if $f''(a) < 0$, then $f(a)$ is a relative maximum value of the function $f(x)$.

Test for relative minimum. If $x = a$ is a critical value for $y = f(x)$ and if $f''(a) > 0$, then $f(a)$ is a relative minimum value of the function $f(x)$.

The test fails when $f''(a) = 0$. When this occurs, the tests of the preceding section must be used. (See Problem 16.3.)

CONCAVITY. Suppose that $f(x)$ is a differentiable function on (a, b). Then, if $f'(x)$ is increasing on (a, b), we call f concave upward on (a, b). See Fig. 16-2(a). If $f'(x)$ is decreasing on (a, b), we say f is concave downward on (a, b). See Fig. 16-2(b).

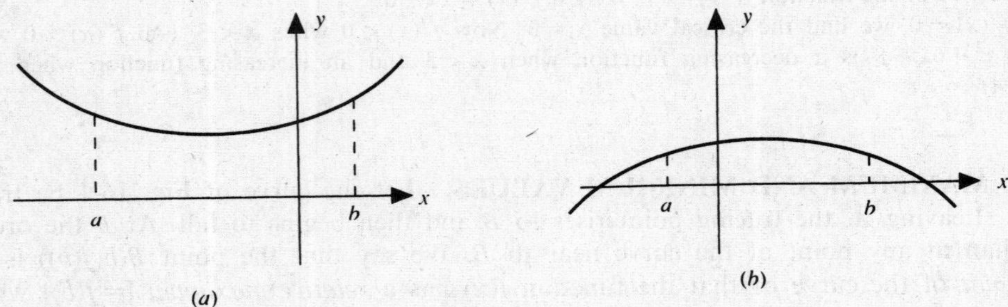

Fig. 16-2

INFLECTION POINT OF A CURVE. If at $x = a$, not necessarily a critical value for $f(x)$, the concavity changes from downward to upward or upward to downward, $(a, f(a))$ is an inflection point of $f(x)$. See Fig. 16-3.

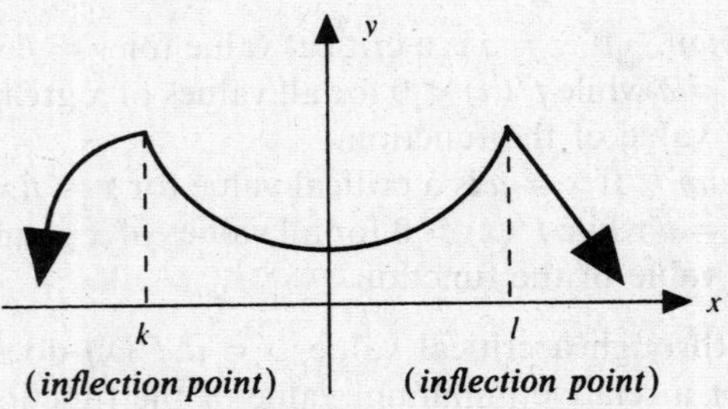

Fig. 16-3

In Fig. 16-1, G and H are inflection points of the curve. Note that at points between A and G the tangent lines to the curve lie above the curve, at points between G and H the tangent lines lie below the curve, and at points between H and E the tangent lines line above the curve. At G and H, the points of inflection, the tangent line *crosses* the curve. Thus, $f''(x)$ must be zero at an inflection point and change sign there.

EXAMPLE 12. For the function $f(x) = x^2 - 6x + 8$ of Example 1, $f'(x) = 2x - 6$ and $f''(x) = 2$. At the critical value $x = 3$, $f''(x) > 0$; hence $f(3) = -1$ is a relative minimum value. Since $f''(x) = 2 \neq 0$, the parabola $y = x^2 - 6x + 8$ has no inflection point. (See Problem 16.4.)

EXAMPLE 13. For the function $f(x) = x^3 + x^2 + x$, $f''(x) = 6x + 2 = 0$ when $x = -\frac{1}{3}$. Since $f'''(-\frac{1}{3}) \neq 0$, concavity must be changing when $x = -\frac{1}{3}$.

VELOCITY AND ACCELERATION. Let a particle move along a horizontal line and let its distance (in meters) at time $t \geq 0$ (in seconds) from a fixed point O of the line be given by $s = f(t)$. Let the positive direction on the line be to the right (that is, the direction of increasing s). A complete description of the motion may be obtained by examining $f(t)$, $f''(t)$, and $f''(t)$. It was noted in Chapter 45 that $f'(t)$ gives the velocity v of the particle. The *acceleration* of the particle is given by a $a = f''(t)$.

EXAMPLE 14. Discuss the motion of a particle which moves along a horizontal line according to the equation $s = t^3 - 6t^2 + 9t - 2$.

When $t = 0$, $s = f(0) = -2$. The particle begins its motion from $A(s = -2)$ See Fig. 16-4.

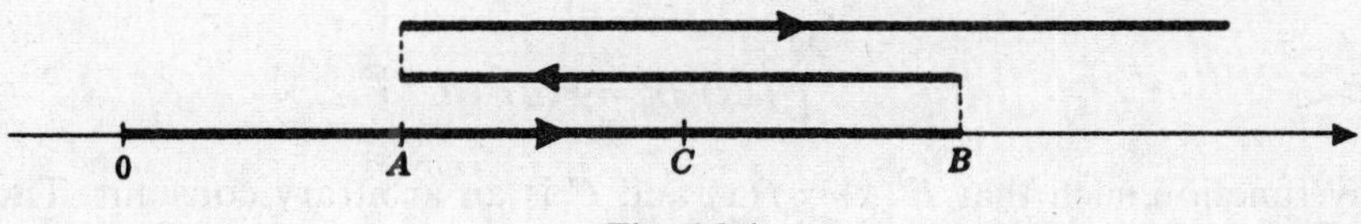

Fig. 16-4

Direction of Motion. Here $v = f'(t) = 3t^2 - 12t + 9 = 3(t-1)(t-3)$. When $t = 0$, $v = f'(0) = 9$); the particle leaves A with initial velocity 9 m/s.

Now $v = 0$ when $t = 1$ and $t = 3$. Thus, the particle moves (from A) for 1 s, stops momentarily ($v = 0$, when $t = 1$), moves off for two more seconds, stops momentarily, and then moves off indefinitely.

On the interval $0 < t < 1$, $v > 0$. Now $v > 0$ indicates that s is increasing; thus the body leaves A with initial velocity 9 ft/s and moves to the right for 1 s to B $[s = f(1) = 6]$ where it stops momentarily.

On the interval $1 < t < 3$, $v < 0$. Now $v < 0$ indicates that s is decreasing; thus the particle leaves B and moves to the left for 2 s to $A[s = f(3) = 2]$ where it stops momentarily.

On the interval $t > 3$, $v > 0$. The particle leaves A for the second time and moves to the right indefinitely.

Velocity and Speed. We have $a = f''(t) = 6t - 12 = 6(t-2)$. The acceleration is 0 when $t = 2$.

On the interval $0 < t < 2$, $a < 0$. Now $a < 0$ indicates that v is decreasing; thus the particle moves for the first 2 s with decreasing velocity. For the first second (from A to B) the velocity decreases from $v = 9$ to $v = 0$. The speed (numerical value of the velocity) decreases from 9 to 0; that is, the particle "slows up." When $t = 2$, $f(t) = 4$ (the particle is at C) and $f'(t) = -3$. Thus from B to $C(t = 1$ to $t = 2)$, the velocity decreases from $v = 0$ to $v = -3$. On the other hand, the speed increases from 0 to 3; that is, the particle "speeds up."

On the interval $t > 2$, $a > 0$; thus the velocity is increasing. From C to A ($t = 2$ to $t = 3$) the velocity increases from $v = -3$ to $v = 0$ while the speed decreases from 3 to 0. Thereafter ($t > 3$) both the velocity and speed increase indefinitely. (See Problem 16.32.)

DIFFERENTIALS. Let $y = f(x)$. Define dx (read, differential x) by the relation $dx = \Delta x$ and define dy (read, differential y) by the relation $dy = f'(x) \cdot dx$. Note $dy \neq \Delta y$.

EXAMPLE 15. If $y = f(x) = x^3$, then

$$\Delta y = (x + \Delta x)^3 - x^3 = 3x^2 \cdot \Delta x + 3x(\Delta x)^2 + (\Delta x)^3 = 3x^2 \cdot dx + 3x(dx)^2 + (dx)^3$$

while $dy = f'(x) \cdot dx = 3x^2 \cdot dx$. Thus, if dx is small numerically, dy is a fairly good approximation of Δy and simple to compute.

Suppose now that $x = 10$ and $dx = \Delta x = .01$. Then for the function above, $\Delta y = 3(10)^2(.01) + 3(10)(.01)^2 + (.01)^3 = 3.0031$ while $dy = 3(10)^2(.01) = 3$.

IF $F(x)$ IS A FUNCTION whose derivative $F'(x) = f(x)$, then $F(x)$ is called an *antiderivative* of $f(x)$.

For example, $F(x) = x^3$ is an antiderivative of $f(x) = 3x^2$ since $F'(x) = 3x^2 = f(x)$. Also, $G(x) = x^3 + 5$ and $H(x) = x^3 - 6$ are antiderivatives of $f(x) = 3x^2$. Why?

If $F(x)$ and $G(x)$ are two distinct antiderivatives of $f(x)$, then $F(x) = G(x) + C$, where C is a constant. (See Problem 16.30)

THE INDEFINITE INTEGRAL OF $f(x)$ is denoted by $\int f(x)\,dx$, and is the most general antiderivative of $f(x)$—that is

$$\int f(x)\,dx = F(x) + C$$

where $F(x)$ is any function such that $F'(x) = f(x)$ and C is an arbitrary constant. Thus the indefinite integral of $f(x) = 3x^2$ is $\int 3x^2 dx = x^3 + C$.

We shall use the following antidifferentiation formulas:

I. $\int x^n\,dx = \dfrac{x^{n+1}}{n+1} + C$, where $n \neq -1$

II. $\int cf(x)\,dx = c\int f(x)\,dx$, where c is a constant

III. $\int [f(x) + g(x)]\,dx = \int f(x)\,dx + \int g(x)\,dx$

EXAMPLE 16

(*a*) $\int x^5 dx = \dfrac{x^{5+1}}{5+1} + C = \dfrac{x^6}{6} + C$

(*b*) $\int 4x^3\,dx = 4\int x^3\,dx = 4\left[\dfrac{x^4}{4} + C\right] = x^4 + 4C$, but if we call $4C$ by the name C, then C_1 still represents an arbitrary constant; so we can simply write $\int 4x^3\,dx = x^4 + C$

(*c*) $\int 3x\,dx = 3\int x\,dx = 3 \cdot \dfrac{x^2}{2} + C = \dfrac{3}{2}x^2 + C$

(*d*) $\int \dfrac{dx}{x^3} = \int x^{-3}\,dx = \dfrac{x^{-2}}{-2} + C = -\dfrac{1}{2x^2} + C$

(*e*) $\int (x^5 + 4x^3 + 3x)\,dx = \int x^5\,dx + \int 4x^3\,dx + \int 3x\,dx = \dfrac{x^6}{6} + x^4 - \dfrac{3x^2}{2} + C$

(See Problems 16.37–16.43.)

AREA BY SUMMATION. Consider the area A bounded by curve $y = f(x) \geq 0$, the x axis, and the ordinates $x = a$ and $x = b$, where $b > a$.

Let the interval $a \leq x \leq b$ be divided into n equal parts each of length Δx. At each point of subdivision, construct the ordinate, thus dividing the area into n strips, as in Fig. 16-5. Since the areas of the strips are unknown, we propose to approximate each strip by a rectangle whose area can be found. In Fig. 16-6, a representative strip and its approximating rectangle are shown.

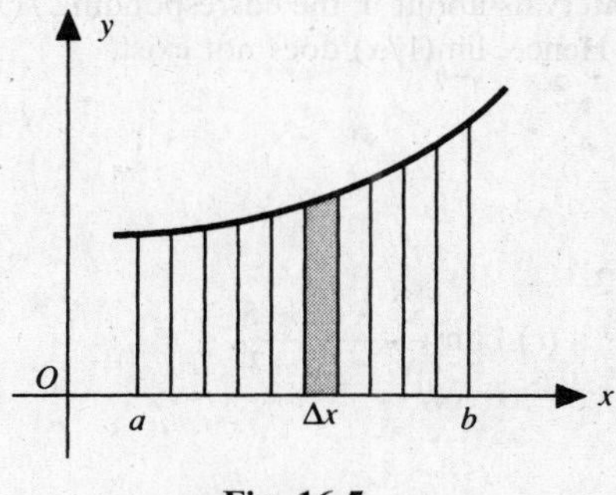

Fig. 16-5

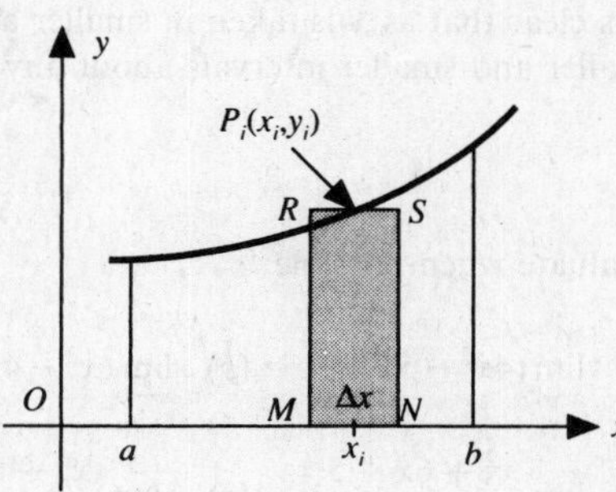

Fig. 16-6

Suppose the representative strip is the ith strip counting from the left, and let $x = x_i$ be the coordinate of the midpoint of its base. Denote by $y_i = f(x_i)$ the ordinate of the point P_i (on the curve) whose abscissa is x_i. Through P_i pass a line parallel to the x axis and complete the rectangle $MRSN$. This rectangle of area $y_i \Delta x$ is the approximating rectangle of the ith strip. When each strip is treated similarly, it seems reasonable to take

$$y_1 \Delta x + y_2 \Delta x + y_3 \Delta x + \cdots + y_n \Delta x = \sum_{i=1}^{n} y_i \Delta x$$

as an approximation of the area sought.

Now suppose that the number of strips (with approximating rectangles) is indefinitely increased so that $\Delta x \rightarrow 0$. It is evident from the figure that by so increasing the number of approximating rectangles the sum of their areas more nearly approximates the area sought; that is,

$$A = \lim_{n \rightarrow \infty} \sum_{i=1}^{n} y_i \Delta x$$

IF WE DEFINE $\int_a^b f(x)\, dx$ (read, the *definite integral* of $f(x)$ between $x = a$ and $x = b$) as

$$\int_a^b f(x)\, dx = F(x)|_a^b = F(b) - F(a), \qquad (F'(x) = f(x))$$

then the area bounded by $y = f(x) \geq 0$, the x axis, and the ordinates $x = a$ and $x = b$ $(b > a)$ is given by

$$A = \int_a^b f(x)\, dx$$

(See Problems 16.44–16.47.)

Solved Problems

16.1 Investigate $f(x) = 1/x$ for values of x near $x = 0$.

If x is near 0, say $-.01 < x < .01$, then $\dfrac{1}{-.01} < \dfrac{1}{x} < \dfrac{1}{.01}$ or $-100 < \frac{1}{x} < 100$.

If x is nearer to 0, say $-.0001 < x < .0001$, then $\dfrac{1}{-.0001} < \dfrac{1}{x} < \dfrac{1}{.0001}$ or $-10\,000 < \frac{1}{x} < 10\,000$.

It is clear that as x is taken in smaller and smaller intervals about 0, the corresponding $f(x)$ does *not* lie in smaller and smaller intervals about any number M. Hence, $\lim_{x\to 0}(1/x)$ does not exist.

16.2 Evaluate when possible:

(*a*) $\lim_{x\to 2}(4x^2 - 5x)$, (*b*) $\lim_{x\to 1}(x^2 - 4x + 10)$, (*c*) $\lim_{x\to 2}\dfrac{x^2+6x+5}{x^2-2x-3}$,

(*d*) $\lim_{x\to 3}\dfrac{x^2+6x+5}{x^2-2x-3}$, (*e*) $\lim_{x\to -1}\dfrac{x^2+6x+5}{x^2-2x-3}$.

(*a*) $\lim_{x\to 2}(4x^2 - 5x) = \lim_{x\to 2}4x^2 - \lim_{x\to 2}5x = 4\lim_{x\to 2}x^2 - 5\lim_{x\to 2}x = 4\cdot 4 - 5\cdot 2 = 6.$

(*b*) $\lim_{x\to 1}(x^2 - 4x + 10) = (1)^2 - 4\cdot 1 + 10 = 7.$

(*c*) $\lim_{x\to 2}(x^2 + 6x + 5) = 21$ and $\lim_{x\to 2}(x^2 - 2x - 3) = -3$; hence

$$\lim_{x\to 2}\frac{x^2+6x+5}{x^2-2x-3} = \frac{\lim_{x\to 2}(x^2+6x+5)}{\lim_{x\to 2}(x^2-2x-3)} = \frac{21}{-3} = -7$$

(*d*) $\lim_{x\to 3}(x^2 + 6x + 5) = 32$ and $\lim_{x\to 3}(x^2 - 2x - 3) = 0$; hence $\lim_{x\to 3}\dfrac{x^2+6x+5}{x^2-2x-3}$ does not exist.

(*e*) $\lim_{x\to -1}(x^2 + 6x + 5) = 0$ and $\lim_{x\to -1}(x^2 - 2x - 3) = 0$. Then, when $x \neq -1$,

$$\frac{x^2+6x+5}{x^2-2x-3} = \frac{(x+5)(x+1)}{(x-3)(x+1)} = \frac{x+5}{x-3} \quad\text{and}\quad \lim_{x\to -1}\frac{x^2+6x+5}{x^2-2x-3} = \lim_{x\to -1}\frac{x+5}{x-3} = \frac{4}{-4} = -1$$

16.3 Tell why each graph in Fig. 16-7 is not continuous at a.

(*a*) $y = g(x)$ exists at a and $\lim_{x\to a}g(x)$ exists, but $\lim_{x\to a}g(x) \neq g(a)$.

(*b*) $\lim_{x\to a}h(x)$ does not exist since $\lim_{x\to a^+}h(x) \neq \lim_{x\to a^-}h(x)$.

(*c*) $f(a)$ is not defused.

16.4 Discuss the continuity of $y = \sqrt{x-1}$. See Fig. 16-8.

Here, $f(1) = \sqrt{1-1} = 0$; $\lim_{x\to 1^+}\sqrt{x-1} = 0$, thus $f(x)$ is continuous on $[1, \infty]$.

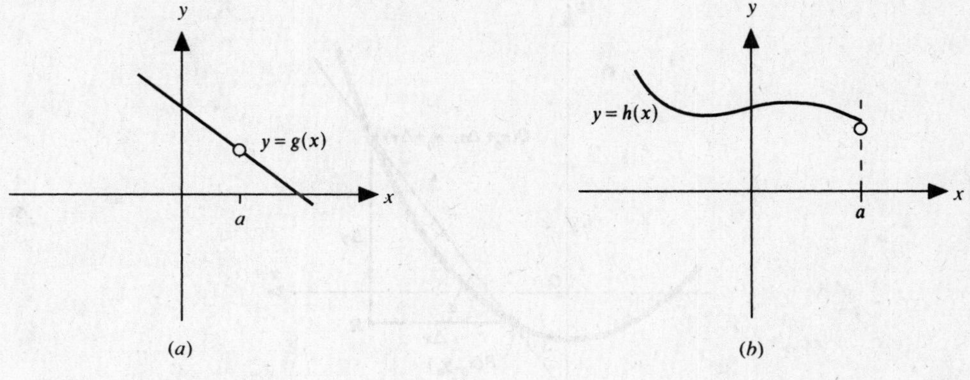

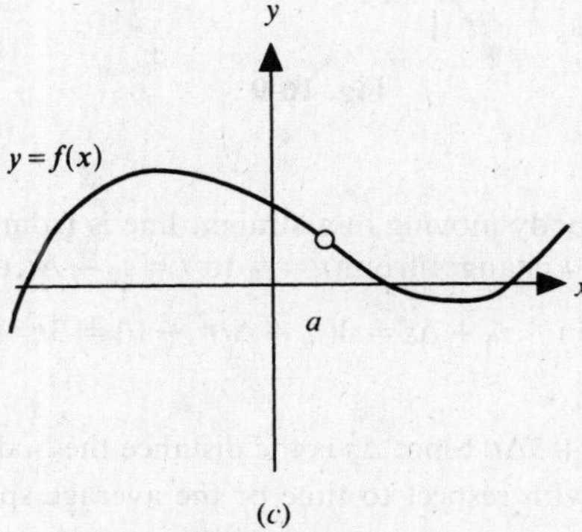

Fig. 16-7

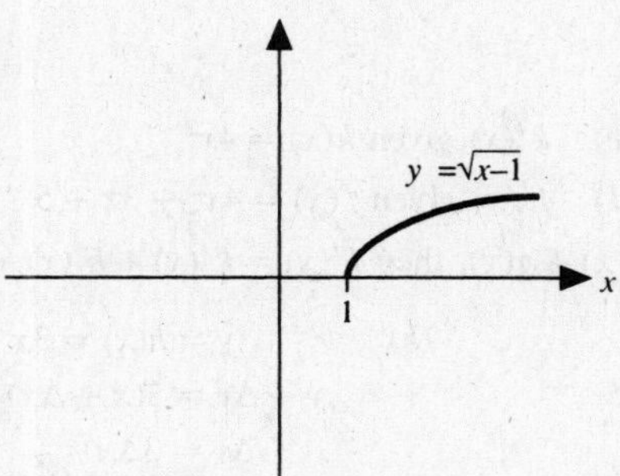

Fig. 16-8

16.5 Let $P(x_0, y_0)$ and $Q(x_0 + \Delta x, y_0 + \Delta y)$ be two distinct points on the parabola $y = x^2 - 3$. Compute $\Delta y/\Delta x$ and interpret.

Here
$$y_0 = x_0^2 - 3$$
$$y_0 + \Delta y = (x_0 + \Delta x)^2 - 3 = x_0^2 + 2x_0 \cdot \Delta x + (\Delta x)^2 - 3$$
$$\Delta y = [x_0^2 + 2x_0 \cdot \Delta x + (\Delta x)^2 - 3] - [x_0^2 - 3] = 2x_0 \cdot \Delta x + (\Delta x)^2$$
and
$$\frac{\Delta y}{\Delta x} = 2x_0 + \Delta x$$

In Fig. 16-9, PR is parallel to the x axis and QR is parallel to the y axis. If α denotes the inclination of the secant line PQ, $\tan \alpha = \Delta y/\Delta x$; thus, $\Delta y/\Delta x$ is the slope of the secant line PQ.

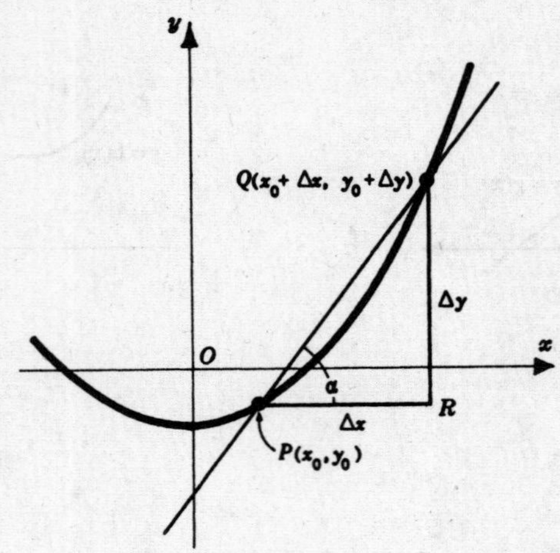

Fig. 16-9

16.6 If $s = 3t^2 + 10$ is the distance a body moving in a straight line is from a fixed point O of the line at time t, (*a*) find the change Δs is s when t changes from $t = t_0$ to $t = t_0 + \Delta t$, (*b*) find $\Delta s/\Delta t$ and interpret.

(*a*) Here $s_0 = 3t_0^2 + 10$. Then $s_0 + \Delta s = 3(t_0 + \Delta t)^2 + 10 = 3t_0^2 + 6t_0 \cdot \Delta t + 3(\Delta t)^2 + 10$ and $\Delta s = 6t_0 \cdot \Delta t + 3(\Delta t)^2$.

(*b*) $\dfrac{\Delta s}{\Delta t} = \dfrac{6t_0 \cdot \Delta t + 3(\Delta t)^2}{\Delta t} = 6t_0 + 3\Delta t$. Since Δs is the distance the body moves in time Δt, $\dfrac{\Delta s}{\Delta t}$ is the average rate of change of distance with respect to time or the average speed of the body in the interval t_0 to $t_0 + \Delta t$.

16.7 Find

(*a*) $g'(x)$, given $g(x) = 5$ (*c*) $k'(x)$, given $k(x) = 4x^2$

(*b*) $h'(x)$, given $h(x) = 3x$ (*d*) $f'(x)$, given $f(x) = 4x^2 + 3x + 5$

Thus verify: If $f(x) = k(x) + h(x) + g(x)$, then $f'(x) = k'(x) + h'(x) + g'(x)$.

(*a*)
$$\begin{aligned} y &= g(x) = 5 \\ y + \Delta y &= g(x + \Delta x) = 5 \\ \Delta y &= 0 \\ \frac{\Delta y}{\Delta x} &= 0 \\ g'(x) &= \lim_{\Delta x \to 0} 0 = 0 \end{aligned}$$

(*b*)
$$\begin{aligned} y &= h(x) = 3x \\ y + \Delta y &= 3(x + \Delta x) = 3x + 3\Delta x \\ \Delta y &= 3\Delta x \\ \frac{\Delta y}{\Delta x} &= 3 \\ h'(x) &= \lim_{\Delta x \to 0} 3 = 3 \end{aligned}$$

(*c*)
$$\begin{aligned} y &= k(x) = 4x^2 \\ y + \Delta y &= 4(x + \Delta x)^2 = 4x^2 + 8x \cdot \Delta x + 4(\Delta x)^2 \\ \Delta y &= 8x \cdot \Delta x + 4(\Delta x)^2 \\ \frac{\Delta y}{\Delta x} &= 8x + 4\Delta x \\ k'(x) &= \lim_{\Delta x \to 0} (8x + 4\Delta x) = 8x \end{aligned}$$

(*d*) $y = f(x) = 4x^2 + 3x + 5$

$$y + \Delta y = 4(x + \Delta x)^2 + 3(x + \Delta x) + 5 = 4x^2 + 8x \cdot \Delta x + 4(\Delta x)^2 + 3x + 3\Delta x + 5$$
$$\Delta y = 8x \cdot \Delta x + 3\Delta x + 4(\Delta x)^2$$
$$\frac{\Delta y}{\Delta x} = 8x + 3 + 4\Delta x$$
$$f'(x) = \lim_{\Delta x \to 0} (8x + 3 + 4\Delta x) = 8x + 3$$

Thus $f'(x) = k'(x) + h'(x) + g'(x) = 8x + 3 + 0$.

16.8 Place a straight edge along PQ in Fig. 45-3. Keeping P fixed, let Q move along the curve toward P and thus verify that the straight edge approaches the tangent line PT as limiting position.

Now as Q moves toward P, $\Delta x \to 0$ and $\lim_{\Delta x \to 0} \frac{\Delta y}{\Delta x} = \lim_{\Delta x \to 0} (2x_0 + \Delta x) = 2x_0$. Thus the slope of the tangent line to $y = f(x) = x^2 - 3$ at the point $P(x_0, y_0)$ is $m = f'(x_0) = 2x_0$.

16.9 Find the slope and equation of the tangent line to the given curve $y = f(x)$ at the given point:
(*a*) $y = 2x^3$ at (1, 2), (*b*) $y = -3x^2 + 4x + 5$ at (3, −10), (*c*) $y = x^2 - 4x + 3$ at (2, −1).

(*a*) By the five-step rule, $f'(x) = 6x^2$; then the slope $m = f'(1) = 6$. The equation of the tangent line at (1,2) is $y - 2 = 6(x - 1)$ or $6x - y - 4 = 0$.

(*b*) Here $f'(x) = -6x + 4$ and $m = f'(3) = -14$. The equation of the tangent line at (3, −10) is $14x + y - 32 = 0$.

(*c*) Here $f'(x) = 2x - 4$ and $m = f'(2) = 0$. The equation of the tangent line at (2, −1) is $y + 1 = 0$. Identify the given point on the parabola.

16.10 Find the equation of the tangent line to the ellipse $4x^2 + 9y^2 = 25$ at (*a*) the point (2, 1), (*b*) the point $(0, \frac{5}{3})$.

Let $P(x, y)$ and $Q(x + \Delta x, y + \Delta y)$ be two nearby points on the ellipse. (Why should P not be taken at an extremity of the major axis?) Then

$$4x^2 + 9y^2 = 25 \tag{1}$$

$$4x^2 + 8x \cdot \Delta x + 4(\Delta x)^2 + 9y^2 + 18y \cdot \Delta y + 9(\Delta y)^2 = 25 \tag{2}$$

Subtracting (*1*) from (*2*), $8x \cdot \Delta x + 4(\Delta x)^2 + 18y \cdot \Delta y + 9(\Delta y)^2 = 0$. Then

$$\Delta y(18y + 9\Delta y) = -\Delta x(8x + 4 \cdot \Delta x) \quad \text{and} \quad \frac{\Delta y}{\Delta x} = -\frac{8x + 4\Delta y}{18y + 9\Delta y}$$

When Q moves along the curve toward P, $\Delta x \to 0$ and $\Delta y \to 0$. Then $m = \lim_{\Delta x \to 0} \frac{\Delta y}{\Delta x} = -\frac{4x}{9y}$.

(*a*) At point (2, 1), $m = -\frac{8}{9}$ and the equation of the tangent line is $8x + 9y - 25 = 0$.

(*b*) At point $(0, \frac{5}{3})$, $m = 0$ and the equation of the tangent line is $y - \frac{5}{3} = 0$.

16.11 The normal line to a curve at a point P on it is perpendicular to the tangent line at P. Find the equation of the normal line to the given curve at the given point of (*a*) Problem 16.9(*b*), (*b*) Problem 16.10.

(*a*) The slope of the tangent line is −14; the slope of the normal line is $\frac{1}{14}$. The equation of the normal line is $y + 10 = \frac{1}{14}(x - 3)$ or $x - 14y - 143 = 0$.

(*b*) The slope of the tangent line at (2, 1) is $-\frac{8}{9}$; the slope of the normal line is $\frac{9}{8}$. The equation of the normal line is $9x - 8y - 10 = 0$.

At the point $(0, \frac{5}{3})$ the normal line is vertical. Its equation is $x = 0$.

16.12 If $s = f(t)$ is the distance of a body, moving in a straight line, from a fixed point O of the line at time t, then (see Problem 16.6) $\dfrac{\Delta s}{\Delta t} = \dfrac{f(t+\Delta t) - f(t)}{\Delta t}$ is the average speed of the body in the interval of time t to $t + \Delta t$ and

$$v = s' = \lim_{\Delta t \to 0} \frac{\Delta s}{\Delta t} = \lim_{\Delta t \to 0} \frac{f(t+\Delta t) - f(t)}{\Delta t}$$

is the *instantaneous speed of the body* at time t. For $s = 3t^2 + 10$ of Problem 16.6, find the (instantaneous) speed of the body at time (*a*) $t = 0$, (*b*) $t = 4$.

Here $v = s' = 6t$. (*a*) When $t = 0, v = 0$. (*b*) When $t = 4, v = 24$.

16.13 The height above the ground of a bullet shot vertically upward with initial velocity of 1152 ft/s is given by $s = 1152t - 16t^2$. Find (*a*) the velocity of the bullet 20 s after was fired and (*b*) the time required for the bullet to reach its maximum height and the maximum height attained.

Here $v = 1152 - 32t$.

(*a*) When $t = 20, v = 1152 - 32(20) = 512$ ft/s.

(*b*) At its maximum height, the velocity of the bullet is 0 ft/s. When $v = 1152 - 32t = 0, t = 36$ s. When $t = 36, s = 1152(36) - 16(36)^2 = 20\,736$ ft, the maximum height.

16.14 Find the derivative of each of the following polynomials:

(*a*) $f(x) = 3x^2 - 6x + 5$, (*b*) $f(x) = 2x^3 - 8x + 4$, (*c*) $f(x) = (x-2)^2(x-3)^2$.

(*a*) $f'(x) = 3 \cdot 2x^{2-1} - 6x^{1-1} + 0 = 6x - 6$.
(*b*) $f'(x) = 2 \cdot 3x^{3-1} - 8x^{1-1} + 0 = 6x^2 - 8$.
(*c*) Here $f(x) = x^4 - 10x^3 + 37x^2 - 60x + 36$. Then $f'(x) = 4x^3 - 30x^2 + 74x - 60 = (x-2)(x-3)(4x-10)$.

16.15 For each of the following functions, find $f'(x), f''(x)$, and $f'''(x)$:

(*a*) $f(x) = 2x^2 + 7x - 5$, (*b*) $f(x) = x^3 - 6x^2$, (*c*) $f(x) = x^5 - x^3 + 3x$.

(*a*) $f'(x) = 4x + 7,\ f''(x) = 4,\ f'''(x) = 0$
(*b*) $f'(x) = 3x^2 - 12x,\ f''(x) = 6x - 12,\ f'''(x) = 6$
(*c*) $f'(x) = 5x^4 - 3x^2 + 3,\ f''(x) = 20x^3 - 6x,\ f'''(x) = 60x^2 - 6$

16.16 Use the five-step rule to obtain y' when $y = 6x^{3/2}$.

We have

$$y = 6x^{3/2}$$
$$y + \Delta y = 6(x + \Delta x)^{3/2}$$
$$\Delta y = 6(x+\Delta x)^{3/2} - 6x^{3/2} = 6[(x+\Delta x)^{3/2} - x^{3/2}]$$

and

$$\frac{\Delta y}{\Delta x} = 6 \cdot \frac{(x+\Delta x)^{3/2} - x^{3/2}}{\Delta x} = 6 \cdot \frac{(x+\Delta x)^{3/2} - x^{3/2}}{\Delta x} \cdot \frac{(x+\Delta x)^{3/2} + x^{3/2}}{(x+\Delta x)^{3/2} + x^{3/2}}$$
$$= 6 \cdot \frac{(x+\Delta x)^3 - x^3}{\Delta x[(x+\Delta x)^{3/2} + x^{3/2}]} = 6 \cdot \frac{3x^2 + 3x \cdot \Delta x + (\Delta x)^2}{(x+\Delta x)^{3/2} + x^{3/2}}$$

Then

$$y' = \lim_{\Delta x \to 0} 6 \cdot \frac{3x^2 + 3x \cdot \Delta x + (\Delta x)^2}{(x+\Delta x)^{3/2} + x^{3/2}} = 6 \cdot \frac{3x^2}{2x^{3/2}} = 9x^{1/2}$$

(NOTE: By Formula I, with $k = 6$ and $n = \frac{3}{2}$, we find $y' = knx^{n-1} = 6 \cdot \frac{3}{2} x^{1/2} = 9x^{1/2}$.)

16.17 Use the five-step rule to find y' when $y = (x^2 + 4)^{1/2}$. Solve also by using Formula II.

We have
$$y = (x^2 + 4)^{1/2}$$
$$y + \Delta y = [(x + \Delta x)^2 + 4]^{1/2}$$
$$\Delta y = [(x + \Delta x)^2 + 4]^{1/2} - (x^2 + 4)^{1/2}$$

and
$$\frac{\Delta y}{\Delta x} = \frac{[(x + \Delta x)^2 + 4]^{1/2} - (x^2 + 4)^{1/2}}{\Delta x} \cdot \frac{[(x + \Delta x)^2 + 4]^{1/2} + (x^2 + 4)^{1/2}}{[(x + \Delta x)^2 + 4]^{1/2} + (x^2 + 4)^{1/2}}$$
$$= \frac{(x + \Delta x)^2 + 4 - (x^2 + 4)}{\Delta x\{[(x + \Delta x)^2 + 4]^{1/2} + (x^2 + 4)^{1/2}\}} = \frac{2x + \Delta x}{[(x + \Delta x)^2 + 4]^{1/2} + (x^2 + 4)^{1/2}}$$

Then
$$y' = \lim_{\Delta x \to 0} \frac{2x + \Delta x}{[(x + \Delta x)^2 + 4]^{1/2} + (x^2 + 4)^{1/2}} = \frac{2x}{2(x^2 + 4)^{1/2}} = \frac{x}{(x^2 + 4)^{1/2}}$$

Let $u = x^2 + 4$ so that $y = u^{1/2}$. Then $u' = 2x$ and $y' = \frac{1}{2}u^{-1/2} \cdot u' = \frac{1}{2}(x^2 + 4)^{-1/2} \cdot 2x = \frac{x}{\sqrt{x^2 + 4}}$.

16.18 Find y', given (*a*) $y = (2x - 5)^3$, (*b*) $y = \frac{2}{3}(x^6 + 4x^3 + 5)^2$, (*c*) $y = \frac{4}{x^2}$, (*d*) $y = \frac{1}{\sqrt{x}}$, (*e*) $y = 2(3x^2 + 2)^{1/2}$.

(*a*) Let $u = 2x - 5$ so that $y = u^3$. Then, differentiating with respect to x, $u' = 2$ and $y' = 3u^2 \cdot u' = 3(2x - 5)^2 \cdot 2 = 6(2x - 5)^2$.

(*b*) Let $u = x^6 + 4x^3 + 5$ so that $y = \frac{2}{3}u^2$. Differentiating with respect to x, $u' = 6x^5 + 12x^2$ and $y' = \frac{4}{3}u \cdot u' = \frac{4}{3}(x^6 + 4x^3 + 5)(6x^5 + 12x^2) = 8(x^6 + 4x^3 + 5)(x^5 + 2x^2)$.

(*c*) Here $y = 4x^{-2}$ and $y' = 4(-2)x^{-3} = -8/x^3$.

(*d*) Here $y = x^{-1/2}$ and $y' = (-\frac{1}{2})x^{-3/2} = -1/(2x\sqrt{x})$.

(*e*) Since $y = 2(3x^2 + 2)^{1/2}$, $y' = 2(\frac{1}{2})(3x^2 + 2)^{-1/2}(6x) = 6x/(3x^2 + 2)^{1/2}$.

16.19 Derive: If $y = f(x) \cdot g(x)$, then $y' = f(x) \cdot g'(x) + g(x) \cdot f'(x)$, provided $f'(x)$ and $g'(x)$ exist.

Let $u = f(x)$ and $v = g(x)$ so that $y = u \cdot v$.

As x changes to $x + \Delta x$, let u change to $u + \Delta u$, v change to $v + \Delta v$, and y change to $y + \Delta y$. Then

$$y + \Delta y = (u + \Delta u)(v + \Delta v) = uv + u \cdot \Delta v + v \cdot \Delta u + \Delta u \cdot \Delta v$$

$$\Delta y = u \cdot \Delta v + v \cdot \Delta u + \Delta u \cdot \Delta v$$

and
$$\frac{\Delta y}{\Delta x} = u \cdot \frac{\Delta v}{\Delta x} + v \cdot \frac{\Delta u}{\Delta x} + \Delta u \cdot \frac{\Delta v}{\Delta x}$$

Then
$$y' = \lim_{\Delta x \to 0}\left(u \cdot \frac{\Delta v}{\Delta x} + v \cdot \frac{\Delta u}{\Delta x} + \Delta u \cdot \frac{\Delta v}{\Delta x}\right)$$
$$= u \cdot v' + v \cdot u' + 0 \cdot v' = u \cdot v' + v \cdot u' = f(x) \cdot g'(x) + g(x) \cdot f'(x)$$

16.20 Find y', given (*a*) $y = x^5(1 - x^2)^4$, (*b*) $y = x^2\sqrt{x^2 + 4}$, (*c*) $y = (3x + 1)^2(2x^3 - 3)^{1/3}$.

(*a*) Set $f(x) = x^5$ and $g(x) = (1 - x^2)^4$. Then $f'(x) = 5x^4$, $g'(x) = 4(1 - x^2)^3(-2x)$, and

$$y' = f(x) \cdot g'(x) + g(x) \cdot f'(x) = x^5 \cdot 4(1 - x^2)^3(-2x) + (1 - x^2)^4 \cdot 5x^4$$
$$= x^4(1 - x^2)^3[-8x^2 + 5(1 - x^2)] = x^4(1 - x^2)^3(5 - 13x^2)$$

(*b*)
$$y = x^2 \cdot \tfrac{1}{2}(x^2+4)^{-1/2} \cdot 2x + (x^2+4)^{1/2} \cdot 2x$$
$$= x^3(x^2+4)^{-1/2} + 2x(x^2+4)^{1/2} = \frac{x^3 + 2x(x^2+4)}{(x^2+4)^{1/2}} = \frac{3x^3+8x}{\sqrt{x^2+4}}$$

(*c*) Here $y = (3x+1)^2(2x^3-3)^{1/3}$ and
$$y' = (3x+1)^2 \cdot \tfrac{1}{3}(2x^3-3)^{-2/3} \cdot 6x^2 + (2x^3-3)^{1/3} \cdot 2(3x+1) \cdot 3$$
$$= 2x^2(3x+1)^2(2x^3-3)^{-2/3} + 6(3x+1)(2x^3-3)^{1/3} = \frac{2x^2(3x+1)^2 + 6(3x+1)(2x^3-3)}{(2x^2-3)^{2/3}}$$
$$= \frac{2(3x+1)[x^2(3x+1)+3(2x^3-3)]}{(2x^3-3)^{2/3}} = \frac{2(3x+1)(9x^3+x^2-9)}{(2x^3-3)^{2/3}}$$

16.21 Prove: If $y = \dfrac{f(x)}{g(x)}$, if $f'(x)$ and $g'(x)$ exist, and if $g(x) \neq 0$, then $y' = \dfrac{g(x)\cdot f'(x) - f(x)\cdot g'(x)}{[g(x)]^2}$.

Let $u = f(x)$ and $v = g(x)$ so that $y = u/v$. Then
$$y + \Delta y = \frac{u+\Delta u}{v+\Delta v} \qquad \Delta y = \frac{u+\Delta u}{v+\Delta v} - \frac{u}{v} = \frac{v\cdot\Delta u - u\cdot\Delta v}{v(v+\Delta v)}$$

and
$$\frac{\Delta y}{\Delta x} = \frac{v\cdot\Delta u - u\cdot\Delta v}{\Delta x\cdot v(v+\Delta v)} = \frac{v\cdot\dfrac{\Delta u}{\Delta x} - u\cdot\dfrac{\Delta v}{\Delta x}}{v(v+\Delta v)}$$

Then
$$y' = \lim_{\Delta x\to 0}\frac{v\cdot\dfrac{\Delta u}{\Delta x} - u\cdot\dfrac{\Delta v}{\Delta x}}{v(v+\Delta v)} = \frac{v\cdot u' - u\cdot v'}{v^2} = \frac{g(x)\cdot f'(x) - f(x)\cdot g'(x)}{[g(x)]^2}$$

16.22 Find y', given (*a*) $y = 1/x^3$, (*b*) $y = \dfrac{2x}{x-3}$, (*c*) $y = \dfrac{x+5}{x^2-1}$, (*d*) $y = \dfrac{x^3}{\sqrt{4-x^2}}$, (*e*) $y = \dfrac{\sqrt{2x-3x^2}}{x+1}$.

(*a*) Take $f(x) = 1$ and $g(x) = x^3$; then
$$y' = \frac{g(x)\cdot f'(x) - f(x)\cdot g'(x)}{[g(x)]^2} = \frac{x^3\cdot 0 - 1\cdot 3x^2}{(x^3)^2} = -\frac{3}{x^4}$$

Note that it is simpler here to write $y = x^{-3}$ and $y' = -3x^{-4} = -3/x^4$.

(*b*) Take $f(x) = 2x$ and $g(x) = x-3$; then
$$y' = \frac{g(x)\cdot f'(x) - f(x)\cdot g'(x)}{[g(x)]^2} = \frac{(x-3)\cdot 2 - 2x\cdot 1}{(x-3)^2} = \frac{-6}{(x-3)^2}$$

Note that $y = \dfrac{2x}{x-3} = 2 + \dfrac{6}{x-3} = 2 + 6(x-3)^{-1}$ and $y' = 6(-1)(x-3)^{-2} = -6/(x-3)^2$.

(*c*)
$$y' = \frac{(x^2-1)(1) - (x+5)(2x)}{(x^2-1)^2} = -\frac{1+10x+x^2}{(x^2-1)^2}$$

(*d*)
$$y' = \frac{(4-x^2)^{1/2}\cdot 3x^2 - x^3\cdot\frac{1}{2}(4-x^2)^{-1/2}(-2x)}{4-x^2} = \frac{(4-x^2)(3x^2) + x^3\cdot x}{(4-x^2)^{3/2}} = \frac{12x^2-2x^4}{(4-x^2)^{3/2}}.$$

The derivative exists for $-2 < x < 2$.

(*e*)
$$y' = \frac{(x+1)\cdot\frac{1}{2}(2x-3x^2)^{-1/2}(2-6x) - (2x-3x^2)^{1/2}(1)}{(x+1)^2}$$
$$= \frac{(x+1)(1-3x) - (2x-3x^2)}{(x+1)^2(2x-3x^2)^{1/2}} = \frac{1-4x}{(x+1)^2(2x-3x^2)^{1/2}}$$

The derivative exists for $0 < x < \frac{2}{3}$.

16.23 Find y', y'', y''' given (*a*) $y = 1/x$, (*b*) $y = \dfrac{1}{x-1} + \dfrac{1}{x+1}$, (*c*) $y = \dfrac{2}{x^2-1}$.

(*a*) Here $y = x^{-1}$; then $y' = -1 \cdot x^{-2} = -x^{-2}$, $y'' = 2x^{-3}$, $y''' = -6x^{-4}$ or $y' = -1/x^2$, $y'' = 2/x^3$, $y''' = -6/x^4$.

(*b*) Here $y = (x-1)^{-1} + (x+1)^{-1}$; then

$$y' = -1(x-1)^{-2} + (-1)(x+1)^{-2} = -\frac{1}{(x-1)^2} - \frac{1}{(x+1)^2}$$

$$y'' = 2(x-1)^{-3} + 2(x+1)^{-3} = \frac{2}{(x-1)^3} + \frac{2}{(x+1)^3}$$

$$y''' = -6(x-1)^{-4} - 6(x+1)^{-4} = -\frac{6}{(x-1)^4} - \frac{6}{(x+1)^4}$$

(*c*) Here $y = 2(x^2-1)^{-1}$; then

$$y' = 2(-1)(x^2-1)^{-2} \cdot (2x) = -4x(x^2-1)^{-2} = \frac{-4x}{(x^2-1)^2}$$

$$y'' = -4(x^2-1)^{-2} - 4x(-2)(x^2-1)^{-3}(2x) = -4(x^2-1)^{-2} + 16x^2(x^2-1)^{-3} = \frac{12x^2+4}{(x^2-1)^3}$$

$$y''' = \frac{(x^2-1)^3(24x) - (12x^2+4) \cdot 3(x^2-1)^2(2x)}{(x^2-1)^6} = \frac{(x^2-1)(24x) - 6x(12x^2+4)}{(x^2-1)^4} = \frac{-48x(x^2+1)}{(x^2-1)^4}$$

16.24 Determine the intervals on which each of the following is an increasing function and the intervals on which it is a decreasing function:

(*a*) $f(x) = x^2 - 8x$ (*c*) $f(x) = x^3 + 3x^2 + 9x + 5$ (*e*) $f(x) = (x-2)^3$
(*b*) $f(x) = 2x^3 - 24x + 5$ (*d*) $f(x) = x^3 + 3x$ (*f*) $f(x) = (x-1)^3(x-2)$

(*a*) Here $f'(x) = 2(x-4)$. Setting this equal to 0 and solving, we find the critical value to be $x = 4$. We locate the point $x = 4$ on the x axis and find that $f'(x) < 0$ for $x < 4$, and $f'(x) > 0$ when $x > 4$. See Fig. 16-10. Thus, $f(x) = x^2 - 8x$ is an increasing function when $x > 4$, and is a decreasing function when $x < 4$.

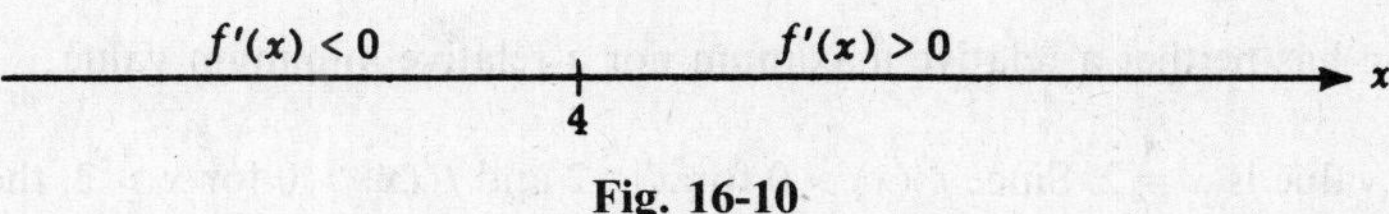

Fig. 16-10

(*b*) $f'(x) = 6x^2 - 24 = 6(x+2)(x-2)$; the critical values are $x = -2$ and $x = 2$. Locating these points and determining the sign of $f'(x)$ on each of the intervals $x < -2, -1 < x < 2$, and $x > 2$ (see Fig. 16-11), we find that $f(x) = 2x^3 - 24x + 5$ is an increasing function on the intervals $x < -2$ and $x > 2$, and is a decreasing function on the interval $-2 < x < 2$.

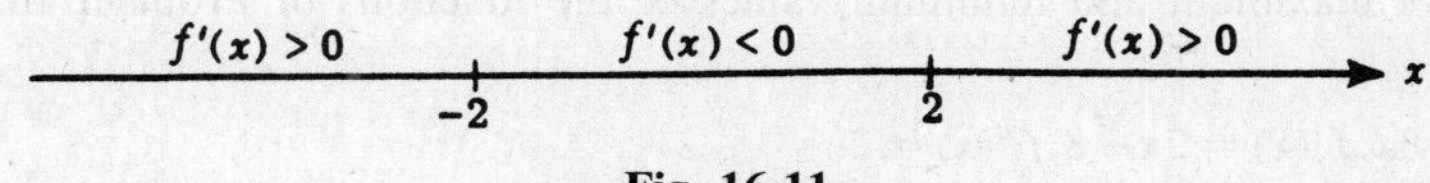

Fig. 16-11

(*c*) $f'(x) = -3x^2 + 6x + 9 = -3(x+1)(x-3)$; the critical values are $x = -1$ and $x = 3$. See Fig. 16-12. Then $f(x)$ is an increasing function on the interval $-1 < x < 3$, and a decreasing function on the intervals $x < -1$ and $x > 3$.

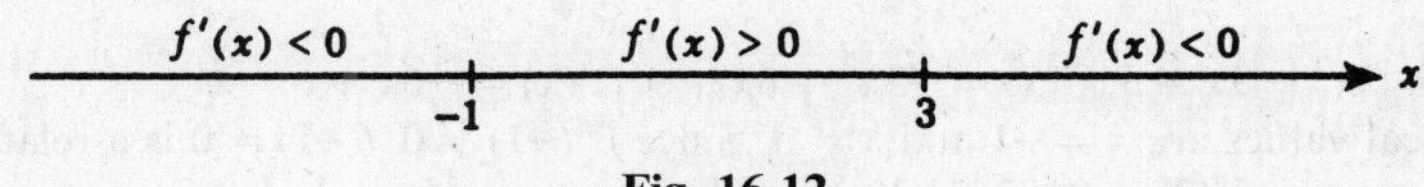

Fig. 16-12

(*d*) $f'(x) = 3x^2 + 3 = 3(x^2 + 1)$; there are no critical values. Since $f'(x) > 0$ for all values of x, $f(x)$ is everywhere an increasing function.

(*e*) $f'(x) = 3(x - 2)^2$; the critical value is $x = 2$. See Fig. 16-13. Then $f(x)$ is an increasing function on the intervals $x < 2$ and $x > 2$.

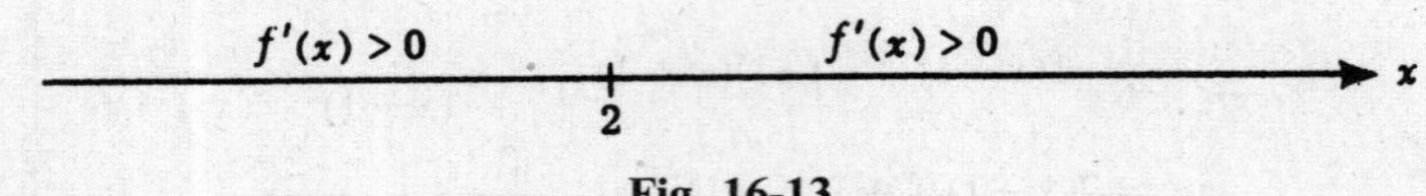

Fig. 16-13

(*f*) $f'(x) = (x - 1)^2(4x - 7)$; the critical values are $x = 1$ and $x = \frac{7}{4}$. See Fig. 16-14. Then $f(x)$ is an increasing function on the interval $x > \frac{7}{4}$ and is a decreasing function on the intervals $x < 1$ and $1 < x < \frac{7}{4}$.

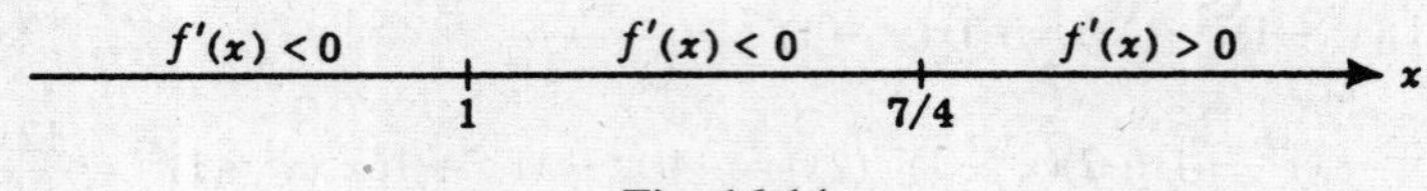

Fig. 16-14

16.25 Find the relative maximum and minimum values of the functions of Problem 16.24

(*a*) The critical value is $x = 4$. Since $f'(x) < 0$ for $x < 4$ and $f'(x) > 0$ for $x > 4$, the function has a relative minimum vale $f(4) = -16$.

(*b*) The critical values are $x = -2$ and $x = 2$. Since $f'(x) > 0$ for $x < -2$ and $f'(x) < 0$ for $-2 < x < 2$, the function has a relative maximum value $f(-2) = 37$. Since $f'(x) < 0$ for $-2 < x < 2$ and $f'(x) > 0$ for $x > 2$, the function has a relative minimum value $f(2) = -27$.

(*c*) The critical values are $x = -1$ and $x = 3$. Since $f'(x) < 0$ for $x < -1$ and $f'(x) > 0$ for $-1 < x < 3$, $f(x)$ has a relative minimum value $f(-1) = 0$. Since $f'(x) > 0$ for $-1 < x < 3$ and $f'(x) < 0$ for $x > 3$, the function has a relative maximum value $f(3) = 32$.

(*d*) The function has neither a relative maximum nor a relative minimum value.

(*e*) The critical value is $x = 2$. Since $f'(x) > 0$ for $x < 2$ and $f'(x) > 0$ for $x > 2$, the function has neither a relative maximum nor minimum value.

(*f*) The critical values are $x = 1$ and $x = \frac{7}{4}$. The function has a relative minimum value $f(\frac{7}{4}) = -\frac{27}{256}$. The critical value $x = 1$ yields neither a relative maximum nor minimum value.

16.26 Find the relative maximum and minimum values of the functions of Problem 16.24, using the second derivative test.

(*a*) $f(x) = x^2 - 8x, f'(x) = 2x - 8, f''(x) = 2$.

The critical value is $x = 4$. Since $f''(4) = 2 \neq 0, f(4) = -16$ is a relative minimum value of the function.

(*b*) $f(x) = 2x^3 - 24x + 5, f'(x) = 6x^2 - 24, f''(x) = 12x$.

The critical values are $x = -2$ and $x = 2$. Since $f''(-2) = -24 < 0, f(-2) = 37$ is a relative maximum value of the function; since $f''(2) = 24 > 0, f(2) = -27$ is a relative minimum value.

(*c*) $f(x) = -x^3 + 3x^2 + 9x + 5, f'(x) = -3x^2 + 6x + 9, f''(x) = -6x + 6$.

The critical values are $x = -1$ and $x = 3$. Since $f''(-1) > 0, f(-1) = 0$ is a relative minimum value of the function; since $f''(3) < 0, f(3) = 32$ is a relative maximum value.

(*d*) $f(x) = x^3 + 3x, f'(x) = 3x^2 + 3, f''(x) = 6x$.

There are no critical values; hence the function has neither a relative minimum nor a relative maximum value.

(*e*) $f(x) = (x-2)^3, f'(x) = 3(x-2)^2, f''(x) = 6(x-2)$.

The critical value is $x = 2$. Since $f''(2) = 0$, the test fails. The test of Problem 16.25 shows that the function has neither a relative maximum nor a relative minimum value.

(*f*) $f(x) = (x-1)^3(x-2), f'(x) = (x-1)^2(4x-7), f''(x) = 6(2x-3)(x-1)$.

The critical value are $x = 1$ and $x = \frac{7}{4}$. Since $f''(1) = 0$, the test fails; the test of Problem 16.25 shows that $f(1)$ is neither a relative maximum nor a relative minimum value of the function. Since $f''(\frac{7}{4}) > 0$, $f(\frac{7}{4}) = -\frac{27}{256}$ is a relative minimum value.

16.27 Find the inflection points and plot the graph of each of the given curves. In sketching the graph, locate the x and y intercepts when they can be found, the relative maximum and minimum points (see Problem 16.25), and the inflection points, if any. Additional points may be found if necessary.

(*a*) $y = f(x) = x^2 - 8x$ (*c*) $y = f(x) = -x^3 + 3x^2 + 9x + 5$ (*e*) $y = f(x) = (x-2)^3$
(*b*) $y = f(x) = 2x^3 - 24x + 5$ (*d*) $y = f(x) = x^3 + 3x$ (*f*) $y = f(x) = (x-1)^3(x-2)$

(*a*) Since $f''(x) = 2$, the parabola does not have an inflection point. It is always concave upward.

The x and y intercepts are $x = 0, x = 8$, and $y = 0$; $(4, -16)$ is a relative minimum point. See Fig. 16-15(*a*).

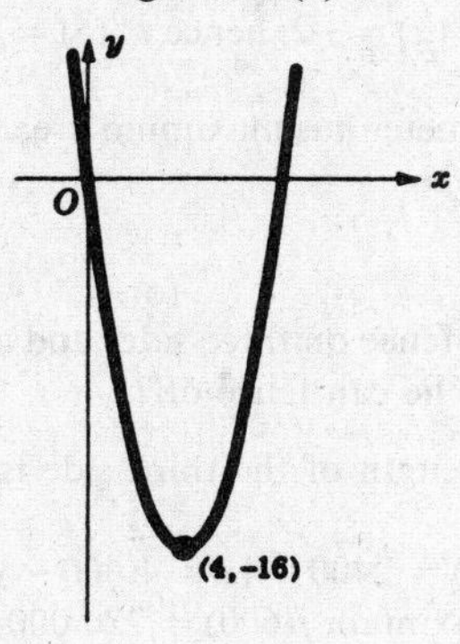

(*a*)

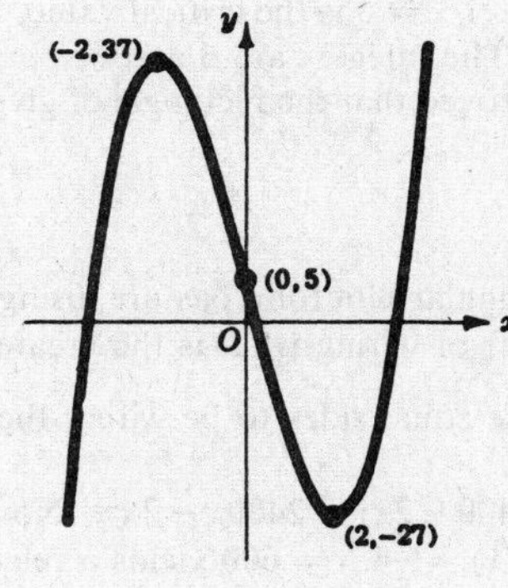

(*b*)

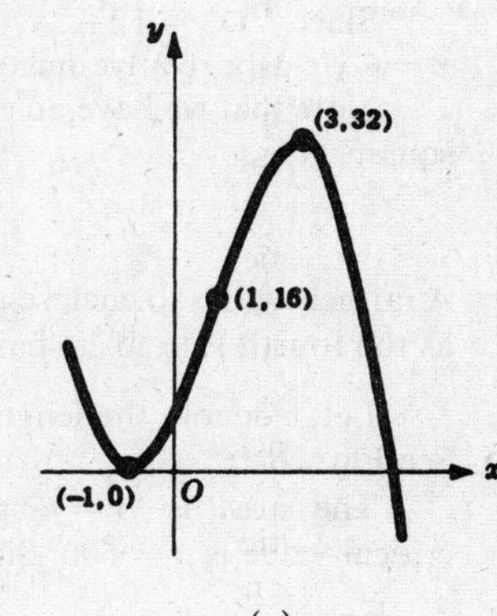

(*c*)

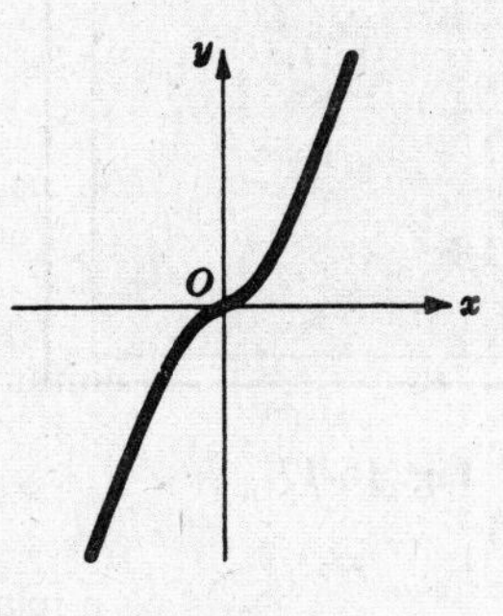

(*d*)

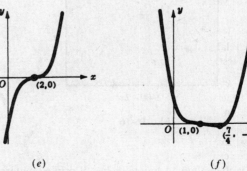

(*e*)

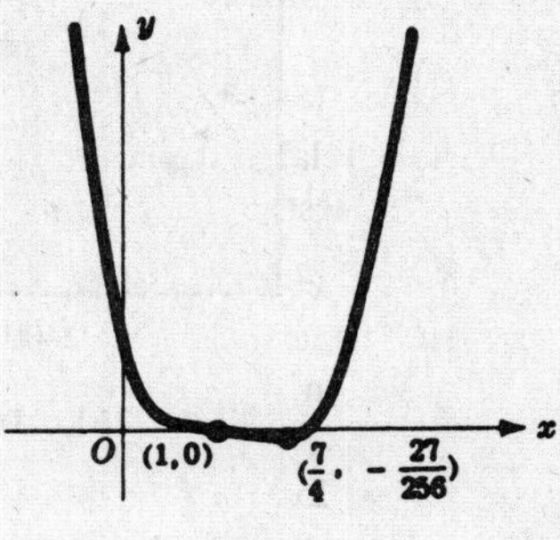

(*f*)

Fig. 16-15

(*b*) $f''(x) = 12x$ and $f'''(x) = 12$. Since $f''(x) = 0$ when $x = 0$ and $f'''(0) = 12 \neq 0, (0, 5)$ is an inflection point. Notice the change in concavity.

They y intercept is $y = 5$, the x intercepts cannot be determined; $(-2, 37)$ is a relative maximum point, $(2, -27)$ is a relative minimum point; (0, 5) is an inflection point. See Fig. 16-15 (*b*).

(*c*) $f''(x) = -6x + 6$ and $f'''(x) = -6$. Since $f''(x) = 0$ when $x = 1$ and $f'''(1) = -6 \neq 0, (1, 16)$ is an inflection point.

The x and y intercepts are $x = -1, x = 5$, and $y = 5$; $(-1, 0)$ is a relative minimum point and (3, 32) is a relative maximum point; (1, 16) is an inflection point. See Fig. 16-15 (*c*).

(*d*) $f''(x) = 6x$ and $f'''(x) = 6$. The point (0, 0) is an inflection point.

The x and y intercepts are $x = 0, y = 0$; (0, 0) is an inflection point. The curve can be sketched after locating the points (1, 4), (2, 14), $(-1, -4)$, and $(-2, -14)$. See Fig. 16-15 (*d*).

(*e*) $f''(x) = 6(x - 2)$ and $f'''(x) = 6$. The point (2, 0) is an inflection point.

The x and y intercepts are $x = 2, y = -8$; (2, 0) is an inflection point. The curve can be sketched after locating the points (3, 1), (4, 8), and $(1, -1)$. See Fig. 16-15 (*e*).

(*f*) $f''(x) = 6(2x - 3)(x - 1)$ and $f'''(x) = 6(4x - 5)$. The inflection points are (1, 0) and $(\frac{3}{2}, -\frac{1}{16})$.

The x and y intercepts are $x = 1$, $x = 2$, and $y = 2$; $(\frac{7}{4}, -\frac{27}{256})$ is a relative minimum point; (1, 0) and $(\frac{3}{2}, -\frac{1}{16})$ are inflection points. For the graph, see Fig. 16-15 (*f*).

16.28 Find two integers whose sum is 12 and whose product is a maximum.

Let x and $12 - x$ be the integers; their product is $P = f(x) = x(12 - x) = 12x - x^2$.

Since $f'(x) = 12 - 2x = 2(6 - x), x = 6$ is the critical value. Now $f''(x) = -2$; hence $f''(6) = -2 < 0$ and $x = 6$ yields a relative maximum. The integers are 6 and 6.

Note that we have, in effect proved that the rectangle of given perimeter has maximum area when it is a square.

16.29 A farmer wishes to enclose a rectangular plot for a pasture, using a wire fence on three sides and a hedge row as the fourth side. If he has 2400 ft of wiring what is the greatest area he can fence off?

Let x denote the length of the equal sides to be wired; then the length of the third side is $2400 - 2x$. See Fig. 16-16.

The area is $A = f(x) = x(2400 - 2x) = 2400x - 2x^2$. Now $f'(x) = 2400 - 4x = 4(600 - x)$ and the critical value is $x = 600$. Since $f''(x) = -4, x = 600$ yields a relative maximum $f(600) = 720\ 000\ \text{ft}^2$.

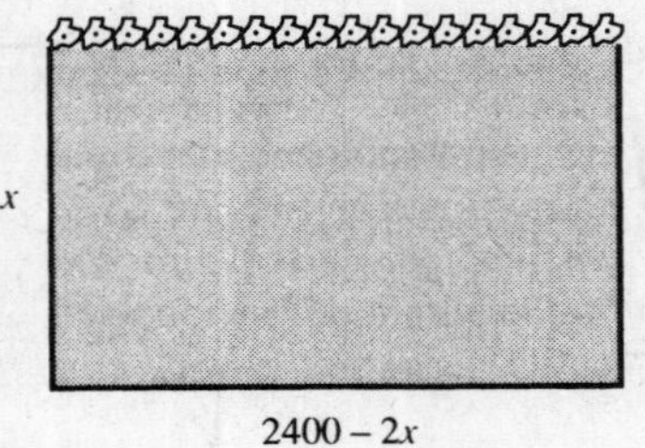

Fig. 16-16

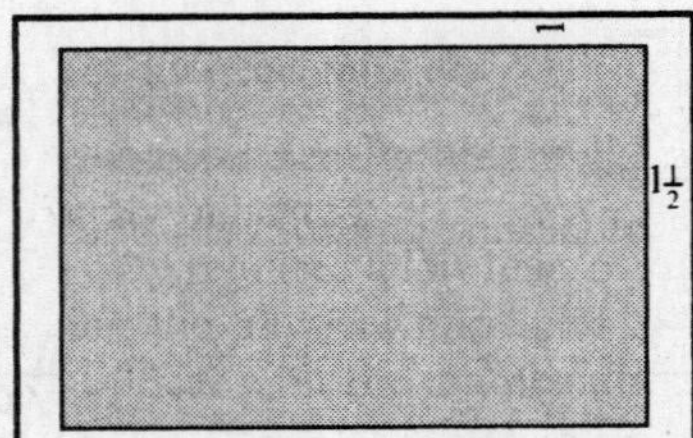

Fig. 16-17

16.30 A page is to contain 54 square inches of printed material. If the margins are 1 in. at top and bottom and $1\frac{1}{2}$ in. at the sides, find the most economical dimensions of the page. See Fig. 16-17.

Let the dimensions of the printed material be denoted by x and y; then $xy = 54$.

The dimensions of the page are $x + 3$ and $y + 2$; the area of the page is $A = (x + 3)(y + 2)$.

Since $y = 54/x, A = f(x) = (x+3)(54/x+2) = 60 + 162/x + 2x$. Then $f'(x) = -162/x^2 + 3$ and the critical values are $x = \pm 9$. Since $f''(x) = 324/x^3$, the relative minimum is given by $x = 9$. The required dimensions of the page are 12 in. wide and 8 in. high.

16.31 A cylindrical container with circular base is to hold 64 cubic centimeters. Find the dimensions so that the amount (surface area) of metal required is a minimum when (*a*) the container is an open cup and (*b*) a closed can.

Let r and h respectively be the radius of the base and height in centimeters, V be the volume of the container, and A be the surface area of the metal required.

(*a*) $V = \pi r^2 h = 64$ and $A = 2\pi rh + \pi r^2$. Solving for $h = 64/\pi r^2$ in the first relation and substituting in the second, we have $A = 2\pi r\left(\frac{64}{\pi r^2}\right) + \pi r^2 = \frac{128}{r} + \pi r^2$.

Then $\frac{dA}{dr} = -\frac{128}{r^2} + 2\pi r = \frac{2(\pi r^3 - 64)}{r^2}$ and the critical value is $r = \frac{4}{\sqrt[3]{\pi}}$.

Now $h = \frac{64}{\pi r^2} = \frac{4}{\sqrt[3]{\pi}}$; thus, $r = h \frac{4}{\sqrt[3]{\pi}}$ cm.

(*b*) $V = \pi r^2 h = 64$ and $A = 2\pi rh + 2\pi r^2 = 2\pi r\left(\frac{64}{\pi r^2}\right) + 2\pi r^2 = \frac{128}{r} + 2\pi r^2$.

Then $\frac{dA}{dr} = -\frac{128}{r^2} + 4\pi r = \frac{4(\pi r^3 - 32)}{r^2}$ and the critical value is $r = 2\sqrt[3]{\frac{4}{\pi}}$.

Now $h = \frac{64}{\pi r^2} = 4\sqrt[3]{\frac{4}{\pi}}$; thus, $h = 2r = 4\sqrt[3]{\frac{4}{\pi}}$ cm.

16.32 Study the motion of a particle which moves along a horizontal line in accordance with

(*a*) $x = t^3 - 6t^2 + 3$, (*b*) $x = t^3 - 5t^2 + 7t - 3$, (*c*) $v = (t-1)^2(t-4)$, (*d*) $v = (t-1)^4$.

(*a*) Here $v = 3t^2 - 12t = 3t(t-4) = 0$ when $t = 0$ and $t = 4$; $a = 6t - 12 = 6(t-2) = 0$ when $t = 2$. The particle leaves $A(s = 3)$ with velocity 0 and moves to $B(-29)$ where it stops momentarily; thereafter it moves to the right. See Fig. 16-18.

The intervals of increasing and decreasing speed are shown in Fig. 16-19.

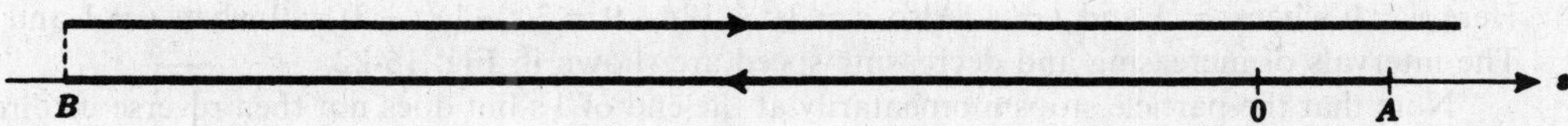

Fig. 16-18

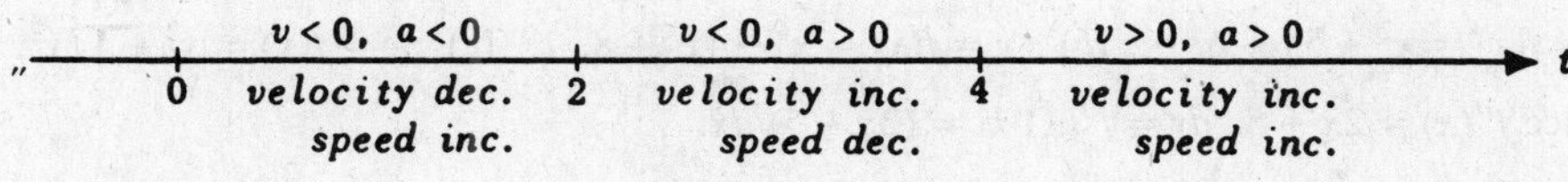

Fig. 16-19

(*b*) Here $v = 3t^2 = 10t + 7 = (t-1)(3t-7) = 0$ when $t = 1$ and $t = \frac{7}{3}$; $a = 6t = 10 = 2(3t-5) = 0$ when $t = \frac{5}{3}$. The particle leaves $A(s = -3)$ with velocity 7 m/s and moves to 0 where it stops momentarily, then it moves to $B(-\frac{32}{27})$ where is stops momentarily. Thereafter it moves to the right. See Fig. 16-20.

The intervals of increasing and decreasing speed are shown in Fig. 16-21.

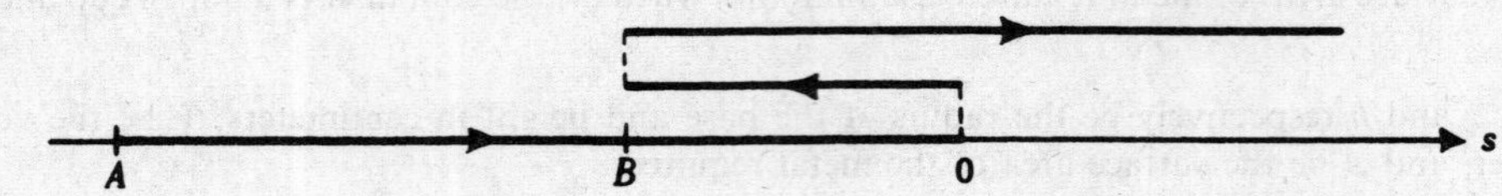

Fig. 16-20

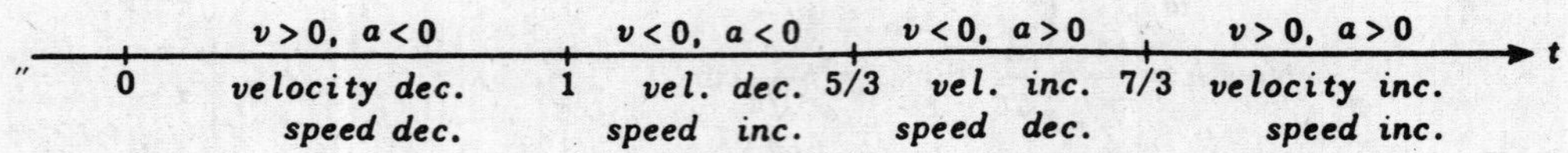

Fig. 16-21

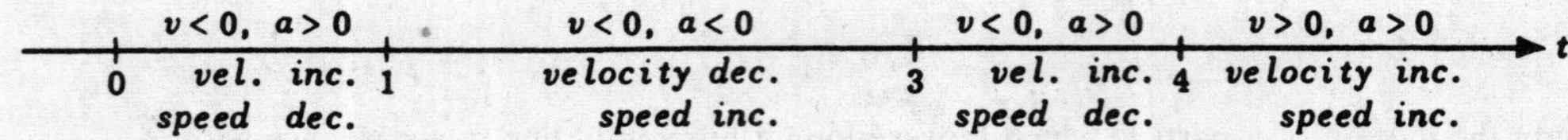

Fig. 16-22

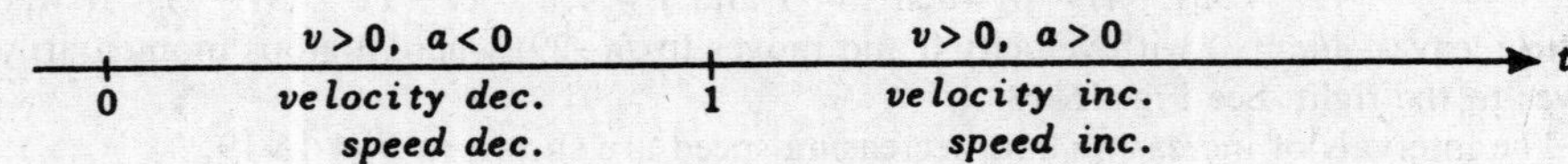

Fig. 16-23

(*c*) Here $v = 0$ when $t = 1$ and $t = 4$. Also $a = 3t^2 - 12t + 9 = 3(t-1)(t-3) = 0$ when $t = 1$ and $t = 3$. The intervals of increasing and decreasing speed are shown in Fig. 16-22.

Note that the particle stops momentarily at the end of 1 s but does not then reverse its direction of motion as, for example, in (*b*).

(*d*) Here $a = 4(t-1)^3$ and $v = a = 0$ when $t = 1$. The intervals of increasing and decreasing speed are shown in Fig. 16-23.

16.33 Find dy in terms of x and dx, given

(*a*) $y = f(x) = x^2 + 5x + 6$, (*b*) $y = f(x) = x^4 - 4x^3 + 8$, (*c*) $y = f(x) = x^2 + 1/x^2$.

(*a*) Since $f'(x) = 2x + 5$, $dy = f'(x)\,dx = (2x+5)\,dx$.

(*b*) Since $f'(x) = 4x^3 - 12x^2$, $dy = (4x^3 - 12x^2)\,dx$.

(*c*) Since $f'(x) = 2x - 2/x^3$, $dy = (2x - 2/x^3)\,dx$.

16.34 Find the approximate displacement of a particle moving along the x axis in accordance with the law $s = t^4 - t^2$, from the time $t = 1.99$ to $t = 2$.

Here $ds = (4t^3 - 2t)dt$. We take $t = 2$ and $dt = -0.01$. Then $ds = (4 \cdot 8 - 2 \cdot 2)(-.01) = -0.28$ and the displacement is 0.28 meters.

16.35 Find using differentials the approximate area of a square whose side is 3.01 cm.

Here $A = x^2$ and $dA = 2x\,dx$. Taking $x = 3$ and $dx = 0.01$, we find $dA = 2 \cdot 3(.01) = 0.06\ \text{cm}^2$. Now the area ($9\ \text{cm}^2$) of a square 3 cm. on a side is increased approximately $0.06\ \text{cm}^2$ when the side is increased to 3.01 cm. Hence the approximate area is $9.06\ \text{cm}^2$. The true area is $9.0601\ \text{cm}^2$.

16.36 Prove: If $F(x)$ and $G(x)$ are distinct integrals of $f(x)$, then $F(x) = G(x) + C$, where C is a constant.

Since $F(x)$ and $G(x)$ are integrals of $f(x)$, $F'(x) = G'(x) = f(x)$.

Suppose $F(x) - G(x) = H(x)$; differentiating with respect to $x, F'(x) - G'(x) = H'(x)$ and $H'(x) = 0$. Thus, $H(x)$ is a constant, say, C, and $F(x) = G(x) + C$.

16.37 (*a*) $\int \sqrt{x}\,dx = \int x^{1/2}\,dx = \dfrac{x^{3/2}}{\frac{3}{2}} + C = \dfrac{2}{3}x^{3/2} + C$

(*b*) $\int (3x^2 + 5)\,dx = x^3 + 5x + C$

(*c*) $\int (5x^6 + 2x^3 - 4x + 3)\,dx = \frac{5}{7}x^7 + \frac{1}{2}x^4 - 2x^2 + 3x + C$

(*d*) $\int (80x^{19} - 32x^{15} - 12x^{-3})\,dx = 4x^{20} - 2x^{16} + \dfrac{6}{x^2} + C$

16.38 At every point (x, y) of a certain curve, the slope is equal to 8 times the abscissa. Find the equation of the curve if it passes through (1,3).

Since $m = \dfrac{dy}{dx} = 8x$, we have $dy = 8x\,dx$. Then $y = \int 8x\,dx = 4x^2 + C$, a family of parabolas. We seek the equation of the parabola of this family which passes through the point (1,3). Then $3 = 4(1)^2 + C$ and $C = -1$. The curve has equation $y = 4x^2 - 1$.

16.39 For a certain curve $y'' = 6x - 10$. Find its equation if it passes through point (1,1) with slope -1.

Since $y'' = 6x - 10$, $y' = 3x^2 - 10x + C_1$; since $y' = -1$ when $x = 1$, we have $-1 = 3 - 10 + C_1$ and $C_1 = 6$. Then $y' = 3x^2 - 10x + 6$.

Now $y = x^3 - 5x^2 + 6x + C_2$ and since $y = 1$ when $x = 1$, $1 = 1 - 5 + 6 + C_2$ and $C_2 = -1$. Thus the equation of the curve is $y = x^3 - 5x^2 + 6x - 1$.

16.40 The velocity at time t of a particle moving along the x axis is given by $v = x' = 2t + 5$. Find the position of the particle at time t, if $x = 2$ when $t = 0$.

Select a point on the x axis as origin and assume positive direction to the right. Then at the beginning of the motion ($t = 0$) the particle is 2 units to the right of the origin.

Since $v = \dfrac{dx}{dt} = 2t + 5, dx = (2t + 5)\,dt$. Then $x = \int (2t + 5)\,dt = t^2 + 5t + C$.

Substituting $x = 2$ and $t = 0$, we have $2 = 0 + 0 + C$ so that $C = 2$. Thus the position of the particle at time t is given by $x = t^2 + 5t + 2$.

16.41 A body moving in a straight line has an acceleration equal to $6t^2$, where time (t) is measured in seconds and distance s is measured in feet. If the body starts from rest, how far will it move during the first 2 s?

Let the body start from the origin; then it is given that when $t = 0$, $v = 0$ and $s = 0$.

Since $a = \dfrac{dv}{dt} = 6t^2, dv = 6t^2 dt$. Then $v \int 6t^2 dt = 2t^3 + C_1$. When $t = 0$, $v = 0$; then $0 = 2 \cdot 0 + C_1$ and $C_1 = 0$. Thus $v = 2t^3$.

Now $v = \frac{ds}{dt} = 2t^3$; then $ds = 2t^3$ and $s = \int 2t^3 dt = \frac{1}{2}t^4 + C_2$. When $t = 0$, $s = 0$; then $C_2 = 0$ and $s = \frac{1}{2}t^4$. When $t = 2$, $s = \frac{1}{2}(2)^4 = 8$. The body moves 8 ft during the first 2 s.

16.42 A ball is thrown upward from the top of a building 320 m high with initial velocity 128 m/s. Determine the velocity with which the ball will strike the street below. (Assume acceleration is 9.81 m/s^2, directed downward.)

First we choose an origin from which all distances are to be measured and a direction (upward or downward) which will be called positive.

First Solution. Take the origin at the top of the building and positive direction as upward.

Then $$a = \frac{dv}{dt} = -32 \quad \text{and} \quad v = -32t + C_1$$

When the ball is released, $t = 0$ and $v = 128$; then $128 = -32(0) + C_1$ and $C_1 = 128$.

Now $v = ds/dt = -32t + 128$ and $s = -16t^2 + 128t + C_2$. When the ball is released, $t = 0$ and $s = 0$; then $C_2 = 0$ and $s = -16t^2 + 128t$.

When the ball strikes the street, it is 320 ft below the origin, that is, $s = -320$; hence, $-320 = -16t^2 + 128t$, $t^2 - 8t - 20 = (t + 2)(t - 10) = 0$, and $t = 10$. Finally, when $t = 10$, $v = -32(10) + 128 = -129$ ft/s.

Second Solution. Take the origin on the street and positive direction as before. Then $a = dv/dt = -32$ and $v = -32t + 128$ as in the first solution.

Now $s = -16t^2 + 128t + C_2$ but when $t = 0$, $s = 320$. Thus $C_2 = 320$ and $s = -16t^2 + 128 + 320$. When the ball strikes the street $s = 0$; then $t = 10$ and $v = -192$ ft/s as before.

16.43 A ball was dropped from a balloon 640 ft above the ground. If the balloon was rising at the rate of 48 ft/s, find (*a*) the greatest distance above the ground attained by the ball, (*b*) the time the ball was in the air, and (*c*) the speed of the ball when it struck the ground.

Assume the origin at the point where the ball strikes the ground and positive distance to be directed upward. Then

$$a = \frac{dv}{dt} = -32 \qquad \text{and} \qquad v = -32t + C_1$$

When $t = 0$, $v = 48$; hence, $C_1 = 48$. Then $v = ds/dt = -32 + 48$ and $s = -16t^2 + 48t + C_2$. When $t = 0$, $s = 640$; hence, $C_2 = 640$ and $s = -16t^2 + 48t + 640$.

(*a*) When $v = 0$, $t = \frac{3}{2}$ and $s = -16(\frac{3}{2})^2 + 48(\frac{3}{2}) + 640 = 676$. The greatest height attained by the ball was 676 ft.

(*b*) When $s = 0, -16t^2 + 48t + 640 = 0$ and $t = -5, 8$. The ball was in the air for 8 s.

(*c*) When $t = 8$, $v = -32(8) + 48 = -208$. The ball struck the ground with speed 208 ft/s.

16.44 Find the area bounded by the line $y = 4x$, the x axis, and the ordinates $x = 0$ and $x = 5$.

Here $y \geq 0$ on the interval $0 \leq x \leq 5$. Then

$$A = \int_0^5 4x\, dx = 2x^2\Big|_0^5 = 50 \text{ sq units}$$

Note that we have found the area of a right triangle whose legs are 5 and 20 units. See Fig. 16-24. The area is $\frac{1}{2}(5)(20) = 50$ sq units.

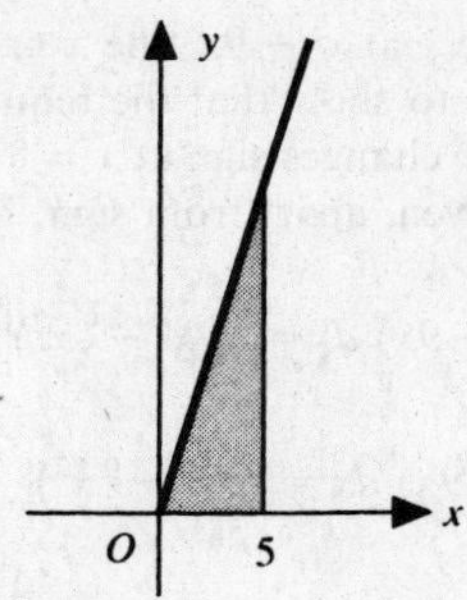

Fig. 16-24

16.45 Find the area bounded by the parabola $y = 8 + 2x - x^2$ and the x axis.

The x intercepts are $x = -2$ and $x = 4$; $y \geq 0$ on the interval $-2 \leq x \leq 4$. See Fig. 16-25. Hence

$$A = \int_{-2}^{4} (8 + 2x - x^2)\,dx = \left(8x + x^2 - \frac{x^3}{3}\right)\Bigg|_{-2}^{4}$$

$$= \left(8 \cdot 4 + 4^2 - \frac{4^3}{3}\right) - \left[8(-2) + (-2)^2 - \frac{(-2)^3}{3}\right] = 36 \text{ sq units}$$

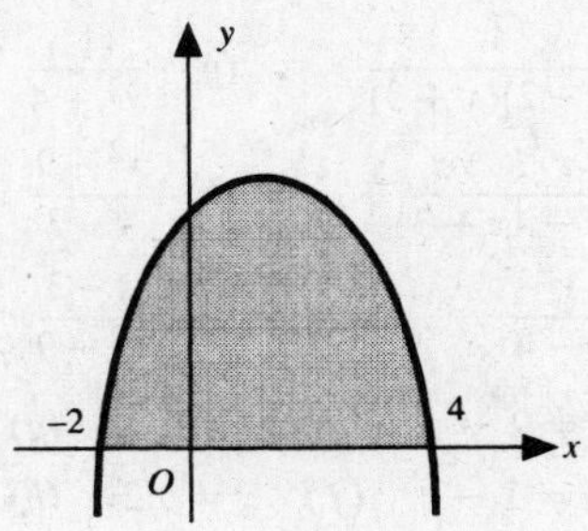

Fig. 16-25

16.46 Find the area bounded by the parabola $y = x^2 + 2x - 3$, the x axis, and the ordinates $c = -2$ and $x = 0$.

On the interval $-2 \leq x \leq 0$, $y \leq 0$. Here

$$A \int_{-2}^{0} (x^2 + 2x - 3)\,dx = \left(\frac{x^3}{3} + x^2 - 3x\right)\Bigg|_{-2}^{0} = 0 - \left[\frac{(-2)^3}{3} + (-2)^2 - 3(-2)\right] = -\frac{22}{3}$$

The negative sign indicates that the area lies entirely below the x axis. The area is $\frac{22}{3}$ sq units. See Fig. 16-26.

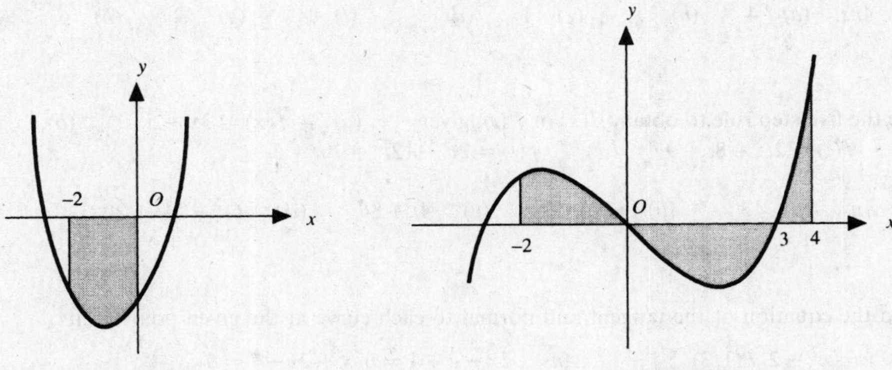

Fig. 16-26

Fig. 16-27

16.47 Find the area bounded by the curve $y = x^3 - 9x$, the x axis, and the ordinates $x = -2$ and $x = 4$.

The purpose of this problem is to show that the required area is *not* given by $\int_{-2}^{4}(x^3 - 9x)\,dx$.

From Fig. 16-27, we note that y changes sign at $x = 0$ and at $x = 3$. The required area consists of three pieces, the individual areas being given, apart from sign, by

$$A_1 = \int_{-2}^{0}(x^3 - 9x)\,dx = \left(\tfrac{1}{4}x^4 - \tfrac{9}{2}x^2\right)\Big|_{-2}^{0} = 0 - (4 - 18) = 14$$

$$A_2 = \int_{0}^{3}(x^3 - 9x)\,dx = \left(\tfrac{1}{4}x^4 - \tfrac{9}{2}x^2\right)\Big|_{0}^{3} = \left(\tfrac{81}{4} - \tfrac{81}{2}\right) - 0 = -\tfrac{81}{4}$$

$$A_3 = \int_{3}^{4}(x^3 - 9x)\,dx = \left(\tfrac{1}{4}x^4 - \tfrac{9}{2}x^2\right)\Big|_{3}^{4} = (64 - 72) - \left(\tfrac{81}{4} - \tfrac{81}{2}\right) = \tfrac{49}{4}$$

Thus, $A = A_1 - A_2 + A_3 = 14 + \frac{81}{4} + \frac{49}{4} = \frac{93}{2}$ sq units.

Note that $\int_{-2}^{4}(x^3 - 9x)\,dx = 6 < A_1$, an absurd result.

Supplementary Problems

16.48 Find all (real) values of x for which each of the following is defined:

(a) $x^2 - 3x + 4$ (d) $\dfrac{1}{(x-2)(x+3)}$ (g) $\dfrac{1}{x^2 + 4}$

(b) $\dfrac{1}{x^2}$ (e) $\dfrac{1}{x^2 - 4x + 3}$ (h) $\dfrac{x^2 - 9}{x - 3}$

(c) $\dfrac{1}{x - 2}$ (f) $\dfrac{1}{x^2 - 4}$ (i) $\dfrac{x - 3}{x^2 - 9}$

Ans. (a) all x (c) $x \neq 2$ (e) $x \neq 1, 3$ (g) all x (i) $x \neq \pm 3$
(b) $x \neq 0$ (d) $x \neq 2, -3$ (f) $x \neq \pm 2$ (h) $x \neq 3$

16.49 For each function $f(x)$ of Problem 16.48 evaluate $\lim_{x \to 1} f(x)$, when it exists.

Ans. (a) 2 (b) 1 (c) -1 (d) $-\frac{1}{4}$ (f) $-\frac{1}{3}$ (g) $\frac{1}{5}$ (h) 4 (i) $\frac{1}{4}$

16.50 For each function $f(x)$ of Problem 16.48 evaluate $\lim_{x \to 3} f(x)$, when it exists.

Ans. (a) 4 (b) $\frac{1}{9}$ (c) 1 (d) $\frac{1}{6}$ (f) $\frac{1}{5}$ (g) $\frac{1}{13}$ (h) 6 (i) $\frac{1}{6}$

16.51 Use the five-step rule to obtain $f'(x)$ or $f'(t)$, given (a) $f(x) = 3x + 5$ (b) $f(x) = x^2 - 3x$ (c) $f(t) = 2t^2 + 8t + 9$ (d) $f(t) = 2t^3 - 12t^2 + 20t + 3$

Ans. (a) 3 (b) $2x - 3$ (c) $4t + 8$ (d) $6t^2 - 24t + 20$

16.52 Find the equation of the tangent and normal to each curve at the given point on it.

(a) $y = x^2 + 2, P(1, 3)$ *Ans.* $2x - y + 1 = 0, x + 2y - 7 = 0$

(b) $y = 2x^2 - 3x, P(1, -1)$ *Ans.* $x - y - 2 = 0, x + y = 0$

(c) $y = x^2 - 4x + 5, P(1, 2)$ *Ans.* $2x + y - 4 = 0, x - 2y + 3 = 0$

(d) $y = x^2 + 3x - 10, P(2, 0)$ *Ans.* $7x - y - 14 = 0, x + 7y - 2 = 0$

(e) $x^2 + y^2 = 25, P(4, 3)$ *Ans.* $4x + 3y - 25 = 0, 3x - 4y = 0$

(f) $y^2 = 4x - 8, P(3, -2)$ *Ans.* $x + y - 1 = 0, x - y - 5 = 0$

(g) $x^2 + 4y^2 = 8, P(-2, -1)$ *Ans.* $x + 2y + 4 = 0, 2x - y + 3 = 0$

(h) $2x^2 - y^2 = 9, P(-3, 3)$ *Ans.* $2x + y + 3 = 0, x - 2y + 9 = 0$

16.53 A particle moves along the x axis according to the law $s = 2t^2 + 8t + 9$ [see Problem 16.51(c)], where s (ft) is the directed distance of the particle from the origin O at time t (seconds). Locate the particle and find its velocity when (a) $t = 0$, (b) $t = 1$.

Ans. (a) 9 ft to the right of O, $v = 8$ ft/s (b) 19 ft to the right of O, $v = 12$ ft/s

16.54 A particle moves along the x axis according to the law $s = 2t^3 - 12t^2 + 20t + 3$ [see Problem 45.20(d)], where s is defined as in Problem 45.22.

(a) Locate the particle and find its velocity when $t = 2$. (b) Locate the particle when $v = 2$ ft/s.

Ans. (a) 11 ft to the right of O, $v = -4$ ft/s

(b) $t = 1$, 13 ft to the right of O; $t = 3$, 9 ft to the right of O

16.55 The height (s m) of a bullet shot vertically upwards is given by $s = 1280t - 16t^2$, with t measured in seconds. (a) What is the initial velocity? (b) For how long will it rise? (c) How high will it rise?

Ans. (a) 1280 m/s (b) 40 s (c) 25 600 m

16.56 Find the coordinates of the points for which the slope of the tangent to $y = x^3 - 12x + 1$ is 0.

Ans. (2, −15), (−2, 17)

16.57 At what point on $y = \frac{1}{2}x^2 - 2x + 3$ is the tangent perpendicular to that at the point (1, 0)?

Ans. $(3, \frac{3}{2})$

16.58 Show that the equation of the tangent to the conic $Ax^2 + 2Bxy + Cy^2 + 2Dx + 2Ey + F = 0$ at the point $P_1(x_1, y_1)$ on it is given by $Ax_1x + B(x_1y + y_1x) + Cy_1y + D(x_1 + x) + E(y_1 + y) + F = 0$. Use this as a formula to solve Problem 16.52.

16.59 Show that the tangents at the extremities of the latus rectum of the parabola $y^2 = 4px$ (a) are mutually perpendicular and (b) intersect on the directrix.

16.60 Show that the tangent of slope $m \neq 0$ to the parabola $y^2 = 4px$ has equation $y = mx + p/m$.

16.61 Show that the slope of the tangent at either end of either latus rectum of the ellipse $b^2x^2 + a^2y^2 = a^2b^2$ is equal numerically to its eccentricity. Investigate the case of the hyperbola.

16.62 Use the differentiation formulas to find y', given

(a) $y = 2x^3 + 4x^2 - 5x + 8$ *Ans.* $y' = 6x^2 + 8x - 5$
(b) $y = -5 + 3x - \frac{3}{2}x^2 - 7x^3$ *Ans.* $y' = 3 - 3x - 21x^2$
(c) $y = (x - 2)^4$ *Ans.* $y' = 4(x - 2)^3$
(d) $y = (x^2 + 2)^3$ *Ans.* $y' = 6x(x^2 + 2)^2$
(e) $y = (4 - x^2)^{10}$ *Ans.* $y' = -20x(4 - x^2)^9$
(f) $y = (2x^2 + 4x - 5)^6$ *Ans.* $y' = 24(x + 1)(2x^2 + 4x - 5)^5$

(g) $y = \frac{1}{5}x^{5/2} + \frac{1}{3}x^{3/2}$ *Ans.* $y' = \frac{1}{2}x^{1/2}(x+1)$
(h) $y = (x^2-4)^{3/2}$ *Ans.* $y' = 3x(x^2-4)^{1/2}$
(i) $y = (1-x^2)^{1/2}$ *Ans.* $y' = -\dfrac{x}{(1-x^2)^{1/2}}$
(j) $y = \dfrac{6}{x} + \dfrac{4}{x^2} - \dfrac{3}{x^3}$ *Ans.* $y' = -\dfrac{6}{x^2} - \dfrac{8}{x^3} + \dfrac{9}{x^4}$
(k) $y = x^3(x+1)^2$ *Ans.* $y' = x^2(x+1)(5x+3)$
(l) $y = (x+1)^3(x-3)^2$ *Ans.* $y' = (x+1)^2(x-3)(5x-7)$
(m) $y = (x+2)^2(2-x)^3$ *Ans.* $y' = -(x+2)(2-x)^2(5x-2)$
(n) $y = \dfrac{x+1}{x-1}$ *Ans.* $y' = -\dfrac{2}{(x-1)^2}$
(o) $y = \dfrac{x^2+2x-3}{x^2}$ *Ans.* $y' = \dfrac{6-2x}{x^3}$
(p) $y = \dfrac{x^2+1}{x^2+2}$ *Ans.* $y' = \dfrac{2x}{(x^2+2)^2}$
(q) $y = \dfrac{1}{(2x+1)^3}$ *Ans.* $y' = -\dfrac{6}{(2x+1)^4}$
(r) $y = \dfrac{1}{(x^2-9)^{1/2}}$ *Ans.* $y' = -\dfrac{x}{(x^2-9)^{3/2}}$
(s) $y = \dfrac{1}{(16-x^2)^{1/2}}$ *Ans.* $y' = \dfrac{x}{(16-x^2)^{3/2}}$
(t) $y = \dfrac{x}{(x+1)^{1/2}}$ *Ans.* $y' = \dfrac{x+2}{2(x+1)^{3/2}}$
(u) $y = \dfrac{(x^2+2)^{1/2}}{x}$ *Ans.* $y' = \dfrac{-2}{x^2(x^2+2)^{1/2}}$

16.63 For each of the following, find $f'(x)$, $f''(x)$, and $f'''(x)$:

(a) $f(x) = 3x^4 - 8x^3 + 12x^2 + 5$ (c) $f(x) = \dfrac{1}{4-x}$
(b) $f(x) = x^3 - 6x^2 + 9x + 18$ (d) $f(x) = (1-x^2)^{3/2}$

Ans. (a) $f'(x) = 12x(x^2-2x+2)$, $f''(x) = 12(3x^2-4x+2)$, $f'''(x) = 24(3x-2)$
(b) $f'(x) = 3(x^2-4x+3)$, $f''(x) = 6(x-2)$, $f'''(x) = 6$
(c) $f'(x) = \dfrac{1}{(4-x)^2}$, $f''(x) = \dfrac{2}{(4-x)^3}$, $f'''(x) = \dfrac{6}{(4-x)^4}$
(d) $f'(x) = -3x(1-x^2)^{1/2}$, $f''(x) = \dfrac{3(2x^2-1)}{(1-x^2)^{1/2}}$, $f'''(x) = \dfrac{3x(3-2x^2)}{(1-x^2)^{3/2}}$

16.64 In each of the following state the values of x for which $f(x)$ is continuous; also find $f'(x)$ and state the values of x for which it is defined.

(a) $f(x) = \dfrac{1}{x^2}$ (b) $f(x) = \dfrac{1}{x-2}$ (c) $f(x) = (x-2)^{4/3}$ (d) $f(x) = (x-2)^{1/3}$

Ans. (a) $x \neq 0$; $f'(x) = \dfrac{-2}{x^3}$, $x \neq 0$ (c) all x; $f'(x) = \frac{4}{3}(x-2)^{1/3}$; all x
(b) $x \neq 2$; $f'(x) = \dfrac{-1}{(x-2)^2}$, $x \neq 2$ (d) all x; $f'(x) = \dfrac{1}{3(x-2)^{2/3}}$, $x \neq 2$

[NOTE: Parts (a) and (b) verify: If $f(x)$ is not continuous at $x = x_0$, then $f'(x)$ does not exist at $x = x_0$. Parts (c) and (d) verify: If $f(x)$ is continuous at $x = x_0$, its derivative $f'(x)$ may or may not exist at $x = x_0$.]

16.65 Determine the intervals on which each of the following is an increasing function and the intervals on which it is a decreasing function.

(*a*) $f(x) = x^2$ *Ans.* Dec. for $x < 0$; inc. for $x > 0$
(*b*) $f(x) = 4 - x^2$ *Ans.* Inc. for $x < 0$; dec. for $x > 0$
(*c*) $f(x) = x^2 + 6x - 5$ *Ans.* Dec. for $x < -3$; inc. for $x > -3$
(*d*) $f(x) = 3x^2 + 6x + 18$ *Ans.* Dec. for $x < -1$; inc. for $x > -1$
(*e*) $f(x) = (x - 2)^4$ *Ans.* Dec. for $x < 2$; inc. for $x > 2$
(*f*) $f(x) = (x - 1)^3(x + 2)^2$ *Ans.* Inc. for $x < -2$; dec. for $-2 < x < -\frac{4}{5}$; inc. for $-\frac{4}{5} < x < x1$ and for $x > 1$

16.66 Find the relative maximum and minimum values of the functions of Problem 47.13.

Ans. (*a*) Min. = 0 (*c*) Min. = −14 (*e*) Min. = 0
(*b*) Max. = 4 (*d*) Min. = 15 (*f*) Max. = 0, Min. = −26 244/3125

16.67 Investigate for relative maximum (minimum) points and points of inflection. Sketch each locus.

(*a*) $y = x^2 - 4x + 8$ (*b*) $y = (x - 1)^3 + 5$ (*c*) $y = x^4 + 32x + 40$ (*d*) $y = x^3 - 3x^2 - 9x + 6$

Ans. (*a*) Min. (2, 4) (*c*) Min. (−2, −8)
(*b*) I.P. (1, 5) (*d*) Max, (−1, 11), Min. (3, −21), I.P. (1, −5)

16.68 The sum of two positive numbers is 12. Find the numbers

(*a*) If the sum of their squares is a minimum

(*b*) If the product of one and the square of the other is a maximum

(*c*) If the product of one and the cube of the other is a maximum

Ans. (*a*) 6 and 6 (*b*) 4 and 8 (*c*) 3 and 9

16.69 Find the dimensions of the largest open box which can be made from a sheet of tin 24 m. square by cutting equal squares from the corners and turning up the sides.

Ans. $16 \times 16 \times 4$ m.

16.70 Find the dimensions of the largest open box which can be made from a sheet of tin 60 in. by 28 in. by cutting equal squares from the corners and turning up the sides.

Ans. $48 \times 16 \times 6$ in.

16.71 A rectangular field is to be enclosed by a fence and divided into two smaller plots by a fence parallel to one of the sides. Find the dimensions of the largest such field which can be enclosed by 1200 ft of fencing.

Ans. 200×300 ft

16.72 If a farmer harvests his crop today, he will have 1200 kg worth $2.00 per kg. Every week he waits, the crop increases by 100 kg but the price drops 10¢ per kg. When should he harvest the crop?

Ans. 4 weeks from today

16.73 The base of an isosceles triangle is 20 m and its altitude is 40 m. Find the dimensions of the largest inscribed rectangle if two of the vertices are on the base of the triangle.

Ans. 10×20 m

16.74 For each of the following compute Δy, dy, and $\Delta y - dy$.

(a) $y = \frac{1}{2}x^2 + x; x = 2, \Delta x = \frac{1}{4}$ *Ans.* $\Delta y = \frac{25}{32}, dy = \frac{3}{4}, \Delta y - dy = \frac{1}{32}$
(b) $y = x^2 - x; x = 3, \Delta x = .01$ *Ans.* $\Delta y = .0501, dy = .05, \Delta y - dy = .0001$

16.75 Approximate using differentials the volume of a cube whose side is 3.005 cm.

Ans. $27.135\,\text{cm}^3$

16.76 Approximate using differentials the area of a circular ring whose inner radius is 5 m. and whose width is $\frac{1}{8}$ m.

Ans. $1.25\pi\,\text{m}^2$

16.77 Find the following indefinite integrals.

(a) $4\int dx = 4x + C$
(b) $\int \frac{1}{2}x\,dx = \frac{1}{4}x^2 + C$
(c) $\int (3x^2 + 4x - 5)\,dx = x^3 + 2x^2 - 5x + C$
(d) $\int x(1-x)\,dx = \frac{1}{2}x^2 - \frac{1}{3}x^3 + C$
(e) $\int 3(x+1)^2 dx = (x+1)^3 + C$
(f) $\int (x-1)(x+2)\,dx = \frac{1}{3}x^3 + \frac{1}{2}x^2 - 2x + C$
(g) $\int \frac{dx}{x^2} = -\frac{1}{x} + C$
(h) $\int \frac{x^2 - 2}{x^2}\,dx = x + \frac{2}{x} + C$

16.78 Find the equation of the family of curves whose slope is the given function of x. Find also the equation of the curve of the family passing through the given point.

(a) $m = 1, (1, -2)$ *Ans.* $y = x + C,\ y = x - 3$
(b) $m = -6x, (0, 0)$ *Ans.* $y = -3x^2 + C,\ y = -3x^2$
(c) $m = 3x^2 + 2x, (1, -3)$ *Ans.* $y = x^3 + x^2 + C,\ y = x^3 + x^2 - 5$
(d) $m = 6x^2, (0, 1)$ *Ans.* $y = 2x^3 + C,\ y = 2x^3 + 1$

16.79 For a certain curve $y'' = 6x + 8$. Find its equation if it passes through (1,2) with slope $m = 6$.

Ans. $y = x^3 + 4x^2 - 5x + 2$

16.80 A stone is dropped from the top of a building 400 ft high. Taking the origin at the top of the building and positive direction downward, find (*a*) the velocity of the stone at time t, (*b*) the position at time t, (*c*) the time it takes for the stone to reach the ground, and (*d*) the velocity when it strikes the ground.

Ans. (*a*) $v = 32t$ (*b*) $s = 16t^2$ (*c*) 5 s (*d*) 160 ft/s

16.81 A stone is thrown downward with initial velocity 20 ft/s from the top of a building 336 ft high. Following the directions of Problem 16.80, find (*a*) the velocity and position of the stone 2 s later, (*b*) the time it takes to reach the ground, and (*c*) the velocity with which it strikes the ground.

Ans. (*a*) 84 ft/s, 232 ft above the ground (*b*) 4 s (*c*) 148 ft/s

16.82 A stone is thrown upward with initial velocity 16 m/s from the top of a building 192 m high. Find (*a*) the greatest height attained by the stone, (*b*) the total time in motion, and (*c*) the speed with which the stone strikes the ground.

Ans. (*a*) 196 m (*b*) 4 s (*c*) 112 m/s

16.83 A boy on top of a building 192 m high throws a rock straight down. What initial velocity did he give it if it strikes the ground after 3 s?

Ans. 16 m/s

16.84 Find the area bounded by the x axis, the given curve, and the indicated ordinates.

(*a*) $y = x^2$ between $x = 2$ and $x = 4$ *Ans.* $\frac{56}{3}$ square units
(*b*) $y = 4 - 3x^2$ between $x = -1$ and $x = 1$ *Ans.* 6 square units
(*c*) $y = x^{1/2}$ between $x = 0$ and $x = 9$ *Ans.* 18 square units
(*d*) $y = x^2 - x - 6$ between $x = 0$ and $x = 2$ *Ans.* $\frac{34}{3}$ square units
(*e*) $y = x^3$ between $x = -2$ and $x = 4$ *Ans.* 68 square units
(*f*) $y = x^3 - x$ between $x = -1$ and $x = 1$ *Ans.* $\frac{1}{2}$ square unit

Chapter 17

The Calculus of Single-Variable Functions: A Physics Approach

OVERVIEW. The physical world is all about change—including, but hardly limited to, changing motions, colors, electromagnetic fields, temperatures, atomic states, and stars. Calculus is the mathematics of change and so plays a central role throughout physics. While the word "calculus" is sometimes used as a kind of popular shorthand to describe particularly challenging mathematics, we shall see that it is based upon simple ideas that ultimately provide a deeper understanding of the workings of the natural world.

VELOCITY AND FIRST DERIVATIVES. Consider the mass at the end of a spring, shown in Fig. 17-1, that is set into motion at time $t = 0$ and thereafter oscillates back and forth, undergoing a displacement as a function of time as shown in the figure. The average velocity of the mass in any time interval can be calculated by dividing its net displacement by the time. For example, the average velocity between points A and D in Fig. 17-2 is $\bar{v}_{\mathrm{AD}} = \frac{x_{\mathrm{D}}-x_{\mathrm{A}}}{t_{\mathrm{D}}-t_{\mathrm{A}}}$. This ratio is simply the slope of the line AD connecting the endpoints.

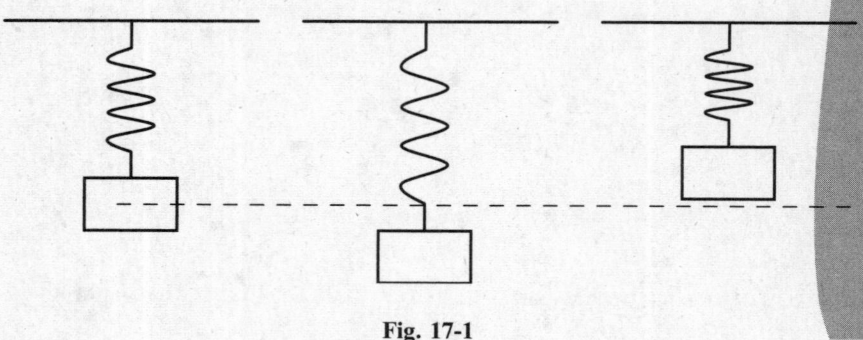

Fig. 17-1

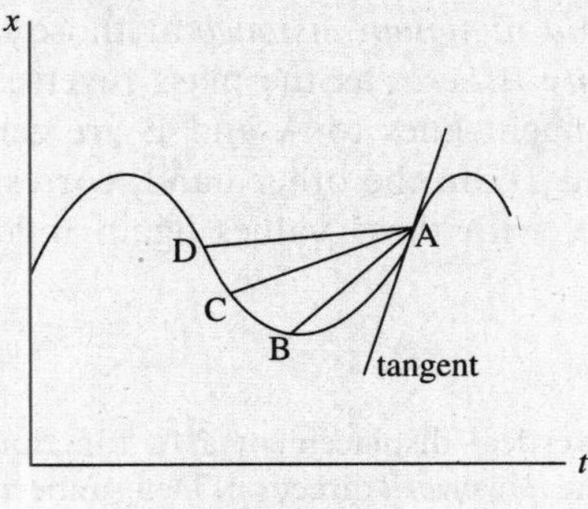

Fig. 17-2

If we want to understand this motion more deeply, then we can consider the average velocity of the mass in a smaller time interval that still includes point A—for example, in the interval between *A* and *C* instead. Suppose we continue shrinking the time interval in this same manner, always including point *A*. The average velocity of the object will approach a constant value, again given by the slope of the (now much smaller) line segment. From the figure, we can see that, as the time interval approaches zero, the line segment eventually will approach the *tangent** to the curve at point *A*. The slope of this tangent is apparently the velocity of the object in the immediate vicinity of point *A*, which we call the *instantaneous velocity* at point *A*. Note that the average velocity is measured over a finite time interval, whereas the instantaneous velocity refers to a particular time. (At this point, we will use the term "velocity" to mean the instantaneous velocity and will separately use the term "average velocity" for that quantity.) Note also that the instantaneous velocity can be either positive or negative, depending upon whether the displacement is increasing or decreasing in the immediate vicinity of this point.

The instantaneous velocity at one point is one of an infinite number of points that describe the instantaneous velocity at any time in the motion of the mass. Taken together, these points comprise a function which, for the example of the oscillating mass, is shown in Fig. 17-3, where both the displacement and the instantaneous velocity as a function of time are shown. Points *A* and *B* correspond to *extreme*

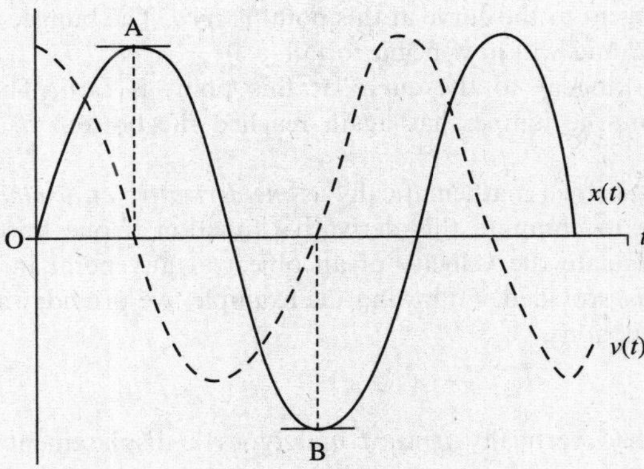

Fig. 17-3

* The tangent to a curve is the line intersecting the curve in exactly one point that has the same direction as the curve at that point.

values: the mass is at a *maximum or a minimum distance* at those locations. *At these points, as shown in the figure, the instantaneous velocity is zero*, as the mass reverses its direction. The figure supports this conclusion: the slopes of the tangent lines to A and B are zero, which equals the instantaneous velocity at those points. Points *C* and *D*, on the other hand, correspond, respectively, to points where the velocity is positive and negative, with these values again reflected in the instantaneous velocity function.

EXAMPLE 1. Figure 17-4 shows the vertical displacement as a function of time for a bungee jumper, where displacement is defined to be *positive* in the *downward* direction. Determine at each of the labeled points whether the displacement or the velocity is at an extreme value.

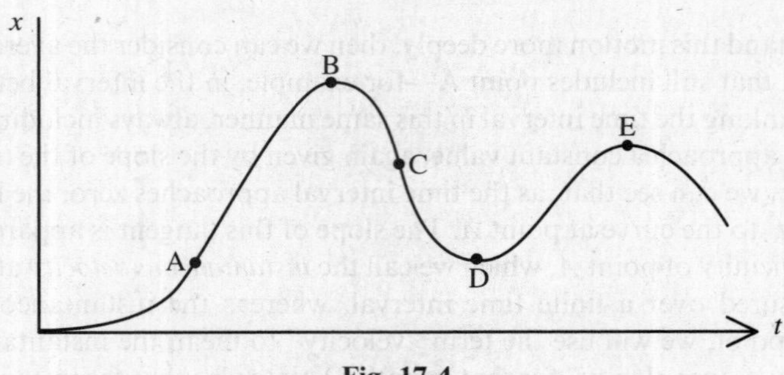

Fig. 17-4

Point A: The slope of the tangent to the curve at this point is a *maximum*. This corresponds to a maximum instantaneous velocity.

Point B: The slope of the tangent to the curve at this point is *zero*. This corresponds to an instantaneous velocity of 0. The displacement is a maximum, and the bungee jumper is reversing direction.

Point C: The slope of the tangent to the curve at this point is more negative than anywhere else. This corresponds to a minimum instantaneous velocity. As time increases and the bungee jumper rises, the magnitude of the jumper's velocity (that is, the *speed*) will decrease as the force of gravity begins to exceed the upward force of the bungee.

Point D: The slope of the tangent to the curve at this point is *zero*. The bungee jumper has reached a *maximum* height following the rebound at *B* and will now begin to fall.

Point E: The slope of the tangent to the curve at this point is *zero*. The displacement is a *maximum* relative to nearby points. The bungee jumper has again reached the bottom of the trajectory and is about to move upward.

The instantaneous velocity is known mathematically as *the derivative of displacement with respect to time*. The techniques of calculus allow one to compute this derivative function if one knows the original function. Such techniques then allow one to calculate the velocity of an object at any point in its trajectory. In the following example, the derivative function is provided. Following the example, we provide the formulae for determining the derivative for several common functions.

EXAMPLE 2. An object launched vertically *upward* undergoes a displacement that varies with time according to

$$y = v_0 t - \frac{1}{2}gt^2$$

where v_0 is the initial velocity, g is the acceleration due to gravity, t is the time and positive y corresponds to the upward direction. According to calculus, the instantaneous velocity of such an object is:

$$v = v_0 - gt$$

(*a*) What is the displacement and velocity of an object 2 s after launch, assuming an initial upward velocity of 20 m/s? (*b*) What is the displacement and velocity of the object when it reaches a maximum height? (*c*) Sketch both the displacement and the velocity as a function of time.

(*a*) The displacement of the object 2 s after launch is

$$y = v_0 t - \frac{1}{2}gt^2 = (20\,\text{m/s})(2\,\text{s}) - \frac{1}{2}(9.8\,\text{m/s}^2)(2\,\text{s})^2 = 20.4\,\text{m}$$

The velocity at this time will be:

$$v = v_0 - gt = 20\,\text{m/s} - (9.8\,\text{m/s}^2)(2\,\text{s}) = .4\,\text{m/s}$$

(*b*) When its displacement is a maximum, the instantaneous velocity will be zero. Therefore, $v = v_0 - gt = 20$ m/s$-9.8t = 0$ at this point. Solving for t, we get $t = 2.04$ s. The maximum height reached is:

$$y = v_0 t - \frac{1}{2}gt^2 = (20\,\text{m/s})(2.04\,\text{s}) - \frac{1}{2}(9.8\,\text{m/s}^2)(2.04\,\text{s})^2 = 20.4\,\text{m}$$

(*c*) Using the above expressions for displacement and velocity, we can draw the following plot:

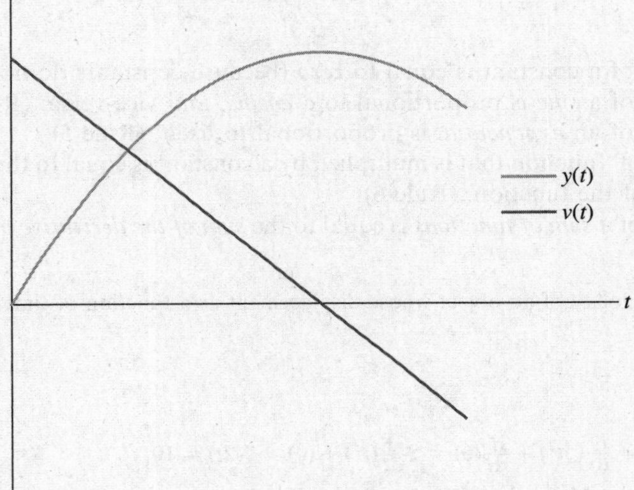

Fig. 17-5

DERIVATIVE FORMULAS The following table provides the derivatives of some common functions along with some common rules for evaluating the derivatives of combinations of functions.*

These rules are all derivable by applying the definition of the derivative to each situation, but we shall not do so here. Instead, we note in somewhat informal language, some of the most useful properties that emerge from these rules:

* A more comprehensive listing can be found in nearly any calculus textbook and in many physics textbooks. The formulas provided here will suffice for most problems likely to be encountered at the introductory physics level and for many beyond that level.

Table 17-1

#	$f(t)$	$f'(t)$	Example
1	constant	0	If $x = 5$, then $\frac{dx}{dt} = 0$
2	t^r ($r \neq 0$)	rt^{r-1}	If $y = t^2$, then $\frac{dy}{dt} = 2t$
3	$\sin(t)$	$\cos(t)$	If $z = \sin x$ $z = \sin(x)$, then $\frac{dz}{dx} = \cos x$
4	$\cos(t)$	$\sin(t)$	If $v = \cos(t)$, then $\frac{dv}{dt} = -\sin(t)$
5	e^t	e^t	If $y = e^x$, then $\frac{dy}{dx} = e^x$
6	$ah(t)$	$ah'(t)$	If $y = A \sin t$, then $\frac{dy}{dt} = A \cos t$
7	$f_1(t) + f_2(t)$	$f_1'(t) + f_2'(t)$	If $x(t) = at + bt^3$, then $v = \frac{dx}{dt} = a + 3bt^2$
8	$f(g(t))$ [function of a function]	$f'(g(t))g'(t)$	If $N = N_0 e^{\lambda t}$, then $\frac{dN}{dt} = \lambda N_0 e^{\lambda t}$
9	$\ln(t)$	$\frac{1}{t}$	If $I = \ln(P)$, then $\frac{dI}{dP} = \frac{1}{P}$
10	$f(t)g(t)$ [product of two functions]	$f'(t)g(t) + f(t)g'(t)$	If $y = 3te^t$, then $\frac{dy}{dt} = 3e^t + 3te^t = 3e^t(1+t)$

1. The derivative of a constant is equal to *zero* (because constants do not change). (Rule 1)
2. The derivative of a *sine* is proportional to a *cosine*, and vice-versa. (Rules 3 and 4)
3. The derivative of an *exponential* is proportional to *itself*. (Rule 5)
4. The derivative of function that is multiplied by a constant is equal to the constant multiplied by the derivative of the function. (Rule 6)
5. The derivative of a *sum of functions* is equal to the *sum of the derivative of each function*. (Rule 7)

EXAMPLE 3. Find the velocity of an object whose displacement as a function of time is given by:

(*a*) $x = 5t^2 + 6$
(*b*) $z(t) = 10 \sin(4\pi t)$
(*c*) $y(t) = (10t^2 + 6t - 4)e^{3t}$

(*a*) $v = \frac{dx}{dt} = \frac{d}{dt}(5t^2 + 6) = \frac{d}{dt}(5t^2) + \frac{d}{dt}(6) = 5\frac{d}{dt}(t^2) + (0) = 5(2t) = 10t$

(*b*) $v = \frac{dz}{dt} = \frac{d}{dt}(10 \sin 4\pi t) = 10\frac{d}{dt}(\sin 4\pi t) = 10(\cos 4\pi t)(4\pi) = 40\pi \cos 4\pi t$

(*c*) $v = \frac{dy}{dt} = \frac{d}{dt}[(10t^2 + 6t - 4)e^{3t}] = \left(\frac{d}{dt}[(10t^2 + 6t - 4)]\right)e^{3t} + (10t^2 + 6t - 4)\frac{d}{dt}(e^{3t})$

where we have used Rule 10, the "product rule" to decompose the product, as well as Rule 8, the "function of a function" rule, to simplify the exponential. For the resulting expression, we now use Rule 2 to decompose the polynomials into individual terms and Rules 5 and 8 to find the derivative of the exponential term. The result is:

$$v = (20t + 6)e^{3t} + (10t^2 + 6t - 4)3e^{3t}$$

EXAMPLE 4. An object is in a "potential well" described by $V(x) = \frac{1}{2}k(x-5)^2$, where $V(x)$ is the potential in joules and x is the position in meters. The force $F(x)$ on such an object is given by the expression $F(x) = -\frac{dV(x)}{dx}$, where F is in newtons. Find the force on the object as a function of position and the place where there is no force on the object.

$$F(x) = -\frac{dV}{dx} = -\frac{1}{2}k \times 2(x-5) = k(x-5)$$

Therefore, $F(x) = k(x-5)$ and so the force will be zero when $x = 5$.

EXAMPLE 5. The temperature along a 3 cm rod, to be used in a research project, varies rapidly according to $T(x) = 5x^3 + 10x^2 - 60x - 1$, where x is the distance in cm from the left-hand end and T is the temperature in degrees Fahrenheit.

(*a*) Find the minimum temperature and the location of the minimum.
(*b*) Find the maximum temperature and the location of the maximum.

As previously seen, an extreme value of a function—whether a minimum or a maximum—can be found by setting the derivative of the function equal to 0. We set $\frac{dT}{dx}$ equal to 0 and then see which values of x correspond to a minimum or a maximum.

(*a*) $\frac{dT}{dx} = \frac{d}{dx}(5x^3 + 10x^2 - 60x - 1) = 15x^2 + 20x - 60 = 0$. Dividing both sides by 5, we get $3x^2 + 4x - 12 = 0$. This is a quadratic equation of the form $ax^2 + bx + c = 0$, where $a = 3$, $b = 4$ and $c = -12$. The solution is given by the quadratic formula:

$$x = \frac{-b \pm \sqrt{b^2 - 4ac}}{2a} = \frac{-4 \pm \sqrt{4^2 - 4(3)(-12)}}{6}$$

This has two solutions, corresponding to the + and − signs above. They are $x_1 = 1.44$ and $x_2 = -2.77$, respectively. These are the values of x at which $T(x)$ has local extrema—i.e., for which $\frac{dT}{dx} = 0$, when all values of x are considered. For our rod, however, x only ranges from 0 to 3, and, because x_2 is outside this range, it is excluded as "unphysical". The value of the temperature at x_1 is $T(x_1) = -48.1$. We also need to examine each of the endpoints of the rod to see if they contain extreme values within the restricted range of the function that we are considering. In the current case, $T(0) = -1$ and $T(3) = 44$. Hence, the minimum temperature of -48.1 occurs at the location $x_1 = 1.44$, and the maximum temperature of 44 occurs at $x_1 = 3$. The variation with temperature along the length of the rod is shown in Fig. 17-6. Notice, in particular, the locations of the minimum and maximum temperature values.

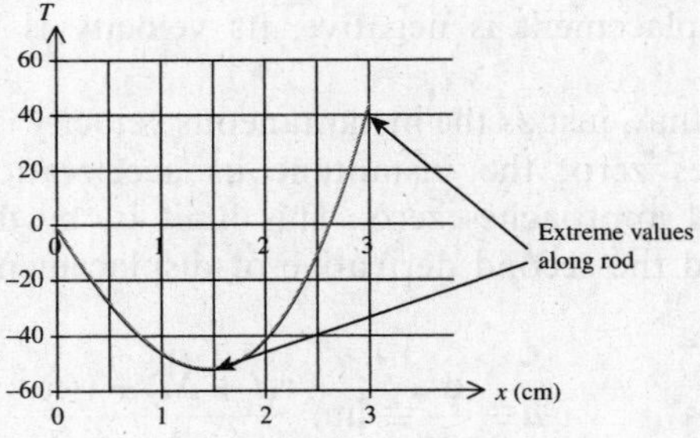

Fig. 17-6

ACCELERATION AND SECOND DERIVATIVES. As we saw in the previous section, velocity is the instantaneous change in displacement with respect to time. It tells us, at any particular time, whether the displacement is increasing, decreasing, or remaining constant. But what about the change in the instantaneous velocity itself? How do we know whether it is rising, falling, or staying constant? As we will see, the answer will have important physical consequences.

Let us consider once more the motion of a mass at the end of a vertical spring.

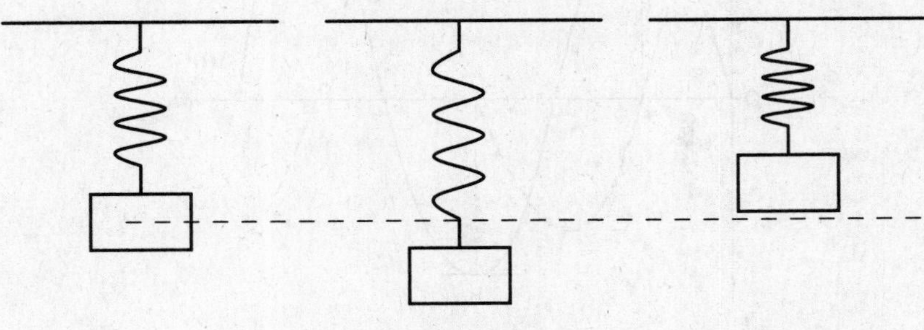

Fig. 17-7

In Fig. 17-8, we reproduce Fig. 17-3. In order to determine the instantaneous velocity near Point A in the figure, we employ a procedure similar to that used in the preceding section. Recall that we defined instantaneous velocity as the limit of the average velocity within a small time interval as that time interval approaches zero. We now look at the *average change in velocity per unit time*—i.e., the *average acceleration*—in the vicinity of Point A, again allowing the time interval to approach zero. We note that the average acceleration—the change in velocity divided by the change in time over a particular time interval—corresponds to the slope of the line connecting the initial and final times (see points D, C and B in Fig. 17-8).

As the time interval shrinks, the slope approaches a limit, which is the slope of the tangent to the curve at point A. It is called the *instantaneous acceleration* (or simply *acceleration*), and is equal to the derivative of the instantaneous velocity with respect to time. Because the instantaneous velocity is the derivative of displacement with respect to time, velocity is sometimes referred to as the *first derivative* of displacement with respect to time, and acceleration—the derivative of a derivative—is referred to as the *second derivative* of displacement with respect to time.

In Fig. 17-9, the displacement, velocity, and acceleration of the mass are plotted as a function of time. Recall that we have defined displacement to be positive when the mass is below the equilibrium point and velocity positive when it is moving in the downward direction. Acceleration is then positive when the velocity is becoming more positive. In Region I, the mass is moving downward but is being slowed by the restraining force of the spring—the velocity is positive but the acceleration is negative. When the mass extends the spring to its maximum extent, then the velocity is momentarily zero—this is shown at point x. After that, in Region II, the mass moves back toward the equilibrium position—the velocity is now negative. Note that the acceleration—the slope of the velocity curve—has remained negative throughout. After the mass rebounds upward through the equilibrium point (Region III), then its displacement is negative, its velocity is negative and its acceleration is positive.

Mathematically, we can say that, just as the instantaneous velocity is the limit of the average velocity as the time interval approaches zero, the instantaneous acceleration is the limit of the average acceleration as the time interval approaches zero. This limit is, by definition, the first derivative of velocity with respect to time and the second derivative of displacement with respect to time. In other words,

$$a \equiv \frac{\mathrm{d}v}{\mathrm{d}t} \equiv \lim_{\delta t \to 0} \frac{v(t + \Delta t) - v(t)}{\Delta t}$$

and, since $v \equiv \frac{\mathrm{d}x}{\mathrm{d}t} \equiv \lim_{\delta t \to 0} \frac{x(t + \Delta t) - x(t)}{\Delta t}$, then it follows that

$$a = \frac{\mathrm{d}v}{\mathrm{d}t} = \frac{\mathrm{d}}{\mathrm{d}t}\left(\frac{\mathrm{d}\,x}{\mathrm{d}t}\right) \equiv \frac{\mathrm{d}^2x}{\mathrm{d}t^2}$$

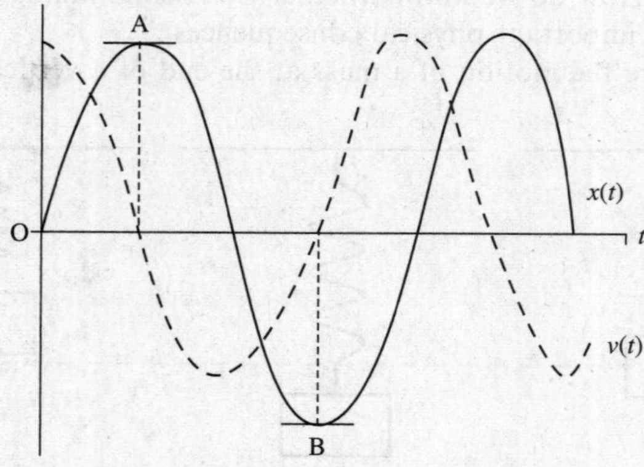

Fig. 17-8

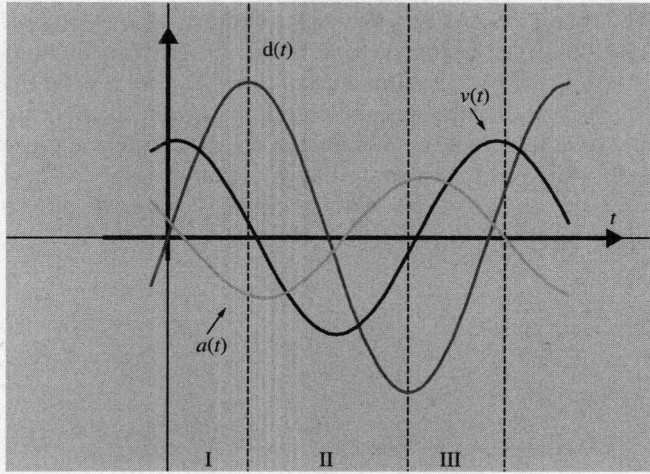

Fig. 17-9

where $\frac{d^2x}{dt^2}$ is the symbol for "the second derivative of x with respect to t" (i.e., the derivative of the derivative of x with respect to t).

We have chosen t as the independent variable and x as the dependent one in the above example. For a function $g(y)$, the second derivative of $g(y)$ with respect to y would be written as $\frac{d^2g}{dy^2}$. An alternative notation is to write $g'(y)$ for the first derivative of g with respect to y and $g''(y)$ for the second derivative.*

EXAMPLE 6. Find the velocity and acceleration as a function of time for a projectile with a displacement given by $y = y_0 + v_0t + \frac{1}{2}gt^2$, where y_0 and v_0 are the initial displacement and velocity, respectively, at $t = 0$, and g is the acceleration of gravity.

From the above equations, we know that:

$$v = \frac{dy}{dt} = \frac{d}{dt}(y_0 + v_0t + \frac{1}{2}gt^2) = v_0 + gt$$

Hence,

$$a = \frac{dv}{dt} = \frac{d}{dt}(v_0 + gt) = g$$

Therefore, the acceleration is a constant and equal to g, the acceleration of gravity. Some of these equations may be familiar to the reader as *kinematic equations of motion* (except that we have used the acceleration of gravity g for the more general acceleration variable a). These expression emerge simply using the tools of calculus (in the following chapter, we shall reverse the process, using the constant acceleration of gravity to extract expressions for displacement and velocity as a function of time).

The use of the second derivative to distinguish among local maxima and minima, as described in the preceding chapter, is of considerable utility.

* In particular, note that neither "2" in $\frac{d^2x}{dt^2}$ is an exponent!

ANTIDERIVATIVES AND INTEGRALS. We saw in the previous section that if we know the displacement as a function of time, then, using the tools of calculus, we can find the velocity—the derivative of the displacement with respect to time—as a function of time. Can we reverse the steps in such a manner that given the velocity as a function of time, we can determine how the displacement varies with time? We shall see that, with certain conditions met, we can indeed do so, and that the results can be applied in a straightforward fashion to find answers to problems that would otherwise be difficult or impossible to solve.

Suppose we know that an object moves with a constant velocity of $v = 1.5$ m/s. What is the displacement x as a function of time? (Fig. 17-10)

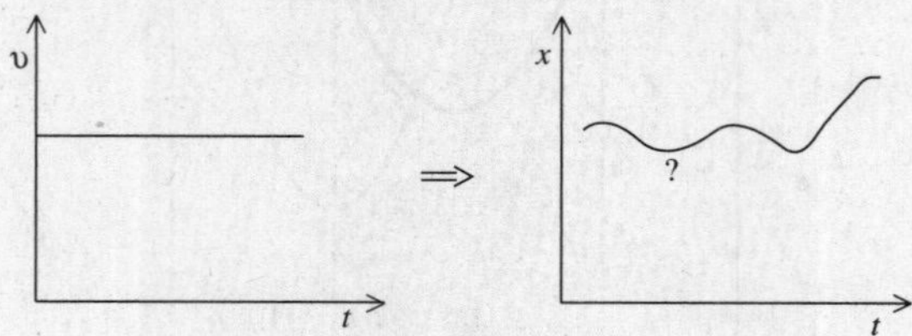

Fig. 17-10

Because the derivative of $x(t)$ is $v(t)$, we know that $v(t)$ is equal to the slope of $x(t)$ for all values at every particular time. Hence, in this case, $x(t)$ has a slope of 5 m/s at all points. Which curve has a constant slope of 5 m/s at all points? It must be a line:

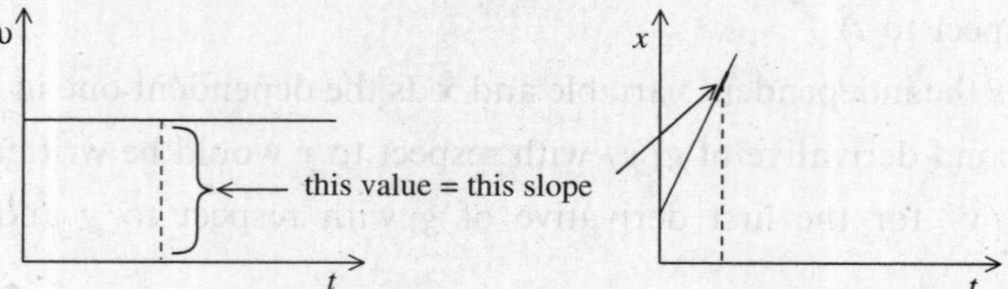

Fig. 17-11

Just as v is the derivative of x with respect to t (or symbolically, $v = \frac{dx}{dt}$), we say that x is the *antiderivative*, or the *indefinite integral*, of v with respect to t—i.e., x is the function whose derivative with respect to t is v. Symbolically, we write that $x = \int v \, dt$.

Note that additional information is needed in order to specify the exact function in Fig. 17-10, since we only know the slope of the line and not its vertical intercept. This additional information is typically given in the form of *initial conditions* or *boundary conditions*. This is shown in the following example.

EXAMPLE 7. Suppose we know that an object moves horizontally along a straight line path. We do not know the function $x(t)$ that describes its position as a function of time. However, suppose we do know that its velocity is given by

$$v(t) = a + bt^2$$

Can we determine the motion—the displacement—of the object as a function of time that is giving rise to this velocity? Using the derivative formulas from the previous chapter, one can directly verify that, in this particular case, $v(t)$ is the derivative of the function $x(t) = at + \frac{b}{3}t^3 + c$, where c is some (as yet unknown constant). Let us indeed verify that $v(t)$ is the time derivative of $x(t)$:

$$\frac{dx(t)}{dt} = \frac{d}{dt}\left(at + \frac{b}{3}t^3 + c\right) = a + \frac{b}{3}(3t^2) + 0 = a + bt^2 = v(t)$$

Notice that c is, at this point, an undetermined constant term; since the derivative of any constant is zero, any constant can be added to $x(t)$ without changing the value of $v(t)$. In order to specify what the displacement $x(t)$ as a function of time actually *is*—to remove the ambiguity of the unknown constant—additional information must be provided. For example, if we are told that the displacement of the particle at $t = 0$ is $x(0) = 5$ then the value of c can be determined from the equation

$$x(0) = a(0) + \frac{b}{3}(0)^3 + c = 5,$$

so that $c = 5$. The displacement is then

$$x(t) = at + \frac{b}{3}t^3 + 5$$

Both $x(t)$ and $v(t)$ are plotted below for the case $a = b = 1$. Note that the value of the slope of the tangent to $x(t)$ at any time t is given by the value of $v(t)$ at that same time. For example, $x(t)$ is positively sloped at point A, is zero at point B and is negative at point C. The values of these slopes are reflected in the values of $v(t)$ at the same corresponding times.

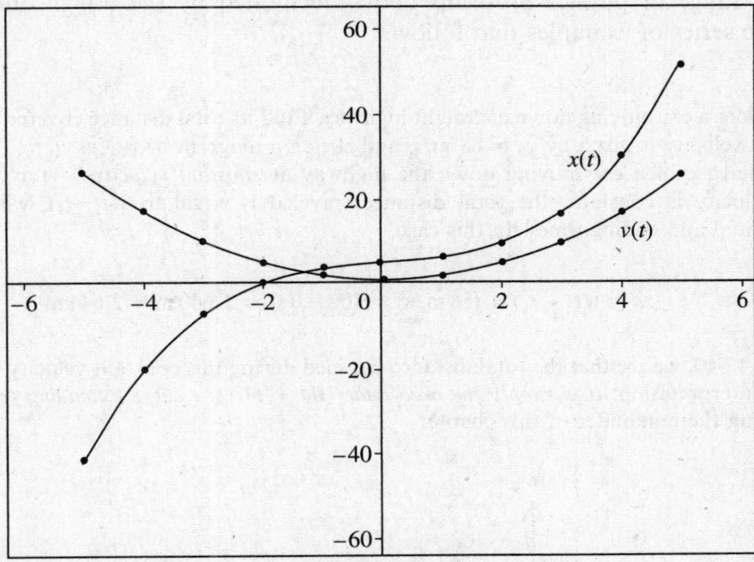

Fig. 17-12

In this case, we say that, given the velocity, we have determined the displacement. Since velocity is the derivative of displacement, we say that the displacement function is the *antiderivative* of velocity. As we shall see below, all of the tools that have been used to find derivatives can be inverted in order to find antiderivatives as well. For nearly all of the functions that we have considered, one could similarly work backwards to figure out the particular function the derivative of which will yield a given function.

Note that, in each case, the constant c is an additive term in the antiderivative. Its value must be determined from the specifics of a given situation, as we will see repeatedly in the problems that follow. Note the use of the expression $\int f(x)\,dx$ (read, alternately as "the antiderivative of $f(x)$", "the integral of $f(x)$" or—more formally—"the indefinite integral of $f(x)$ over x")

Table 17-2

Rule	$f(t)$	Antiderivative $\int f(t)\,dt$	Example
1	$t^n\ (n \neq -1)$	$\frac{t^{n+1}}{n+1}$	If $f(t) = t^2$, then $\int f(t)\,dt = \frac{1}{3}t^3 + c$
2	$\sin(t)$	$-\cos(t)$	If $z = \sin(x)$, then $\int \sin(x)\,dx = -\cos(x) + c$
3	$\cos(t)$	$\sin(t)$	If $a(t) = \cos(t)$, then $\int a(t)\,dt = \sin(t) + c$
4	e^t	e^t	If $y = e^x$, then $\int e^x\,dx = e^x + c$
5	$a\,f(t)$	$a \int f(t)\,dt$	If $v(t) = gt$, then $\int v(t)\,dt = \frac{1}{2}gt^2 + c$
6	$f_1(t) + f_2(t)$	$\int f_1(t)\,dt + \int f_2(t)\,dt$	If $E = E_1 + E_2$, then $\int E\,dl = \int E_1\,dl + \int E_2\,dl$
7	$\frac{1}{t}$	$\ln(t)$	If $f(x) = \frac{1}{x}$, then $\int f(x)\,dx = \ln(x) + c$

EXAMPLE 8. Find the antiderivative of $g(x)$ when $g(x)$ is (*a*) $6x$; (*b*) $A\cos x$; (*c*) e^x and (*d*) $\frac{1}{x^3}$

Using the above formulas, we get:

(*a*) $\int g(x)\,dx = \int 6x\,dx = 6\int x\,dx = 6\left(\frac{x^2}{2}\right) + c = 3x^2 + c$ [using Rule 1];

(*b*) $\int g(x)\,dx = \int A\cos x\,dx = A\int \cos x\,dx = A\sin x + c$ [using Rules 3 and 5]

(*c*) $\int g(x)\,dx = \int e^x dx = e^x + c$ [using Rule 4]; and

(*d*) $\int g(x)\,dx = \int \frac{1}{x^3}dx = -\frac{1}{2x^2}$

THE AREA UNDER A CURVE. The ability to calculate the area under a curve is critical to the solution of a wide range of physics problems and is facilitated by the use of antiderivatives, as is demonstrated in the series of examples that follow.

EXAMPLE 9. Consider a car moving down a straight highway. Find its total distance covered during a 30-s period starting at $t = 0$ if its velocity is given by $v_0 = 88$ m/s, and sketch a diagram of $v(t)$ vs. t.

This is the simple case of a car moving down the highway at *constant velocity* v—for which no calculus is needed. Since the velocity is constant, the total distance traveled is equal to $v(t_2 - t_1)$, where t_1 and t_2 refer, respectively, to the initial and ending times. In this case,

$$d = v(t_2 - t_1) = (88\,\text{m/s}) \times (30\,\text{s} - 0\,\text{s}) = 2,640\,\text{m} = 2.64\,\text{km}$$

Referring to Fig. 17-13, we see that the total distance traveled during this constant velocity trip has a simple but important graphical interpretation: *it is simply the area under the $v(t)$ vs. t curve extending from t_1 to t_2.* We will utilize this result during the remainder of this chapter.

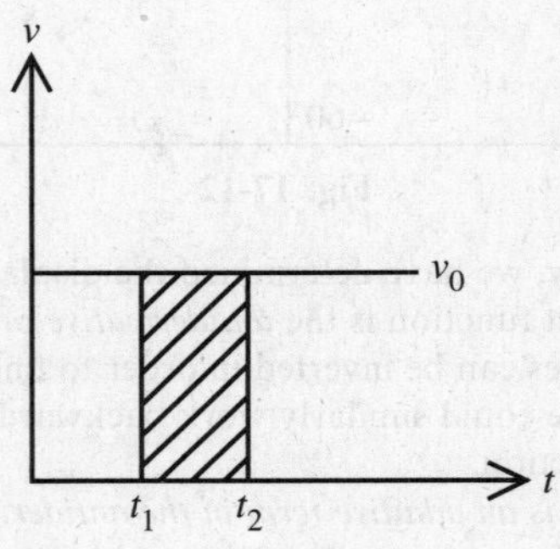

Fig. 17-13

EXAMPLE 10. Find the relationship between the distance covered by an object moving in 1D with a velocity $v(t)$ and the area under its $v-t$ curve.

Suppose the velocity varies with time according to Fig. 17-14*a*. What is the total distance covered by an object during the first 30 s? To answer this question, let us approximate the curve by a set of rectangles (Fig. 17-14*b*). As we replace the curve by more and more rectangles—each representing a smaller time slice—we more closely approximate the original curve. Extending the results of Example 11, we can approximate the total distance covered by the sum of the areas of the rectangles.

If we extend this logic, dividing the trip up into more and more segments of shorter and shorter time intervals, then we eventually will get a graph somewhat like Fig. 17-14*c*, in which the velocity varies continuously over time. Hence, we can consider the area under the original curve to consist of an infinite number of infinitely thin rectangles, each spanning an instant of time. Using the same reasoning as above, we conclude that *the total distance covered is equal to the area under the velocity–time curve*. If we can find a means for finding the area under this curve, then we can find the total distance covered.

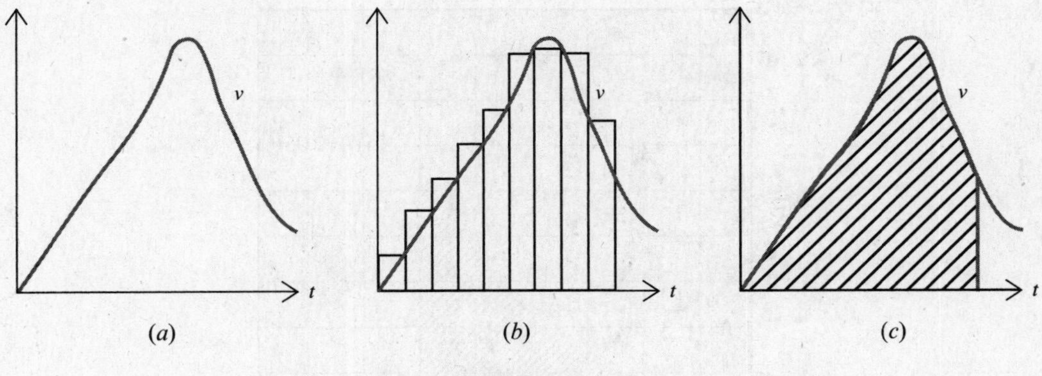

Fig. 17-14

THE FUNDAMENTAL THEOREM OF CALCULUS. There is an important relationship between the antiderivative of a function $v(t)$ and the area under $v(t)$:

$$\int_{t_1}^{t_2} v(t)\,\mathrm{d}t = \text{Area under } v(t) \text{ between } t_1 \text{ and } t_2.$$

where $\int_{t_1}^{t_2} v(t)\,\mathrm{d}t$ is defined as $x(t_1)-x(t_2)$ and is called the **definite integral** of $v(t)$ between t_1 and t_1.

APPLICATIONS OF THE FUNDAMENTAL THEOREM. The fundamental theorem of calculus makes possible the solution of many problems involving continuously changing quantities. This is demonstrated in the following example and in the solved problems that follow.

EXAMPLE 11. Find the total distance covered in the first 30 s by an object undergoing 1-D motion with a velocity that varies according to:

$$v(t) = 0.05t^2$$

According to the Fundamental Theorem of Calculus, the area under the shaded portion of the velocity vs. time curve in Fig. 17-15 — and therefore the total displacement, is

$$x = \int_0^{30} 0.05t^2\,\mathrm{d}t = \frac{0.05}{3}t^3\Big|_0^{30} = \frac{0.05}{3}30^3 - \frac{0.05}{3}0^3 = 450\,\mathrm{m}$$

This problem provides an example of the remarkable power of calculus to address problems involving continuous change. The same technique is used in the problems of the previous chapter as well as those that follow, finds wide application throughout all of physics and, indeed, the natural and social sciences.

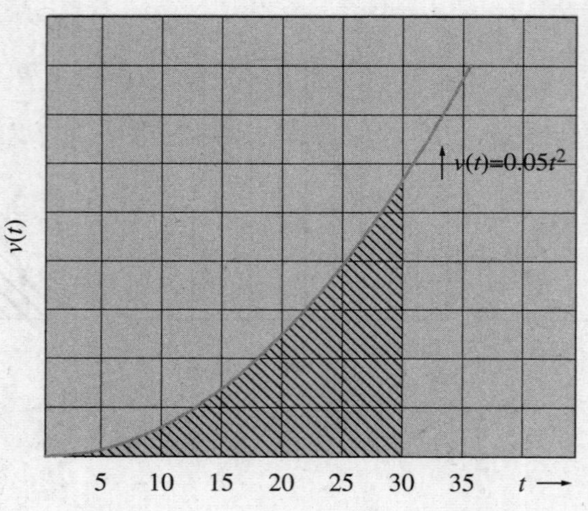

Fig. 17-15

Table 10-2 provides guidance to the integration of several of the most common functions. A number of useful techniques exist for integration in more advanced situations, particularly, when combinations of such functions are involved (e.g., $\int x^2 e^{-\alpha x^2}\,dx$).

Standard techniques are covered in any number of books on calculus but are beyond the scope of the current text. However, because of its wide utility, one technique, integration by parts, is described in the Appendix.

Solved Problems

17.1 Find the velocity of a rocket 4 s after it is launched vertically upwards if its displacement is given by $y = 15 + 3t + \frac{1}{2}t^2$, where t is in seconds and y is in meters.

$$y = 15 + 3t + \frac{1}{2}gt^2$$

$$v = \frac{dy}{dt} = \frac{d}{dt}(15 + 3t + \frac{1}{2}t^2)$$

$$v = \frac{dy}{dt} = \frac{d}{dt}(15) + \frac{d}{dt}(3t) + \frac{d}{dt}\left(\frac{1}{2}t^2\right)$$

$$= 0 + 3 + t$$

$$\text{At, } t = 4,\ v = 3(1) + (.4)(9.8)(2) = 10.8\,\text{m/s}$$

17.2 An object undergoes displacement described by $x(t) = 3t^3 + 5t^2 - 6$, where the displacement x is in meters and the time t is in seconds. Find (*a*) its average velocity during its first 10 s of motion; (*b*) its instantaneous velocity at $t = 5$; (*c*) its average acceleration during this same time; and (*d*) its instantaneous acceleration at $t = 4$ s.

(*a*) $\bar{v}_{21} = \frac{x_2 - x_1}{t_2 - t_1}$, where the subscripts 1 and 2 refer, respectively, to the starting and ending points of the motion.

At $t = 0$, $x(0) = 3 \times 0^3 + 5 \times 0^2 - 6 = -6$

At $t = 10$, $x(10) = 3 \times 10^3 + 5 \times 10^2 - 6 = 3,496$

Therefore, $\bar{v}_{21} = \frac{x_2 - x_1}{t_2 - t_1} = \frac{3,496 - (-6)}{10 - 0} = 350.2\,\text{m/s}$

(*b*) The instantaneous value of v at time t is given by

$$\begin{aligned} v &= \frac{dx}{dt} = \frac{d}{dt}(3t^3 + 5t^2 - 6) \\ &= \frac{d}{dt}(3t^3) + \frac{d}{dt}(5t^2) - \frac{d}{dt}(6) \\ &= 9t^2 + 10t \end{aligned}$$

Therefore, the value of the instantaneous velocity at $t = 5$ is

$$v = 9 \cdot 5^2 + 10 \cdot 5 = 275\,\text{m/s}$$

(*c*) The average acceleration of the object between points 1 and 2 is

$$a = \frac{v_2 - v_1}{t_2 - t_1}$$

In order to evaluate this expression, we need to know how the object velocity varies with time. Fortunately, we know that the displacement varies with time according to

$$x = 3t^3 + 5t^2 - 6$$

Because we are talking about the variation of velocity at each instant in time, what we want is the instantaneous velocity. As we have seen, this is just the derivative of the displacement with respect to time:

$$v = \frac{dx}{dt} = \frac{d}{dt}(3t^3 + 5t^2 - 6) = 9t^2 + 10t$$

At $t = 0$, $v_1 = 0$ and at $t = 10$, $v_2 = 1000$

So, during the first 10 s, the average acceleration is $\bar{a} = \dfrac{(1000 - 0)\,\text{m/s}}{(10 - 0)\text{s}} = 100\,\text{m/s}^2$

(*d*) The instantaneous acceleration is just the derivative of the instantaneous velocity with respect to time:

$$a = \frac{dv}{dt} = \frac{d}{dt}(9t^2 + 10t) = 18t + 10$$

At $t = 4$, the instantaneous acceleration is $a = 18(4) + 10 = 82\,\text{m/s}^2$.

17.3 An object subject only to the force of gravity has a displacement relative to the ground of $y = y_0 + v_0 t + \frac{1}{2}gt^2$, where, in English units, y is measured in feet and $g = 32$ ft/s^2.

(*a*) If the object dropped from rest from a height of 35,000 ft, how fast will it be moving after 15 s?

(*b*) What will be its height above the ground at this point?

(*c*) With what velocity will it strike the ground?

(*a*) We choose as our origin the point at 35,000 ft where the object is initially dropped. This corresponds to $y_0 = 0$. Note that, because the object is dropped with no initial vertical velocity, that $v_0 = 0$ as well.

The velocity of the object is given by

$$v = \frac{dy}{dt} = \frac{d}{dt}\left(y_0 + v_0 t + \frac{1}{2}gt^2\right) = 0 + 0 + \frac{d}{dt}\left(\frac{1}{2}gt^2\right) = \frac{1}{2}g(2t) = gt$$

Therefore $v = gt$

After 15 s, the velocity of the object is $v = (32\,\text{ft/s}^2) \times (15\,\text{s}) = 480$ ft/s

(*b*) Its height above the ground at this point can be calculated by determining its displacement from the original location from which it was dropped, which we choose at the origin. We then have that $y = y_0 + v_0 t + \frac{1}{2}gt^2$, where y measures the displacement from the origin, $y_0 = 0$ and $v_0 = 0$. We choose the vertical axis so that y increases positively in the downward direction.

Therefore, $y = \frac{1}{2}(32\,\text{ft/s}^2)(30\,\text{s})^2 = 16,384\,\text{ft}$

This is the displacement from the ejection point at 35,000 ft.

Therefore, the height above the ground at this point is:

$$h = 35,000\,\text{ft} - 19,384\,\text{ft} = 15,616\,\text{ft}$$

(*c*) At what time and at what velocity will it strike the ground?

Using the same origin defined above, we know that

$$y = y_0 + v_0 t + \frac{1}{2}gt^2$$

Since $v_0 = 0$, we have $y = y_0 + \frac{1}{2}gt^2$.

Upon rearrangement, we get, for the elapsed time:

$$t = \sqrt{\frac{2(y - y_0)}{g}} = \sqrt{\frac{2(35,000\,\text{ft})}{32\,\text{ft/s}^2}} = 46.8\,\text{s}$$

17.4 A mass on a spring undergoes an oscillatory motion, with its displacement given by $x = A\sin(kx - \omega t)$.

(*a*) Find the velocity of the spring as a function of time.

(*b*) Find the acceleration of the spring as a function of time.

(*a*) $v = \frac{dx}{dt} = \frac{d}{dt}(A\sin(kx - \omega t)) = A\omega\cos(kx - \omega t)$

(*b*) $a = \frac{dv}{dt} = \frac{d}{dt}(A\omega\cos(kx - \omega t) = -A\omega^2\sin(kx - \omega t)$

17.5 The temperature of a 50 cm metal rod varies along its length according to $T(x) = .01 \times (\frac{1}{3}x^3 + 3x^2 - 40x)$, where x is the distance in centimeter from the left end and the temperature is in degrees Celsius. Find the maximum and minimum temperatures of the rod and the positions where these extremes are reached.

To find the maximum and minimum temperature, we look for the places where the derivative of the temperature function with respect to position is equal to 0:

$$\frac{dT}{dx} = \frac{d}{dx}\left(.01 \times \left(\frac{1}{3}x^3 + 3x^2 - 40x\right)\right) = .01 \times (x^2 + 6x - 40) = 0$$

By factoring the last expression, we get:

$$(x + 10)(x - 4) = 0 \Rightarrow x = -10 \text{ or } x = 4$$

Because the rod only extends from $x = 0$ to $x = 50$, we reject the first solution as unphysical.

At the other solution, $x = 4$, substitution into the temperature function yields $T = -.91°\text{C}$. This is a local minimum or maximum. However, we need to the temperature at this point with that at the ends of the rod—because of the possibility that the "bump" that we have found is only a local extreme point and not a minimum or maximum of the entire function.

At the ends of the rod, the values of the temperature are:

$$T(0) = .01 \times \left(\frac{1}{3}(0)^3 + 3(0)^2 - 40(0)\right) = 0$$

and

$$T(50) = .01 \times \left(\frac{1}{3}(50)^3 + 3(50)^2 - 40(50)\right) = 472\,°\text{C}$$

Therefore, the minimum temperature of $-.91°\text{C}$ is reached at $x = 4$ cm, while the maximum of 472°C is reached at x = 50 cm.

17.6 The current i through a 50 h inductor has the form $i = 3t^2 + 4\sin\left(\frac{\pi t}{120}\right) + 6$, where t is the time in seconds. What is the potential difference across the inductor at $t = 30$?

A potential difference across a circuit element in which magnetic flux is linked is a consequence of Faraday's Law of Induction. For our purposes here, we only note that the potential difference across such an element depends on the derivative of the current with respect to time according to the formula

$$V = -L\frac{\mathrm{d}i}{\mathrm{d}t},$$

where the potential difference V is measured in volts, the self-inductance L is measured in henrys and $\frac{\mathrm{d}i}{\mathrm{d}t}$ gives the rate of change of current with respect to time (in units of A/s).

In our case,

$$V = -L\frac{\mathrm{d}}{\mathrm{d}t}\left[3t^2 + 4\sin\left(\frac{\pi t}{120}\right) + 6\right]$$
$$= -L\left(6t + \frac{\pi}{30}\cos\left(\frac{\pi t}{120}\right)\right)$$

Substituting $L = 50$ h and $t = 30$ s, we get:

$$V = -50\left(6 \times 30 + \frac{3.14}{30}\cos\left(\frac{\pi}{4}\right)\right)$$
$$= -9.00\,\mathrm{kV}$$

17.7 If the velocity of an object moving through a viscous fluid decreases according to $v = v_0 e^{-bt^2}$, what is its acceleration at $t = 5$ s? Assume an initial velocity at $t = 0$ of .5 m/s and that $b = .3\ \mathrm{s}^{-2}$

$$v = v_0 e^{-bt^2}$$

The acceleration

$$a = \frac{\mathrm{d}v}{\mathrm{d}t} = \frac{\mathrm{d}}{\mathrm{d}t}(v_0 e^{-bt^2}) = v_0\frac{\mathrm{d}}{\mathrm{d}t}(e^{-bt^2}) = v_0 e^{-bt^2}(-2bt) = -2btv_0 e^{-bt^2} = (-2)(.3)(.5\mathrm{m/s})e^{(-.3\times.25)} = -.278\,\mathrm{m/s^2}$$

17.8 The current through a resistor varies with time according to $i(t) = t^2 + 6t + 4$, where the current $i(t)$ is measured in amps. What is the instantaneous rate of change of the current at $t = 10$ s?

The current through a resistor is the rate at which charge is flowing through the resistor. If the charge is measured in the standard SI units of coulombs, then the current is measured in amps. The rate of change of the current, $\frac{\mathrm{d}i}{\mathrm{d}t}$, tells how quickly the current is growing or declining with time. In this case,

$$\frac{\mathrm{d}i}{\mathrm{d}t} = \frac{\mathrm{d}}{\mathrm{d}t}(t^2 + 6t + 4) = \frac{\mathrm{d}}{\mathrm{d}t}(t^2) + \frac{\mathrm{d}}{\mathrm{d}t}(6t) + \frac{\mathrm{d}}{\mathrm{d}t}(4) = 2t + 6$$

At $t = 10$ s, $\frac{\mathrm{d}i}{\mathrm{d}t} = 2(10) + 6 = 26$ A/s. So the current is increasing at the rate of 26 A/s.

17.9 According to quantum theory, when X-rays are scattered from electrons they undergo a change in wavelength given by the Compton formula:

$$\Delta\lambda = \frac{h}{m_e c}(1 - \cos\theta)$$

where h is Planck's constant, m_e is the mass of the electron, c is the speed of light and θ is the x-ray scattering angle (θ can range between 0° and 180°)

(*a*) What scattering angle results in the greatest change in wavelength?

(*b*) Compute the change in wavelength that results in this case.

(*a*) $\Delta\lambda = \frac{h}{m_e c}(1 - \cos\theta)$

To find the maximum of the function, we set its derivative equal to 0:

$$\frac{d\Delta\lambda}{d\theta} = \frac{d}{d\theta}\left(\frac{h}{m_e c}(1 - \cos\theta)\right) = \frac{h}{m_e c}\sin\theta = 0$$

$\sin\theta = 0$ then yields points at which $\Delta\lambda$ has an extreme value—a maximum or a minimum. Inspection of the sine function shows that, for angles between 0° and 180°, the function has extreme values at 0° and 180°. Substitution into the Compton formula above shows that $\theta = 0°$ corresponds to a minimum while $\theta = 180°$ corresponds to a maximum. Since we are seeking the greatest change in wavelength, we conclude that when a photon scatters from an electron with a scattering angle of 180°, the change in wavelength is maximal.

(*b*) The change in wavelength is given by

$$\Delta\lambda = \frac{h}{m_e c}(1 - \cos\theta) = \frac{h}{m_e c}(1 - (-1)) = \frac{6.6 \times 10^{-34}\ \text{J} - \text{s}}{(9.11 \times 10^{-31}\ \text{kg})(3 \times 10^8\ \text{m/s})}(1 - (-1)) = 4.8 \times 10^{-10}\ \text{m}$$

17.10 How long will a ball dropped at rest from the top of the Empire State Building (approximate height: 1500 ft) take to reach the ground if we assume a constant gravitational acceleration of 32 ft/s^2 and ignore the effects of air resistance?

If we define the velocity of the ball $v(t)$ to be positive in the downward direction, then we know that the acceleration of gravity $g = \frac{dv}{dt}$. Then $dv = g\,dt$ and $v = \int dv = \int g\,dt = g\int dt = gt + C$, where C is a constant arising from the integration that needs to be determined from a consideration of the initial conditions. In this case, the initial conditions are that, at $t = 0$, $v = gt + C = g(0) + C = 0$, because the ball is initially at rest, and so $C = 0$. Hence $v = gt$.

Recognizing that velocity is the derivative of position with respect to time, we then have: $v = \frac{dx}{dt} \rightarrow dx = v\,dt \rightarrow x = \int dx = \int v\,dt = \int gt\,dt = g\int t\,dt = \frac{1}{2}gt^2 + D$. Hence, $x = \frac{1}{2}gt^2 + D$.

If we define the origin of position to be at the top of the building, then $x = 0$ at $t = 0$, and, according to the preceding equation, we conclude that $D = 0$ and, therefore, $x = \frac{1}{2}gt^2$.

Since $x = \frac{1}{2}gt^2$, $t = \sqrt{\frac{2x}{g}} = \sqrt{\frac{2 \times 1500\ \text{ft}}{32\ \text{ft/s}^2}} = 9.7\ \text{s}$

17.11 A subatomic particle begins at rest from $x = 0$ and moves with an acceleration $a(t) = 4t^3 - 6t^2 + 30\cos(\pi t)$, where $a(t)$ is in units of m/s^2. After 10 s, find (*a*) the velocity of the particle; and (*b*) the displacement of the particle.

(*a*) The velocity is:

$$\begin{aligned} v &= \int a\,dt = \int [4t^3 - 6t^2 + 30\cos(\pi t)]\,dt = 4\int t^3\,dt - 6\int t^2\,dt + 30\int \cos(\pi t)\,dt \\ &= t^4 - 2t^3 - \frac{60}{\pi}\sin(\pi t) + c = \{t^4 - 2t^3 - \frac{60}{\pi}\sin(\pi t)\}|_0^{10} \\ &= 10^4 - 2 \times 10^3 - \frac{60}{\pi}\sin(10\pi) - 0^4 - 2 \times 0^3 - 60\sin 0 \\ &= 8,000\ \text{m/s} = 8\ \text{km/s} \end{aligned}$$

(*b*) $x(t) = \int v\,dt = \int \left[t^4 - 2t^3 - \frac{60}{\pi}\sin(\pi t)\right]dt = \frac{t^5}{5} - \frac{t^4}{2} + 60\cos(\pi t) + c$

Since $x(0) = 0 + 0 + 60\cos(0) + c = 60 + c$ and $x(0) = 0$ at $t = 0$, $c = -60$

Therefore $x(t) = \frac{t^5}{5} - \frac{t^4}{2} + 60\cos(\pi t) - 60$ and after 10 s the displacement is:

$$\Delta x = x_{10} - x_0 = \frac{10^5}{5} - \frac{10^4}{2} + 60\cos(10\pi) - 60 - 0 = 14,880\ \text{m} = 14.8\ \text{km}$$

17.12 The current through a 100 ohm resistor increases by 5 mA/s. How much energy is dissipated by the resistor in the first 30 s?

The energy dissipated through a resistor is related to the power generated through:

$W = \int_0^t P(t)\,\mathrm{d}t$, where the power generated in a resistor is given by $P(t) = i^2(t)R$, where i is the current through the resistor and R is the resistance.

Therefore, $W = \int i^2 R\,\mathrm{d}t$. In this case $I = 0.005t$, where I is in amps, and $R = 100\,\Omega$.

Therefore, $W = \int i^2 R\,\mathrm{d}t = R\int i^2\,\mathrm{d}t$. Since $I = .005t$, we have

$$W = R\int_0^{30} i^2(t)\,\mathrm{d}t = R\int_0^{30}(.005t)^2\,\mathrm{d}t = R(.005)^2\int_0^{30} t^2\,\mathrm{d}t = R(.005)^2\left.\frac{t^3}{3}\right|_0^{30} = (100)\times(.005)^2(30^3 - 0^3) = 67.5\,\mathrm{J}$$

17.13 What is the work done on a particle subject to a force that varies as $F = ax^2 + bx - cx^{2/3}$ as it moves from $x = 5$ to $x = 25$?

The work done on the particle as it moves from x_1 to x_2 is defined by $W = \int_{x_1}^{x_2} F(x)\,\mathrm{d}x$.

Therefore,

$$W = \int_0^{25}(ax^2 + bx - cx^{2/3})\,\mathrm{d}x = \left(\frac{a}{3}x^3 + \frac{b}{2}x^2 - \frac{3}{5}cx^{5/3}\right)\Big|_0^{25}$$

$$= \left(\frac{a}{3}25^3 + \frac{b}{2}25^2 - \frac{3c}{5}25^{5/3}\right) - \left(\frac{a}{3}0^3 + \frac{b}{2}0^2 - \frac{3c}{5}0^{5/3}\right)$$

$$= \frac{a}{3}25^3 + \frac{b}{2}25^2 = \frac{15625}{3}a + \frac{625}{2}b - \frac{3c}{5}5^{10/3}$$

17.14 What is the impulse imparted to an object subject to a force $F = \frac{3}{t+4}$ as it moves during the first 10 s? Assume that, in this formula, the time t is in seconds and F is in newtons. (The impulse I imparted to an object is given by the expression $I = \int_{t_i}^{t_f} F\,\mathrm{d}t$, where F is the force applied to the object between times t_i and t_f. The units of Impulse are $[I] = [F]\,[t] = N{-}s = (kg\frac{m}{s^2})(s) = (kg\frac{m}{s})$

$$I = \int_{t_i}^{t_f} F\,\mathrm{d}t$$

$$I = \int_{t_i}^{t_f}\frac{3}{t+4}dt = 3\ln(t+4)\Big|_{t_i}^{t_f} = 3\ln(t_f + 4) - 3\ln(t_i + 4)$$

$$= 3\ln\frac{t_f + 4}{t_i + 4} = 3\ln\frac{14}{4} = 3.76\,\mathrm{kg\,m/s}$$

17.15 For relatively small changes in temperature, the length of a rod changes with temperature according to the formula $L(T) = L_0 + \alpha(T - T_0)$, where L(T) is its length (in meters) at temperature T, L_0 is its length at temperature T_0 and α is the coefficient of thermal expansion. If the change in length is small compared with the original length, find (*a*) the change in length of a 1 m steel rod when it is heated 10°C; (*b*) an expression for the fractional change in surface area of a square steel plate of side L. (For steel, $\alpha \approx 1.1 \times 10^{-5}/°\mathrm{C}$).

(*a*) For the rod, we know that

$$L(T) = L_0 + \alpha(T - T_0)$$

The fractional change in length is

$$\frac{\Delta L}{L_0} = \frac{L - L_0}{L_0} = \frac{\alpha(T - T_0)}{L_0}$$

For a change in temperature of 10°C in a steel rod, the fractional change in length is

$$\frac{\Delta L}{L_0} == \frac{\alpha(T-T_0)}{L_0} = \frac{(1.1\times10^{-5}/\text{C})(10\text{C})}{1\text{ m}} \approx 10^{-4}$$

In other words, a 1 m steel rod will expand by approximately 10^{-4}—i.e., one tenth of a millimeter—for every 10° temperature rise.

(*b*) For a square plate with side L_0 at temperature T_0, the area A_0 is given by:

$$A_0 = L_0^2$$

For a different temperature T, the area is given by:

$$A = L^2 = [L_0 + \alpha(T-T_0)]^2 = L_0^2 + 2\alpha L_0(T-T_0) + \alpha^2(T-T_0)^2$$
$$\approx L_0^2 + 2\alpha L_0(T-T_0)$$

where we have used the fact that $\alpha \ll 1$ to ignore the term quadratic in α.
The fractional difference in the areas is therefore:

$$\frac{A-A_0}{A_0} = \frac{2\alpha L_0(T-T_0)}{L_0^2} = \frac{2\alpha(T-T_0)}{L_0}$$

This is exactly twice the value of the expression for the fractional change in linear expansion $\frac{\Delta L}{L_0}$ above. Therefore, the area of a 1 m square steel plate will increase by $\approx 2\times10^{-4}$ m^2 when it is raised by 10°C.

Supplementary Problems

17.16 A particle moves with a displacement $x(t)$, where $x(t) = t^2 \sin^3\theta + 6t - 3$. Find the velocity and the acceleration of the particle as a function of time.

Ans. $v = 2t\sin^3\theta + 6$; $a = 2\sin^3\theta$

17.17 Find the induced current in a conducting loop of wire of resistance R through which a magnetic flux of magnitude Φ passes if the induced current $i = -\frac{1}{R}\frac{d\Phi}{dt}$ and $\Phi = 12t^2e^{-3t}$

Ans. $i = \frac{1}{R}(36t^2e^{-3t} - 24te^{-3t})$

17.18 The velocity of a projectile varies according to $v = v_0e^{-bt}$, where $v_0 = 10$ m/s and $b = .5\ \text{s}^{-1}$. Find its acceleration at $t = 3$ s.

Ans. -1.11 m/s^2

17.19 What is the maximum magnitude of the acceleration of a car having a velocity $v(t) = Ae^{-t^2}$, and at what time is it reached?

Ans. .429 m/s^2 (a deceleration) at $t = .707$ s

17.20 A mass on a spring undergoes a displacement $x = A\sin(\omega t + \phi)$ where ω and ϕ are constants. (*a*) Find the velocity of the mass as a function of time (*b*) Find the acceleration of the mass as a function of time.

Ans. (*a*) $v = A\omega\cos(\omega t + \phi)$; (*b*) $a = -A\omega^2\sin(\omega t + \phi)$

17.21 An object undergoes an angular displacement $x(t)$ according to $\theta(t) = \left(\frac{\pi}{2} - t^2\right)(1 + e^{-5t})$

(*a*) Find its angular velocity ω, where $\omega = \frac{d\theta}{dt}$. (*b*) Find its angular acceleration α, where $\alpha = \frac{d^2\theta}{dt^2}$

Ans. (*a*) $\omega = \frac{-5\pi}{2}e^{-5t} + 5t^2e^{-5t} - 2t - 2te^{-5t}$ (*b*) $\alpha = \left(\frac{25\pi}{2} - 2\right)e^{-5t} + 20te^{-5t} - 25t^2e^{-5t} - 2$

17.22 A pendulum undergoes simple harmonic motion, undergoing an angular displacement $\theta(t) = A\sin(\omega t)$. Find its angular acceleration $\alpha = \frac{d^2\theta}{dt^2}$

Ans. $\alpha = -A\omega^2 \sin \omega t$

17.23 An object is subject to a force $F = 3x^2 + 4x$ that acts along its direction of motion. What is the work done on the object as it moves from $x = 3$ to $x = 17$? (The work done on an object along its direction of motion is equal to $\int_a^b F\,dx$, where F is the force along the direction of motion x and a and b are the initial and final values of x. Assume SI units throughout.)

Ans. 5446 N m

17.24 (*a*) What is the potential difference between two point charges A and B, each with charge q, located at x_A and x_B, if the potential difference $V_{AB} = -\int_A^B E\,dx$ and the electric field $E = k\frac{q}{x^2}$, where k is Coulomb's constant ($\approx 9 \times 10^9$ in SI units)?

Ans. $V = kq\left(\frac{1}{x_A} - \frac{1}{x_B}\right)$

17.25 An ball bearing fired into a viscous substance with an initial velocity of $v_0 = 1$ m/s descends with a velocity $v = v_0\, e^{-bt}$ where $b = .4\ s^{-1}$. (*a*) Find an expression for the distance traveled as a function of time. (*b*) How far has the bearing fallen after 3 s? (*c*) after 10 s?

Ans. (*a*) $x = \frac{v_0}{b}(1 - e^{-bt})$; (*b*) 1.74 m; (*c*) 2.45 m

17.26 (*a*) Find the energy U, in joules, stored in a capacitor C, initially uncharged, as it is charged to a charge q, using the expression $W = \int_0^q V\,dq$ and the relationship $q = CV$. (*b*) Find the energy stored in a 10 pF capacitor that is charged to 10^{-6} C. (1 pF = 10^{-12} F (farads), where farads are the SI units of capacitance.)

Ans. (*a*) $U = \frac{q^2}{2C}$; (*b*) 0.05 J

17.27 (*a*) Find an expression for the work $W = \int_{V_1}^{V_2} p\,dV$ done by an ideal gas as it expands in volume from V_1 to V_2 at constant temperature T. Use the ideal gas law $pV = nRT$, where n is the number of moles, R is the universal gas constant and T is the temperature in Kelvin. (*b*) Find the work done by one mole of an ideal gas as it doubles in volume at a constant temperature of 300 K.

Ans. (*a*) $W = nRT \ln\left(\frac{V_2}{V_1}\right)$; (*b*) $W = 17.05$ N m

17.28 A retarding force $F = -3N$ is applied to an object of mass 5 kg as it moves along a horizontal line. If its initial velocity is 3 m/s, what is the time and displacement at which it begins to reverse its motion?

Ans. 1 s, 2.7 m

Chapter 18

Vectors

VECTORS. There exist a number of physical quantities—including velocity, force and acceleration—that are described by the same underlying mathematics. These particular entities are examples of *vectors*. A vector includes a *magnitude* (or number) and a *direction* and combines with other vectors in well-defined ways that we describe below. Vectors and related concepts are used in areas as disparate as mechanics, electromagnetism, and quantum mechanics to solve a variety of problems. The trigonometric relationships that relate the sides of a right triangle are often used in solving problems involving vectors, and so some of the results of the previous chapter will be useful in this chapter as well.

In this chapter we discuss the graphical representation of vectors, the addition and subtraction of vectors, their commutative and associative properties and different forms of vector multiplication. We illustrate the manipulation of vectors and the analysis of problems involving vectors using a variety of examples drawn primarily from the areas of mechanics. We begin with the symbolic and geometric representation of vectors and then proceed to the other topics.

SYMBOLIC REPRESENTATION OF VECTORS. Vectors are typically written in equations or labeled on diagrams by a letter with an arrow on top (e.g., $\vec{A}$). Some books instead boldface the letter (e.g., **A**) and others do both (e.g., $\vec{\mathbf{A}}$). The magnitude of a vector—i.e., the number associated with the vector—is represented by the vector symbol bracketed by two vertical lines (e.g., $|\vec{A}|$).

GEOMETRIC REPRESENTATION OF VECTORS. Vectors are represented as arrows with a tip that indicates the direction of the vector and a length that is proportional to the magnitude. For example, a force of 20 newtons (20 N) would be represented with a line that is twice as long as one representing a force of 10 newtons, with the arrow pointing in the direction of the force. The actual length of the line is not fixed but can be chosen for convenience (e.g., 1 cm on paper could represent 10 N of force).

EXAMPLE 1. A box is pushed horizontally along a smooth floor. Sketch its velocity vector.

The situation is shown in Fig. 18-1.

Fig. 18-1

EXAMPLE 2. Draw a vector representing a wind blowing in the northeast direction at 10 m/s.

The vector is drawn in Fig. 18-2. The vector makes an angle of 20° with the "east" axis.

The wind blowing northeast at 10 m/s represents a velocity vector, involving both a magnitude (10 m/s) and a direction (northeast). The wind velocity vector is indicated by the diagonal arrow, with the arrow pointing in the direction of the wind and the length of the arrow proportional to the wind speed.

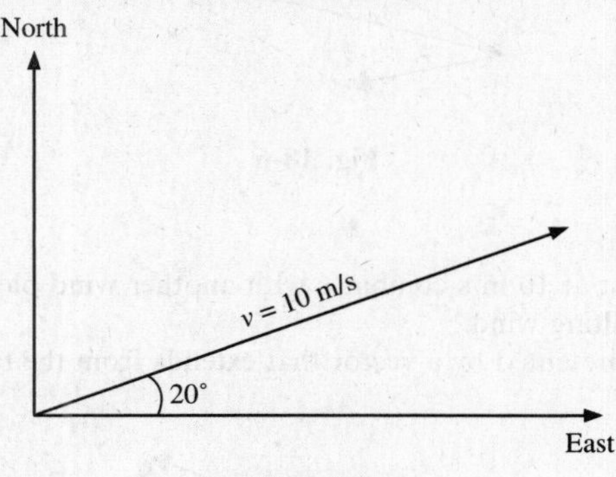

Fig. 18-2

EXAMPLE 3. Draw vectors representing wind speeds of 5, 10 and 15 m/s.

Three different wind speeds, all pointing in the same direction, are indicated below. Note, as indicated in Fig. 18-3 that the length of each arrow is proportional to the magnitude of the velocity (the speed):

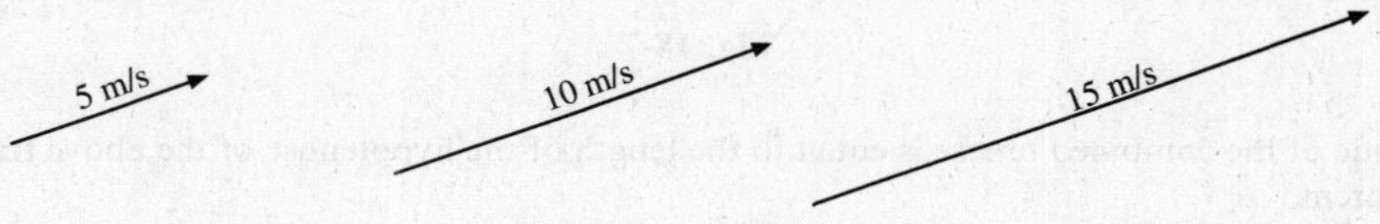

Fig. 18-3

GRAPHICAL ADDITION OF VECTORS. Vector quantities, such as wind velocity, combine with each other in a common manner. As indicated in Fig. 18-4, two vectors $\vec{A}$ and $\vec{B}$ can be "added" by moving one of the vectors parallel to itself so that the "head" of one vector is at the tail of the other vector. The sum of the vectors, $\vec{A} + \vec{B}$, is then defined to be the vector that extends from the tail of one vector to the head of the other and is represented by $\vec{C}$ in Fig. 18-4. The vector resulting from the combining of two or more vectors is called the resultant and its magnitude is equal to its length in the units in which the vector is measured (e.g., newtons for force, meter per second for velocity, etc.)

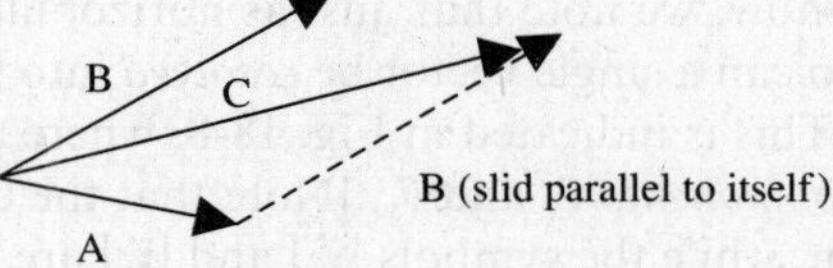

Fig. 18-4

In the above figure, we have moved *B* parallel to itself so that its tail lines up with *A*'s head. However, we could equally well have slid *A* parallel to itself so that it lines up with *B*'s head, as indicated in Fig. 18-5:

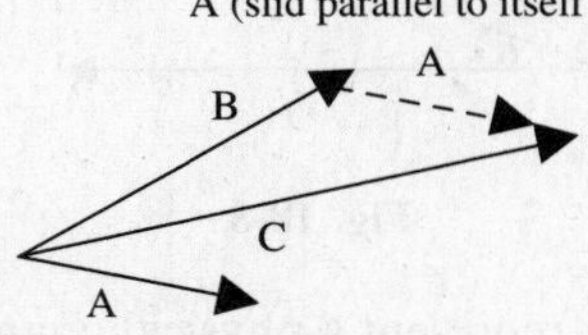

Fig. 18-5

Figures 18-4 and 18-5 together form a parallelogram, as indicated in Fig. 18-6, and so this means of combining vectors is sometimes referred to as the "parallelogram method".

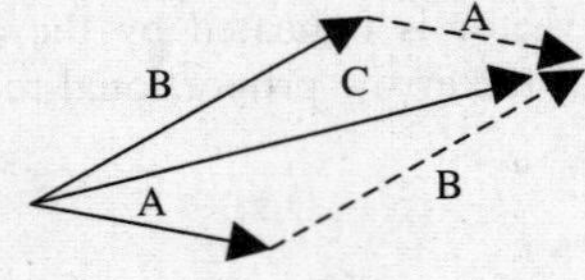

Fig. 18-6

EXAMPLE 4. A wind blowing east at 10 m/s combines with another wind blowing north at 5 m/s. Find the magnitude and direction of the resulting wind.

The combined winds can be represented by a vector that extends from the tail of one vector to the head of the other, as indicated in Fig. 18-7:

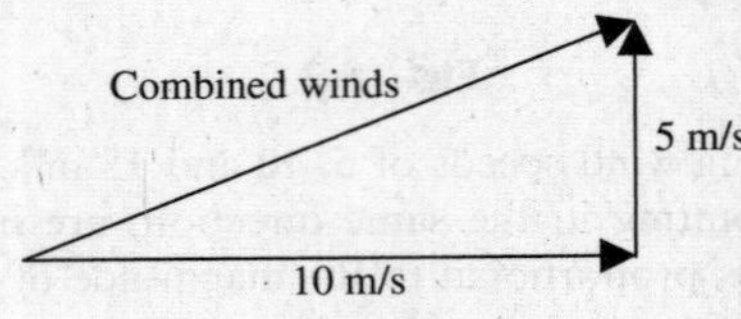

Fig. 18-7

The magnitude of the combined breeze is equal to the length of the hypotenuse of the above triangle. From the Pythagorean theorem,

$$|\vec{C}|^2 = |\vec{A}|^2 + |\vec{B}|^2$$

$$C = \sqrt{|\vec{A}|^2 + |\vec{B}|^2} = \sqrt{|10|^2 + |5|^2} = 11.18\,\text{m/s}.$$

The direction of the breeze can be found by noting that, according to the definition of the tangent,

$$\tan\theta = \frac{5\,\text{m/s}}{10\,\text{m/s}} = .5$$

$$\theta = \arctan(.5) = 26.56^\circ$$

VECTOR COMPONENTS. For now, we note that, just as horizontal and vertical vectors can be combined into one combined vector, so can a single vector be *resolved* into separate component vectors—one horizontal and the other vertical. This is indicated in Fig. 18-8, where a velocity vector $\vec{v}$ is resolved into separate horizontal and vertical *components* $\vec{v}_x$ and $\vec{v}_y$. [Note that the components $\vec{v}_x$ and $\vec{v}_y$ are vectors, with both magnitude and direction, while the symbols $|\vec{v}_x|$ and $|\vec{v}_y|$ are *scalar quantities*, referring only to the magnitudes of the corresponding vector components.]

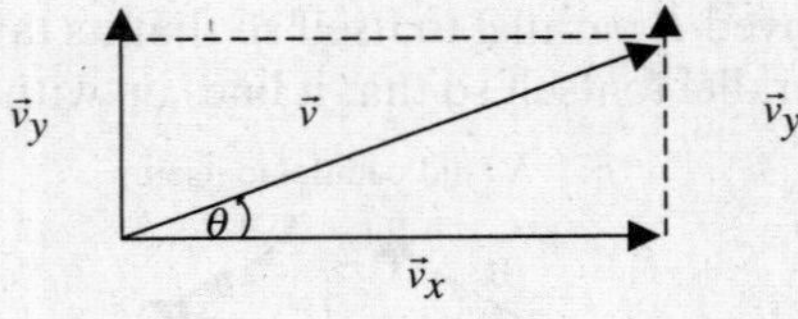

Fig. 18-8

The lengths of the sides, which often represent a physical quantity of interest (such as the magnitude of a force or of an electric field), can be determined using the properties of triangles. We summarize only a few primary relationships here, shown in Fig. 18-9 and the relations that follow, referring the reader to Chapter 13 for a more complete review.

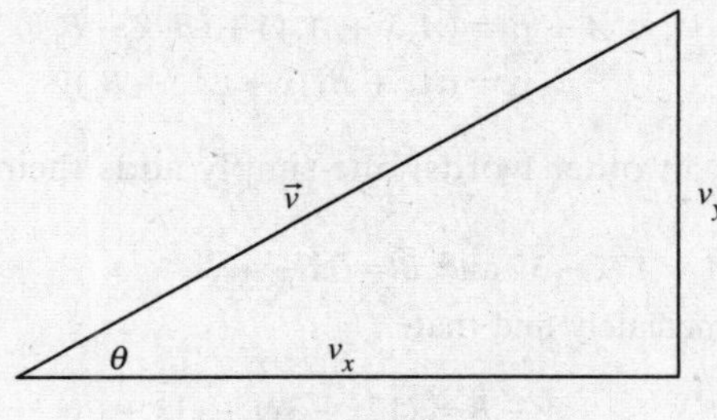

Fig. 18-9

$$\cos\theta = \frac{\text{Adjacent side}}{\text{Hypotenuse}} = \frac{|\vec{v}_x|}{|\vec{v}|} \rightarrow v_x = |\vec{v}|\cos\theta$$

$$\sin\theta = \frac{\text{Opposite side}}{\text{Hypotenuse}} = \frac{|\vec{v}_y|}{|\vec{v}|} \rightarrow v_y = |\vec{v}|\sin\theta$$

$$\tan\theta = \frac{\text{Opposite side}}{\text{Adjacent side}} = \frac{|\vec{v}_y|}{|\vec{v}_x|} \rightarrow v_y = |\vec{v}_x|\tan\theta$$

THE UNIT VECTOR. A unit vector is a vector of magnitude one. In a Cartesian coordinate system, the symbols $\hat{x}$, $\hat{y}$, and $\hat{z}$ represent unit vectors along each of the three coordinate axes. An alternative representation is to use the symbols $\hat{i}$, $\hat{j}$, and $\hat{k}$, respectively, for $\hat{x}$, $\hat{y}$, and $\hat{z}$. (See Fig. 18-10.)

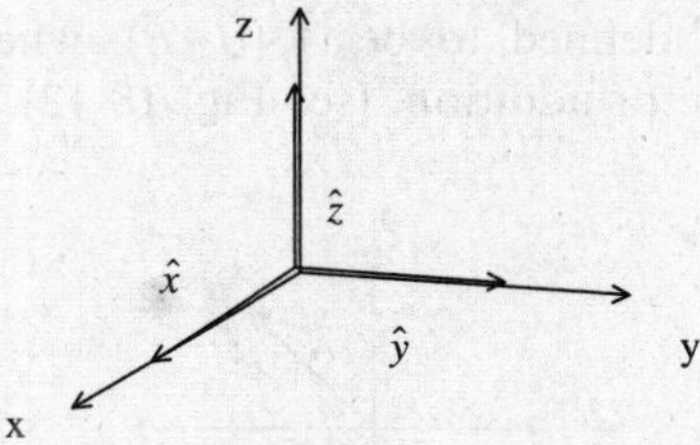

Fig. 18-10

ALGEBRAIC REPRESENTATION OF VECTORS. In addition to the geometric representation described above, vectors can also be represented algebraically. Using the concepts developed above, we can represent the vector $\vec{v}$ in Fig. 18-8 as:

$$\vec{v} = v_x\hat{x} + v_y\hat{y}$$

From the preceding discussion, we can see that the vector can also be written in the following forms:

$$\vec{v} = v_x\hat{i} + v_y\hat{j}$$
$$\vec{v} = v\cos\theta\hat{x} + v\sin\theta\hat{y}$$

We can therefore write any vector $\vec{R}$ in the form $\vec{R} = R_x\hat{x} + R_y\hat{y} + R_z\hat{z}$, where R_x, R_y, and R_z represent the respective projections of each vector along each of the coordinate axes and where $\hat{x}$, $\hat{y}$, and $\hat{z}$ are the unit vectors along each of these axes.

ALGEBRAIC ADDITION OF VECTORS. The addition of two vectors $\vec{A}$ and $\vec{B}$ to form a third vector $\vec{C}$ was described geometrically above. The same operation can be expressed in algebraic form as follows:

$$\begin{aligned}\vec{A}+\vec{B} &= (A_x\hat{x}+A_y\hat{y})+(B_x\hat{x}+B_y\hat{y})\\ &= (A_x+B_x)\hat{x}+(A_y+B_y)\hat{y}\end{aligned}$$

To add two vectors together, in other words, one simply adds their respective components.

EXAMPLE 5. Find $\vec{A}+\vec{B}$, where $\vec{A}=17\hat{x}-3\hat{y}$ and $\vec{B}=3\hat{x}+4\hat{y}$

Adding the components, we immediately find that:

$$\begin{aligned}\vec{A}+\vec{B} &= (17\hat{x}-3\hat{y})+(3\hat{x}+6\hat{y})\\ &= (17+3)\hat{x}+(-3+6)\hat{y}\\ &= 20\hat{x}+3\hat{y}\end{aligned}$$

MULTIPLICATION OF A VECTOR BY A NUMBER. In addition to adding together different vectors, one can also add a vector to itself. A shorthand way of writing $\vec{A}+\vec{A}$ is to write $2\vec{A}$. In general, $n\vec{A}$ is read as "n multiplied by $\vec{A}$". It is defined to be the vector that points in the same direction as $\vec{A}$ with a magnitude equal to $n|\vec{A}|$ (n need not be an integer).

THE NULL VECTOR. The null vector $\vec{0}$ is defined to be one for which $\vec{A}+\vec{0}=\vec{A}$ for any vector $\vec{A}$. It is a vector with a magnitude of 0. It is used below to help define the operation of subtracting one vector from another.

SUBTRACTION OF VECTORS. The vector $-\vec{A}$ ("negative A") is defined to be one for which $\vec{A}+(-\vec{A})=0$. This is a vector with a magnitude equal to $\vec{A}$, but oppositely directed. (See Fig. 18-11). The subtraction operation $\vec{A}-\vec{B}$ is defined to be $\vec{A}+(-\vec{B})$, where the two vectors $\vec{A}$ and $-\vec{B}$ are combined using the usual laws of vector addition. (see Fig. 18-12).

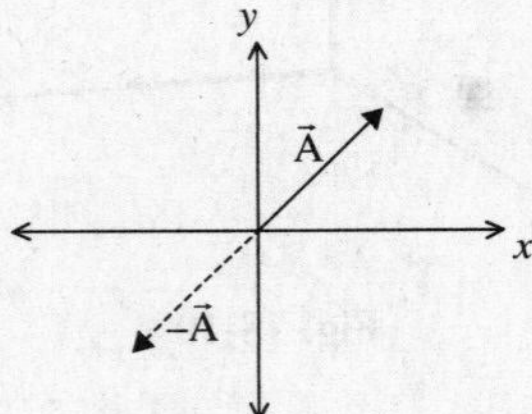

Fig. 18-11

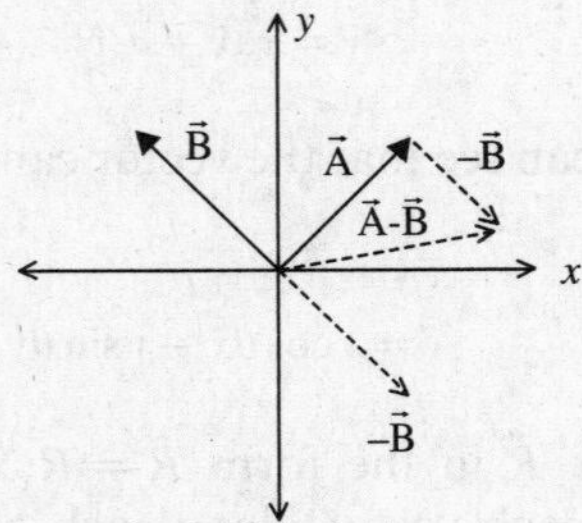

Fig. 18-12

EXAMPLE 6. What is the net result of the following two winds pictured in Fig. 18-13?

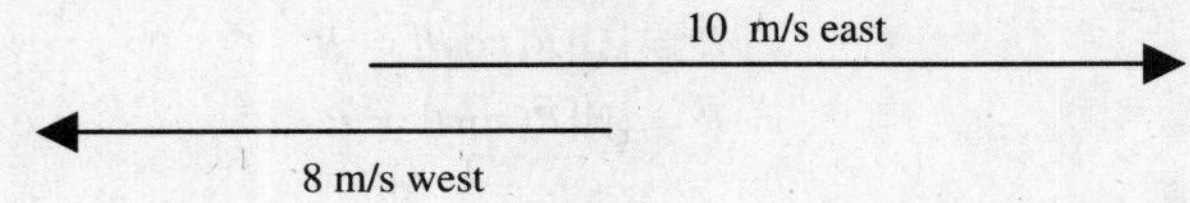

Fig. 18-13

The combined wind is formed using the law of vector addition described above. It moves to the east at 2 (= 10–8) m/s.

EXAMPLE 7. What is $\vec{R}-\vec{S}$ if $\vec{R}=7\hat{x}+14\hat{y}$ and $\vec{S}=2\hat{x}+5\hat{y}$?

$$\vec{R}-\vec{S}=(7\hat{x}+14\hat{y})-(2\hat{x}+5\hat{y})$$

VECTOR COMMUTATIVITY AND ASSOCIATIVITY. From Fig. 18-8, it is clear that the order in which two vectors are added does not affect the result. Formally, we say that vector addition is commutative: i.e., $\vec{A}+\vec{B}=\vec{B}+\vec{A}$. In addition, three (or more) vectors add in an associative fashion: i.e., $(\vec{A}+\vec{B})+\vec{C}=\vec{A}+(\vec{B}+\vec{C})$

VECTORS IN THREE DIMENSIONS. We can extend the preceding discussion from two dimensions to three. In three dimensions, $\vec{R}=R_x\hat{x}+R_y\hat{y}+R_z\hat{z}$ where $\hat{x}$, $\hat{y}$, and $\hat{z}$ are unit vectors along the three mutually perpendicular coordinate axes.

For example, in 3D the addition of $\vec{A}$ and $\vec{B}$ is:

$$\begin{aligned}\vec{A}+\vec{B}&=(A_x\hat{x}+A_y\hat{y}+A_z\hat{z})+(B_x\hat{x}+B_y\hat{y}+B_z\hat{z})\\&=(A_x+B_x)\hat{x}+(A_y+B_y)\hat{y}+(A_z+B_z)\hat{z}\end{aligned}$$

EXAMPLE 8. What is $3\vec{A}+4\vec{B}-2\vec{C}$ if $\vec{A}=7\hat{x}-3\hat{y}+23\hat{z}$, $\vec{B}=30\hat{x}+4\hat{y}+8\hat{z}$, and $\vec{C}=3\hat{x}+4\hat{y}-\hat{z}$?

$$\begin{aligned}3\vec{A}+4\vec{B}-2\vec{C}&=3(7\hat{x}-3\hat{y}+23\hat{z})+4(30\hat{x}+4\hat{y}+8\hat{z})-2(3\hat{x}+4\hat{y}-\hat{z})\\&=(21+120-6)\hat{x}+(-9+16-8)\hat{y}+(69+32+2)\hat{z}\\&=135\hat{x}-\hat{y}+103\hat{z}\end{aligned}$$

VECTOR MULTIPLICATION. Vectors differ from pure numbers and cannot be multiplied together in the usual sense. In addition to the multiplication of a vector by a number discussed above, there are two kinds of vector "multiplication" that are commonly used in physics problems. They are the "scalar product" (also known as the "dot product") and the cross product (also known as the "vector product"). We first define these terms and then provide some examples that illustrate their usefulness.

THE SCALAR PRODUCT. The scalar product (or "dot product") of two vectors $\vec{A}$ and $\vec{B}$ is a number (rather than another vector) that is defined by:

$$\vec{A}\cdot\vec{B}=|\vec{A}||\vec{B}|\cos\theta_{AB}$$

where $|\vec{A}|$ and $|\vec{B}|$ indicate the magnitudes of those vectors and θ is the angle between the two vectors. This definition is valid for both vectors in a 2D plane or in 3D space.

For two vectors $\vec{A}$ and $\vec{B}$, with components $\vec{A}=A_x\hat{x}+A_y\hat{y}+A_z\hat{z}$ and $\vec{B}=B_x\hat{x}+B_y\hat{y}+B_z\hat{z}$, the scalar product of the two vectors is written in terms of the components as:

$$\vec{A}\cdot\vec{B}=A_xB_x+A_yB_y+A_zB_z$$

EXAMPLE 9. Find the scalar products of the unit vectors $\hat{x}$ and $\hat{y}$ with the vector $\vec{R} = x\hat{x} + y\hat{y} + z\hat{z}$.

$$\hat{x} \cdot \vec{R} = |\hat{x}||\vec{R}| \cos\theta = R_x$$
$$\hat{y} \cdot \vec{R} = |\hat{y}||\vec{R}| \sin\theta = R_y$$

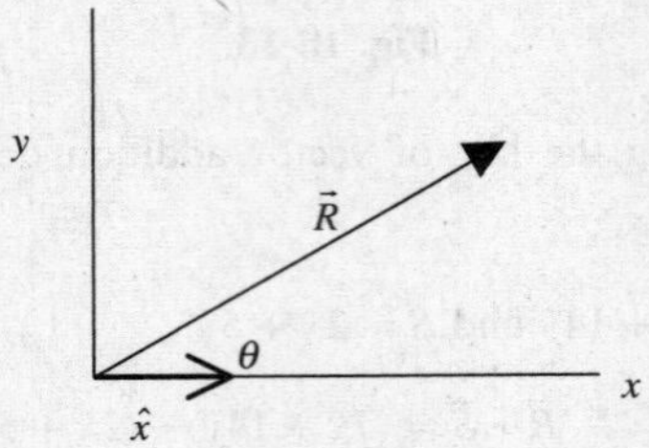

Fig. 18-14

Note that by taking the dot product of $\vec{R}$ with a unit vector, we obtain the component of $\vec{R}$ along the unit vector. In general, the scalar product can be used to determine the magnitude of the component of a vector that lies along a desired direction.

EXAMPLE 10. The power P associated with an object moving with a velocity $\vec{v}$ and subject to a force is given by the expression:

$P = \vec{F} \cdot \vec{v}$, where P is the power driving the object (in W), $\vec{F}$ is the force (in N), and $\vec{v}$ is the velocity of the object (Fig. 18-15).

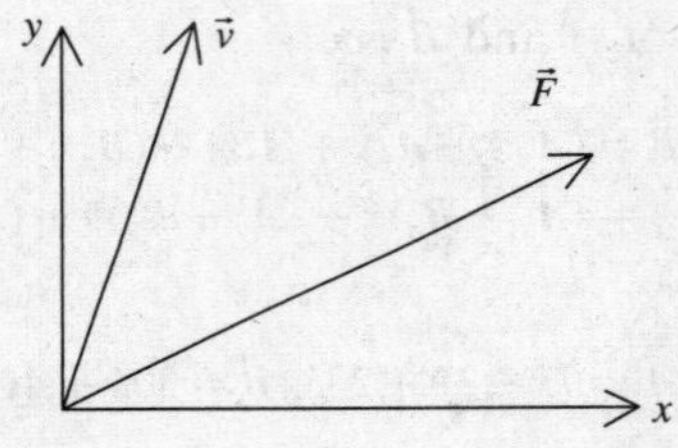

Fig. 18-15

Suppose that the object is moving at 3 m/s at 57° with respect to the horizontal and that the force is acting with 25 N at 25° to the horizontal. What is the power delivered?

Using the above formula,

$$P = \vec{F} \cdot \vec{v} = |\vec{F}||\vec{v}| \cos\theta_{Fv} = (25\,\text{N})(3\,\text{m/s})\cos(57° - 32°) = 62.18\,\text{W}$$

Note that, from the definition above, taking the scalar product of two vectors is a commutative operation: i.e., $\vec{A} \cdot \vec{B} = \vec{B} \cdot \vec{A}$

THE CROSS PRODUCT. The cross product (or "vector product") of two vectors is another operation that comes up frequently throughout physics—rotational motion and electrodynamics are two notable examples. In contrast to the scalar product, the cross product of two vectors is another vector, rather than a number. The cross product of $\vec{A}$ and $\vec{B}$ is written as $\vec{A} \times \vec{B}$ and, as a vector, has both a magnitude and a direction. Its *magnitude* is defined to be:

$$|\vec{A} \times \vec{B}| = |\vec{A}||\vec{B}| \sin\theta_{\text{AB}}$$

where, once again, $|\vec{A}|$ and $|\vec{B}|$ are the magnitudes of the individual vectors and θ_{AB} is the angle between them.

The *direction* of $\vec{A} \times \vec{B}$ is defined to be mutually perpendicular to both $\vec{A}$ and $\vec{B}$ and points in a direction given by the *right-hand rule*: when the fingers of the right hand are waved from $\vec{A}$ to $\vec{B}$, thumb extending outwards so that it is perpendicular to both $\vec{A}$ and $\vec{B}$, the thumb will point in the direction of the cross product. (Note that, according to the definition of the cross product, $\vec{A} \times \vec{B}$ is not equal to $\vec{B} \times \vec{A}$; these two vectors are of equal magnitude, but point in opposite directions.)

Note also that when $\vec{A}$ and $\vec{B}$ are parallel to each other, the cross product is equal to zero because the angle θ_{AB} between the vectors is 0 and so $\sin\theta_{AB} = \sin 0 = 0$. It follows through similar reasoning that for two vectors of particular magnitudes, their cross product will be maximum when they are mutually perpendicular to one another (because $\sin\theta_{AB}$ reaches its maximum value of 1 when $\theta_{AB} = 90°$).

The usefulness of this definition will become clearer in the examples that follow.

EXAMPLE 11. Find the torque exerted when a wrench is pulled with a horizontal force of 25 N on a vertically oriented bolt from a distance of 40 cm, with an angle of 74° between the direction of the force and the lever arm.

The torque on an object is given by $\vec{\tau} = \vec{R} \times \vec{F}$, where $\vec{\tau}$ is the torque in N m, $\vec{R}$ is the lever arm* and $\vec{F}$ is the applied force in N.

The magnitude of the torque is given by:

$$|\vec{\tau}| = |\vec{R}||\vec{F}| \sin\theta$$

$$= 40\,\text{cm} \cdot \frac{1\,\text{m}}{100\,\text{cm}} \cdot 25\,\text{N} \cdot \sin(74°)$$

$$= 9.61\,\text{N m}$$

The direction of the torque is mutually perpendicular to the directions of the force and the lever arm. Using the right-hand rule described above, it is mutually perpendicular to both the force vector and the lever arm in the direction indicated in Fig. 18-16.

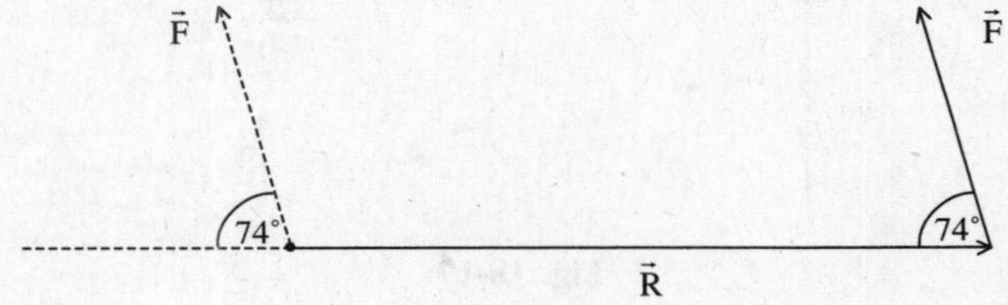

Fig. 18-16

EXAMPLE 12. Find the motion of a proton moving at 2×10^6 m/s in a uniform magnetic field of 1 T when:

(*a*) the proton velocity is parallel to the field;
(*b*) the proton velocity is perpendicular to the field; and
(*c*) the proton velocity is traveling at an angle of 37° with respect to the field.

The force acting on a charged particle in a uniform magnetic field is given by the formula $\vec{F} = q\vec{v} \times \vec{B}$, where the force $\vec{F}$ is given in newtons (N), the particle charge q is given in coulombs (C) and the magnetic field $\vec{B}$ is given in teslas (T). The magnitude of the magnetic force is therefore $F = qvB\sin\theta$, where θ is the angle between the velocity vector of the proton and the direction of the magnetic field. See Fig. 18-15.

* As pictured in Fig. 18-16, the lever arm $\vec{R}$ vector with a magnitude equal to the distance from the center of the bolt (in m) and a direction along its axis extending from the center of the bolt to the point at which the force is applied.

For (a): Because the proton is traveling parallel to the field, θ is 0 and therefore $F = qvB\sin\theta = qvB\sin 0 = 0$. Because no force acts upon the particle, it will move—according to Newton's First Law—in a straight line at the indicated velocity of 2×10^6 m/s.

For (b): In this case, because the proton is traveling perpendicular to the field, θ is 90° and therefore $F = qvB\sin\theta = qvB\sin 90° = qvB(1) = qvB$. The direction of the force is always perpendicular to the particle motion and the magnetic field; one can convince oneself that this results in uniform circular motion of the particle.

One can go further by using the physics of uniform circular motion, combining the centripetal force relation $F = mv^2/r$ with the magnetic force law $F = qvB$ to find an expression for the radius of the circle: $r = mv/qB$. Using the known values of the proton mass and charge and the given values of the other quantities, we get

$$r = \frac{(1.67 \times 10^{-27}\,\text{kg})(2 \times 10^6\,\text{m/s})}{(1.6 \times 10^{-19}\,\text{C})(1\,\text{T})} = 2 \times 10^{-2}\,\text{m} = 2\,\text{cm}$$

For (c): The velocity vector is now oriented at an angle θ with respect to the magnetic field. As before, the magnitude of the force acting upon the proton is given by $F = qvB\sin\theta$. Because the sine function increases in the region from 0° to 90°, the magnitude of the force acting upon the proton does as well. To find the direction of motion, it is easiest to decompose the velocity vector into two vectors: one parallel and one perpendicular to the magnetic field. We have solved for these motions in (a) and (b). The combined motion is therefore a circular motion superimposed upon a straight line motion—a *helix*, which is illustrated in Fig. 18-17.

Using reasoning similar to that in (b), we leave it as an exercise for the reader to show that the radius of the helix is $r = mv/qB\sin\theta$.

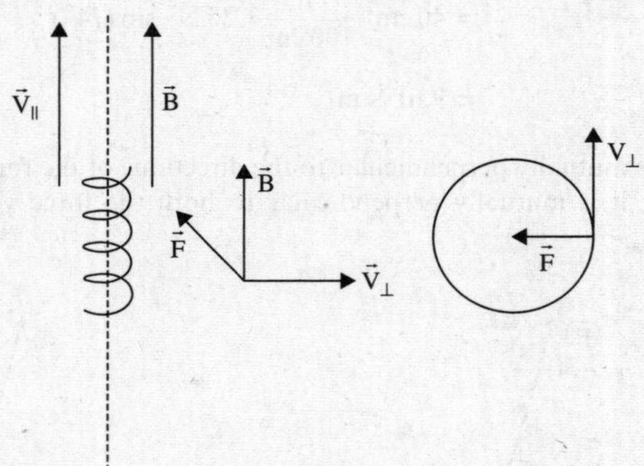

Fig. 18-17

VECTOR PRODUCT IDENTITIES. There are a number of identities involving vector products that recur throughout physics. Among the most useful is the expansion of a "triple cross product":

$$\vec{A} \times (\vec{B} \times \vec{C}) = \vec{B}(\vec{A} \cdot \vec{C}) - \vec{C}(\vec{A} \cdot \vec{B}) \qquad (18.1)$$

(because of the expression on the right-hand side, a useful mnemonic is to refer to this to as the "BAC minus CAB" rule). Another commonly used identity is one involving a combination of scalar and cross products:

$$\vec{A} \cdot (\vec{B} \times \vec{C}) = \vec{B} \cdot (\vec{C} \times \vec{A}) = \vec{C} \cdot (\vec{A} \times \vec{B}) \qquad (18.2)$$

EXAMPLE 13. Show that $\vec{A} \cdot (\vec{A} \times \vec{B}) = 0$ for any vectors $\vec{A}$ and $\vec{B}$.

We know that $\vec{A} \times \vec{B}$ is a vector perpendicular to $\vec{A}$ (as well as $\vec{B}$). Therefore $\vec{A} \cdot (\vec{A} \times \vec{B}) = |\vec{A}||\vec{A} \times \vec{B}|\cos 90 = |\vec{A}||\vec{A} \times \vec{B}|(0) = 0$

EXAMPLE 14. Evaluate $\hat{x} \times (\hat{x} \times \hat{z})$

Method 1: Using the right-hand rule, we note that $\hat{x} \times \hat{z} = -\hat{y}$. Then $\hat{x} \times (\hat{x} \times \hat{z}) = \hat{x} \times (-\hat{y}) = -(\hat{x} \times \hat{y}) = -\hat{z}$ where, in the last step, we have used the right-hand rule in evaluating $(\hat{x} \times \hat{y})$

Method 2: Using the vector identity of (18.1)

$$\hat{x} \times (\hat{x} \times \hat{z}) = \hat{x}(\hat{x} \cdot \hat{z}) - \hat{z}(\hat{x} \cdot \hat{x})$$

We can simplify the right-hand side of this expression by noting that, because $\hat{x}$ is perpendicular to $\hat{z}$, $\hat{x} \cdot \hat{z} = |\hat{x}||\hat{z}| \cos 90 = (1)(1)(0) = 0$.

Also, $\hat{x} \cdot \hat{x} = |\hat{x}| \cdot |\hat{x}| \cos(0) = 1 \times 1 \times 1 = 1$

Hence, $\hat{x} \times (\hat{x} \times \hat{z}) = \hat{x}(0) - \hat{z}(1) = -\hat{z}$

EXPANSION OF CROSS PRODUCT COMPONENTS. The cross product $\vec{A} \times \vec{B}$ can be expressed in terms of the components of $\vec{A}$ and $\vec{B}$ according to the following formula*

$$\vec{A} \times \vec{B} = (A_yB_z - r_zF_y)\hat{x} + (A_zB_x - A_xB_z)\hat{y} + (A_xB_y - A_yB_x)\hat{z} \qquad (18.3)$$

EXAMPLE 15. Evaluate the torque $\vec{\tau}$ in terms of the components of the lever arm $\vec{r}$ and of the applied force $\vec{F}$ using the relationship and (18.3)

$$\vec{\tau} = \vec{r} \times \vec{F} = (r_yF_z - r_zF_y)\hat{x} + (r_zF_x - r_xF_z)\hat{y} + (r_xF_y - r_yF_x)\hat{z}$$

Solved Problems

18.1 Find the speed and angle of a baseball whose horizontal component of velocity is 15 m/s and whose vertical component of velocity is 25 m/s (Fig. 18-18).

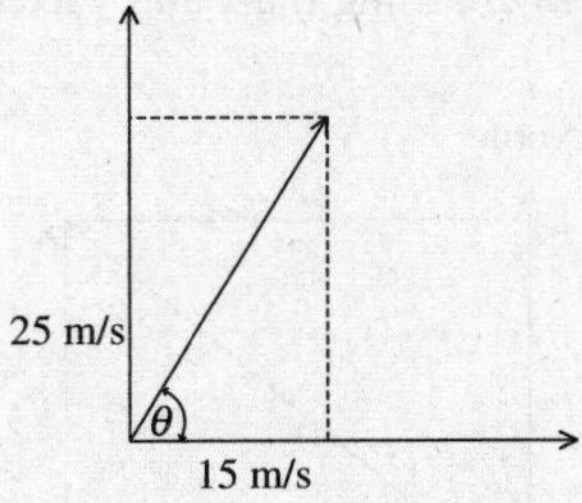

Fig. 18-18

The speed of an object is equal to the magnitude of the total velocity: $v = |\vec{v}| = v_x^2 + v_y^2 = \sqrt{15^2 + 25^2}$. The angle θ is related to the sides through the relation

$$\tan\theta = \frac{v_y}{v_x} = \frac{25}{15}$$

If $\tan\theta = 25/15$, then $\theta = \arctan(25/15)$, where arctan (the arctangent) is the inverse tangent function. Like the other functions, this one is also tabulated and commonly available as a calculator function for scientific or mathematical calculators.

In this case, $\theta = \arctan(25/15) = 1.19 \text{ rad} = 59.01°$

* The cross product can also be conveniently expressed using *determinants*, but we do not pursue that topic here.

18.2 What is the combined effect of a wind blowing to the north at 15 m/s and, at the same time, another wind blowing to the east at 10 m/s?

The combined effect of these breezes is indicated in Fig. 18-19. Each of these winds is regarded as a velocity vector, because each involves a magnitude and a direction and because we assume that each wind adds to the total velocity in an independent manner.

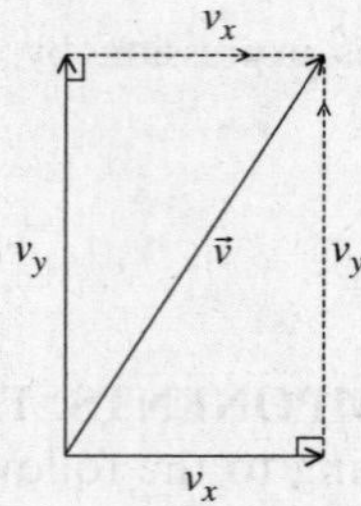

Fig. 18-19

The wind speed and its direction can be calculated as follows.
Using the Pythagorean theorem, $v^2 = v_x^2 + v_y^2$
In order to determine the angle at which it blows, one uses the property of the tangent:

$$\tan\theta = \frac{v_y}{v_x} = \frac{15}{10} = 1.5$$

Hence, the combined wind blows at $\sqrt{15^2 + 10^2} = 18.03\,\text{m/s}$ at an angle of 43.6° relative to the eastern direction.

18.3 What are the vector components of a wind blowing at 10 m/s in a direction 40° north of east? (Fig. 18-20)

The wind is represented in Fig. 18-20, using the x and y axes to represent, respectively, the eastern and northern directions.

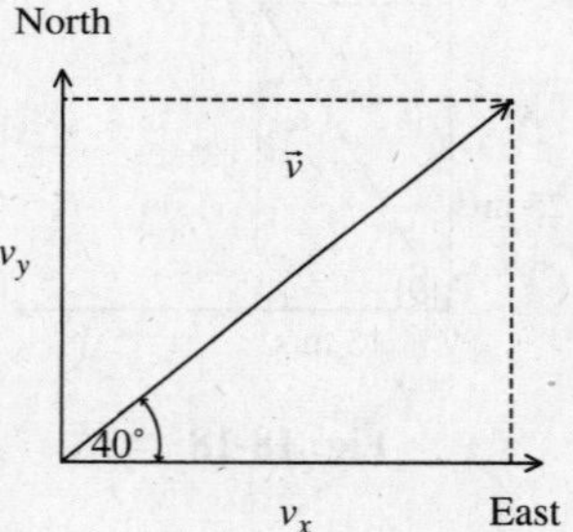

Fig. 18-20

From trigonometry and Fig. 18-20,

$$\vec{v} = \vec{v}_x + \vec{v}_y = |v|\cos\theta\hat{x} + |v|\sin\theta\hat{y}$$

$$\cos 40° = \frac{v_x}{|\vec{v}|} \cdot \sin 40° = \frac{v_y}{|\vec{v}|}$$

Hence,

$$\vec{v}_x = v\cos 40°\,\hat{x} = 10(.46)\,\hat{x} = 4.6\,\text{m/s}\,\hat{x}$$
$$\vec{v}_y = v\sin 40° = 10(.64) = 6.4\,\text{m/s}\,\hat{y}$$

18.4 A wind blowing at 10 m/s at 35° combines with a second wind blowing at 20 m/s at 150°. What is the magnitude and direction of the combined wind?

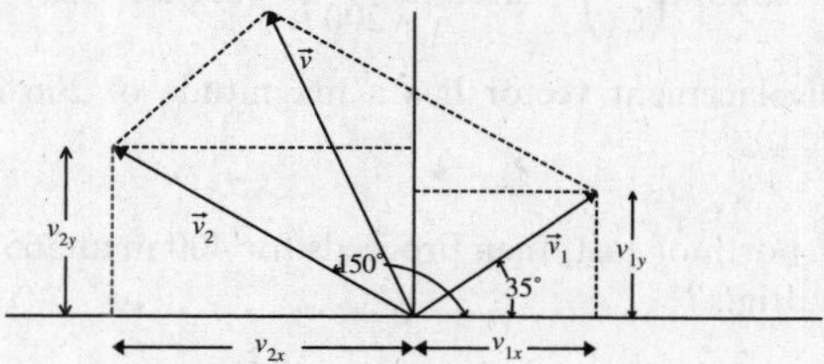

Fig. 18-21

The total velocity vector (Fig. 18-21) is related to its components by $\vec{v} = \vec{v}_1 + \vec{v}_2$ Resolving this into components gives

$$v_x = v_{1x} + v_{2x}$$
$$v_y = v_{1y} + v_{2y}$$

Using

$$v_{1x} = v_1 \cos\theta_1 \quad v_{1y} = v_1 \sin\theta_1$$
$$v_{2x} = v_2 \cos\theta_2 \quad v_{2y} = v_2 \sin\theta_2$$

we get:

$$v_x = v_1 \cos\theta_1 + v_2 \cos\theta_2 = 10\cos 35 + 20\cos 150 = -9.129$$
$$v_y = v_1 \sin\theta_1 + v_2 \sin\theta_2 = 10\sin 35 + 20\sin 150 = 15.736$$

The magnitude of the total velocity is, therefore,

$$v = \sqrt{v_x^2 + v_y^2} = \sqrt{(-9.129)^2 + (15.736)^2} = 18.19\,\text{m/s}$$

Using $\tan\theta = \frac{v_y}{v_x}$, we then get $\theta = \arctan(v_y/v_x) = \arctan(15.736/-9.129) = 120.12°$

The resulting wind, therefore, blows at 18.19 m/s at an angle of 120.1°.

18.5 George walks 200 m east and 50 m south. What is his net displacement?

The displacement describes the relative position of an object as both a distance from a starting point and a direction. For example, " 30 km from home" refers to a distance, but does not include a direction. "30 km in a direction 48° north of east" provides both a magnitude (the 30 mi distance) and a direction (48°).

The following problem shows how to combine two vectors where one is entirely horizontal and the other is entirely vertical.

The displacement is given by $\vec{C} = \vec{A} + \vec{B}$. Hence, $C_x = A_x + B_x$ and $C_y = A_y + B_y$. From Fig. 18-22, the angle is given by $\tan\theta = C_y/C_x$

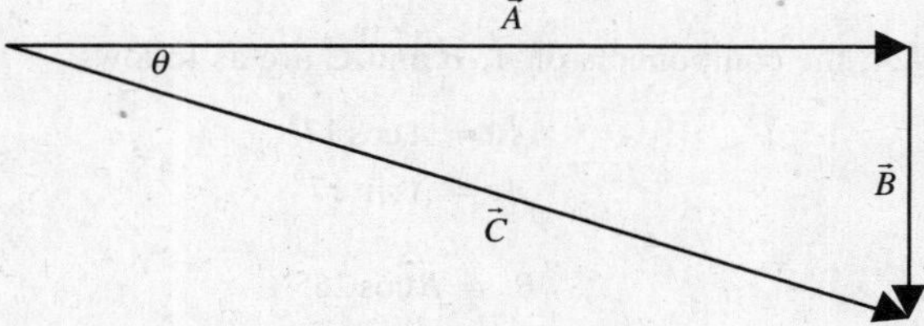

Fig. 18-22

The magnitude of the displacement is given directly by the Pythagorean theorem:

$$|\vec{C}| = \sqrt{|\vec{A}|^2 + |\vec{B}|^2} = \sqrt{(200\,\text{m})^2 + (50\,\text{m})^2} = 206\,\text{m}$$

while the direction of the displacement vector is given by

$$\theta = \arctan\left(\frac{C_y}{C_x}\right) = \arctan\left(\frac{-50}{200}\right) = \arctan(-.25) = -.245.$$

Hence, George's net displacement vector has a magnitude of 206 m and points in a direction .245 radians south of east.

18.6 Natasha walks 250 m at 17° north of east, then proceeds for 400 m at 265° and finally heads 90° for 300 m. Where is she relative to the origin?

This problem shows how to combine three vectors that are not oriented in "convenient" directions—i.e., they are not entirely horizontal or vertical. As indicated, one first resolves each vector into horizontal and vertical components and then combines those individual components.

The displacement vectors are indicated in the Fig. 18-23.

A: 250 m at 17° with respect to the x axis
B: 400 m at 265° with respect to the x axis
C: 300 m at 90° with respect to the x axis

The total displacement $\vec{D} = \vec{A} + \vec{B} + \vec{C}$

$$D_x = A_x + B_x + C_x$$
$$D_y = A_y + B_y + C_y$$

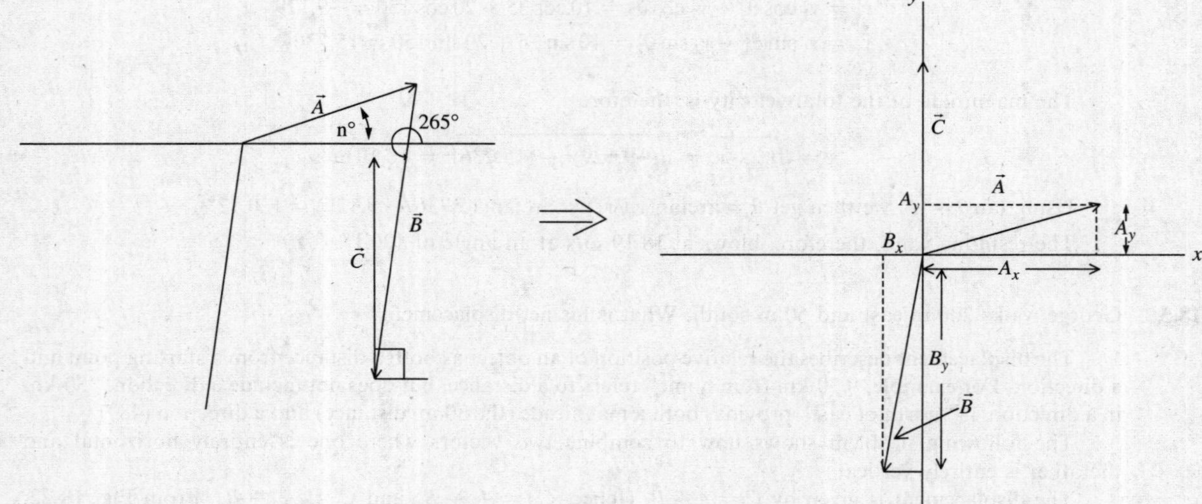

Fig. 18-23

As seen in Fig. 18-24, the components of $\vec{A}$, $\vec{B}$ and $\vec{C}$ are as follows:

$$A_x = A\cos 17°$$
$$A_y = A\sin 17°$$

$$B_x = B\cos 265°$$
$$B_y = B\sin 265°$$

$$C_x = C\cos 90°$$
$$C_y = C\sin 90°$$

where $|\vec{A}| = 250$ m, $|\vec{B}| = 400$ m and $|\vec{C}| = 300$ m

Therefore,

$$D_x = A\cos 17 + B\cos 265 + C\cos 90 = 250(.956) + 400(-.0871) + 300(0) = 204.16\,\text{m}$$

$$D_y = A\sin 17 + B\sin 265 + C\sin 90 = 250(.2924) + 400(-0.9962) + 300(1) = -25.38\,\text{m}$$

(All quantities above are assumed to be in m.)
From the Pythagorean theorem, the magnitude of the displacement is

$$D = \sqrt{D_x^2 + D_y^2} = \sqrt{(204.16\,\text{m})^2 + (-25.38\,\text{m})^2} = 205.7\,\text{m}$$

This occurs at a particular angle with respect to the origin. From the figure, we see that $\tan\theta = \frac{D_y}{D_x}$, so that $\theta = \arctan\left(\frac{D_y}{D_x}\right) = -82.9070°$

This problem happens to involve displacements, but the techniques shown for resolving individual vectors into components and then combining the components is a general one and can be used in similar situations involving other vectors—e.g., vectors of force, momentum, electric, and magnetic fields, and so on.

18.7 A boat travels across a river flowing at 1 m/s. The speed of the boat with respect to the water is 3 m/s. What is the net speed and direction of the boat? (Fig. 18-24)

The physical principle is that the boat will acquire a component of velocity equal to that of the downstream movement of the river, while its velocity in the direction perpendicular to the river movement will be unchanged. From the figure, we can use the Pythagorean theorem to find the net speed of the boat: $v = \sqrt{v_x^2 + v_y^2}$ where v_x is the river speed and v_y is the boat speed. Substituting the above numbers, we get

$$v = \sqrt{v_x^2 + v_y^2} = \sqrt{1^2 + 3^2} = \sqrt{10}\,\text{m/s} = 3.2\,\text{m/s}$$

To find the direction with respect to the river, we use the relationship $\tan\theta = v_y/v_x$ and get $\theta = \arctan(v_y/v_x) = \arctan(3/1) = 72°$

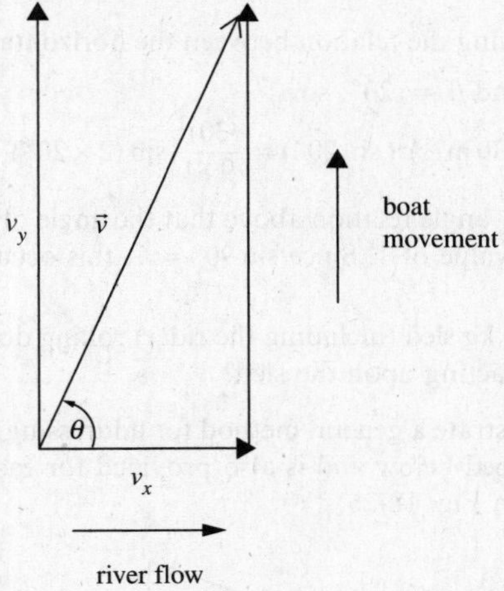

Fig. 18-24

18.8 Find the horizontal range of a golf ball hit at 30 m/s at an angle of 20° upward with respect to the horizontal. (For the flat golfing green we will assume here, the horizontal range is simply equal to the horizontal distance traversed during the flight of the ball.)

The situation is sketched in Fig. 18-25, with horizontal and vertical axes suitably labeled.

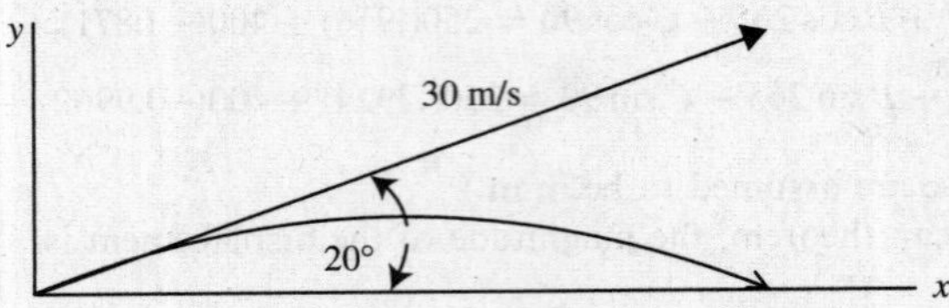

Fig. 18-25

According to the kinematic equations for two dimensional motion,

$$x = x_0 + v_0\,t + \frac{1}{2}at^2$$

Choosing $x_0 = 0$ and noting that the horizontal acceleration a_x is 0 (because only the vertical acceleration of gravity is present), we have

$$x = v_0 t$$

In the vertical, direction we have:

$$y = y_0 + v_{0y}\,t - \frac{1}{2}gt^2$$

To find the horizontal range, we must determine the location where the ball returns to the ground—i.e., where the vertical position y is zero.

Setting $y = 0$ in the above equation, we get

$$0 = v_{0y}\,t - \frac{1}{2}gt^2$$
$$= t\left(v_{0y} - \frac{1}{2}gt\right)$$

Therefore, $v_{0y} = \frac{1}{2}gt$

So, $t = \frac{2v_{0y}}{g} = 2v_0 \sin\theta/g$

Hence, $x = v_{0x}\,t = (v\cos\theta)(2v_0\sin\theta/g) = 2v_0^2 \sin\theta\cos\theta/g$

We can simplify this last expression using the relation $\sin 2\theta = 2\sin\theta\cos\theta$

Therefore $x = \frac{v_0^2}{g}\sin 2\theta$, giving the relation between the horizontal range of the ball and its initial angle.

In our case, $v_0 = 30\,\text{m/s}$ and $\theta = 20°$, so

$$x = (30\,\text{m/s})^2(\sin 20°) = \frac{(30)^2}{9.81}\sin(2\times 20°) = 58.97\,\text{m}$$

Note that, from the range—angle relation above that the angle of maximum range occurs when the sine function reaches its maximum value of 1. Since $\sin 90° = 1$, this occurs when $2\theta = 90°$, or $\theta = 45°$.

18.9 What is the acceleration of a 90 kg sled (including the rider) rolling down a frictionless incline at an angle of 30°? What is the normal force acting upon the sled?

We use this problem to illustrate a general method for addressing problems involving multiple forces on an object. Its solution is described below and is also provided for easy reference in Appendix C.

The situation is sketched in Fig. 18-26.

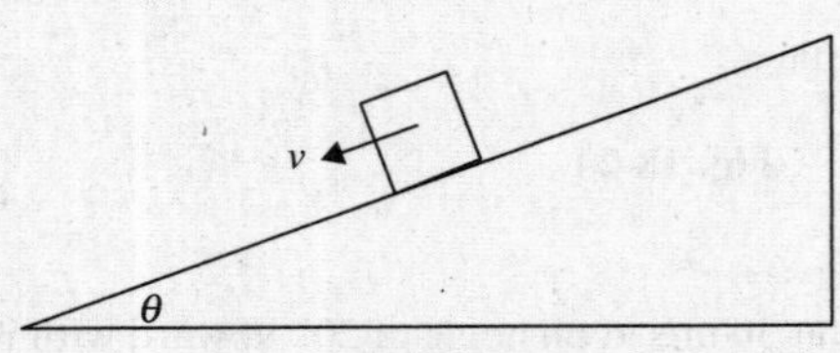

Fig. 18-26

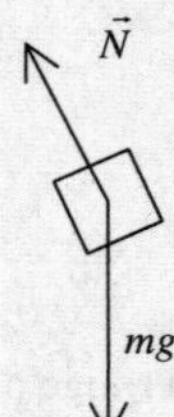

Fig. 18-27

Problems of this sort—involving multiple forces and motion that is neither horizontal nor vertical—can often be effectively addressed through the following sequence of steps:

(1) Read the problem carefully—twice; and try to make an estimate (or even a wild guess) of what might be a rough answer;
(2) Draw a clear diagram;
(3) Indicate all forces acting upon the object of interest;
(4) Choose and label coordinate axes carefully;
(5) Decompose the forces into vector components along these axes;
(6) Apply Newton's Second Law (force = mass · acceleration) to each of the components;
(7) Solve the problem, avoiding the substitution of numbers as long as possible;
(8) Compare to your initial estimate or guess, and check that the units are correct.

Let us go through these steps for this case:

Step (1): Read the problem carefully—twice; and try to make an estimate (or even a wild guess) of what might be a rough answer.

This step could apply to any physics problem. Take your time to understand what you are being asked to find, what information is given and which physics principles may be relevant. Related concepts, examples from class, from the textbook or related problems can all be helpful. (Ultimately, there is no substitute for simply gaining experience in doing as many problems as possible—many problems will then begin to look familiar.)

In this case, we are asked to find *the acceleration of the sled down the incline.*

A rough guess as to the acceleration? Well, if it were falling vertically, the acceleration would be the acceleration of gravity, which is 9.8 m/s^2. It is moving down a slope instead, so the acceleration will be some fraction of that—let us guess about 3 m/s^2. We certainly do not expect the acceleration to be *greater* than if it were falling vertically!

Steps (2 and 3): Draw a clear diagram; indicate all forces of interest.

Draw a simple but clear diagram (Fig. 18-26) to represent all the important features of the problem, including lengths, angles, velocities, etc.

It is often helpful to draw a *free-body diagram* (Fig. 18-27) showing all of the forces acting on the sled. For our problem, this includes the force of gravity and the "normal force" of the incline against the sled (i.e., the force of the incline acting perpendicular to its surface).

Step (4): Choose and label coordinate axes carefully.

In situations like these, it is often convenient to choose coordinate axes that are, respectively, parallel and perpendicular to the direction of motion of the object; and to then label these axes in the diagram (Fig. 18-24).

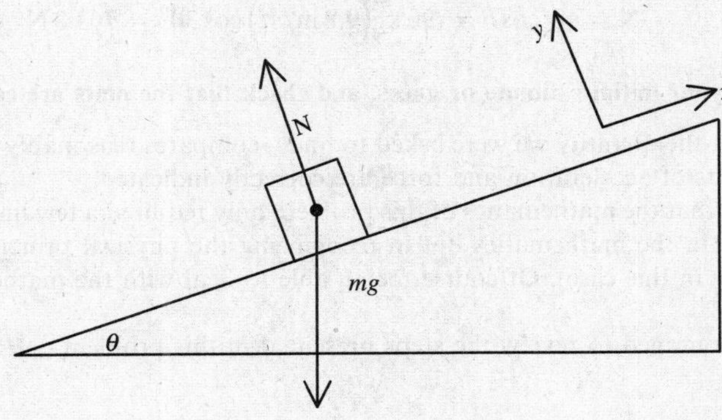

Fig. 18-28

Step (5): Decompose forces along these coordinate axes.

Try to resolve each force vector into components acting along these axes. The normal force is already perpendicular to that motion, but the vertical force of gravity must be resolved into components as indicated in Fig. 18-28.

In order to find these components of force along the chosen axes, one needs to recognize that the angle between the vertical and the downward going normal is equal to θ, the incline angle.* Recognizing this, and being able to draw the triangle in Fig. 18-28 with the vertical force of gravity as the hypotenuse, is critical to the solution of many sloped incline problems.

Step (6): Apply Newton's Second Law (force = mass · acceleration) to each of the components

We now apply Newton's Second Law of Motion, which states that the sum of the external forces acting upon a body is equal to the product of its mass and its acceleration. In equation form, this reads:

$$\vec{F}^{\text{total}} = m\vec{a}$$

where $\vec{F}^{\text{total}}$ is the sum of all external forces acting on the sled of mass m and $\vec{a}$ is the acceleration resulting from that force.

Resolving this equation into its components then yields†:

$$\vec{F}_x^{\text{total}} = ma_x$$

$$\vec{F}_y^{\text{total}} = ma_y$$

Recall that we are trying to find a_x—the acceleration of the sled down the incline. On the other hand, we know that a_y must equal zero, because the sled does not move in a direction perpendicular to the incline surface.

In the direction parallel to the incline, we have the component of gravity shown. In the direction perpendicular to the incline, we have both a component of gravity and the *normal force* of the incline on the ball, which, by definition, acts in a direction perpendicular to the incline.

Hence,

$$\vec{F}_x^{\text{total}} = -mg\sin\theta = ma_x$$

$$\vec{F}_y^{\text{total}} = \text{N} - mg\cos\theta = ma_y = 0$$

Step (7): Solve the problem, avoiding the substitution of numbers as long as possible.

$$-mg\sin\theta = ma_x \Rightarrow a_x = -g\sin\theta$$

$$\text{N} - mg\cos\theta = ma_y = 0 \Rightarrow \text{N} = mg\cos\theta$$

The acceleration of the sled down the incline, and the normal force on the sled, are therefore

$$a_x = g\sin\theta = \left(9.8\,\text{m/s}^2\right)(\sin 30°) = 4.9\,\text{m/s}^2$$

$$\text{N} = mg\cos\theta = (90\,\text{kg}\left(9.8\,\text{m/s}^2\right)\cos 30° = 763.8\,\text{N}$$

Step (8): Compare to your initial estimate or guess, and check that the units are correct.

The acceleration—the quantity we were asked to find—compares reasonably well to our initial guess of 3 m/s^2, and the SI units of acceleration and force are correctly indicated.

It is worth noting that the mathematics of this problem only required a few lines of algebra. The heart of the challenge was not in the mathematics but in recognizing the physical principles that apply (primarily Newton's Second Law in this case). Of course, being able to deal with the mathematics—the focus of this book—is also essential.

The reader is encouraged to review the steps presented in this problem.

---

* This is not difficult to see: the original angle θ is the angle between a horizontal line and the sloped surface. If one rotates these lines by 90°, then one gets the angle in question.

† The force equation is also commonly expressed using $\sum_i \vec{F}_i = m\vec{a}$, where the summation symbol $\sum_i$ represents the sum over contributions from a total of "i" different forces.

18.10 A man pushes a wheelbarrow down a street with a force of 45 N directed at 30° with respect to the horizontal. Find the work done by the force on the wheelbarrow as it moves 10 m down the street in a straight line.

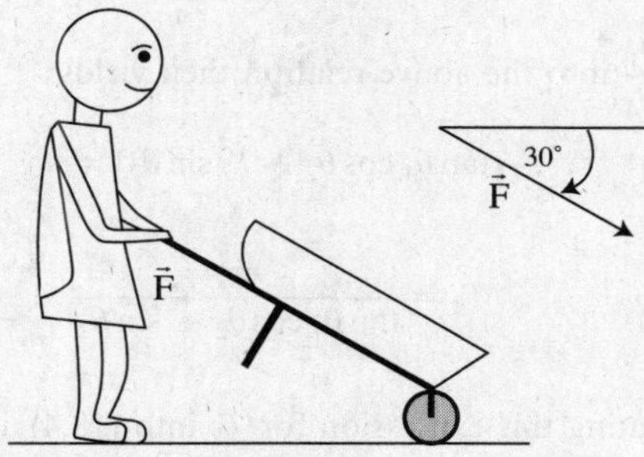

Fig. 18-29

The general definition of work in physics is beyond the scope of this problem. However, in the problem described here, in which a constant force is exerted along a straight line path, we may express work as

$$W = \vec{F} \cdot \vec{d}$$

where W is the work done, $\vec{F}$ is the force exerted, and $\vec{d}$ is the displacement.

In this case $W = |\vec{F}||\vec{d}|\cos\theta = (45\text{ N})(10\text{ m})\cos 30° = 390\text{ N m}$. In this expression, θ is the angle between the direction of the force and the displacement.

18.11 A 10 kg planter is suspended by two cables at angles of 43° and 58° with respect to the vertical (Fig. 18-30). What is the tension in each cable?

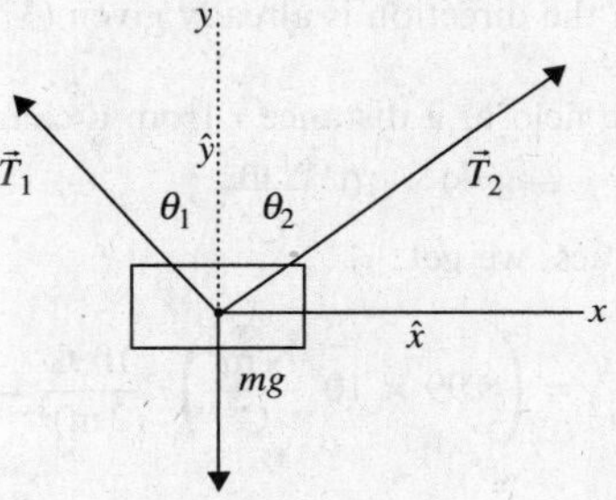

Fig. 18-30

According to Newton's Second Law, the total (or net) force acting upon the planter is equal to its acceleration: $\vec{F}_{net} =$ ma. In this case, the forces acting upon the cable include the tensions in the cables as well as the vertical force of gravity. Because the planter does not move, its acceleration (change of velocity) is 0. Hence we have:

$$\vec{T}_1 + \vec{T}_2 - mg\hat{y} = 0$$

Resolving this into horizontal and vertical components yields:

$$T_1 \cos\theta_1 + T_2 \cos\theta_2 = 0 \rightarrow T_1 = -T_2 \cos\theta_2 / \cos\theta_1 \tag{18.4}$$

$$T_1 \sin\theta_1 + T_2 \sin\theta_2 = mg \tag{18.5}$$

Substituting (18.4) into (18.5) yields

$$T_2 \frac{\cos\theta_2}{\cos\theta_1} \sin\theta_1 + T_2 \sin\theta_2 = mg$$

Substituting $\tan\theta_1 = \dfrac{\sin\theta_1}{\cos\theta_1}$ into the above relation then yields:

$$T_2\,[\tan\theta_1 \cos\theta_2 + T_2 \sin\theta_2] = mg$$

Isolating T_2 then gives:

$$T_2 = \frac{mg}{\tan\theta_1 \cos\theta_2 + \sin\theta_2}$$

We can find T_1 by substituting this expression for T_2 into (18.4) above, yielding:

$$T_1 = \frac{mg}{(\tan\theta_1 \cos\theta_2 + \sin\theta_2)} \frac{\cos\theta_2}{\cos\theta_1}$$

Substituting $\theta_1 = 43°$, $\theta_2 = 58°$, and m = 10 kg, into these expressions, we find:

$$T_2 = \frac{(10\,\text{kg})\,(9.80\,\text{m/s}^2)}{\tan 43° \cos 58° + \sin 58°} = 73.02\,\text{N}$$

$$T_1 = \frac{(10\,\text{kg})\,(9.80\,\text{m/s}^2)}{(\tan 43° \cos 58° + \sin 58°)} \frac{\cos 58°}{\cos 43°} = 53\,\text{N}$$

18.12 What are the x and y components of electric field at a point 2 m and at an angle of 353° with respect to the positive x axis, assuming that the a 10 C point charge is located at the origin?

An electric field is an example of a vector field, which associates a vector with each point in space. We need to find both the magnitude and direction of the electric field at the point described. If we take as our origin the 10 C point charge, then the direction is already given (37° southeast), so we only need to find the magnitude at the point.

The magnitude of the electric field at a distance r from a charge q in vacuum is given by $|\vec{E}| = k\frac{q}{r^2}$ where, in SI units, $k = 8.99 \times 10^9 \frac{\text{N}\,\text{m}^2}{\text{C}^2}$

Substituting in the above values, we get

$$|\vec{E}| = k\frac{q}{r^2} = \left(8.99 \times 10^9 \frac{\text{N}\,\text{m}^2}{\text{C}^2}\right) \cdot \frac{10\ \text{C}}{(2\,\text{m})^2} = 2.25 \times 10^{10} \frac{\text{N}}{\text{C}}$$

Therefore, the electric field components at the point are:

$$E_x = |\vec{E}| \cos 37° = \left(2.25 \times 10^{10} \frac{\text{N}}{\text{C}}\right)(.798) = 1.79 \times 10^{10} \frac{\text{N}}{\text{C}}$$

$$E_y = |\vec{E}| \sin 37° = \left(2.25 \times 10^{10} \frac{\text{N}}{\text{C}}\right)(.602) = 1.35 \times 10^{10} \frac{\text{N}}{\text{C}}$$

18.13 Find the magnitude and direction of the total electric field E when $\vec{E}_1 = 5\hat{x} + 10\hat{y}$ and $\vec{E}_2 = 12\hat{x} - 3\hat{y}$

Because the electric field is a vector, $\vec{E} = \vec{E}_1 + \vec{E}_2 = (5\hat{x} + 10\hat{y}) + (12\hat{x} - 3\hat{y}) = 17\hat{x} + 7\hat{y}\,\text{N/C}$

Therefore $|\vec{E}| = \sqrt{E_x^2 + E_y^2} = \sqrt{17^2 + 7^2} = 18.38\,\text{N/C}$

$\tan\theta = \dfrac{E_y}{E_x}$, so that $\theta = \arctan\left(\dfrac{E_y}{E_x}\right) = \arctan(7/17) = 22.4°$

Supplementary Problems

18.14 Find the magnitude and direction of the following vectors:

(*a*) $\vec{F} = 5\hat{x} + 7\hat{y}$
(*b*) $\vec{E} = 10\hat{\imath} + 2\hat{\jmath}$
(*c*) $\vec{R} = 25\hat{x} + 4\hat{y}$

Ans: (*a*) $|F| = 8.60;\ \theta = 54°$
(*b*) $|E| = 10.19;\ \theta = 119°$
(*c*) $|R| = 10.77;\ \theta = 11°$

18.15 Find the x and y components of the following vectors:

(*a*) $v = 25$ m/s at 30° with respect to the horizontal
(*b*) $a = 13.5$ m/s^2 at 53° with respect to the horizontal
(*c*) $R = 12$m at 1.6 rad with respect to the horizontal

Ans: (*a*) $v_x = 25\cos 30° = 22\,\text{m/s}$
$v_y = 25\sin 30° = 13\,\text{m/s}$

(*b*) $a_x = 13.5\cos 53° = 8.12\,\text{m/s}^2$
$a_y = 13.5\sin 53° = 10.78\,\text{m/s}^2$

(*c*) $R_x = 12\cos(1.6) = -.35\,\text{m}$
$R_y = 12\sin(1.6) = 11.99\,\text{m}$

18.16 A boy subject to a gravitational force of 300 N slides down a slide inclined at 40° to the horizontal. What is the normal force $\vec{N}$ of the slide acting on the boy?

Ans: $\vec{N} = 230\,\text{N}$

18.17 $|\vec{A}| = 10$ and points horizontally to the right. $|\vec{B}| = 8$ and points 35° south of east. Compute:
(*a*) $\vec{A} + \vec{B}$ (*b*) $\vec{A} - \vec{B}$ (*c*) $\vec{A} \cdot \vec{B}$ (*d*) $\vec{A} \times \vec{B}$

Ans: (*a*) $16.6\hat{x} - 4.6\hat{y}$ (*b*) $3.5\hat{x} + 4.6\hat{y}$
(*c*) 65.5 (*d*) the resultant vector is directed into the page and has a magnitude of 45.9

18.18 A plane flies south at 600 mi/h for 2 h, then turns east at 500 mi/h for 30 min and finally heads north at 700 mi/h for 45 min. Find the displacement vector $\vec{R}$ and the straight-line distance from the starting point

Ans: $\vec{R} = 250\hat{x} - 675\hat{y}$; the plane is 719.8 mi from its starting point.

18.19 A neutron initially moving horizontally at 5000 m/s is subject to a gravitational force. This force does not affect the horizontal velocity but results in a vertical component of velocity v_y given by $v_y = gt$ (where t is the time and g is the acceleration due to gravity ($g = 9.8$ m/s^2). At what angle with respect to the horizontal is the neutron traveling:

(*a*) after 1 s? (*b*) after 10 s? (*c*) after 1000 s?

Ans: (*a*) .11° (*b*) 1.12° (*c*) 62.9°

18. 20 When the sum of all external forces acting on an object is equal to zero, the object is said to be in equilibrium. The shipping crate shown in Fig. 18-31 is suspended by two ropes, one of which exerts a force of

50 N at angle of 75° from the horizontal, while gravity exerts a downward force of 120 N. What force is exerted by the second rope if it acts at an angle of 65° from the horizontal? (Fig. 18-28)

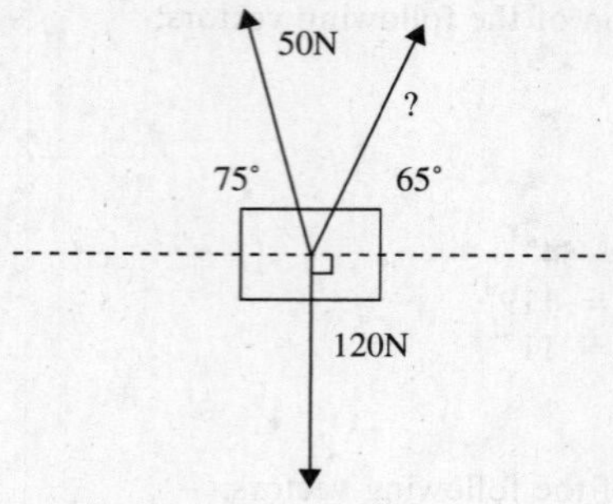

Fig. 18-31

Ans: $|\vec{F}| = 61.37$ N at an angle of 57.27°

18.21 Find the components F_x and F_y if $\vec{F}$:

(*a*) has a magnitude of 10 and is at 45°

(*b*) has a magnitude of 10 and is at 60°

(*c*) has a magnitude of 10 and is at 90°

Ans: (*a*) $F_x = 7.07 \quad F_y = 7.07$ (*b*) $F_x = 5 \quad F_y = 8.66$ (*c*) $F_x = 0 \quad F_y = 10$

18.22 If $\vec{R} = 4\hat{x} + 3\hat{y}$ and $\vec{S} = \hat{x} + 2\hat{y}$, find:

(*a*) $4\vec{R} + 3\vec{S}$ (b) $4\vec{R} - 3\vec{S}$

Ans: (*a*) $19\hat{x} + 18\hat{y}$ (*b*) $13\hat{x} + 6\hat{y}$

18.23 If $\vec{R} = 40\hat{x} + 6\hat{y} - 3\hat{z}$ and $\vec{S} = \hat{x} + 2\hat{y} + 4\hat{z}$, calculate:

(*a*) $5\,\vec{R}\cdot\vec{S}$ (*b*) $2\,\vec{R}\times\vec{S}$

Ans: (*a*) 200; (*b*) the resulting vector has a magnitude of 181 and points in a direction mutually perpendicular to R and S according to the right-hand rule

18.24 Show that, for any vectors T and V, that $\vec{T}\cdot(\vec{V}\times\vec{T}) = 0$

PART III

ADVANCED TOPICS IN MATHEMATICS

Chapter 19

Polar, Spherical, and Cylindrical Coordinate Systems

OVERVIEW OF COORDINATE SYSTEMS. There are several different coordinate systems commonly used in physics. While a satellite orbiting the Earth may follow a unique trajectory, different coordinate systems can be used to describe that trajectory in different ways. The judicious choice of a coordinate system can be crucial in determining the ease with which a problem can be solved. One often needs to be able to freely *transform* a position, velocity or other quantity expressed in one coordinate system into another one. The treatment of derivatives, particularly with respect to time, requires careful attention, particularly in advanced applications. Having previously discussed Cartesian coordinates in two and three dimensions, we now consider three additional coordinate systems that are frequently encountered throughout physics: *polar, spherical and cylindrical coordinate systems.* We define each system, including its unit vectors, and indicate how various quantities—including unit vectors, displacements, time derivatives, vector and scalar products—are expressed when transforming between each of these systems and the Cartesian coordinate system. Later in this book, we consider more sophisticated applications using these systems—including, in particular, the calculation of velocities and accelerations and the integration over areas and volumes.

IN THE POLAR COORDINATE SYSTEM a point in the plane is located by giving its position relative to a fixed point and a fixed line (direction) through the fixed point. The fixed point O (see Fig. 19-1) is called the *pole* and the fixed half line OA is called the *polar axis.*

Let θ denote the smallest positive angle measured counterclockwise in degrees or radians from OA to OB, and let r denote the (positively) directed distance OP. Then P is uniquely determined when r and θ are known. These two measures constitute the *polar coordinates* of P and we write $P(r, \theta)$. The quantity r is called the *radius vector* and θ is called the *vectorial angle* of P. Note that a positive direction, indicated by the arrow, has been assigned on the half line $\overrightarrow{OB}$.

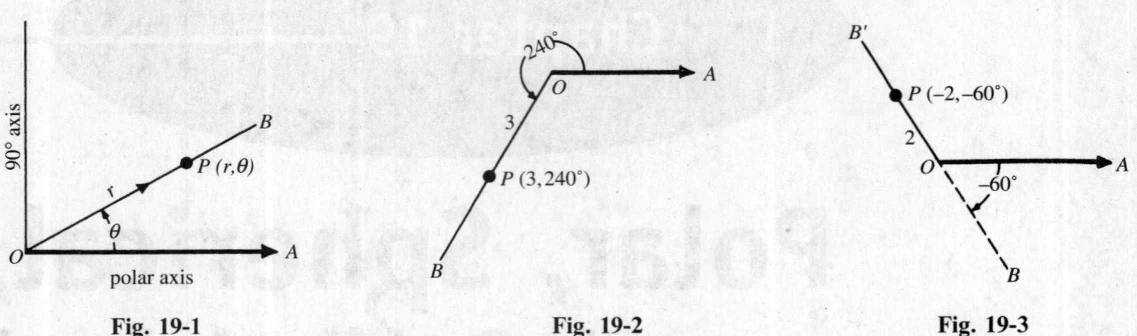

Fig. 19-1 **Fig. 19-2** **Fig. 19-3**

EXAMPLE 1. Locate the point $P(3, 240°)$ or $P(3, 4\pi/3)$. Refer to Fig. 19-2.

Lay off the vectorial angle $\theta = m\measuredangle AOB = 240°$, measured counterclockwise from $\overrightarrow{OA}$, and on $\overrightarrow{OB}$ locate P such that $r = OP = 3$.

In the paragraph above we have restricted r and θ so that $r \geq 0$ and $0° \leq \theta < 360°$. In general, these restrictions will be observed; however, at times it will be more convenient to permit r and θ to have positive or negative values. If θ is negative and r is positive, we lay off the angle $\theta = \angle AOB$, measured clockwise from $\overrightarrow{OA}$, and locate P on $\overrightarrow{OB}$ so that $OP = r$. If r is negative, we lay off $\theta = \angle AOB$, extend $\overrightarrow{OB}$ through the pole to B′, and locate P on $\overrightarrow{OB'}$ a distance $|r|$ from O.

EXAMPLE 2. Locate the point $P(-2, 60°)$ or $P(-2, -\pi/3)$. Refer to Fig. 19-3.

Lay off the vectorial angle $\theta = \angle AOB = 60°$, measured clockwise from OA, extend OB through the pole to B', and on OB' locate P a distance 2 units from O. (See Problems 19.1–19.2.)

Although not a part of the polar system, it will be helpful at times to make use of a half line, called the 90° axis, which issues from the pole perpendicular to the polar axis.

TRANSFORMATIONS BETWEEN POLAR AND RECTANGULAR COORDINATES. If the pole and polar axis of the polar system coincide respectively with the origin and positive x axis of the rectangular system, and if P has rectangular coordinates (x, y) and polar coordinates (r, θ), then the following relations hold:

(*1*) $x = r\cos\theta$ (*4*) $\theta = \arctan y/x$
(*2*) $y = r\sin\theta$ (*5*) $\sin\theta = y/r$ and $\cos\theta = x/r$
(*3*) $r = \sqrt{x^2 + y^2}$

If relations (*3*)–(*5*) are to yield the restricted set of coordinates of the section above, θ is to be taken as the smallest positive angle satisfying (*5*) or, what is equivalent, θ is the smallest positive angle satisfying (*4*) and terminating in the quadrant in which $P(x, y)$ lies.

EXAMPLE 3. Find the rectangular coordinates of $P(3, 300°)$.

Here $r = 3$ and $\theta = 300°$; then $x = r\cos\theta = 3\cos 300° = 3(\frac{1}{2}) = \frac{3}{2}$, $y = r\sin\theta = 3\sin 300° = 3(-\frac{1}{2}\sqrt{3}) = -3\sqrt{3}/2$, and the rectangular coordinates are $(\frac{3}{2}, -3\sqrt{3}/2)$.

EXAMPLE 4. Find the polar equation of the circle whose rectangular equation is $x^2 + y^2 - 8x + 6y - 2 = 0$.

Since $x = r\cos\theta$, $y = r\sin\theta$, and $x^2 + y^2 = r^2$, the polar equation is $r^2 - 8r\cos\theta + 6r\sin\theta - 2 = 0$. (See Problems 19.3–19.5.)

CURVE SKETCHING IN POLAR COORDINATES. Preliminary to sketching the locus of a polar equation, we discuss symmetry, extent, etc., as in the case of rectangular equations. However, there are certain complications at times due to the fact that in polar coordinates a given curve may have more than one equation.

EXAMPLE 5. Let $P(r, \theta)$ be an arbitrary point on the curve $r = 4\cos\theta - 2$. Now P has other representations: $(-r, \theta+\pi), (-r, \theta-\pi), (r, \theta-2\pi), \ldots$.

Since (r, θ) satisfies the equation $r = 4\cos\theta - 2, (-r, \theta+\pi)$ satisfies the equation $-r = 4\cos(\theta+\pi) - 2 = -4\cos\theta - 2$ or $r = 4\cos\theta + 2$. Thus, $r = 4\cos\theta - 2$ and $r = 4\cos\theta + 2$ are equations of the same curve. Such equations are called *equivalent*. The reader will show that $(-r, \theta-\pi)$ satisfies $r = 4\cos\theta + 2$ and $(r, \theta-2\pi)$ satisfies $r = 4\cos\theta - 2$.

EXAMPLE 6. Show that point $A(-1, \pi/6)$ is on the ellipse $r = \dfrac{3}{4+2\sin\theta}$.

Note that the given coordinates do not satisfy the given equation.

First Solution. Another set of coordinates for A is $(1, 7\pi/6)$. Since these coordinates satisfy the equation, A is on the ellipse.

Second Solution. An equivalent equation for the ellipse is

$$-r = \frac{3}{4+2\sin(\theta-\pi)} \qquad \text{or} \qquad r = \frac{-3}{4-2\sin\theta}$$

Since the given coordinates satisfy this equation, A is on the ellipse.

SYMMETRY. A locus is symmetric with respect to the polar axis if an equivalent equation is obtained when

(*a*) θ is replaced by $-\theta$, or
(*b*) θ is replaced by $\pi - \theta$ and r by $-r$ in the given equation.

A locus is symmetric with respect to the 90° axis if an equivalent equation is obtained when

(*a*) θ is replaced by $\pi - \theta$, or
(*b*) θ is replaced by $-\theta$ and r by $-r$ in the given equation.

A locus is symmetric with respect to the pole if an equivalent equation is obtained when

(*a*) θ is replaced by $\pi + \theta$, or
(*b*) r is replaced by $-r$ in the given equation.

EXTENT. The locus whose polar equation is $r = f(\theta)$ is a closed curve if r is real and finite for all values of θ, but is not a closed curve if there are values of one variable which make the other become infinite.

The equation should also be examined for values of one variable which make the other imaginary.

At times, as in the equation $r = a(1 + \sin\theta)$, the values of θ which give r its maximum values can be readily determined. Since the maximum value of $\sin\theta$ is 1, the maximum value of r is $2a$ which it assumes when $\theta = \frac{1}{2}\pi$.

DIRECTIONS AT THE POLE. Unlike all other points, the pole has infinitely many pairs of coordinates $(0, \theta)$ when θ is restricted to $0° \le \theta < 360°$. While two such pairs $(0, \theta_1)$

and $(0, \theta_2)$ define the pole, they indicate different directions (measured from the polar axis) there. Thus, the values of θ for which $r = f(\theta) = 0$ give the directions of the tangents to the locus $r = f(\theta)$ at the pole.

POINTS ON THE LOCUS. We may find as many points on a locus as desired by assigning values to θ in the given equation and solving for the corresponding values of r.

EXAMPLE 7. Discuss and sketch the locus of the cardioid $r = a(1 - \sin\theta)$.

Symmetry. An equivalent equation is obtained when θ is replaced by $\pi - \theta$; the locus is symmetric with respect to the 90° axis.

Extent. Since r is real and $\leq 2a$ for all values of θ, the locus is a closed curve, lying within a circle of radius $2a$ with center at the pole. Since $\sin\theta$ is of period 2π, the complete locus is described as θ varies from 0 to 2π.

Direction at the pole. When $r = 0$, $\sin\theta = 1$ and $\theta = \frac{1}{2}\pi$. Thus, the locus is tangent to the 90° axis at the pole.

After locating the points in Table 19.1 and making use of symmetry of the locus with respect to the 90° axis, we obtain the required curve as shown in Fig. 19-4. (See Problems 19.11–19.17.)

Table 19.1

θ	$\frac{1}{2}\pi$	$2\pi/3$	$3\pi/4$	$5\pi/6$	π	$7\pi/6$	$5\pi/4$	$4\pi/3$	$3\pi/2$
r	0	$0.13a$	$0.29a$	$0.5a$	a	$1.5a$	$1.71a$	$1.87a$	$2a$

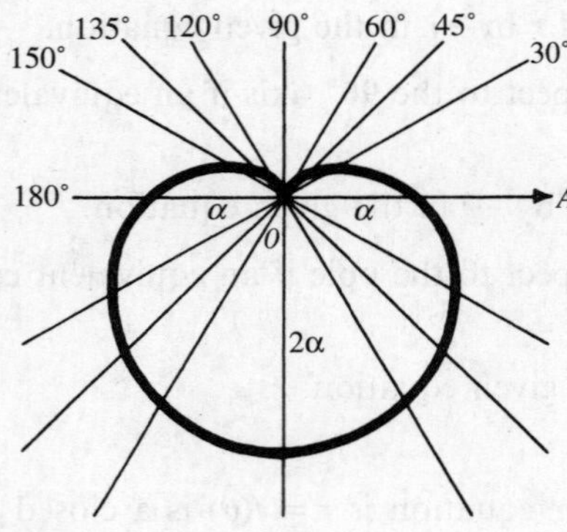

Fig. 19-4

INTERSECTIONS OF POLAR CURVES. It is to be expected that in finding the points of intersection of two curves with polar equations $r = f_1(\theta)$ and $r = f_2(\theta)$, we set $f_1(\theta) = f_2(\theta)$ and solve for θ. However, because of the multiplicity of representations both of the coordinates of a point and the equation of a curve, this procedure will fail at times to account for all of the intersections. Thus, it is a better policy to determine from a figure the exact number of intersections before attempting to find them.

EXAMPLE 8. Since each of the circles $r = 2\sin\theta$ and $r = 2\cos\theta$ passes through the pole, the circles intersect in the pole and in one other point. See Fig. 19-5. Since each locus is completely described on the interval 0 to π, we set $2\sin\theta = 2\cos\theta$ and solve for θ on this interval. The solution $\theta = \frac{1}{4}\pi$ yields the point $(\sqrt{2}, \frac{1}{4}\pi)$.

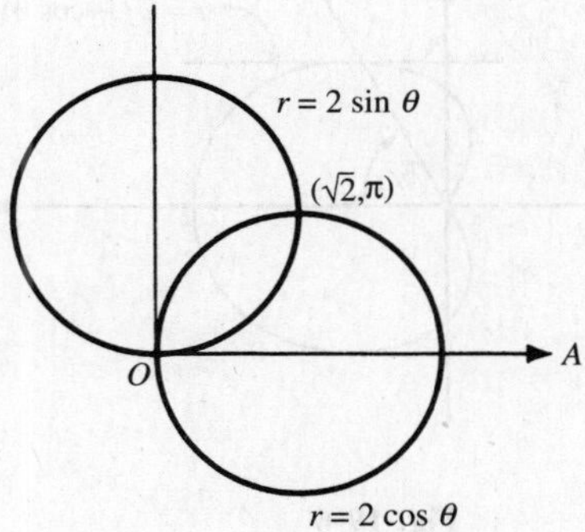

Fig. 19-5

Analytically we may determine whether or not the pole is a point of intersection by setting $r = 0$ in each of the equations and solving for θ. Setting $\sin\theta = 0$ we find $\theta = 0$, and setting $\cos\theta = 0$ we find $\theta = \frac{1}{2}\pi$. Since both equations have solutions, the pole is a point of intersection. The procedure above did not yield this solution since the coordinates of the pole $(0, 0)$ satisfy $r = 2\sin\theta$ while the coordinates $(0, \frac{1}{2}\pi)$ satisfy $r = 2\cos\theta$. (See Problems 19.18–19.19.)

SLOPE OF A POLAR CURVE. We state the following results without proof:

Given a polar function, $r = f(\theta)$,

$$\frac{dy}{dx} = \frac{r'\sin\theta + r\cos\theta}{r'\cos\theta - r\sin\theta} \tag{19.1}$$

If $f(x) = \sin x$, then

$$f'(x) = \cos x \tag{19.2}$$

and if $g(x) = \cos x$, then

$$g'(x) = -\sin x \tag{19.3}$$

Using (*19.1*), (*19.2*), and (*19.3*) above, we can find derivatives and thus slopes for polar curves.

EXAMPLE 9. Find the slope of the cardioid

$$r = 2(1 + \cos\theta) \text{ at } \theta = \frac{\pi}{3}.$$

$$\begin{aligned}\frac{dy}{dx} &= \text{slope of curve} \\ &= \frac{r'\sin\theta + r\cos\theta}{r'\cos\theta - r\sin\theta} \\ &= \frac{(-2\sin\theta)(\sin\theta) + 2(1 + \cos\theta)(\cos\theta)}{(-2\sin\theta)(\cos\theta) - 2(1 + \cos\theta)(\sin\theta)}\end{aligned}$$

At $\theta = \frac{\pi}{3}, \frac{dy}{dx} = 0$. This indicates that $r = 2(1 + \cos\theta)$ has a horizontal tangent line at $\theta = \frac{\pi}{3}$. See Fig. 19-6 and see Problem 19.20.

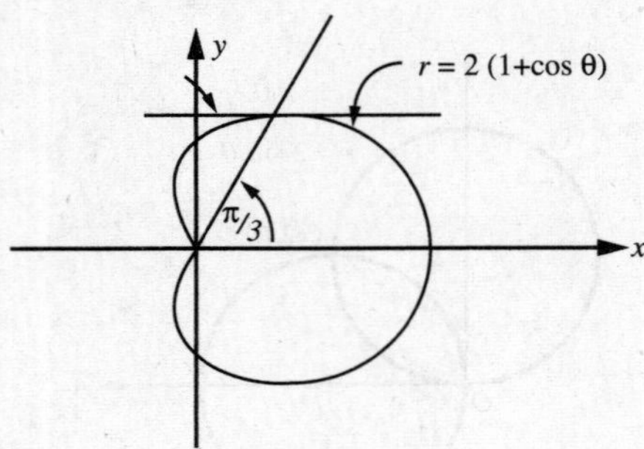

Fig. 19-6

VECTOR REPRESENTATIONS IN POLAR COORDINATES. A vector, regardless of the coordinate system in which it is described, is something that has a direction and a magnitude. In polar coordinates, vectors that point out radially from the origin (or that point radially inward, such as the tension in a string with a marble at the end moving in a circle) can be described using the unit vector $\hat{r}$, which is shown in Fig. 19-3. As indicated, $\hat{r}$ points from the origin outward and has a magnitude of one unit. Not all vectors are radial, however. The vector that describes the change in velocity as the rock swings from position 1 to position 2 is at right angles to $\hat{r}$ and is expressed in terms of the unit vector $\hat{\theta}$, which is defined to be at $+90°$ with respect to $\hat{r}$. (Note that $\hat{r}$ and $\hat{\theta}$, in contrast to the Cartesian unit vectors $\hat{x}$ and $\hat{y}$, each has a direction that is not constant but that varies from point to point.)

In general, a vector may contain components that lie along both $\hat{r}$ and $\hat{\theta}$ and can be expressed as

$$\vec{f}(r, \theta) = f_r\hat{r} + f_\theta\hat{\theta} \tag{19.4}$$

where, in the most general case, f_r and f_θ are each functions of r and θ and the magnitude of f is

$$\left|\vec{f}(r, \theta)\right| = \sqrt{|f_r|^2 + |f_\theta|^2} \tag{19.5}$$

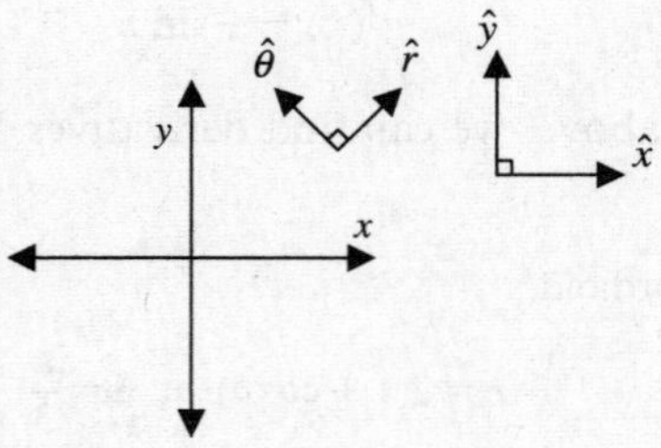

Fig. 19-7

EXAMPLE 10. Find the magnitude of the force field $\vec{F} = 3r^2\theta\hat{r} + 5r\theta^2\hat{\theta}$ at the point $r = 5$ and $\theta = \pi/4$. Assume that each of the components of force is expressed in newtons.

At the value of (r, θ) specified,

$$\vec{F} = 3r^2\theta\hat{r} + 5r\theta^2\hat{\theta} = 3(5)^2\left(\frac{\pi}{4}\right)\hat{r} + 5(5)\left(\frac{\pi}{4}\right)^2\hat{\theta} = \frac{75\pi}{4}\hat{r} + \frac{25\pi^2}{16}\hat{\theta},$$

where each of the components is expressed in newtons.

VELOCITY AND ACCELERATION IN POLAR COORDINATES. It is frequently of interest to calculate the velocity and acceleration of an object, the position of which varies with time. In Cartesian coordinates, the velocity of an object with a displacement

$$\vec{r} = x\hat{x} + y\hat{y} + z\hat{z}$$

is simply:

$$\vec{v} = v_x\hat{x} + v_y\hat{y} + v_z\hat{z}$$

where $v_x = \frac{dx}{dt}$, $v_y = \frac{dy}{dt}$, and $v_z = \frac{dz}{dt}$

In polar coordinates, we need to be slightly more careful in taking the time derivative of the displacement vector $\vec{r}(= r\hat{r})$, because the orientation of the unit vector $\hat{r}$ varies with position. One can show that $\hat{r}$ varies with time as follows:

$$\frac{d\hat{r}}{dt} = \frac{d\theta}{dt}\hat{\theta} \tag{19.6}$$

Hence, as a particle changes its position over time, so, in general will $\hat{r}$ and $\hat{\theta}$. In fact, the reader can show (Problem 19.21) that the total time derivative $\frac{d\vec{r}}{dt}$ is given by:

$$\frac{d\vec{r}}{dt} = \frac{dr}{dt}\hat{r} + r\frac{d\theta}{dt}\hat{\theta} \tag{19.7}$$

Hence, while the position vector $\vec{r}$ is simply $\vec{r} = r\hat{r}$, its time derivative has components that, in general, include components along $\hat{r}$ (the radial component of the velocity) and along $\hat{\theta}$ (the angular component).

Further consideration of the derivatives of scalar and vector functions, which may vary with both position and time, is deferred to Chapters 22 and 23.

SPHERICAL COORDINATES. A spherical coordinate system is used to describe natural phenomena in three dimensions using one radial coordinate and two angular coordinates. (Note that, in contrast, plane polar coordinates describe a two-dimensional space, also using a radial coordinate but only a single angular coordinate.) The trajectory of a satellite orbiting the Earth or of the electromagnetic field surrounding a radio tower would naturally be described in spherical coordinates.

We first remind the reader of the three-dimensional Cartesian coordinate system, shown in Fig. 19-9, in which we refer to the location of an object by identifying its location with respect to an origin at which we envision three mutually perpendicular coordinate axes. A point in such a system is typically labeled (x,y,z), where the variables refer to the individual axes.

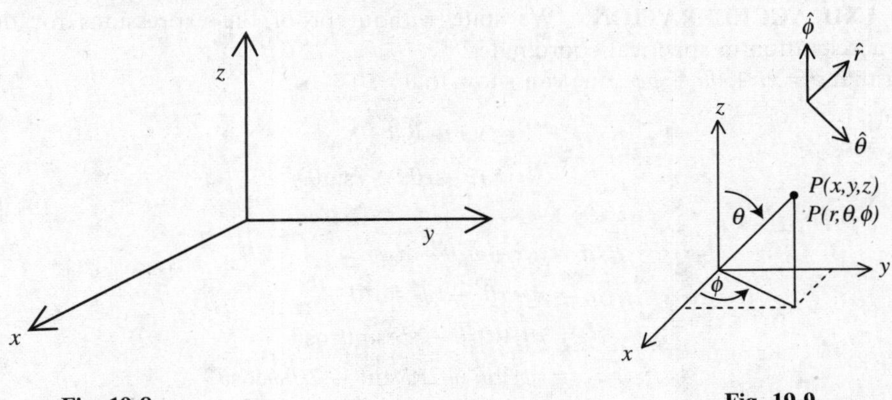

Fig. 19-8

Fig. 19-9

The system of spherical coordinates is illustrated in Fig. 19-9. In spherical coordinates, a location in space is determined by a radial distance r from an origin and the value of two angles θ and ϕ. The first spherical coordinate, *the radial distance r*, is defined as the magnitude of the vector extending from the origin to the point in question. A line (called the polar axis and often labeled the z axis) is perpendicular to a plane that contains an origin and the usual Cartesian x-y coordinate system. The second spherical coordinate, the *polar angle* θ, is defined as the angle between the radial vector and the positive z axis. In geographical terms, it corresponds to the angle of latitude. The third spherical coordinate, *the azimuthal angle* ϕ, is the angle between the projection of r onto the xy plane and the x axis. It corresponds to an object's longitudinal position.* Note that these coordinate variables are restricted to $0 \leq r \leq \infty$, $0 \leq \theta \leq \pi$ and $0 \leq \phi \leq 2\pi$.

TRANSFORMATION BETWEEN SPHERICAL AND CARTESIAN COORDINATES. Applying the basic trigonometric relations to Fig. 19-8, Cartesian coordinates can be expressed in spherical coordinates through the following relations:

$$\begin{aligned} x &= r\sin\theta\cos\phi \\ y &= r\sin\theta\sin\phi \\ z &= r\cos\theta \end{aligned} \tag{19.8}$$

EXAMPLE 11. Find the Cartesian coordinates of a point located 3.50 m from the origin at a polar angle of 1.26 rad and an azimuthal angle of 4.73 rad.

Using equation (*19.8*), we immediately find

$$\begin{aligned} x &= 3.50\sin(1.26)\cos(4.73) = .059 \\ y &= 3.50\sin(1.26)\sin(4.73) = -3.33 \\ z &= 3.50\cos(4.73) = .062 \end{aligned}$$

UNIT VECTORS IN SPHERICAL COORDINATES. In order to describe vectors in spherical coordinates, we define a set of mutually perpendicular unit vectors labeled $\hat{r}$, $\hat{\theta}$ and $\hat{\phi}$. The unit vectors (which, by definition, each have a magnitude of one) are shown in Fig. 19-10 along with the Cartesian unit vectors $\hat{x}$, $\hat{y}$ and $\hat{z}$. $\hat{r}$ lies along the radial vector. $\hat{\theta}$ points in the direction of increasing θ and is mutually perpendicular to $\hat{r}$ and $\hat{z}$.

Similarly, $\hat{\phi}$ points in the direction of increasing ϕ and is mutually perpendicular to $\hat{r}$ and $\hat{\theta}$.

VELOCITY AND ACCELERATION. We state without proof the expressions for displacement, velocity, and acceleration in spherical coordinates:

Recalling that $\vec{r} = r\hat{r} + \theta\hat{\theta} + \phi\hat{\phi}$, one can show that

$$\begin{aligned} v &= v_r\hat{r} + v_\theta\hat{\theta} + v_\phi\hat{\phi} \\ &= \dot{r}\hat{r} + r\dot{\theta}\hat{\theta} + r\sin\theta\dot{\phi} \end{aligned} \tag{19.9}$$

$$\begin{aligned} \vec{a} &= a_r\hat{r} + a_\theta\hat{\theta} + a_\phi\hat{\phi} \\ a_r &= \ddot{r} - r\dot{\theta}^2 - r\dot{\phi}^2\sin^2\theta \\ a_\theta &= r\ddot{\theta} + 2\dot{r}\dot{\theta} - r\dot{\phi}^2\sin\theta\cos\theta \\ a_\phi &= r\ddot{\phi}\sin\theta + 2\dot{r}\dot{\phi}\sin\theta + 2r\dot{\theta}\dot{\phi}\cos\theta \end{aligned}$$

* The use of the symbols θ and ϕ for, respectively, the polar and the azimuthal angles is common but not universal. In particular, the reverse usage is often seen, with ϕ describing the polar angle and θ describing the azimuthal one.

CYLINDRICAL COORDINATES. A system of cylindrical coordinates, shown in Fig. 19-10, adds the z axis to the system of plane polar coordinates in order to describe something in three dimensions. A point P is therefore described with respect to the radial distance r of its projection onto the xy plane; the angle ϕ that projection makes with respect to the x axis; and the vertical projection z onto that plane. Physical situations that naturally make use of such a system are those, not surprisingly, which involve a cylindrical geometry: finding the water pressure in a pipe or the electrical field in the vicinity of a cylindrical conductor are just two examples that lend themselves naturally to the use of cylindrical coordinates. Note that the coordinate variables are restricted to $0 \leq r \leq \infty$, $0 \leq \phi \leq \pi$, and $-\infty < z < \infty$.

While spherical coordinates also describe an object in three dimensions, it uses a length (r) and two angles (θ and ϕ) to do so, while cylindrical coordinates uses two lengths (r and z) and an angle (ϕ) to convey the same information*.

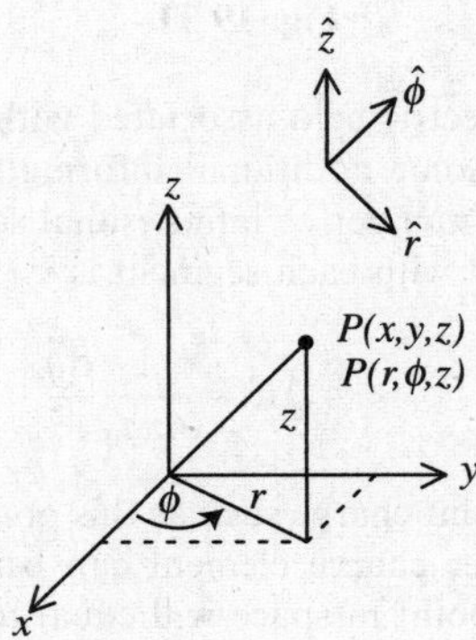

Fig. 19-10

TRANSFORMATION BETWEEN CYLINDRICAL AND CARTESIAN COORDINATES. Once again, we can apply the basic trigonometric relations to Fig. 19-10 to determine the relationship between cylindrical and Cartesian coordinates:

$$x = r\cos\phi$$
$$y = r\sin\phi$$
$$z = z$$

EXAMPLE 12. Find the Cartesian coordinates of point P in Fig. 19-10, if $r = 3.0$ m, $\phi = \pi/3$ and $z = 0.5$.

Substituting these values into the equation above immediately yields:

$$x = 3\cos(\pi/3) = 1.5\,\text{m}$$
$$y = 3\sin(\pi/3) = 2.6\,\text{m}$$
$$z = 0.50\,\text{m}$$

VECTORS IN CYLINDRICAL COORDINATES. Vectors in cylindrical coordinates can be expressed as

$$\vec{f}(r, \phi, z) = f_r\hat{r} + f_\phi\hat{\phi} + f_z\hat{z} \tag{19.10}$$

where f_r, f_ϕ, and f_z can, in the most general case, each be a function of r, ϕ, and z. Note that the vector has components that lie along the unit vectors $\hat{r}$, $\hat{\phi}$, and $\hat{z}$, the three mutually perpendicular vectors

* As with other coordinate systems, the notation used here is common but not universal—one frequently sees the symbol ρ rather than r, used for the radial coordinate, and θ, rather than ϕ, used for the angular one.

shown in Fig. 19-7. The unit vectors $\hat{r}$ and $\hat{\phi}$ are identical to those of polar coordinates, while $\hat{z}$ is familiar from the Cartesian system. Differential elements dr and $d\phi$ are shown in Fig. 19-11.

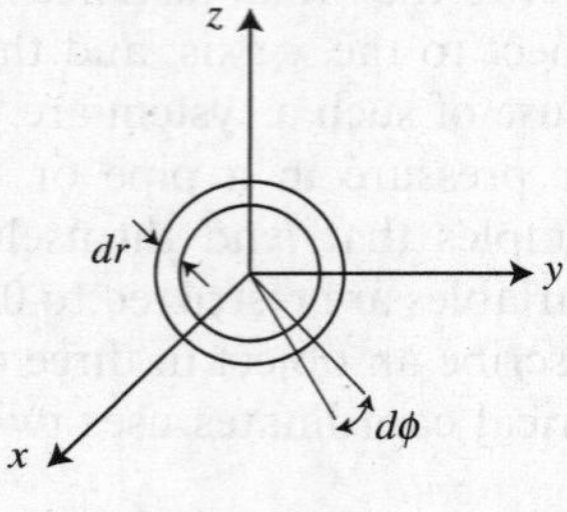

Fig. 19-11

EXAMPLE 13. What is the shape of the electric field associated with an infinitely long uniformly charged wire?

The situation is shown in Fig. 19-11. Some additional information is needed to address this question. We consider the wire as consisting of an infinite number of infinitesimal segments, each with charge dq and length dl. The magnitude of the electric field associated with each segment is

$$\mathrm{d}E = \frac{1}{4\pi\varepsilon_0}\frac{\mathrm{d}q}{r^2}$$

The electric field from a positive (negative) point charge such as this point radially outward (inward) in all directions and possesses spherical symmetry. Any other charge element can be similarly described. From the principle of superposition, the total electric field at any point in space will equal to the sum of the electric fields from each of these charge elements. Now, we utilize the symmetry of this situation: for every charge element on the wire that contributes an electric field component with a component parallel to the wire, there is an exact cancellation from another charge element (see Fig. 19-8).

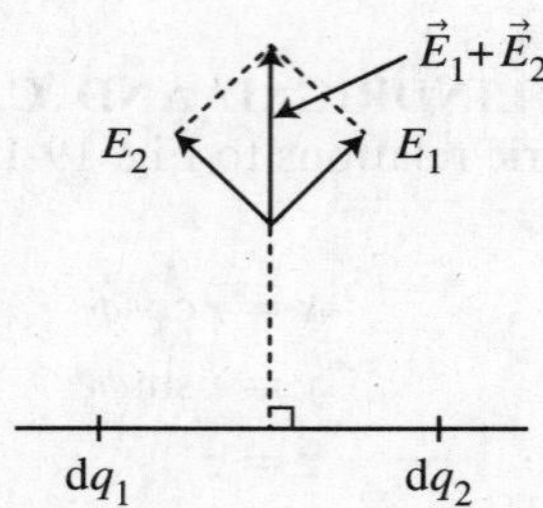

Fig. 19-12

As a result, the only nonzero components of the electric field are those that are perpendicular to the wire. We therefore conclude that

$$\vec{E} = E(r)\hat{r}$$

with no components in either the $\hat{\phi}$ or the $\hat{z}$ directions. (The form of the function $E(r)$ can be found using Gauss' Law, but that is not our focus here.)

EXAMPLE 14. The temperature distribution in a cylindrical air shaft is

$$T(r, \phi, z) = 11r^2 e^{-0.1\phi^2}(z + 300) \tag{19.11}$$

where the units are MKS and ϕ is in radians. How quickly is the temperature changing in the vertical direction at the point 1 m from the axis of the air shaft at the point where $\phi = \pi/6$ and $z = 0.75$ m?

Differentiating (19.11) with respect to z, we get:

$$\frac{\partial T}{\partial z} = 11r^2 e^{-0.1\phi^2} = 11(1)^2 e^{-0.1(\pi/6)^2} = 10.7\,\mathrm{K/m}$$

VELOCITY AND ACCELERATION IN CYLINDRICAL COORDINATES. The velocity and acceleration of the displacement vector $r = r\hat{r} + z\hat{z}$ are expressed in cylindrical coordinates as:

$$\vec{v} = \dot{r}\hat{r} + r\dot{\phi}\hat{\phi} + \dot{z}\hat{z} \tag{19.12}$$

$$\vec{a} = (\ddot{r} - r\dot{\phi}^2)\hat{r} + (2\dot{r}\dot{\phi} + r\ddot{\phi})\hat{\phi} + \ddot{z}\hat{z} \tag{19.13}$$

EXAMPLE 15. An airplane in the vicinity of an airport control tower circles it once every 6 min in a radius of 10 km. As it does so, it descends steadily by 10 m/s. What is its velocity in cylindrical coordinates?

The angular velocity $\dot{\phi} = \frac{2\pi}{t} = \frac{2(3.14)}{6\,\text{min} \times 60\,\text{s/min}} = .017\,\text{s}^{-1}$.

Because r is constant at 10 km, $\dot{r} = 0$. Therefore

$$\begin{aligned}\vec{v} &= \dot{r}\hat{r} + r\dot{\phi}\hat{\phi} + \dot{z}\hat{z}\\ &= (10\,\text{km})(.017\,\text{s}^{-1})\hat{\phi} - 10\,\text{ms}^{-1}\hat{z}\\ &= (170\hat{\phi} - 10\hat{z})\,\text{m/s}\end{aligned}$$

COORDINATE SYSTEMS IN TRANSLATION AND ROTATION. There are many situations in which one needs to compare measurements made in two frames of reference that are moving and/or rotating with respect to each other. For example, it is important to be able to compare the velocity components of a rocket from either the surface of the Earth or a monitoring satellite. The comparison of reference frames that are moving linearly with respect to one another (i.e., in relative translation) or which are in relative rotation is a topic, the full treatment of which is beyond the scope of this text. Nevertheless, we briefly consider some common situations involving such systems.

We do so in the case of *translating systems* by recognizing that for two Cartesian coordinate systems, where one moves at a relative velocity $\vec{v}$ with respect to the other, have coordinates related by:

$$x' = x - v_x t$$

$$y' = y - v_y t$$

$$z' = z - v_z t$$

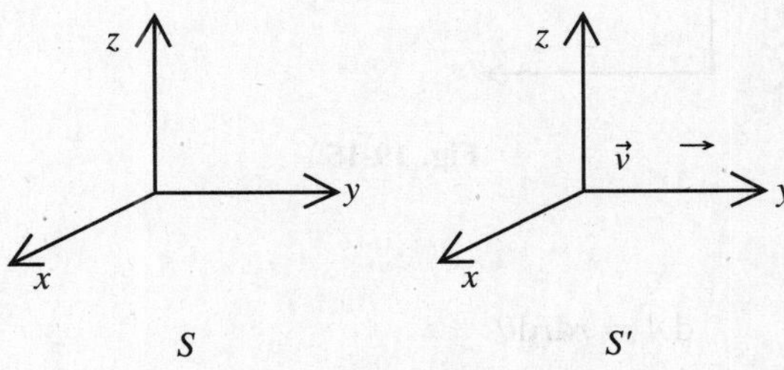

Fig. 19-13

In this case, the relative motion is along the x axis, and system S' (the "primed system") is moving at a velocity v with respect to S (the "unprimed system").

For *rotating systems* S and S', moving at a constant rotational speed ω with respect to one another, the same angle, measured to be θ in system S and θ' in system S', the angular measurements are related by:

$$\theta' = \theta - \omega t$$

where θ and θ' are in the same plane and ω gives the relative rotation between the planes.

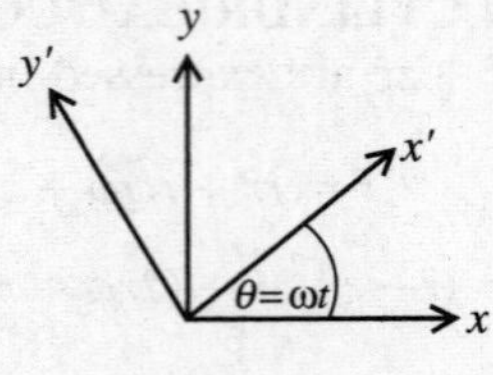

Fig. 19-14

where we assume that the rotational motion occurs in the plane of the measured angle.

For example, consider a system that rotates at an angular velocity $2\pi s^{-1}$. The angle it has rotated through after t seconds is ωt. Hence, $\theta = \theta_0 + \omega t$, where θ_0 and θ are, respectively, the starting and ending angles. As a result, we can account for rotation by tacking on the factor of ωt to the appropriate angle.

These are very specific cases, with the full treatment of translational and rotational motion beyond the scope of this book.

ELEMENTS OF AREA AND VOLUME. Just as dx is a differential element of distance in one dimension, so too are there differential elements of area and volume that can be expressed in each of the coordinate systems presented in this chapter and that are needed when integrating over area or volume in one of these systems. The full treatment of this topic is beyond the scope of this book. Instead, we present only the forms that these differential elements take in the different coordinate systems, providing some simple examples of their use, particularly in physics, in some of the following problems as well as in subsequent chapters.

Area elements:

Cartesian Coordinates: $\mathrm{d}A = \mathrm{d}x\mathrm{d}y$

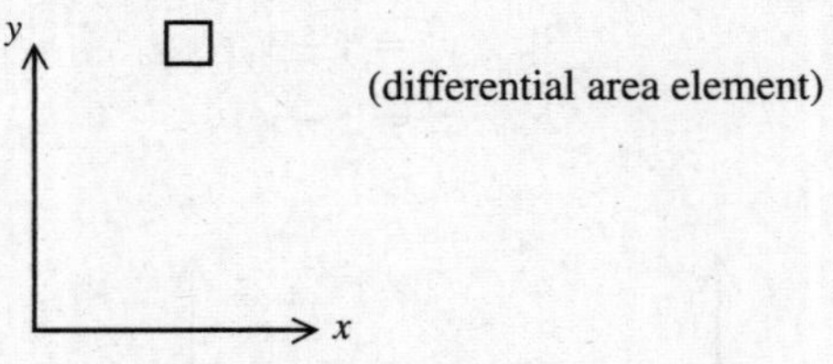

Fig. 19-15

Polar Coordinates: $\mathrm{d}A = r\mathrm{d}r\mathrm{d}\theta$

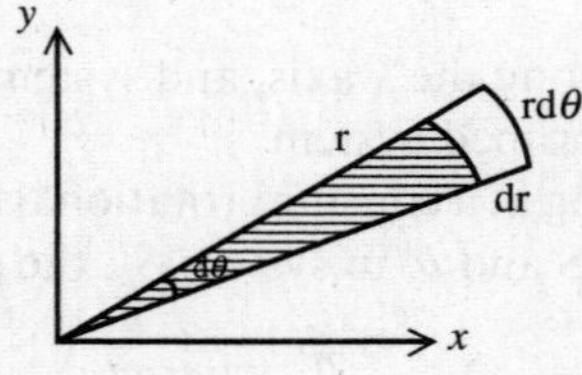

Fig. 19-16

Volume elements:

Cartesian Coordinates: $dV = dx dy dz$

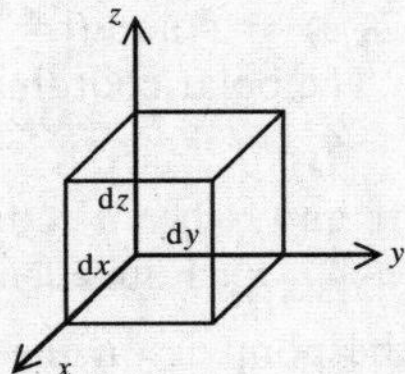

Fig. 19-17

Spherical Coordinates: $dA = \sin\theta\, d\theta d\phi$
$dV = r^2 \sin\theta\, dr d\theta d\phi$

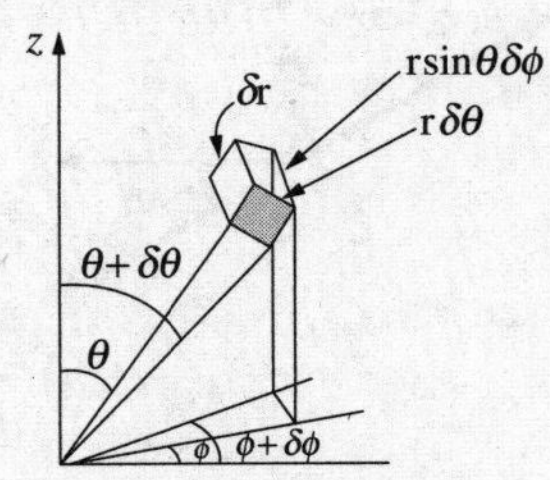

Fig. 19-18

Cylindrical Coordinates: $dA = r\, d\phi dz$
$dV = r\, dr d\theta dz$

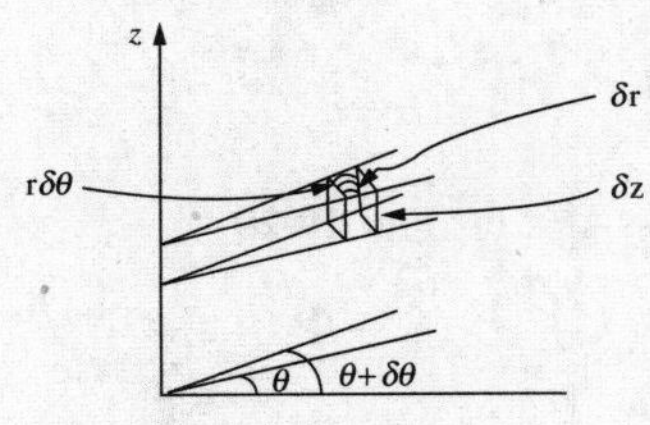

Fig. 19-19

Solved Problems

19.1 Locate the following points and determine which coincide with $P(2, 150°)$ and which with $Q(2, 30°)$:

(*a*) $A(2, 750°)$ (*b*) $B(-2, -30°)$ (*c*) $C(-2, 330°)$ (*d*) $D(-2, -150°)$ (*e*) $E(2, -210°)$

The points B, C, and E coincide with P; the points A and D coincide with Q. See Figs. 19-7(*a*)–(*e*).

19.2 Find the distance between the points

(*a*) $P_1(5, 20°)$ and $P_2(3, 140°)$ (*b*) $P_1(4, 50°)$ and $P_2(3, 140°)$ (*c*) $P_1(r_1, \theta_1)$ and $P_2(r_2, \theta_2)$

In any triangle $OP_1P_2, (P_1P_2)^2 = (OP_1)^2 + (OP_2)^2 - 2(OP_1)(OP_2)\cos \angle P_1OP_2$.

(*a*) From Fig. 19-8 (*a*), $(P_1P_2)^2 = (5)^2 + (3)^2 - 2 \cdot 5 \cdot 3 \cos 120° = 49$; hence, $P_1P_2 = 7$.

(*b*) From Fig. 19-8 (*b*), $P_1P_2 = \sqrt{(4)^2 + (3)^2 - 2 \cdot 4 \cdot 3 \cos 90°} = 5$.

(*c*) From Fig. 19-8 (*c*), $P_1P_2 = \sqrt{r_1^2 + r_2^2 - 2r_1r_2 \cos(\theta_1 - \theta_2)}$.

19.3 Find the set of polar coordinates, satisfying $r \geq 0, 0° \leq \theta < 360°$, of P whose rectangular coordinates are (*a*) $(2, -2\sqrt{3})$, (*b*) (a, a), (*c*) $(-3, 0)$, (*d*) $(0, 2)$. Find two other sets of polar coordinates for each point.

(*a*) We have $r = \sqrt{x^2 + y^2} = \sqrt{(2)^2 + (-2\sqrt{3})^2} = 4$ and $\theta = \arctan y/x = \arctan(-\sqrt{3})$. Since the point is in the first quadrant, we take $\theta = 300°$. The polar coordinates are $(4, 300°)$ or $(4, 5\pi/3)$. Equivalent sets of polar coordinates are $(4, -60°)$ and $(-4, 2\pi/3)$.

(*b*) Here $r = \sqrt{a^2 + a^2} = a\sqrt{2}$ and $\theta = \arctan 1$, when $a > 0$. Since the point is in the first quadrant, we take $\theta = \frac{1}{4}\pi$. The polar coordinates are $(a\sqrt{2}, \frac{1}{4}\pi)$. Equivalent sets are $(a\sqrt{2}, -7\pi/4)$ and $(-a\sqrt{2}, -3\pi/4)$.

(*c*) Here $r = \sqrt{(-3)^2 + (0)^2} = 3$. Since the point is on the negative x axis, we take $\theta = \pi$ and the polar coordinates are $(3, \pi)$. Equivalent sets are $(-3, 0)$ and $(3, -\pi)$.

(*d*) Here $r = \sqrt{(0)^2 + (2)^2} = 2$. Since the point is on the positive y axis, we take $\theta = \pi/2$ and the polar coordinates are $(2, \pi/2)$. Equivalent sets are $(2, -3\pi/2)$ and $(-2, 3\pi/2)$.

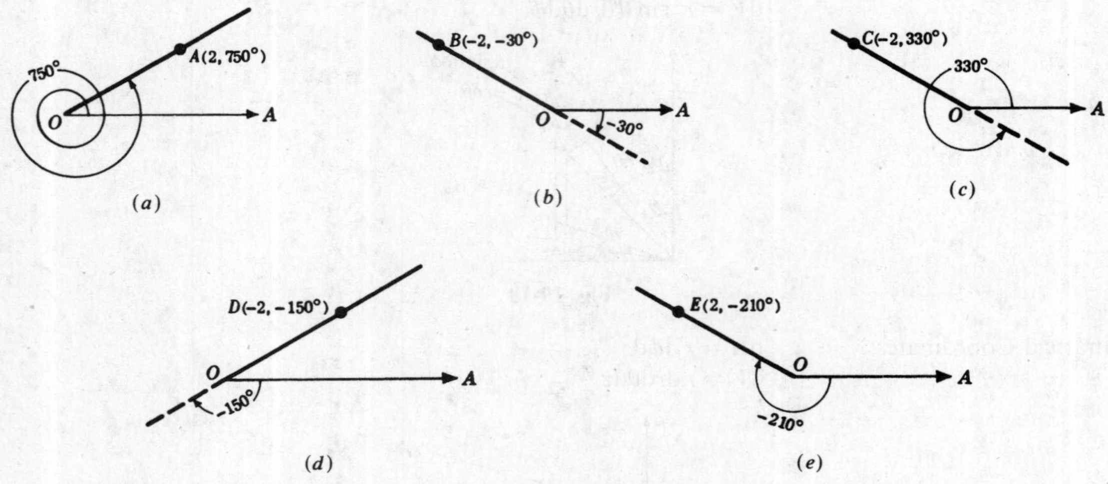

Fig. 19-20

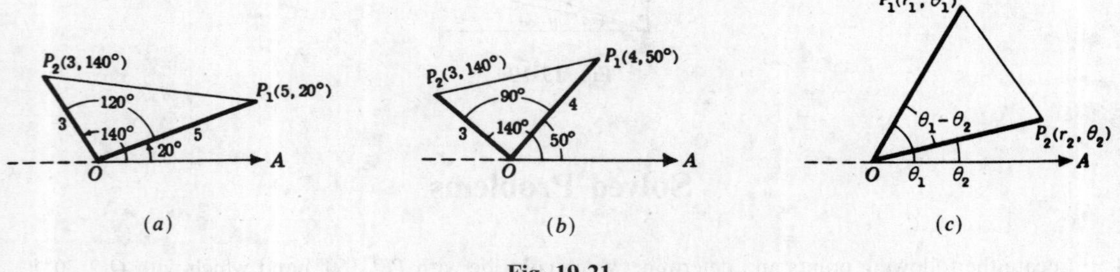

Fig. 19-21

19.4 Transform each of the following rectangular equations into their polar form:

(*a*) $x^2 + y^2 = 25$ (*c*) $3x - y = 0$ (*e*) $(x^2 + y^2 - ax)^2 = a^2(x^2 + y^2)$

(*b*) $x^2 - y^2 = 4$ (*d*) $x^2 + y^2 = 4x$ (*f*) $x^3 + xy^2 + 6x^2 - 2y^2 = 0$

We make use of the transformation: $x = r\cos\theta, y = r\sin\theta, x^2 + y^2 = r^2$.

(*a*) By direct substitution we obtain $r^2 = 25$ or $r = \pm 5$. Now $r = 5$ and $r = -5$ are equivalent equations since they represent the same locus, a circle with center at the origin and radius 5.

(*b*) We have $(r\cos\theta)^2 - (r\sin\theta)^2 = r^2(\cos^2\theta - \sin^2\theta) = r^2\cos 2\theta = 4$.

(c) Here $3r\cos\theta - r\sin\theta = 0$ or $\tan\theta = 3$. The polar equation is $\theta = \arctan 3$.

(d) We have $r^2 = 4r\cos\theta$ or $r = 4\cos\theta$ as the equation of the circle of radius 2 which passes through the origin and has its center on the polar axis.

(e) Here $(r^2 - ar\cos\theta)^2 = a^2r^2$; then $(r - a\cos\theta)^2 = a^2$ and $r - a\cos\theta = \pm a$. Thus we may take $r = a(1+\cos\theta)$ or $r = -a(1-\cos\theta)$ as the polar equation of the locus.

(f) Writing it as $x(x^2+y^2) + 6x^2 - 2y^2 = 0$, we have $r^3\cos\theta + 6r^2\cos^2\theta - 2r^2\sin^2\theta = 0$. Then $r\cos\theta = 2\sin^2\theta - 6\cos^2\theta - 2(\sin^2\theta + \cos^2\theta) - 8\cos^2\theta = 2 - 8\cos^2\theta$ and $r = 2(\sec\theta - 4\cos\theta)$ is the polar equation.

19.5 Transform each of the following equations into its rectangular form:

(a) $r = -2$ (c) $r\cos\theta = -6$ (e) $r = 4(1+\sin\theta)$ (f) $r = \dfrac{4}{2-\cos\theta}$

(b) $\theta = 3\pi/4$ (d) $r = 2\sin\theta$

In general, we attempt to put the polar equation in a form so that the substitutions x^2+y^2 for r^2, x for $r\cos\theta$, and y for $r\sin\theta$ can be made.

(a) Squaring, we have $r^2 = 4$; the rectangular equation is $x^2+y^2 = 4$.

(b) Here $\theta = \arctan y/x = 3\pi/4$; then $y/x = \tan 3\pi/4 = -1$ and the rectangular equation is $x + y = 0$.

(c) The rectangular form is $x = -6$.

(d) We first multiply the given equation by r to obtain $r^2 = 2r\sin\theta$. The rectangular form is $x^2+y^2 = 2y$.

(e) After multiplying by r, we have $r^2 = 4r + 4r\sin\theta$ or $r^2 - 4r\sin\theta = 4r$; then $(r^2 - 4r\sin\theta)^2 = 16r^2$ and the rectangular equation is $(x^2+y^2-4y)^2 = 16(x^2+y^2)$.

(f) Here $2r - r\cos\theta = 4$ or $2r = r\cos\theta + 4$; then $4r^2 = (r\cos\theta+4)^2$ and the rectangular form of the ellipse is $4(x^2+y^2) = (x+4)^2$ or $3x^2+4y^2-8x-16=0$.

19.6 Derive the polar equation of the straight line:

(a) Passing through the pole with vectorial angle k

(b) Perpendicular to the polar axis and $p > 0$ units from the pole

(c) Parallel to the polar axis and $p > 0$ units from the pole

Let $P(r,\theta)$ be an arbitrary point on the line.

(a) From Fig. 19-22 (a) the required equation is $\theta = k$.

(b) From Fig. 19-22 (b) the equation is $r\cos\theta = p$ or $r\cos\theta = -p$ according as the line is to the right or left of the pole.

(c) From Fig. 19-22 (c) the equation is $r\sin\theta = p$ or $r\sin\theta = -p$ according as the line is above or below the pole.

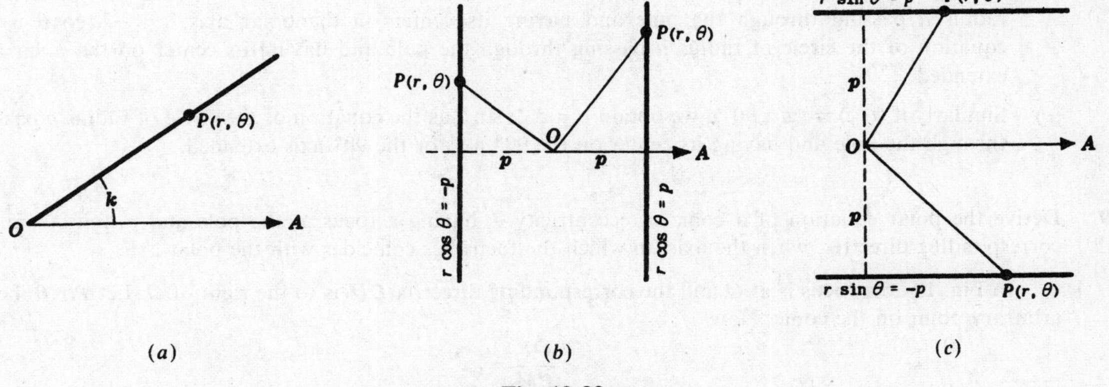

Fig. 19-22

19.7 Derive the polar equivalent of the normal form of the rectangular equation of the straight line not passing through the pole.

Let $P(r, \theta)$ be an arbitrary point on the line. Then the foot of the normal from the pole has coordinates $N(p, w)$. Using triangle ONP, the required equation is $r\cos(\theta - \omega) = p$. See Fig. 19-23.

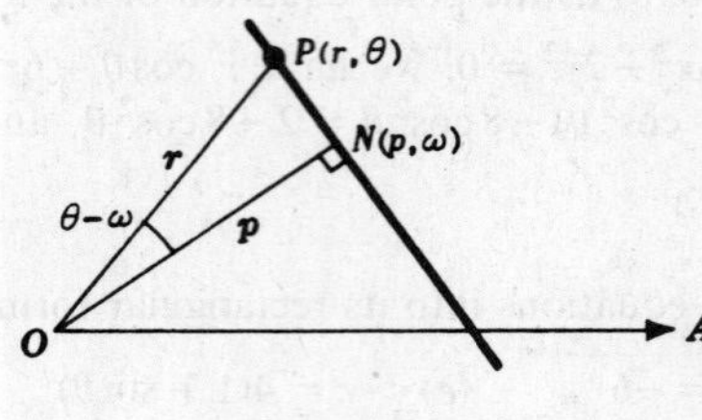

Fig. 19-23

19.8 Derive the polar equation of the circle of radius a whose center is at (c, γ).

Let $P(r, \theta)$ be an arbitrary point on the circle. See Fig. 19-24. Then [see Problem 19-2(*c*)]

$$r^2 + c^2 - 2rc\cos(\gamma - \theta) = a^2$$

or

$$r^2 - 2rc\cos(\gamma - \theta) + c^2 - a^2 = 0 \tag{1}$$

is the required equation.

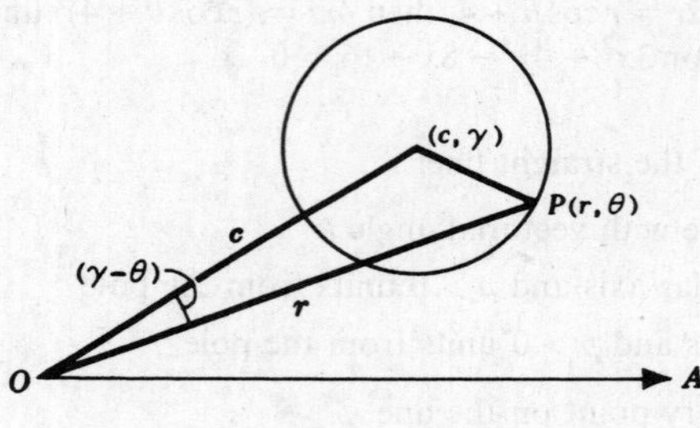

Fig. 19.24

The following special cases are of interest:

(*a*) If the center is at the pole, (*19.14*) becomes $r^2 = a^2$. Then $r = a$ or $r = -a$ is the equation of the circle of radius a with center at the pole.

(*b*) If $(c, \gamma) = (\pm a, 0°)$, (*19.14*) becomes $r = \pm 2a\cos\theta$. Thus, $r = 2a\cos\theta$ is the equation of the circle of radius a passing through the pole and having its center on the polar axis; $r = -2a\cos\theta$ is the equation of the circle of radius a passing through the pole and having its center on the polar axis extended.

(*c*) Similarly if $(c, \gamma) = (\pm a, 90°)$, we obtain $r = \pm 2a\sin\theta$ as the equation of the circle of radius a passing through the pole and having its center on the 90° axis or the 90° axis extended.

19.9 Derive the polar equation of a conic of eccentricity e, having a focus at the pole and p units from the corresponding directrix, when the axis on which the focus lies coincides with the polar axis.

In Fig. 19-25 a focus is at O and the corresponding directrix DD' is to the right of O. Let $P(r, \theta)$ be an arbitrary point on the conic. Now

$$\frac{OP}{PM} = e$$

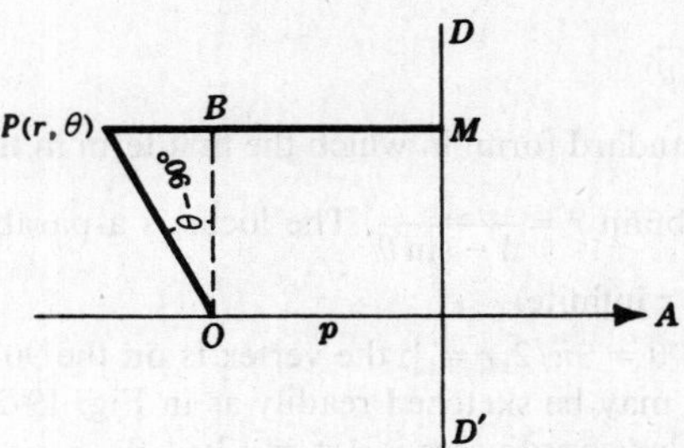

Fig. 19-25

where $OP = r$ and $PM = PB + BM = r\sin(\theta - 90°) + p = p - r\cos\theta$. Thus

$$\frac{r}{p - r\cos\theta} = e, \qquad r(1 + e\cos\theta) = ep, \qquad \text{and} \qquad r = \frac{ep}{1 + e\cos\theta}.$$

It is left for the reader to derive the equation $r = \dfrac{ep}{1 - e\cos\theta}$ when the directrix DD' lies to the left of O.

Similarly it may be shown that the polar equation of a conic of eccentricity e, having a focus at the pole and p units from the corresponding directrix, is

$$r = \frac{ep}{1 \pm e\sin\theta}$$

where the positive sign (negative sign) is used when the directrix lies above (below) the pole.

19.10 Find the locus of the third vertex of a triangle whose base is a fixed line of length $2a$ and the product of the other two sides is the constant b^2.

Take the base of the triangle along the polar axis with the midpoint of the base at the pole. The coordinates of the end points of the base are $B(a, 0)$ and $C(a, \pi)$. Denote the third (variable) vertex by $P(r, \theta)$. See Fig. 19-26.

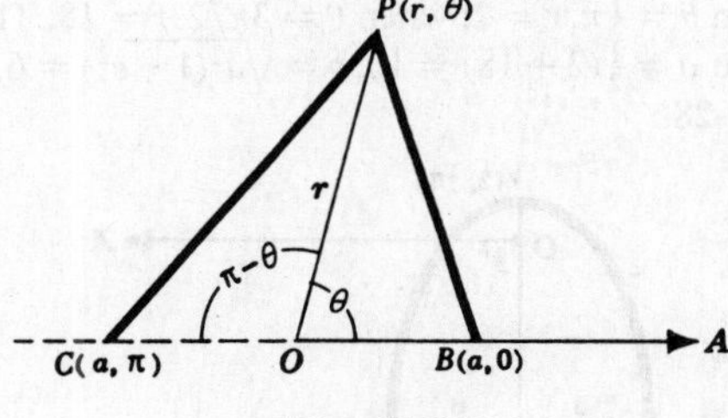

Fig. 19-26

From the triangle BOP, $(BP)^2 = r^2 + a^2 - 2ar\cos\theta$ and from the triangle COP, $(CP)^2 = r^2 + a^2 - 2ar\cos(\pi - \theta) = r^2 + a^2 + 2ar\cos\theta$. Now $(BP)(CP) = b^2$; hence

$$(r^2 + a^2 - 2ar\cos\theta)(r^2 + a^2 + 2ar\cos\theta) = (b^2)^2 = b^4$$

Then

$$(r^2 + a^2)^2 - 4a^2r^2\cos^2\theta = b^4$$

$$r^4 + 2a^2r^2(1 - 2\cos^2\theta) = r^4 - 2a^2r^2\cos 2\theta = b^4 - a^4$$

$$r^4 - 2a^2r^2\cos 2\theta + a^4\cos^2 2\theta = b^4 - a^4 + a^4\cos^4 2\theta = b^4 - a^4\sin^2 2\theta$$

and the required equation is $r^2 = a^2\cos 2\theta \pm \sqrt{b^4 - a^4\sin^2 2\theta}$.

19.11 Sketch the conic $r = \dfrac{3}{2 - 2\sin\theta}$.

To put the equation in standard form, in which the first term in the denominator is 1, divide numerator and denominator by 2 and obtain $r = \dfrac{\frac{3}{2}}{1 - \sin\theta}$. The locus is a parabola ($e = 1$) with focus at the pole. It opens upward ($\theta = \frac{1}{2}\pi$ makes r infinite).

When $\theta = 0, r = \frac{3}{2}$. When $\theta = 3\pi/2, r = \frac{3}{4}$; the vertex is on the 90° axis extended $\frac{3}{4}$ unit below the pole. With these facts the parabola may be sketched readily as in Fig. 19-27.

The equation in rectangular coordinates is $4x^2 = 12y + 9$.

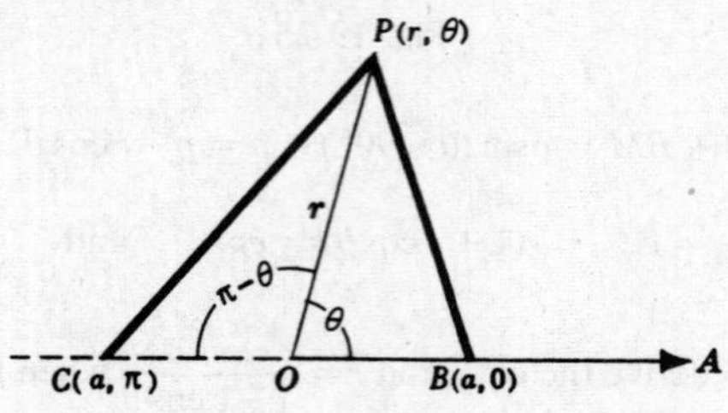

Fig. 19-27

19.12 Sketch the conic $r = \dfrac{18}{5 + 4\sin\theta}$.

After dividing numerator and denominator by 5, we have

$$r = \frac{\frac{18}{5}}{1 + \frac{4}{5}\sin\theta}$$

The locus is an ellipse ($e = \frac{4}{5}$) with a focus at the pole.

Since an equivalent equation is obtained when θ is replaced by $\pi - \theta$, the ellipse is symmetric with respect to the 90° axis; thus, the major axis is along the 90° axis. Since $ep = \frac{18}{5}$ and $ep = \frac{4}{5}, p = \frac{9}{2}$; the directrix is $\frac{9}{2}$ units above the pole. When $\theta = \frac{1}{2}\pi, r = 2$; when $\theta = 3\pi/2, r = 18$. Thus the vertices are 2 units above and 18 units below the pole. Since $a = \frac{1}{2}(2 + 18) = 10, b = \sqrt{a^2(1 - e^2)} = 6$. With these facts the ellipse may be readily sketched as in Fig. 19-28.

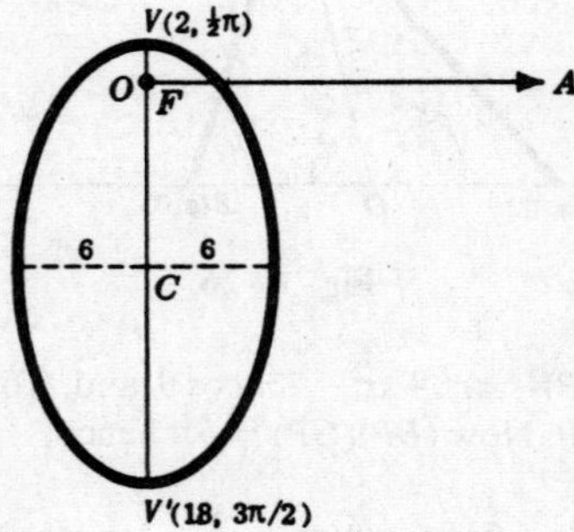

Fig. 19-28

In rectangular coordinates, the equation is $25x^2 + 9y^2 + 144y - 324 = 0$.

19.13 Sketch the conic $r = \dfrac{8}{3 - 5\cos\theta}$.

After dividing the numerator and denominator by 3, we have $r = \dfrac{\frac{8}{3}}{1 - \frac{5}{3}\cos\theta}$. The locus is a hyperbola ($e = \frac{5}{3}$) with a focus at the pole.

An equivalent equation is obtained when θ is replaced by $-\theta$; hence, the hyperbola is symmetric with respect to the polar axis and its transverse axis is on the polar axis. When $\theta = 0, r = -4$ and when $\theta = \pi, r = 1$; the vertices are respectively 4 units and 1 unit to the left of the pole. Then $a = \frac{1}{2}(4-1) = \frac{3}{2}$ and $b = \sqrt{a^2(e^2-1)} = 2$. The asymptotes, having slopes $\pm b/a = \pm\frac{4}{3}$, intersect at the center $\frac{1}{2}(1+4) = \frac{5}{2}$ units to the left of the pole. Since $ep = \frac{8}{3}$ and $e = \frac{5}{3}$, $p = \frac{8}{5}$; the directrix is $\frac{8}{5}$ units to the left of the pole.

In rectangular coordinates, the equation is $16x^2 - 9y^2 + 80x + 64 = 0$. See Fig. 19-29.

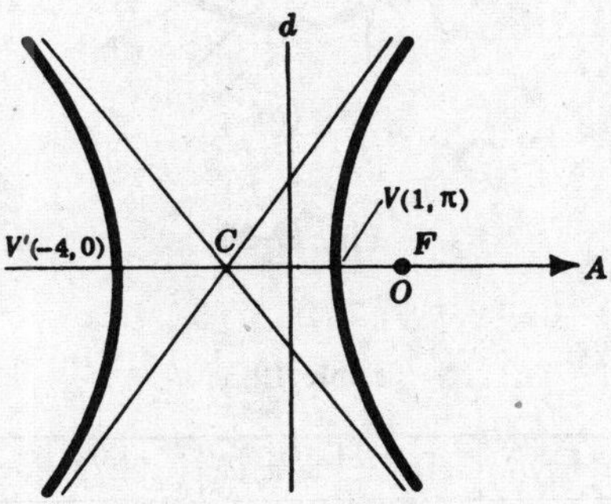

Fig. 19-29

19.14 Sketch the limacon $r = 2a\cos\theta + b$ when (*a*) $a = 2, b = 5$, (*b*) $a = 2, b = 4$, (*c*) $a = 2, b = 3$.

(*a*) The equation is $r = 4\cos\theta + 5$.

Symmetry. An equivalent equation is obtained when θ is replaced by $-\theta$; the locus is symmetric with respect to the polar axis.

Extent. Since r is real and finite for all values of θ, the locus is a closed curve. Since $\cos\theta$ is of period 2π, the complete locus is described as θ varies from 0 to 2π.

Directions at the Pole. When $r = 0, \cos\theta = -\frac{5}{4}$; the locus does not pass through the pole. After locating the points in Table 19.2 and making use of symmetry with respect to the polaris axis, we obtain the required curve shown in Fig. 19-30(*a*). The equation in rectangular coordinates is $(x^2 + y^2 - 4x)^2 = 25(x^2 + y^2)$.

Table 19.2

θ	0	$\pi/6$	$\pi/4$	$\pi/5$	$\pi/2$	$2\pi/3$	$3\pi/4$	$5\pi/6$	π
r	9.00	8.48	7.84	7.00	5.00	3.00	2.16	1.52	1.00

(*b*) The equation is $r = 4(1 + \cos\theta)$.

The locus a closed curve, symmetric with respect to the polar axis, and is completely described as θ varies from 0 to 2π.

When $r = 0, \cos\theta = -1$ and $\theta = \pi$. The locus passes through the pole and is tangent to the polar axis there.

After locating the points in Table 19.3 and making use of symmetry, we obtain the required curve shown in Fig. 19.30(*b*).

In rectangular coordinates the equation of the cardioid is $(x^2 + y^2 - 4x)^2 = 16(x^2 + y^2)$.

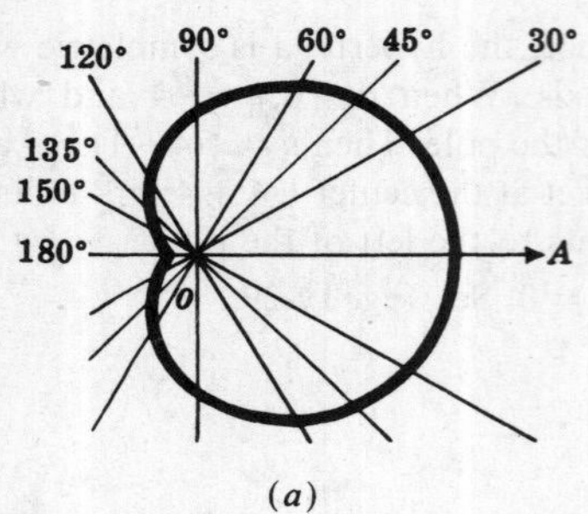

(a)

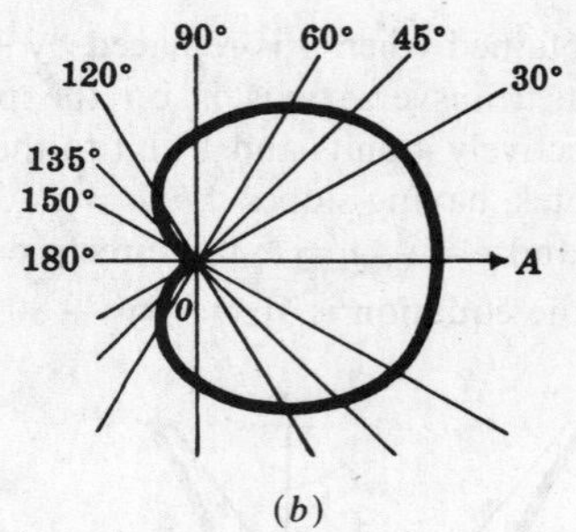

(b)

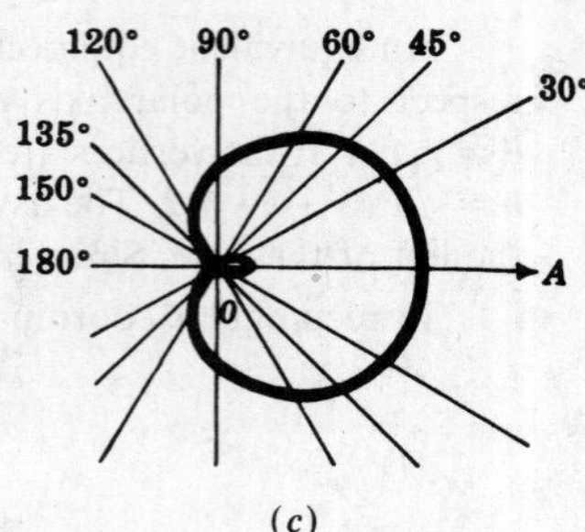

(c)

Fig. 19-30

Table 19.3

θ	0	$\pi/6$	$\pi/4$	$\pi/3$	$\pi/2$	$2\pi/3$	$3\pi/4$	$5\pi/6$	π
r	8.00	7.48	6.84	6.00	4.00	2.00	1.16	0.52	0

(c) The equation is $r = 4\cos\theta + 3$.

The locus is a closed curve, symmetric with respect to the polar axis, and is completely described as θ varies from 0 to 2π.

When $r = 0, \cos\theta = -\frac{3}{4} = -0.750$ and $\theta = 138°40', 221°20'$. The locus passes through the pole with tangents $\theta = 138°40'$ and $\theta = 221°20'$.

After putting in the these tangents as guide lines, locating the points in Table 19.4, and making use of symmetry, we obtain the required curve shown in Fig. 19.17(c).

The equation in rectangular coordinates is $(x^2 + y^2 - 4x)^2 = 9(x^2 + y^2)$.

Table 19.4

θ	0	$\pi/6$	$\pi/4$	$\pi/3$	$\pi/2$	$2\pi/3$	$3\pi/4$	$5\pi/6$	π
r	7.00	6.48	5.84	5.00	3.00	1.00	0.16	−0.48	−1.00

19.15 Sketch the rose $r = a\cos 3\theta$.

The locus is a closed curve, symmetric with respect to the polar axis. When $r = 0, \cos 3\theta = 0$ and $\theta = \pi/6, \pi/2, 5\pi/6, 7\pi/5, \ldots$; the locus passes through the pole with tangent lines $\theta = \pi/6, \theta = \pi/2$, and $\theta = 5\pi/6$ there.

The variation of r as θ changes is shown in Table 19.5.

Table 19.5

θ	3θ	r
0 to $\pi/6$	0 to $\pi/2$	a to 0
$\pi/6$ to $\pi/3$	$\pi/2$ to π	0 to $-a$
$\pi/3$ to $\pi/2$	π to $3\pi/2$	$-a$ to 0
$\pi/2$ to $2\pi/3$	$3\pi/2$ to 2π	0 to a
$2\pi/3$ to $5\pi/6$	2π to $5\pi/2$	a to 0
$5\pi/6$ to π	$5\pi/2$ to 3π	0 to $-a$

Caution. The values plotted are (r, θ) not $(r, 3\theta)$. The curve starts at a distance a to the right of the pole on the polar axis, passes through the pole tangent to the line $\theta = \pi/6$, reaches the tip of a loop when $\theta = \pi/3$, passes through the pole tangent to the line $\theta = \pi/2$, and so on. The locus is known as a three-leaved rose.

The rectangular equation is $(x^2 + y^2)^2 = ax(x^2 + 3y^2)$. See Fig. 19-31.

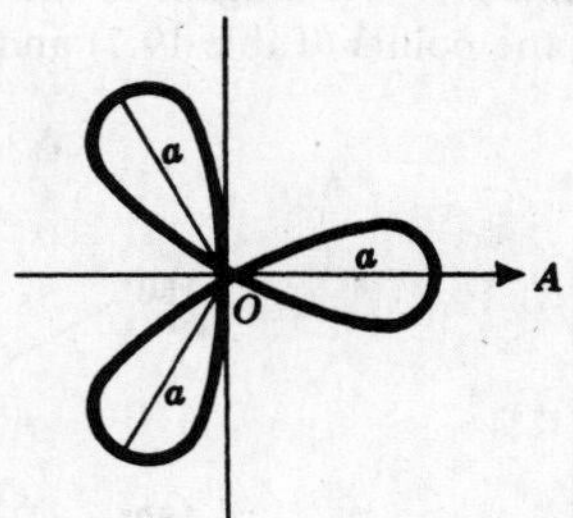

Fig. 19-31

In general, the roses $r = a \sin n\theta$ and $r = a \cos n\theta$ consist of n leaves when n is an *odd* integer.

19.16 Sketch the rose $r = a \sin 4\theta$.

The locus is a closed curve, symmetric with respect to the polar axis (an equivalent equation is obtained when θ is replaced by $\pi - \theta$ and r by $-r$), with respect to the 90° axis (an equivalent equation is obtained when θ is replaced by $-\theta$ and r by $-r$), and with respect to the pole (an equivalent equation is obtained when θ is replaced by $\pi + \theta$).

When $r = 0$, $\sin 4\theta = 0$ and $\theta = 0, \pi/4, \pi/2, 3\pi/4, \ldots$; the locus passes through the pole with tangent lines $\theta = 0, \theta = \pi/4, \theta = \pi/2$, and $\theta = 3\pi/4$ there.

The variation of r as θ changes from 0 to $\pi/2$ is shown in Table 19.6.

Table 19.6

θ	4θ	r
0 to $\pi/8$	0 to $\pi/2$	0 to a
$\pi/8$ to $\pi/4$	$\pi/2$ to π	a to 0
$\pi/4$ to $3\pi/8$	π to $3\pi/2$	0 to $-a$
$3\pi/8$ to $\pi/2$	$3\pi/2$ to 2π	$-a$ to 0

The complete curve, consisting of 8 leaves, can be traced by making use of the symmetry. See Fig. 19-32.

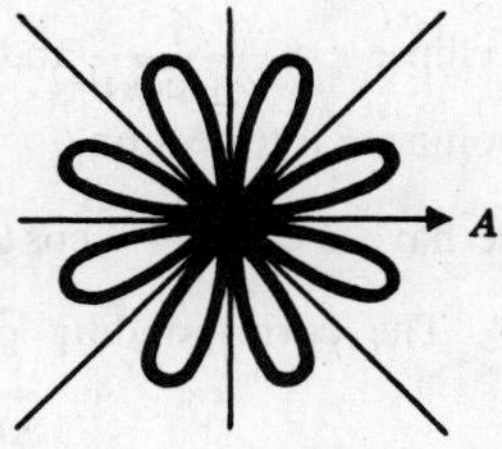

Fig. 19-32

In rectangular coordinates, the equation of the locus is $(x^2 + y^2)^5 = 16a^2(x^3y - xy^3)^2$.

In general, the roses $r = a \sin n\theta$ and $r = a \cos n\theta$ consist of $2n$ leaves when n is an *even* integer.

19.17 Sketch the locus of $r = \cos\frac{1}{2}\theta$.

Other equations of the locus are $-r = \cos\frac{1}{2}(\theta + \pi) = -\sin\frac{1}{2}\theta$ or $r = \sin\frac{1}{2}\theta, -r = \cos\frac{1}{2}(\theta - \pi) = \sin\frac{1}{2}\theta$ or $r = -\sin\frac{1}{2}\theta$, and $r = \cos\frac{1}{2}(\theta - 2\pi) = -\cos\frac{1}{2}\theta$.

The locus is a closed curve, symmetric with respect to the polar axis, the 90° axis, and the pole. It is completely described as θ varies from 0 to 4π.

When $r = 0, \theta = \pi, 3\pi, \ldots$; the line $\theta = \pi$ is tangent to the locus at the pole.

The curve is traced by locating the points (Table 19.7) and making use of symmetry. See Fig. 19-33.

Table 19.7

θ	$\frac{1}{2}\theta$	r
0	0	1.00
$\pi/6$	$\pi/12$	0.97
$\pi/3$	$\pi/6$	0.87
$\pi/2$	$\pi/4$	0.71
$2\pi/3$	$\pi/3$	0.50
$5\pi/6$	$5\pi/12$	0.26
π	$\pi/2$	0.00

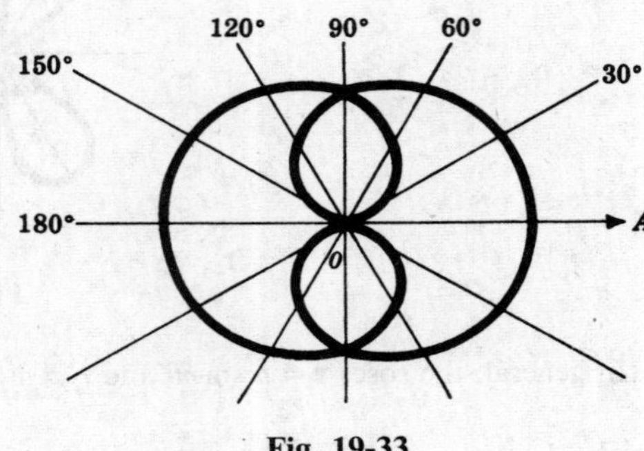

Fig. 19-33

19.18 Find the points of intersection of the limacon $r = 2\cos\theta + 4$ and the circle $r = 8\cos\theta$

From Fig. 19-34 there are two points of intersection.

Setting $2\cos\theta + 4 = 8\cos\theta$, we obtain $\cos\theta = \frac{2}{3}$; then $\theta = 48°10'$ and $311°50'$. (We solve for θ on the range $0 \le \theta < 2\pi$ since the limacon is completely described on this range.) The points of intersection are $(\frac{16}{3}, 48°10')$ and $(\frac{16}{3}, 311°50')$.

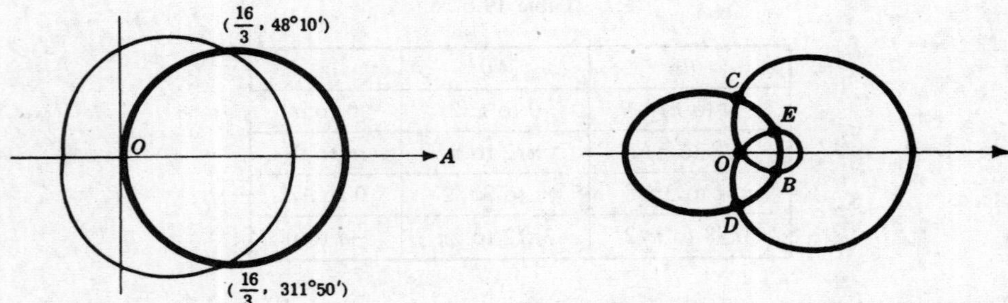

Fig. 19-34 **Fig. 19-35**

19.19 Find the points of intersection of the ellipse $r = \dfrac{4}{2 + \cos\theta}$ and the limacon $r = 4\cos\theta - 2$.

From Fig. 19-35 there are four points of intersection.

Setting $\dfrac{4}{2 + \cos\theta} = 4\cos\theta - 2$, we have $2\cos^2\theta + 3\cos\theta - 4 = 0$. Then $\cos\theta = \dfrac{-3 \pm \sqrt{41}}{4} = 0.851$ or -2.351, and $\theta = 31°40'$ and $328°20'$. The corresponding points are $E(-5 + \sqrt{41}, 31°40')$ and $B(-5 + \sqrt{41}, 328°20')$.

To obtain the other two points, we solve the equation of the ellipse with another equation $r = 4\cos\theta + 2$ (see Example 5) of the limacon. From $4\cos\theta + 2 = \dfrac{4}{2 + \cos\theta}$, we obtain $2\cos^2\theta + 5\cos\theta = (\cos\theta)(2\cos\theta + 5) = 0$. Then $\cos\theta = 0$ and $\theta = \frac{1}{2}\pi$ and $3\pi/2$. The corresponding points are $C(2, \pi/2)$ and $D(2, 3\pi/2)$.

[NOTE: When sketching $r = 4\cos\theta - 2$, the coordinates of C were found as $(-2, 3\pi/2)$ and those of D as $(-2, \pi/2)$.]

19.20 Find $\dfrac{dy}{dx}$ at $\theta = \dfrac{\pi}{6}$ for $r = 2(1 + \cos\theta)$.

$$\frac{dy}{dx} = \frac{(-2\sin\theta)(\sin\theta) + 2(1+\cos\theta)(\cos\theta)}{(-2\sin\theta)(\cos\theta) - 2(1+\cos\theta)(\sin\theta)} \text{ when } \theta = \tfrac{\pi}{6}.$$

Ans. -1

19.21 Derive Eq. (*19.6*)

Using the product rule of differentiation, we have

$$\frac{d\vec{r}}{dt} = \frac{d(r\hat{r})}{dt} = \frac{dr}{dt}\hat{r} + r\frac{d\hat{r}}{dt} = \frac{dr}{dt}\hat{r} + r\frac{d\theta}{dt}\hat{\theta}$$

Substituting (*19.6*) for $\frac{d\hat{r}}{dt}$, we conclude that:

$$\frac{d\vec{r}}{dt} = \frac{dr}{dt}\hat{r} + r\frac{d\theta}{dt}\hat{\theta}$$

19.22 Express a circle of radius 5 that is centered on the point (10,8) in both Cartesian and polar coordinates.

In Cartesian coordinates the circle can be expressed as $5 = \sqrt{(x-10)^2 + (y-8)^2}$ or $25 = (x-10)^2 + (y-8)^2$. Rearranging, this becomes $y = 8 \pm \sqrt{25 - (x-10)^2}$, where x ranges over values for which the radical yields real values. By inspection, this corresponds to values $5 \le x \le 15$. The previous equation then yields the corresponding values of y, which corresponds to points above and below the horizontal line bisecting the circle.

19.23 Express the rectangular coordinate (140,325,500) in spherical coordinates.

The transformation between a rectangular coordinate (x_1, y_1, z_1) and the corresponding spherical coordinate (r, θ, ϕ) is

$$x_1 = r\sin\theta\cos\phi$$
$$y_1 = r\sin\theta\sin\phi$$
$$z_1 = r\cos\theta$$

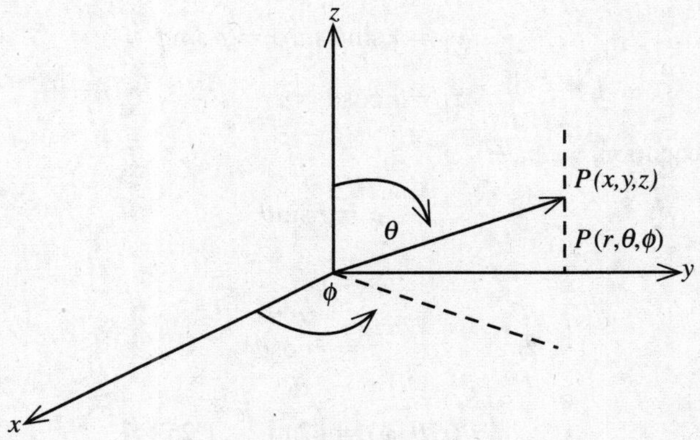

Fig. 19-36

Combining the above first two equations, we get:

$$\frac{y_1}{x_1} = \tan\phi \Rightarrow \phi = \tan^{-1}\frac{y_1}{x_1}$$

From the figure, we see that $r = \sqrt{x_1^2 + y_1^2 + z_1^2}$

From the figure, we can also see that $\tan\theta = \frac{\sqrt{x_1^2 + y_1^2}}{z_1} \Rightarrow \theta = \tan^{-1}\frac{\sqrt{x_1^2 + y_1^2}}{z_1}$.

Therefore, the point can be expressed in spherical coordinates as

$$(r, \theta, \phi) = \left(\sqrt{x_1^2 + y_1^2 + z_1^2}, \tan^{-1}\frac{\sqrt{x_1^2 + y_1^2}}{z_1}, \phi = \tan^{-1}\frac{y_1}{x_1}\right)$$

Substituting in $x_1 = 140$, $y_1 = 325$, and $z_1 = 500$, we get:

$$(r, \theta, \phi) = \left(\sqrt{140^2 + 325^2 + 500^2}, \tan^{-1}\frac{\sqrt{140^2 + 325^2}}{500}, \phi = \tan^{-1}\frac{325}{140}\right)$$

$$= (613, .616, 1.16)$$

19.24 Convert $(r, \theta, \phi) = \left(2m, \frac{\pi}{6}, 1.25\right)$ from spherical to cylindrical coordinates.

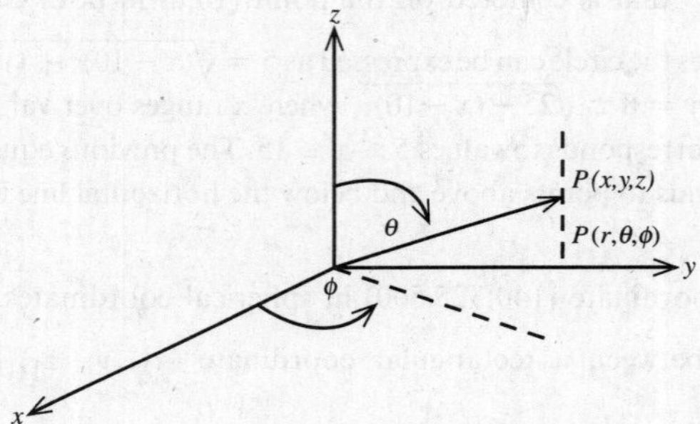

Fig. 19-37

The angle ϕ is defined the same in both spherical and cylindrical coordinates. Using the previous equations defining those systems, we have:

$$x_1 = r\sin\theta\cos\phi = \rho\cos\phi$$

$$y_1 = r\sin\theta\sin\phi = \rho\sin\phi$$

$$z_1 = r\cos\theta = z$$

From the first equation, we have

$$\rho = r\sin\theta$$

$$\phi = \phi$$

$$z = r\cos\theta$$

$$(r, \theta, \phi) = \left(2\,\text{m}, \frac{\pi}{6}, 1.25\right)$$

Substituting the values of the point in spherical coordinates, we get:

$$\rho = (2\,\text{m})\sin\frac{\pi}{6} = 1\,\text{m}$$
$$\phi = 1.25$$
$$z = r\cos\theta = (2\,\text{m})\cos\frac{\pi}{6} = 1.732\,\text{m}$$

19.25 Express the point $(r, \phi, z) = (12.8, 127°, -4.3)$ in spherical coordinates.

Combining the equations for spherical and cylindrical coordinates, we get:

$$x_1 = \rho\cos\phi = r\sin\theta\cos\phi$$
$$y_1 = \rho\sin\phi = \sin\theta\sin\phi$$
$$z_1 = z = r\cos\theta$$

We want to express the point—originally expressed in cylindrical coordinates (ρ, ϕ, z)—in terms of spherical coordinates (r, θ, ϕ).

From the definition of r and ρ, we have:

$$r = \sqrt{\rho^2 + z^2}$$

Combining or rearranging these equations, we get:

$$r\sin\theta = \rho$$
$$r\cos\theta = z$$

Dividing the first equation above by the second, we get:

$$\tan\theta = \frac{\rho}{z} \Rightarrow \theta = \tan^{-1}\left(\frac{\rho}{z}\right)$$

Using $(\rho, \phi, z) = (12.8, 127°, -4.3)$ in these equations then yields:

$$r = \sqrt{\rho^2 + z^2} = \sqrt{(12.8)^2 + (-4.3)^2} = 13.50\,\text{m}$$
$$\phi = 127°$$
$$\theta = \tan^{-1}\left(\frac{\rho}{z}\right) = \tan^{-1}\left(\frac{12.8}{-4.3}\right) = -71°$$

19.26 Convert the point $(r, \theta, \phi) = \left(10, \frac{\pi}{2}, 0\right)$ from spherical to Cartesian coordinates.

$$x = r\sin\theta\cos\phi = 10\sin\frac{\pi}{2}\cos 0 = 10(1)(1) = 10$$
$$y = r\sin\theta\sin\phi = 10\sin\frac{\pi}{2}\sin 0 = 0$$
$$z = r\cos\theta = 10\cos\frac{\pi}{2} = 10(0) = 0$$

19.27 Convert the point $(r, \phi, z) = (12.2, 1.5, 1.8)$ from cylindrical to Cartesian coordinates.

Using the transformation equation between the two coordinate systems, we have immediately:

$$x = r\cos\phi = 12.2\cos(1.5) = .9$$
$$y = r\sin\phi = 12.2\sin(1.5) = 12.2$$
$$z = z = 1.8$$

19.28 Find the surface area of a sphere of radius R using spherical coordinates

$$A = \int \mathrm{d}A = \int R^2\,\mathrm{d}\cos\theta\,\mathrm{d}\phi = R^2 \int_0^{\pi} \mathrm{d}\theta\sin\theta \int_0^{2\pi} \mathrm{d}\phi = R^2(-\cos\theta|_0^{\pi})2\pi = 2\pi R^2(1-(-1)) = 4\pi R^2$$

19.29 Find the volume of a sphere of radius R using spherical coordinates.

$$V = \int \mathrm{d}^3 r = \int r^2\,\mathrm{d}\cos\theta\,\mathrm{d}\phi = \int_0^R r^2\,\mathrm{d}r \int_0^{\pi} \sin\theta\,\mathrm{d}\theta \int_0^{2\pi} \mathrm{d}\phi = \left(\frac{R^3}{3}\right)(-\cos\theta|_0^{\pi})(2\pi) = \left(\frac{R^3}{3}\right)(1-(-1))2\pi$$
$$= \frac{4}{3}\pi R^3$$

19.30 Find the speed of a particle that moves in an elliptic orbit according to

$$\vec{r} = \hat{x}\mathrm{d}\cos\omega t + \hat{y}\mathrm{b}\sin\omega t$$

The speed of the particle is

$$v = \sqrt{v_x^2 + v_y^2} = \sqrt{\mathrm{d}^2\cos^2\omega t + \mathrm{b}^2\sin^2\omega t}$$

19.31 Find the speed of the particle in the previous problem in the case that the particle moves in a circle— i.e., d = b.

In this case,

$$v = \sqrt{v_x^2 + v_y^2} = \sqrt{\mathrm{d}^2\cos^2\omega t + \mathrm{d}^2\sin^2\omega t} = \mathrm{d}\sqrt{\cos^2\omega t + \sin^2\omega t} = \mathrm{d}$$

19.32 A particle moves along the straight line path $x = 10$ in a plane at a speed of 5 m/s. What are its polar coordinates?

We know that

$$x = 10$$
$$y = 5\,\mathrm{m/s}$$
$$r = \sqrt{x^2 + y^2} = \sqrt{10^2 + (5t)^2} = \sqrt{100 + 25t^2}$$
$$\tan\theta = \frac{y}{x} \Rightarrow \theta = \tan^{-1}\frac{y}{x} = \tan^{-1}\frac{5t}{10} = \tan^{-1}\left(\frac{t}{2}\right)$$

19.33 A particle moves in a helix that, in cylindrical coordinates, is expressed as:

$$r = R$$
$$\phi = \omega t$$
$$z = kt$$

Find the velocity and acceleration of the particle in cylindrical coordinates?

The velocity and acceleration are given by

$$\vec{v} = \dot{r}\hat{r} + r\dot{\phi}\hat{\phi} + \dot{z}\hat{z} \qquad (19.15)$$

From the equations preceding (19.15), we have:

$$r = R \Rightarrow \dot{r} = 0$$
$$\phi = \omega t \Rightarrow \dot{\phi} = \omega$$
$$z = kt \Rightarrow \dot{z} = k$$

Substituting the above expressions into (19.15), then yields

$$\vec{v} = \dot{r}\hat{r} + r\dot{\phi}\hat{\phi} + \dot{z}\hat{z} = R\omega\hat{\phi} + k\hat{z}$$

and

$$\vec{a} = (\ddot{r} - r\dot{\phi}^2)\hat{r} + (2\dot{r}\dot{\phi} + r\ddot{\phi})\hat{\phi} + \ddot{z}\hat{z} = -R\omega^2\hat{r}$$

Supplementary Problems

19.34 Find the rectangular coordinates of P whose polar coordinates are

(a) $(-2, 45°)$ (b) $(3, \pi)$ (c) $(2, \pi/2)$ (d) $(4, 2\pi/3)$

Ans. (a) $(-\sqrt{2}, -\sqrt{2})$ (b) $(-3, 0)$ (c) $(0, 2)$ (d) $(-2, 2\sqrt{3})$

19.35 Find a set of polar coordinates of P whose rectangular coordinates are

(a) $(1, \sqrt{3})$ (b) $(0, -5)$ (c) $(1, -1)$ (d) $(-12, 5)$

Ans. (a) $(2, \pi/3)$ (b) $(5, 3\pi/2)$ (c) $(\sqrt{2}, 7\pi/4)$ (d) $(13, \pi - \text{Arctan}\,\frac{5}{12})$

19.36 Transform each of the following rectangular equations into polar form:

(a) $x^2 + y^2 = 16$ (b) $y^2 - x^2 = 9$ (c) $x = 4$ (d) $y = \sqrt{3}x$ (e) $xy = 12$
(f) $(x^2 + y^2)x = 4y^2$

Ans. (a) $r = 4$ (c) $r \cos\theta = 4$ (e) $r^2 \sin 2\theta = 24$
(b) $r^2 \cos 2\theta + 9 = 0$ (d) $\theta = \pi/3$ (f) $r = 4 \tan\theta \sin\theta$

19.37 Transform each of the following polar equations into rectangular form:

(a) $r \sin\theta = -4$ (c) $r = 2\cos\theta$ (e) $r = 1 - 2\cos\theta$ (f) $r = \dfrac{4}{1 - 2\sin\theta}$
(b) $r = -4$ (d) $r = \sin 2\theta$

Ans. (a) $y = -4$ (c) $x^2 + y^2 - 2x = 0$ (e) $(x^2 + y^2 + 2x)^2 = x^2 + y^2$
(b) $x^2 + y^2 = 16$ (d) $(x^2 + y^2)^3 = 4x^2y^2$ (f) $x^2 - 3y^2 - 16y - 16 = 0$

19.38 Derive the polar equation $r = \dfrac{ep}{1 \pm e \sin\theta}$ of the conic of eccentricity e with a focus at the pole and with corresponding directrix p units from the focus.

19.39 Write the polar equation of each of the following:

(a) Straight line bisecting the second and fourth quadrants
(b) Straight line through $(4, 2\pi/3)$ and perpendicular to the polar axis
(c) Straight line through $N(3, \pi/6)$ and perpendicular to the radius vector of N

(*d*) Circle with center at $C(4, 3\pi/2)$ and radius = 4

(*e*) Circle with center at $C(-4,0)$ and radius = 4

(*f*) Circle with center at $C(4, \pi/3)$ and radius = 4

(*g*) Parabola with focus at the pole and directrix $r = -\Delta \sec\theta$

(*h*) Parabola with focus at the pole and vertex at $V(3, \pi/2)$

(*i*) Ellipse with eccentricity $\frac{3}{4}$, one focus at the pole, and the corresponding directrix 5 units above the polar axis

(*j*) Ellipse with one focus at the pole, the other focus at $(8, \pi)$, and eccentricity = $\frac{2}{3}$

(*k*) Hyperbola with eccentricity = $\frac{3}{2}$, one focus at the pole, and the corresponding directrix 5 units to the left of the 90° axis

(*l*) Hyperbola, conjugate axis = 24 parallel to and below the polar axis, transverse axis = 10, and one focus at the pole

Ans. (*a*) $\theta = 3\pi/4$ (*c*) $r\cos(\theta - \pi/6) = 3$ (*e*) $r = -8\cos\theta$
(*b*) $r\cos\theta = -2$ (*d*) $r = -8\sin\theta$ (*f*) $r = 8\cos(\theta - \pi/3)$
(*g*) $r = \dfrac{4}{1-\cos\theta}$ (*i*) $r = \dfrac{15}{4+3\sin\theta}$ (*k*) $r = \dfrac{15}{2-3\cos\theta}$
(*h*) $r = \dfrac{6}{1+\sin\theta}$ (*j*) $r = \dfrac{10}{3+2\cos\theta}$ (*l*) $r = \dfrac{144}{5-13\sin\theta}$

19.40 Discuss and sketch:

(*a*) $r = \sin(\theta - 45°) = -2$
(*b*) $r = 10\sin\theta$
(*c*) $r = -6\cos\theta$
(*d*) $r = \dfrac{8}{2-\sin\theta}$
(*e*) $r = \dfrac{6}{1+2\cos\theta}$
(*f*) $r = \dfrac{2}{1-\cos\theta}$
(*g*) $r = 2 - 4\cos\theta$
(*h*) $r = 4 - 2\cos\theta$
(*i*) $r^2 = 9\cos 2\theta$
(*j*) $r^2 = 16\sin 2\theta$
(*k*) $r = 2\cos 2\theta$
(*l*) $r = 4\sin 2\theta$
(*m*) $r = 2a\tan\theta\sin\theta$
(*n*) $r = 4\tan^2\theta\sec\theta$
(*o*) $r = \cos\frac{3}{2}\theta$
(*p*) $r = 2\theta$
(*q*) $r = a/\theta$

19.41 Find the complete intersection of

(*a*) $r = 2\cos\theta, r = 1$ (*b*) $r^2 = 4\cos 2\theta, r = 2\sqrt{2}\sin\theta$ (*c*) $r = 1 + \sin\theta, r = \sqrt{3}\cos\theta$

Ans. (a) $(1, \pi/3), (1, 5\pi/3)$ (b) $(0,0), (\sqrt{2}, \pi/6), (\sqrt{2}, 5\pi/6)$ (c) $(0,0), (\frac{3}{2}, \pi/6)$

19.42 Find $\dfrac{dy}{dx}$ for $r = \sin\theta$ at $\theta = \pi/2$.

Ans. 0

19.43 Consider a circle of radius 10 centered on the Cartesian coordinates (40,6) Express the circle in plane polar coordinates.

Ans. $r^2 - 92r\cos\theta + 1536 = 0$

19.44 Express the rectangular coordinates (54.0,13.0) in plane polar coordinates.

Ans. $(r, \theta) = (55.5, .236)$

19.45 Express the rectangular coordinates (13,7,-4) in spherical coordinates.

Ans. $(r, \theta, \phi) = (15.3, 1.84, .494)$

19.46 Express the rectangular coordinates (65,35,8) in cylindrical coordinates.

Ans. $(r, \phi, z) = (74, 0.5, 8)$

19.47 Express the cylindrical coordinates $(r,\ \phi,\ z)\ =\ (45.,\ 3.2,\ 1.8)$ in spherical coordinates.

Ans. $(r, \theta, \phi) = (48, .97, 3.2)$

Chapter 20

Multivariate Calculus

FUNCTIONS OF MULTIPLE VARIABLES. In the preceding chapters, we have focused on relationships among physical quantities in which a quantity—such as the gravitational force between two objects—depends upon another quantity, such as the distance between the objects. In such a situation, we speak of the distance as the *independent variable* and the force between the objects as the *dependent variable*. In this particular case, there is only one independent variable (the distance) determining the value of the dependent variable.

In fact, physical quantities typically depend on more than one variable. For example, the gravitational force between the Sun and the Earth depends not only on the distance between the two masses but also upon the masses themselves:

$$F = \frac{-GM\mathrm{m}}{r^2} \tag{20.1}$$

where F is the gravitational force, M is the mass of the Sun, m is the mass of the Earth and r is the distance between their centers.

Because the force depends on more than one quantity, it is an example of a *multivariate function*. In preceding chapters, in problems involving gravitational force, we assumed that the masses were constant and only considered the variation of force with distance.

DIFFERENTIATION OF MULTIVARIATE FUNCTIONS. If we want to understand, for example, how the gravitational force is affected by changes in the Sun–Earth distance—we would consider only the variation of the force F with distance r, treating all other quantities as if they were constant. We take the derivative of F with respect to the particular variable r (called the partial derivative of F with respect to r and written $\frac{\partial F}{\partial r}$):

$$\frac{\partial F}{\partial r} = \frac{-2GM\mathrm{m}}{r^3} = \frac{-2\left(6.67\times 10^{-11}\,\frac{\mathrm{N\,m^2}}{\mathrm{kg^2}}\right)(2\times 10^{30}\,\mathrm{kg})(6\times 10^{24}\,\mathrm{kg})}{(150\times 10^{6}\,\mathrm{m})^3} = 4.7\times 10^{20}\,\frac{\mathrm{N}}{\mathrm{m}} \tag{20.2}$$

This tells us that the total gravitational force between the Earth and the Sun varies by about $4.7 \cdot 10^{20}$ N for every meter change in the distance between the two bodies (provided the change in distance is not too great).

But F does not depend solely upon distance. As indicated in equation (*20.1*), it also depends, for example, upon the mass of the Sun. That mass steadily decreases, due to the conversion of mass into energy through nuclear fusion deep in its interior, eventually radiating from the Sun as light.

EXAMPLE 1. What is the differential change in the gravitation force with time due to the differential loss of mass from the Sun $\frac{dM}{dt}$ (approximately $4 \cdot 10^9$ kg/s)?

To compute the differential change in the gravitational force $\frac{\partial F}{\partial t}$ knowing the differential mass loss per unit time $\frac{dM}{dt}$, we first recognize that

$$\frac{\text{change in force}}{\text{time}} = \frac{\text{change in force}}{\text{change in mass}} \cdot \frac{\text{change in mass}}{\text{time}}$$

Putting this into equation form, we have:

$$\frac{\partial F}{\partial t} = \frac{\partial F}{\partial M}\frac{dM}{dt}$$

(Note that the partial derivative symbols are used with F, which depends upon more than one variable, while the mass of the Sun, M, is considered to be solely a function of time so that $\frac{dM}{dt}$ is an ordinary derivative.)

We need to find $\frac{\partial F}{\partial M}$, the differential change of the force F with respect to the solar mass M. We find it by differentiating F in equation (*20.1*), now regarding M, rather than the distance r, as the independent variable and regarding all other quantities as constant. Using the tools of differentiation developed in previous chapters, we simply find:

$$\frac{\partial F}{\partial M} = \frac{G\,m}{r^2}$$

and, using the above expression for $\frac{\partial F}{\partial t}$, we have finally that

$$\frac{\partial F}{\partial t} = \frac{\partial F}{\partial M}\frac{\partial M}{\partial t} = \frac{G\,m}{r^2}\left(4 \cdot 10^9\,\frac{\text{kg}}{\text{s}}\right) = \frac{\left(6.67 \times 10^{-11}\frac{\text{Nm}^2}{\text{kg}^2}\right)(5.98 \cdot 10^{24}\,\text{kg})\left(4 \cdot 10^9\,\frac{\text{kg}}{\text{s}}\right)}{(1.50 \cdot 10^6\,\text{m})^2} = 7.1 \times 10^{11}\,\frac{\text{N}}{\text{s}}$$

Hence, the total gravitational force between the Earth and the Sun currently decreases by approximately this rate. (This may seem large but is a small reduction per second to the total force of $F = \frac{-GMm}{r^2} \approx 3.6 \times 10^{32}$ N between the Sun and the Earth.)

INTEGRATION OF MULTIVARIATE FUNCTIONS. The above discussion focused on small changes—with respect to distance or mass in the above case—of the gravitational force law and, by extension, on the differentiation of multivariate functions. There are other situations where one needs to add up—or integrate—the contributions from such a function.

For example, suppose we want to know the mass M of a rectangular block with a density that varies within the volume V of the block. Recall that the *average* density ρ of an object of mass M and volume V is M/V. However, the density of our block need not be constant throughout its volume. The density could vary throughout the volume of the block, depending upon its position along the length of the block, along its width, and along its depth. For a small chunk of the block, we know that the density ρ is defined as the mass per unit volume. Hence, $M = \rho V$, when the density is constant. When the density is not constant but varies with position within the block, then we must add up the separate contributions.

Within a small volume ΔV centered about the coordinates (x, y, z) and containing a mass Δm, the *local* density $\rho = \rho(x, y, z) = \frac{\Delta m}{\Delta V}$. If we consider subdividing the block into small chunks of mass Δm_i, volume ΔV_i, each with its own local density ρ_i, then

$$M = \sum_{i=1}^{n} m_i = \sum_{i=1}^{n} \rho_i \Delta V_i$$

where ρ_i is the density of the ith mass element and ΔV_i is the volume of that element.*

**A word about units:* Note that because we are computing a mass, the answer needs to have the dimensions of a mass. The mass is equal to the local density (in this case, measured in units of $\frac{\text{kg}}{\text{m}^3}$) multiplied by the volume element (measured in units of m^3). Hence, each chunk of mass is measured in units of:

$$[\Delta m] = [\rho(x, y, z)][dV] = [\rho(x, y, z)][dx][dy][dz] = \frac{\text{kg}}{\text{m}^3}\text{m} \cdot \text{m} \cdot \text{m} = \text{kg}$$

In order to add up these small elements, we can choose Cartesian coordinates x, y, and z as indicated in Fig. 20-1, with the origin at one of the corners of the block.

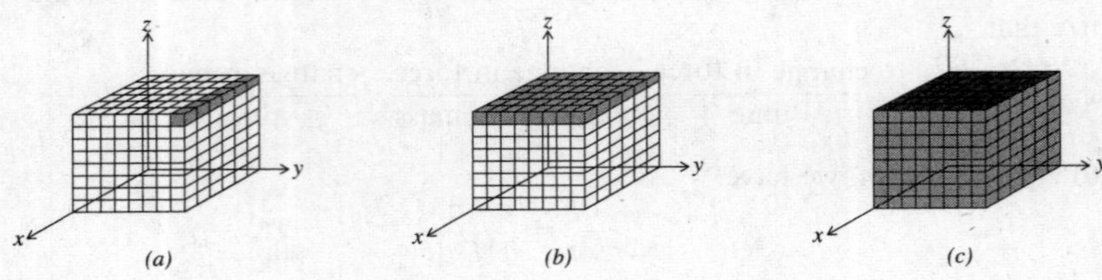

Fig. 20-1

To find the total mass of the block, we add up all of these small elements by adding up the separate contributions along the x, y and z directions. In principle, this can be done in any order. For convenience, we add up the contributions in the order indicated in Fig. 20-1 above: first as small blocks that are part of a thin rectangular solid (Fig. 20-1(*a*)), then adding in neighboring contributions to form a thin slab (Fig. 20-1(*b*)) and finally including all thin slabs to comprise the total volume of the mass (Fig. 20-1(*c*)).

More specifically, then, we have the following steps:

(*1*) We first add up all elements along the length x of the block for some constant value of the width y and depth z. That is consider just the row indicated in Fig. 20-1(*a*). The mass of this narrow rectangular solid is apparently:

$$M_{rectangular\ pipe} = \sum_{i=1}^{n} \rho_{ijk} \Delta V$$

where ρ_{ijk} is the mass density of a particular cube element in location (i, j, k) and $\Delta V = \Delta x \Delta y \Delta z$ is the volume of each element (all assumed to be equal).

(*2*) We now add together all the strips in the same range of x but now including strips that span the full range of y (Fig. 20-1(*b*)):

$$M_{thin\ slab} = \sum_{k=1}^{p} \sum_{i=1}^{n} \rho_{ijk} \Delta V$$

(*3*) Finally, we add together the volumes of all of the thin slabs that lie at different values of z in order to arrive at the mass of the rectangular cube (Fig. 20-1(*c*)):

$$M = \lim_{\delta x \to 0} \lim_{\delta y \to 0} \lim_{\delta z \to 0} \left(\sum_{k=1}^{p} \sum_{j=1}^{n} \sum_{i=1}^{m} \rho_{ijk} \Delta x \Delta y \Delta z \right) \tag{20.3}$$

If we replace the sums in equation (*20.3*) by integrals over a differential volume element $dV (= dx dy dz)$ and the discrete indices by continuous variables, then we get

$$M = \int dM = \int_{\text{volume}} \rho dV = \int_{x_i}^{x_f} dx \int_{y_i}^{y_f} dy \int_{z_i}^{z_f} dz\, \rho(x, y, z)$$

where $dM = \rho(V) dV = \rho(x, y, z)\, dx dy dz$ and the density ρ is a continuous function of the location in the volume. We are, in effect, performing the same three-step process of adding up strips and ribbons but using an infinite number of infinitesimal pieces.

EXAMPLE 2. Find the total mass of a rectangular solid bounded by $1 \le x \le 7$, $-2 \le y \le 6$, and $4 \le z \le 12$, assuming the following density function:

$$\rho(x, y, z) = 3x^2 + 2y^3 + 4z.$$

Then,

$$M = \int dM = \int_{x_i}^{x_f} dx \int_{y_i}^{y_f} dy \int_{z_i}^{z_f} dz\, \rho(x, y, z) = \int_{x_i}^{x_f} dx \int_{y_i}^{y_f} dy \int_{z_i}^{z_f} dz\, 3x^2 + 2y^3 + 4z \qquad (20.4)$$

So we must now perform three integrations—over the variables x, y, and z in the ranges specified. The order in which we perform these integrations will not affect our final result (just as the order in which one adds up the chunks of mass will not affect the total mass). As a result, the answer does not depend upon whether we write $\int_{x_i}^{x_f} dx \int_{y_i}^{y_f} dy \int_{z_i}^{z_f} dz$ or $\int_{z_i}^{z_f} dz \int_{x_i}^{x_f} dx \int_{y_i}^{y_f} dy$ or some other ordering of this expression.

Typically, one evaluates an expression like equation (*20.4*) using the variables in the order from right to left, so that one would first integrate over dz—treating x and y as constants and evaluating the integral at the endpoints of z. One would then take the resulting expression and integrate it over y, now treating x as a constant and evaluating the integral at the endpoints of y. Finally, one would perform the integration over x, evaluating the integral at the endpoints of x.

Proceeding along these lines, we first integrate over the variable z, treating x and y as constant:

$$= \int_{x_i}^{x_f} dx \int_{y_i}^{y_f} dy \left[\int_{z_i}^{z_f} dz\,(3x^2 + 2y^3 + 4z) \right] = \int_{x_i}^{x_f} dx \int_{y_i}^{y_f} dy \left[3x^2 z + 2y^3 z + 2z^2 \right] \Big|_4^{12}$$

$$= \int_{x_i}^{x_f} dx \int_{y_i}^{y_f} dy \left[3x^2(12) + 2y^3(12) + 2(12)^2 \right] - \int_{x_i}^{x_f} dx \int_{y_i}^{y_f} dy \left[3x^2(4) + 2y^3(4) + 2(4)^2 \right]$$

$$= \int_{x_i}^{x_f} dx \int_{y_i}^{y_f} dy \left[3x^2(12) + 2y^3(12) + 2(12)^2 \right] - \left[3x^2(4) + 2y^3 4 + 2(4)^2 \right]$$

$$= \int_{x_i}^{x_f} dx \int_{y_i}^{y_f} dy \left[24x^2 + 16y^3 + 256 \right]$$

Having now integrated over x, we now proceed with the integration over y:

$$= \int_{x_i}^{x_f} dx \int_{y_i}^{y_f} dy \left[24x^2 + 16y^3 + 256 \right]$$

$$= \int_{x_i}^{x_f} dx \int_{-2}^{6} dy \left[24x^2 + 16y^3 + 256 \right] = \int_{x_i}^{x_f} dx \left[24x^2 y + 4y^4 + 256y \right] \Big|_{-2}^{6}$$

$$= \int_{x_i}^{x_f} dx \left[(24x^2(6) + 4(6)^4 + 256(6)) - (24x^2(-2) + 4(-2)^4 + 256(-2)) \right]$$

$$= \int_{x_i}^{x_f} dx\,(192x^2 + 7,296)$$

Having completed the integration over y, we now integrate over x to find the total mass of the object:

$$= \int_{x_i}^{x_f} dx\,(192x^2 + 7,296) = \left[\frac{192x^3}{3} + 7,296x \right] \Big|_1^7$$

$$= \left(\frac{192(7)^3}{3} + 7,296(7) \right) - \left(\frac{192(1)^3}{3} + 7,296(1) \right)$$

$$= 21,952 + 51,072 - 64 - 7,296 = 65,664 \text{ kg}$$

where we assume that the density function $\rho(x, y, z)$ and coordinates are expressed in SI units.

We provide these examples only as a starting point to indicate the power of being able to differentiate and integrate multivariate functions. The larger point is that physical relationships typically involve several variables, and that we need to understand how to deal with them. The topic is a vast one, and the problems that follow provide only a glimpse into this deep and complex realm, the full range of which is beyond the scope of this book. In particular, the specific techniques used for multiple integration depend upon the physical situation, what information is being sought, the particular functional dependence(s) and the choice of coordinate systems.

Solved Problems

20.1. Find the partial derivatives of f with respect to x, y, z, and t for each of the following functions.

(a) $f(x, y, z, t) = 3x^2y + 9xzt + 12;$
(b) $f(x, y, z, t) = (x-5)^{42} + 3\cos(\omega t + \phi) + \sqrt{z^2y + 4};$
(c) $f(x, y, z, t) = 12e^{(x-5)^2}(y + zt)$

(a) $\dfrac{\partial f}{\partial x} = 6xy + 9zt$

$\dfrac{\partial f}{\partial y} = 3x^2$

$\dfrac{\partial f}{\partial z} = 9xt$

$\dfrac{\partial f}{\partial t} = 9xz$

(b) $\dfrac{\partial f}{\partial x} = 42(x-5)^{41}$

$\dfrac{\partial f}{\partial y} = \frac{1}{2}(z^2y + 4)^{-\frac{1}{2}}z^2$

$\dfrac{\partial f}{\partial z} = \frac{1}{2}(z^2y + 4)^{-\frac{1}{2}}(2z) = z(z^2y + 4)^{-\frac{1}{2}}$

$\dfrac{\partial f}{\partial t} = -3\,\omega\sin(\omega t + \phi)$

(c) $\dfrac{\partial f}{\partial x} = 12\,e^{(x-5)^2}(2(x-5))(y + zt) = 24\,e^{(x-5)^2}(x-5)(y + zt)$

$\dfrac{\partial f}{\partial y} = 12\,e^{(x-5)^2}$

$\dfrac{\partial f}{\partial z} = 12\,t\,e^{(x-5)^2}$

$\dfrac{\partial f}{\partial t} = 12\,z\,e^{(x-5)^2}$

20.2 An electric field has the form $\vec{E}(x, y, z) = 10\sin(3x + 5t)\,\hat{x} + x^2y\,\hat{y} + (3y^2z - 4xz)\,\hat{z}$, where $\vec{E}$ is measured in volts per meter (V/m). Consider a point in the field located at (10, 2, 3), where the coordinates are in centimeters. What is the rate of change at this point of *(a)* the z component of electric field along the z direction?; *(b)* the y component of electric field along the z direction?; and *(c)* the y component of electric field along the z direction?

(a) $\dfrac{\partial E_z}{\partial z} = \dfrac{\partial}{\partial z}(3y^2z - 4xz) = 3y^2 - 4x = 3\cdot 2^2 - 4\cdot 10 = -28\,\text{V/m}^2$

(note that the units are V/m^2—that is volts per meter—indicating the instantaneous rate of *change* along the z direction of the electric field (which is measured in volts per meter))

(b) $\dfrac{\partial E_x}{\partial y} = \dfrac{\partial}{\partial y}(10\sin(3x + 5t)) = 0$ (since E_x does not depend upon y)

(c) $\dfrac{\partial E_y}{\partial x} = \dfrac{\partial}{\partial y}(x^2y) = x^2 = 100\,\text{V/m}^2$

20.3 Find $\int_0^5 \mathrm{d}x \int_5^9 \mathrm{d}y \int_4^6 \mathrm{d}z\, g(x,y,z)$ when

(*a*) $g(x,y,z) = 10x^2 + y^2 + 5$
(*b*) $g(x,y,z) = y\,e^x + 4z^2$
(*c*) $g(x,y,z) = A\sin(2\pi x)y^5(z-5)$

(*a*) $$\int_0^5 \mathrm{d}x \int_5^9 \mathrm{d}y \int_4^6 \mathrm{d}z\, g(x,y,z) = \int_0^5 \mathrm{d}x \int_5^9 \mathrm{d}y \int_4^6 \mathrm{d}z\,(10x^2 + y^2 + 5)$$

$$= \int_0^5 \mathrm{d}x \int_5^9 \mathrm{d}y\,(10x^2 + y^2 + 5)z\Big|_{z=4}^{z=6}$$

$$= 2\int_0^5 \mathrm{d}x \int_5^9 \mathrm{d}y\,(10x^2 + y^2 + 5) = 2\int_0^5 \mathrm{d}x\left(10x^2y + \frac{y^3}{3} + 5y\right)\Bigg|_{y=5}^{y=9}$$

$$= 2\int_0^5 \mathrm{d}x\left(90x^2 + \frac{9^3}{3} + 45\right) - 2\int_0^5 \mathrm{d}x\left(50x^2 - \frac{5^3}{3} - 25\right)$$

$$= \int_0^5 \mathrm{d}x\,(80x^2 + 486 + 40 - 83.33) = \left(80\frac{x^3}{3} + 442.66x\right)\Bigg|_0^5$$

$$= 80\cdot\frac{5^3}{3} + 442.66\cdot 5 = 5{,}547$$

(*b*) $$\int_0^5 \mathrm{d}x \int_5^9 \mathrm{d}y \int_4^6 \mathrm{d}z\, g(x,y,z) = \int_0^5 \mathrm{d}x \int_5^9 \mathrm{d}y \int_4^6 \mathrm{d}z\,(ye^x + 4z^2) = \int_0^5 \mathrm{d}x \int_5^9 \mathrm{d}y\left(zye^x + 4\frac{z^3}{3}\right)\Bigg|_{z=4}^{z=6}$$

$$= \int_0^5 \mathrm{d}x \int_5^9 \mathrm{d}y\left(2ye^x + 4\frac{6^3}{3} - 4\frac{4^3}{3}\right) = \int_0^5 dx\left[y^2e^x + \left(4\frac{6^3}{3} - 4\frac{4^3}{3}\right)y\right]\Bigg|_{y=5}^{y=9}$$

$$= \int_0^5 \mathrm{d}x\left[56e^x + 4\left(4\frac{6^3}{3} - 4\frac{4^3}{3}\right)\right]$$

$$= \left[56e^x + 4\left(4\frac{6^3}{3} - 4\frac{4^3}{3}\right)x\right]\Bigg|_0^5 = 56(e^5 - 1) + 80\left(\frac{6^3}{3} - \frac{4^3}{3}\right)$$

$$= 8311.14 - 1 + 7466.7 = 15{,}777$$

(*c*) $$\int_0^5 \mathrm{d}x \int_5^9 \mathrm{d}y \int_4^6 \mathrm{d}z\, g(x,y,z) = \int_0^5 \mathrm{d}x \int_5^9 \mathrm{d}y \int_4^6 \mathrm{d}z\left(x + y^2 + \sqrt{z}\right) = \int_0^5 \mathrm{d}x \int_5^9 \mathrm{d}y\left(xz + y^2z + \frac{2}{3}z^{3/2}\right)\Bigg|_{z=4}^{z=6}$$

$$= \int_0^5 \mathrm{d}x \int_5^9 \mathrm{d}y\left(6x + 6y^2 + \frac{2}{3}6^{3/2}\right) - \int_0^5 \mathrm{d}x \int_5^9 \mathrm{d}y\left(4x + 4y^2 + \frac{2}{3}4^{3/2}\right)$$

$$= \int_0^5 \mathrm{d}x \int_5^9 \mathrm{d}y\left[2x + 2y^2 + \frac{2}{3}(6^{3/2} - 4^{3/2})\right] = \int_0^5 \mathrm{d}x\left[2xy + \frac{2}{3}y^3 + \frac{2}{3}(6^{3/2} - 4^{3/2})y\right]\Bigg|_{y=5}^{y=9}$$

$$= \int_0^5 dx\left[18x + \frac{2}{3}9^3 + 6(6^{3/2} - 4^{3/2})\right] - \int_0^5 dx\left[10x + \frac{2}{3}5^3 + \frac{10}{3}(6^{3/2} - 4^{3/2})\right]$$

$$= \int_0^5 dx\left[8x + \frac{2}{3}(9^3 - 5^3) + \frac{8}{3}(6^{3/2} - 4^{3/2})\right] = 8\frac{x^2}{2} + \left[\frac{2}{3}(9^3 - 5^3) + \frac{8}{3}(6^{3/2} - 4^{3/2})\right]x\Big|_0^5$$

$$= 8\frac{5^2}{2} + \left[\frac{2}{3}(9^3 - 5^3) + \frac{8}{3}(6^{3/2} - 4^{3/2})\right]5 = 100 + [2013.33 + 89.29] = 2,202$$

20.4 The temperature in a rectangular room is described by the function $T(x, y, z) = 68 + 2e^{(-3(x^2+y^2+4z^2)}$. The room occupies the volume $-3 \leq x \leq 3$, $-5 \leq y \leq 5$, $-6 \leq z \leq 6$, where the dimensions are in meters. What is the instantaneous rate of change of temperature along each of the coordinate axes at the center of the room?

We need to first find the partial derivatives of T with respect to x, y, and z and then substitute in the coordinates of the center of the room. This is straightforward:

$$\frac{\partial T}{\partial x} = \frac{\partial}{\partial x}\left(68 + 2e^{(-3(x^2+y^2+4z^2))}\right) = 2e^{(-3(x^2+y^2+4z^2))}(-6x) = -12x\,e^{(-3(x^2+y^2+4z^2))} = 0$$

$$\frac{\partial T}{\partial y} = \frac{\partial}{\partial y}\left(68 + 2e^{(-3(x^2+y^2+4z^2))}\right) = 2e^{(-3(x^2+y^2+4z^2))}(-6y) = -12x\,e^{(-3(x^2+y^2+4z^2))} = 0$$

$$\frac{\partial T}{\partial z} = 68 + 2e^{(-3(x^2+y^2+4z^2))} = 2e^{(-3(x^2+y^2+4z^2))}(-24z) = -48z\,e^{(-3(x^2+y^2+4z^2))} = 0$$

Therefore, the instantaneous temperature change along each of the coordinate axes in the center of the room is zero. (This is also apparent by examining the symmetry of the temperature function, which does not depend upon the sign of any of the coordinates.)

20.5 An object of mass m travels in a potential of the form $V(x, y, z) = x^2 + xy - yz^3 - 2z$, where x, y, and z are in meters.

(*a*) Find the force on the particle given that $\vec{F}_x = -\frac{\partial V}{\partial x}\hat{x}$, $\vec{F}_y = -\frac{\partial V}{\partial y}\hat{y}$, $\vec{F}_z = -\frac{\partial V}{\partial z}\hat{z}$

$$\vec{F}_x = -\frac{\partial V}{\partial x}\hat{x} = (2x + y)\hat{x}$$

$$\vec{F}_y = -\frac{\partial V}{\partial y}\hat{y} = (x + z^3)\hat{y}$$

$$\vec{F}_z = -\frac{\partial V}{\partial z}\hat{z} = (-y - 2)\hat{z}$$

(*b*) Find the acceleration of the particle.
The components of the acceleration $\vec{a}$ are:

$$a_x = \frac{F_x}{m} \quad a_y = \frac{F_y}{m} \quad a_z = \frac{F_z}{m}$$

(c) Where is the force equal to zero?

$$F_x = 0 \rightarrow 2x + y = 0 \tag{1}$$

$$F_y = 0 \rightarrow x + z^3 = 0 \tag{2}$$

$$F_z = 0 \rightarrow -y - 2 = 0 \tag{3}$$

From (3), $y = -2$. Substituting this into (1) yields $x = 1$. Therefore, $1 + z^3 = 0$. So, $z = (-1)^{1/3} = -1$. Therefore, the force is zero at $(x, y, z) = (1, -2, -1)$.

20.6 The displacement y of a string from its equilibrium position is described by

$$y = A\sin(kx - \omega t), \qquad (1)$$

where y is in cm, the wave number k is in cm^{-1}, x is the position along the string in cm, ω is the angular speed in s^{-1}, and t is the time in s.

(*a*) What is the physical interpretation of $\partial y/\partial x$? Find its value at $x = 5$ cm and $t = 30$ s.
(*b*) What is the physical interpretation of $\partial y/\partial t$? Find its value at the same horizontal position x and time t given in (*a*).
(*c*) Determine what relationship between the angular speed ω, the wave number k and the wave speed c must exist in order for the wave equation $\frac{\partial^2 y}{\partial x^2} - \frac{1}{c^2}\frac{\partial^2 y}{\partial t^2} = 0$ to be obeyed.

(*a*) The displacement y of the string depends upon both the time t and the horizontal position x according to equation (*1*). If we consider a "snapshot" of the string at a particular time t, then we see a sine wave that stretches horizontally due to the presence of x in $\sin(kx - \omega t)$.

The partial derivative $\frac{\partial y}{\partial x}$ is simply the change in y with respect to the change in x for a particular time t. That is, it gives the slope of the wave with respect to the horizontal (x) axis at time t (which can also be interpreted as the slope of the tangent to the wave at that point).

Differentiating equation (*1*) with respect to x while regarding t as fixed, we get:

$$\frac{\partial y}{\partial x} = kA\cos(kx - \omega t)$$

(*b*) On the other hand, if we consider some particular location x along the string and focus on its vertical motion, then we will see it oscillate up and down. Mathematically, this is due to the presence of the time t in $\sin(kx - \omega t)$. $\frac{\partial y}{\partial t}$ is then the change in the vertical displacement of the wave with respect to time at a particular horizontal location x. Once again, we differentiate equation (*1*) but this time with respect to the time t:

$$\frac{\partial y}{\partial t} = -\omega A\cos(kx - \omega t)$$

In summary, the partial derivatives $\partial y/\partial x$ and $\partial y/\partial t$ tell, respectively, how fast the displacement is changing with respect to the horizontal position at a particular time, and how fast it is changing with respect to time at a particular horizontal position.

(*c*) Using the above expressions, we get $\frac{\partial^2 y}{\partial x^2} = Ak^2 A\sin(kx - \omega t)$ and $\frac{\partial^2 y}{\partial t^2} = \omega^2 A\sin(kx - \omega t)$

Therefore, $\frac{\partial^2 y}{\partial x^2} - \frac{1}{c^2}\frac{\partial^2 y}{\partial t^2} = Ak^2\sin(kx - \omega t) - \frac{1}{c^2}A\omega^2\sin(kx - \omega t) = A\sin(kx - \omega t)\left(k^2 - \frac{\omega^2}{c^2}\right) = 0$

Since $\sin(kx - \omega t)$ will generally be nonzero, $k^2 - \frac{\omega^2}{c^2} = 0$ so that $\omega = ck$ is the necessary condition for the wave equation to be obeyed. This relationship among the angular speed ω, the wave speed c, and the wave number k occurs throughout wave phenomena.

20.7 The temperature in a room is described by the function $T(x, y, z) = 68 + .005\,x^2z^4 - .004\,xy^2 + .010\,yz^3$. Find the rate of change of the temperature at $(x, y, z) = (3, 5, 6)$ in the x, y, and z directions.

Using the above expression for $T(x, y, z)$, we can directly find the partial derivatives $\frac{\partial T}{\partial x}, \frac{\partial T}{\partial y}$, and $\frac{\partial T}{\partial z}$:

$$\frac{\partial T}{\partial x} = .010x\,z^4 - .004y^2$$

$$\frac{\partial T}{\partial y} = -.008y + .010z^3$$

$$\frac{\partial T}{\partial z} = .020x^2z^3 + .030yz^2$$

Since these expressions give the change of temperature, respectively, in the x, y, and z directions, the temperature change at the indicated location is:

$$\frac{\partial T}{\partial x} = .010(3)6^4 - .004(5)^2 = 38.8$$
$$\frac{\partial T}{\partial y} = -.008(5) + .010(6)^3 = -.04 + 2.16 = 2.1$$
$$\frac{\partial T}{\partial z} = .020(3)^2(6)^3 + .030(5)(6)^2 = 38.9 + 5.4 = 44.3$$

20.8 Find the total charge distributed throughout a sphere of radius R with a volume charge density of $\rho(r) = \rho_0 \frac{e^{-r/r_0}}{r^2}$

$$Q = \int \rho(r)\,dV = \int \rho_0 \frac{e^{-r/r_0}}{r^2}\,dV = \rho_0 \int \frac{e^{-r/r_0}}{r^2}\,dV$$

To integrate over the volume, we need to choose a convenient set of coordinates. For a problem such as this, involving a sphere, spherical coordinates provide a natural choice. In this coordinate system, recall (Chapter 19) that the volume element is

$$dV = \sin(\theta) r^2\, dr d\theta d\phi$$

Hence,

$$Q = \rho_0 \int \frac{e^{-r/r_0}}{r^2}\,dV = \rho_0 \int r^2 dr \int \sin(\theta)\,d\theta \int \frac{e^{-r/r_0}}{r^2}\,d\phi = \rho_0 \int dr \int \sin(\theta)\,d\theta \int e^{-r/r_0}\,d\phi$$

We need to integrate over three dimensions in order to find the total charge on this three-dimensional object. In this case, however, we now integrate over the variables r, θ, and ϕ instead of, in the Cartesian case, x, y, and z. However, we note that the charge density ρ only depends upon the radius r and not upon the angular variables θ or ϕ. As a result, the integration over those variables is straightforward (remember that the polar angle θ is integrated from $\theta = 0°$ to $180°$ (corresponding to $-1 \le \cos\theta \le 1$) and that the azimuthal angle ϕ is integrated from 0 to 2π).
Performing first the integration over ϕ, we get:

$$Q = \rho_0 \int e^{-r/r_0}\,dV = 2\pi\,\rho_0 \int dr \int \sin(\theta)\,d\theta\, e^{-r/r_0}$$

(since the integrand does not depend upon ϕ and $\int_0^{2\pi} d\phi = 2\pi$).
Next, we integrate over the polar angle θ:

$$Q = \rho_0 \int e^{-r/r_0}\,dV = 2\pi\rho_0 \int dr \int \sin(\theta)\,d\theta\, e^{-r/r_0} = -2\pi\rho_0 \int dr\, e^{-r/r_0} \cos(\theta)\Big|_0^{\pi}$$
$$= 4\pi\rho_0 \int dr\, e^{-r/r_0} = -r_0 4\pi\rho_0 e^{-r/r_0}\Big|_0^R = -4\pi r\rho_0(e^{-R/r_0} - 1)$$

So, the total charge distributed throughout the sphere is $Q = -4\pi r\rho_0(e^{-R/r_0} - 1)$

20.9 The location of the center of mass of a three-dimensional object is $\vec{R} = \frac{1}{M} \int_{\text{Volume}} \vec{r}\rho(\vec{r})\,d^3r$, where $M = \int_{\text{Volume}} \rho(\vec{r})\,d^3r$, where ρ is the mass density of the object.

Find the x coordinate of the center of mass of a rectangular solid with a density $\rho(x, y, z) = x^2 + 3z\sin(5\pi y)$. Assume SI units throughout.

$$x_{\mathrm{CM}} = \frac{1}{M} \int\limits_{\text{Volume}} x\rho(\vec{r})\,\mathrm{d}x\mathrm{d}y\mathrm{d}z = \frac{1}{M} \int\limits_{\text{Volume}} x(x^2 + 3z\sin(5\pi y))\,\mathrm{d}x\mathrm{d}y\mathrm{d}z$$

$$= \int \mathrm{d}x \int \mathrm{d}y \int \mathrm{d}z\, x(x^2 + 3z\sin(5\pi y)) = \int \mathrm{d}x \int \mathrm{d}y \int \mathrm{d}z\, x(x^2 + 3z\sin(5\pi y))$$

$$= \int \mathrm{d}x \int \mathrm{d}y\, x\left(x^2 z + 3\frac{z^2}{2}\sin(5\pi y)\right)\Bigg|_3^{10}$$

$$= \int \mathrm{d}x \int \mathrm{d}y\, x\left(7x^2 + 3\frac{91}{2}\sin(5\pi y)\right)$$

$$= \int \mathrm{d}x \int \mathrm{d}y\, x\left(7x^2 y - \frac{3}{5}\frac{91}{2}\cos(5\pi y)\right)\Bigg|_{-3}^{7}$$

$$= \int \mathrm{d}x \int x\left(7x^2(7-(-3)) - \frac{3}{5}\frac{91}{2}\cos(5\pi(7)) + \frac{3}{5}\frac{91}{2}\cos(5\pi(-3))\right)$$

$$= \int \mathrm{d}x\, x\left(70x^2 + \frac{3}{5}\frac{91}{2} - \frac{3}{5}\frac{91}{2}\right) = 70\int x^3 \mathrm{d}x$$

$$= \frac{70x^4}{4}\Bigg|_6^{13} = \frac{70(13^4 - 6^4)}{4} = 477{,}100 \text{ m}$$

where we have inserted dimensions in meters in accordance with the assumption that SI units are used.

20.10 What is the probability of finding an electron in the octant of the cube defined by $\{0<x<\frac{L}{2},\ 0<y<\frac{L}{2},\ 0<z<\frac{L}{2}\}$? Assume that the electron is confined to a three-dimensional cube of length L located at $0<x<L$, $0<y<L$, $0<z<L$ and that it has an electron probability density

$$|\psi|^2 = (2/L)^3 \sin^2\frac{\pi x}{L}\sin^2\frac{\pi y}{L}\sin^2\frac{\pi z}{L}.$$

In quantum mechanics, the probability of finding a particle within a particular volume of space is equal to the integral of its probability density $|\psi|^2$ over that volume.

In classical physics, we would expect the electron to have a 1/8 probability of being in any of the eight octants of the box. In quantum mechanics, this is not necessarily the case, and we must use the probability density given above. Doing so, we get:

$$P = \int\limits_{\text{Volume}} |\psi|^2\,\mathrm{d}V = (2/L)^3 \int_0^{L/2} \mathrm{d}x \int_0^{L/2} \mathrm{d}y \int_0^{L/2} \mathrm{d}z \sin^2\frac{\pi x}{L}\sin^2\frac{\pi y}{L}\sin^2\frac{\pi z}{L}$$

$$= (2/L)^3\left[\int_0^{L/2} \mathrm{d}x\sin^2\frac{\pi x}{L}\right]\left[\int_0^{L/2} \mathrm{d}y\sin^2\frac{\pi y}{L}\right]\left[\int_0^{L/2} \mathrm{d}z\sin^2\frac{\pi z}{L}\right]$$

Inspection of the above equation shows that the three integrals over x, y, and z are really the same integral, and so we can replace the above equation with:

$$P = \int\limits_{\text{Volume}} |\psi|^2\,\mathrm{d}V = (2/L)^3\left[\int_0^{L/2} \mathrm{d}x\sin^2\left(\frac{\pi x}{L}\right)\right]^3$$

Evaluating the integral, we get:

$$\int_0^{.5} \mathrm{d}x\sin^2\left(\frac{\pi x}{L}\right) = \frac{1}{2}\int_0^{L/2} \mathrm{d}x\left(1 - \cos\left(\frac{2\pi x}{L}\right)\right) = \frac{1}{2}\left(x - \frac{L}{2\pi}\sin\left(\frac{2\pi x}{L}\right)\right)\Bigg|_0^{L/2} = \frac{1}{2}\left(\frac{L}{2} - \frac{L}{2\pi}\sin(\pi)\right) = L/4$$

Therefore,

$$P = \int_{\text{Volume}} |\psi|^2 \, \mathrm{d}V = (2/L)^3 \left(\frac{L}{4}\right)^3 = \frac{1}{8}$$

Remarkably, this quantum mechanical result is identical to the classical result we predicted above, although in general this will not be the case.

20.11 Find the volume of a sphere.

$$V = \int \mathrm{d}V = \int \mathrm{d}^3 r = \int_0^R r^2 \, \mathrm{d}r \int_0^\pi \sin(\theta) \, \mathrm{d}\theta \int_0^{2\pi} \mathrm{d}\phi = 2\pi \int_0^R r^2 \, \mathrm{d}r \int_0^\pi \sin(\theta) \, \mathrm{d}\theta = -2\pi \int_0^R r^2 \mathrm{d}r \cos(\theta)\Big|_0^\pi$$

$$= -2\pi \int_0^R r^2 \, \mathrm{d}r(-1-1) = 4\pi \int_0^R r^2 \, \mathrm{d}r = 4\pi \frac{r^3}{3}\Big|_0^R = 4\pi\left(\frac{R^3}{3} - \frac{0^3}{3}\right) = \frac{4}{3}\pi R^3$$

20.12 Find the weight of dielectric material needed to be sandwiched into a set of cylindrical capacitors, each of which is 2 cm long, assuming that 500,000 capacitors are to be produced, that the material is to be sandwiched into a region that lies between 1.00 and 1.25 cm from the cylindrical axis of each unit, and that the density of material at the inner boundary is $\rho_0 = 2.5\,\text{kg/m}^3$ and increases in direct proportion to the distance from the axis.

The weight of dielectric material is given by
$W = Nmg$, where N is the number of capacitors and mg is the weight of dielectric material in each capacitor.

The mass of dielectric material m in each capacitor is given by integrating the density of material over the occupied portion of the cylindrical volume:

$$m = \int \rho(r) \, \mathrm{d}V$$

The geometry of the problem naturally lends itself to cylindrical coordinates, for which the volume element is

$$\mathrm{d}V = r \, \mathrm{d}r \, \mathrm{d}\theta \, \mathrm{d}z$$

Since the density is directly proportional to the distance from the axis and is equal to ρ_0 at the inner boundary, it is given by:
$\rho(r) = \rho_0 \frac{r}{r_1}$, where r_1 is the radius at the inner boundary.
Putting it all together, we get

$$W = Nmg = Ng \int \rho(r) \, \mathrm{d}V = Ng \int \rho_0 \frac{r}{r_1} \mathrm{d}V = N \int_0^L \mathrm{d}z \int_{r_1}^{r_2} \mathrm{d}r \int_0^{2\pi} \mathrm{d}\phi \, \rho_0 r \left(\rho_0 \frac{r}{r_i}\right)$$

where the limits of integration correspond to the volume occupied by the dielectric material.
Performing the integration, choosing first to integrate over the angular coordinate (although any other order would do as well), we get:

$$W = Ng \int_0^L \mathrm{d}z \int_{r_1}^{r_2} \mathrm{d}r \int_0^{2\pi} \mathrm{d}\phi \rho_0 r \left(\rho_0 \frac{r}{r_1}\right) = \frac{N\rho_0 g}{r_1} \int_0^L \mathrm{d}z \int_{r_1}^{r_2} \mathrm{d}r \int_0^{2\pi} d\phi \, r^2$$

$$= \frac{2\pi N \rho_0 g}{r_1} \int_0^L \mathrm{d}z \int_{r_1}^{r_2} \mathrm{d}r \, r^2 = \frac{2\pi N \rho_0 g}{r_1} \int_0^L \mathrm{d}z \int_{r_1}^{r_2} \mathrm{d}r \, r^2$$

Performing the remaining integrations over r and z, we get:

$$W = \frac{2\pi N\rho_0 g}{r_1}\int_0^L dz\int_{r_1}^{r_2} dr\, r^2 = \frac{2\pi N\rho_0 g}{r_1}\int_0^L dz\left(\frac{r_2^3}{3}-\frac{r_1^3}{3}\right) = \frac{2\pi N\rho_0 g}{r_1}\left(\frac{r_2^3}{3}-\frac{r_1^3}{3}\right)\int_0^L dz = \frac{2\pi N\rho_0 g L}{r_1}\left(\frac{r_2^3}{3}-\frac{r_1^3}{3}\right)$$

Finally, substituting in the values above, we get:

$$W = \frac{2(3.14)(500{,}000)\left(2.5\,\frac{\text{kg}}{\text{m}^3}\right)\left(9.8\,\frac{\text{m}}{\text{s}^2}\right)(.02\text{ m})}{.01\text{ m}}\left(\frac{(.0125\text{ m})^3}{3}-\frac{(.01\text{ m})^3}{3}\right) = 48.9\,N$$

20.13 Find the area of a circle.

$$A = \int_o^R r\,dr\int_0^{2\pi} d\theta = \int_o^R r\,dr\,(2\pi) = 2\pi\frac{R^2}{2} = \pi R^2$$

20.14 The radioactivity of a substance varies over a flat surface according to the law $A(r,\theta) = 32r^2 + 4e^{-3\theta}$, where the activity A is the number of disintegrations per second and r and θ are measured with respect to a fixed origin. Find the rate of change of activity (*a*) as a function of radius for a constant angle and (*b*) as a function of angle for a constant radius

(*a*) $\frac{\partial A}{\partial r} = 64\,r$

(*b*) $\frac{\partial A}{\partial \theta} = -12\,e^{-3\theta}$

20.15 The temperature over part of a hot surface is described by the function $T(r,\theta) = 325 + 3r^2\sin 4\theta$, where T is in SI units.

(*a*) Find the temperature at $r = 3$ m *and* $\theta = 25°$.
(*b*) Find the rate of change of temperature with radius at an angle of 25°.
(*c*) Find the rate of change of temperature with angle at a radius of 3 m.

(*a*) $T(r,\theta) = 325 + 3(3)^2\sin 4(25°) = 334$ K.
(*b*) Using the usual rules for taking partial derivatives, we take the derivative with respect to r while holding q constant:

$$\frac{\partial T}{\partial r} = \frac{\partial[3r^2\sin 4\theta]}{\partial r} = 6r\sin 4\theta = 6r\sin 100 = 5.91r\text{ K/ m.}$$

(*c*) We now take the partial derivative with respect to q while holding r constant, yielding: $\frac{\partial T}{\partial \theta} = \frac{\partial[3r^2\sin 4\theta]}{\partial \theta} = 12\,r^2\cos\theta = 12\cdot 3^2\cos\theta = 108\cos\theta$ K/°.

Supplementary Problems

20.16 The temperature in a room varies according to $T(x,y,z) = 3x^2yz + xy^3 + 4x^2z^2$. Find $\frac{\partial T}{\partial x}$, $\frac{\partial T}{\partial y}$, and $\frac{\partial T}{\partial z}$ at $(x,y,z) = (3,4,5)$.

Ans. $\frac{\partial T}{\partial x} = 1024$; $\frac{\partial T}{\partial y} = 279$; $\frac{\partial T}{\partial z} = 144$

20.17 The density of a gas varies according to $\rho(x,y,z) = 3x^2yz + xy^3 + 4x^2z^2$. Find $\frac{\partial \rho}{\partial x}$, $\frac{\partial \rho}{\partial y}$, and $\frac{\partial \rho}{\partial z}$ at $(x,y,z) = (3,4,5)$

Ans. $\frac{\partial \rho}{\partial x} = -4800$; $\frac{\partial \rho}{\partial y} = -1080$; $\frac{\partial \rho}{\partial z} = -144$

20.18 The pressure on a flat surface varies according to $p(r, \theta) = 4r^2(1 + 3\theta)$. Find $\frac{\partial p}{\partial r}$ and $\frac{\partial p}{\partial \theta}$.

Ans. $\frac{\partial p}{\partial r} = 8r(1 + 3\theta)$; $\frac{\partial p}{\partial \theta} = 12r^2$

20.19 The components of the electric field are related to the potential according to

$E_x = -\frac{\partial V}{\partial x}$; $E_y = -\frac{\partial V}{\partial y}$; $E_z = -\frac{\partial V}{\partial z}$

where V is the potential (in volts).

What are the components of the electric field given a potential that varies according to $V(x, y, z) = 3x^2(1 + 2y) \sin 2\pi z$?

Ans. $E_x = 3x^2$; $E_y = 6x^2 \sin 2\pi z$; $E_z = 6\pi x^2(1 + 2y) \cos 2\pi z$

20.20 Find the mass of a sphere of radius R with a density that varies as

$$\rho(r, \theta, \phi) = \frac{A}{r^2}.$$

Ans. $4\pi AR$

20.21 Find the total charge on a rectangular sheet on which the surface charge density varies according to $\sigma(x, y) = 12x^2(1 + y)$, where σ is measured in nanocoulombs (=10^{-9}C) per cm^2.

Ans. 10.4 μC

20.22 Find the moment of inertia of a cylinder with a uniform density ρ about its axis. The moment of inertia relates the angular momentum of an object to its rotational velocity. In the current context, the moment of inertia is defined by $I = \int \rho r^2 dV$, where the mass density ρ may vary throughout the volume and where r is the perpendicular distance from a volume element to the axis with respect to which I is calculated.

Ans. $\frac{1}{2}MR^2$

20.23 Find the moment of inertia of a sphere of radius R and uniform density ρ (see previous problem) about a diameter. (*Hint:* Decompose the sphere into infinitesimal disks, each consisting of a set of infinitesimal annular rings. Then integrate.)

Ans. $\frac{2}{5}MR^2$

20.24 Find the total charge, in coulombs, traveling down a wire of radius 0.5 cm in 10 s if the current density J (current per square meter cross-sectional area) varies with radius according to

$J = J_0(1 + \alpha r)$, where $J_0 = 1A/m^2$ and $\alpha = .3m^{-1}$ (the total charge is given by $Q = \int I\,dt$, where I is the current crossing a cross-sectional area A. I is related to the current density according to

$I = \int J\,dA$. Assume that the charge associated with both J and I is traveling parallel to the axis of the wire.

Ans. 900 μC

Chapter 21

Elementary Linear Algebra

DETERMINANTS OF ORDER TWO. The symbol $\begin{vmatrix} a_1 & b_1 \\ a_2 & b_2 \end{vmatrix}$, consisting of 2^2 numbers called *elements* arranged in two rows and two columns, is called a *determinant of order two*. The elements a_1 and b_2 are said to lie along the *principal diagonal*; the elements a_2 and b_1 are said to lie along the *secondary diagonal*. Row 1 consists of a_1 and b_1. Row 2 consists of a_2 and b_2. Column 1 consists of a_1 and a_2, and column 2 consists of b_1 and b_2.

The *value* of the determinant is obtained by forming the product of the elements along the principal diagonal and subtracting from it the product of the elements along the secondary diagonal; thus,

$$\begin{vmatrix} a_1 & b_1 \\ a_2 & b_2 \end{vmatrix} = a_1b_2 - a_2b_1$$

(See Problem 21.1.)

THE SOLUTION of the consistent and independent equations

$$\begin{cases} a_1x + b_1y = c_1 \\ a_2x + b_2y = c_2 \end{cases} \tag{21.1}$$

may be expressed as quotients of determinants of order two:

$$x = \frac{c_1b_2 - c_2b_1}{a_1b_2 - a_2b_1} = \frac{\begin{vmatrix} c_1 & b_1 \\ c_2 & b_2 \end{vmatrix}}{\begin{vmatrix} a_1 & b_1 \\ a_2 & b_2 \end{vmatrix}}, \qquad y = \frac{a_1c_2 - a_2c_1}{a_1b_2 - a_2b_1} = \frac{\begin{vmatrix} a_1 & c_1 \\ a_2 & c_2 \end{vmatrix}}{\begin{vmatrix} a_1 & b_1 \\ a_2 & b_2 \end{vmatrix}}$$

These equations are consistent and independent if and only if $\begin{vmatrix} a_1 & b_1 \\ a_2 & b_2 \end{vmatrix} \neq 0$. See Chapter 5.

EXAMPLE 1. Solve $\begin{cases} y = 3x + 1 \\ 4x + 2y - 7 = 0 \end{cases}$ using determinants.

Arrange the equations in the form (*19.1*): $\begin{cases} 3x - \ \ y = -1 \\ 4x + 2y = \ \ \ 7 \end{cases}$. The solution requires the values of three determinants:

The denominator, D, formed by writing the coefficients of x and y in order

$$D = \begin{vmatrix} 3 & -1 \\ 4 & 2 \end{vmatrix} = 3 \cdot 2 - 4(-1) = 6 + 4 = 10$$

The numerator of x, N_x, formed from D by replacing the coefficients of x by the constant terms

$$N_x = \begin{vmatrix} -1 & -1 \\ 7 & 2 \end{vmatrix} = -1 \cdot 2 - 7(-1) = -2 + 7 = 5$$

The numerator of y, N_y, formed from D by replacing the coefficients of y by the constant terms

$$N_y = \begin{vmatrix} 3 & -1 \\ 4 & 7 \end{vmatrix} = 3 \cdot 7 - 4(-1) = 21 + 4 = 25$$

Then $x = \dfrac{N_x}{D} = \dfrac{5}{10} = \dfrac{1}{2}$ and $y = \dfrac{N_y}{D} = \dfrac{25}{10} = \dfrac{5}{2}$. (See Problem 21.2.)

DETERMINANTS OF ORDER THREE. The symbol

$$\begin{vmatrix} a_1 & b_1 & c_1 \\ a_2 & b_1 & c_2 \\ a_3 & b_3 & c_3 \end{vmatrix},$$

consisting of 3^2 elements arranged in three rows and three columns, is called a *determinant of order three*. Its value is

$$a_1b_2c_3 + a_2b_3c_1 + a_3b_1c_2 - a_1b_3c_2 - a_2b_1c_3 - a_3b_2c_1$$

This may be written as

$$a_1(b_2c_3 - b_3c_2) - b_1(a_2c_3 - a_3c_2) + c_1(a_2b_3 - a_3b_2)$$

or

$$a_1\begin{vmatrix} b_2 & c_2 \\ b_3 & c_3 \end{vmatrix} - b_1\begin{vmatrix} a_2 & c_2 \\ a_3 & c_3 \end{vmatrix} + c_1\begin{vmatrix} a_2 & b_2 \\ a_3 & b_3 \end{vmatrix} \tag{21.2}$$

to involve three determinants of order two. Note that the elements which multiply the determinants of order two are the elements of the first row of the given determinant. (See Problem 21.3.)

THE SOLUTION of the system of consistent and independent equations

$$\begin{cases} a_1x + b_1y + c_1z = d_1 \\ a_2x + b_2y + c_2z = d_2 \\ a_3x + b_3y + c_3z = d_3 \end{cases}$$

in determinant form is given by

$$x = \frac{N_x}{D} = \frac{\begin{vmatrix} d_1 & b_1 & c_1 \\ d_2 & b_2 & c_2 \\ d_3 & b_3 & c_3 \end{vmatrix}}{\begin{vmatrix} a_1 & b_1 & c_1 \\ a_2 & b_2 & c_2 \\ a_3 & b_3 & c_3 \end{vmatrix}}, \qquad y = \frac{N_y}{D} = \frac{\begin{vmatrix} a_1 & d_1 & c_1 \\ a_2 & d_2 & c_2 \\ a_3 & d_3 & c_3 \end{vmatrix}}{D}, \qquad z = \frac{N_z}{D} = \frac{\begin{vmatrix} a_1 & b_1 & d_1 \\ a_2 & b_2 & d_2 \\ a_3 & b_3 & d_3 \end{vmatrix}}{D}.$$

The determinant D is formed by writing the coefficients of x, y, z in order, while the determinant appearing in the numerator for any unknown is obtained from D by replacing the column of coefficients of that unknown by the column of constants.

The system is consistent and independent if and only if $D \neq 0$.

EXAMPLE 2. Solve, using determinants: $\begin{cases} x + 3y + 2z = -13 \\ 2x - 6y + 3z = 32 \\ 3x - 4y - z = 12 \end{cases}$

The solution requires the values of four determinants:
The denominator,

$$D = \begin{vmatrix} 1 & 3 & 2 \\ 2 & -6 & 3 \\ 3 & -4 & -1 \end{vmatrix} = 1(6 + 12) - 3(-2 - 9) + 2(-8 + 18)$$

$$= 18 + 33 + 20 = 71$$

The numerator of x,

$$N_x = \begin{vmatrix} -13 & 3 & 2 \\ 32 & -6 & 3 \\ 12 & -4 & -1 \end{vmatrix} = -13(6 + 12) - 3(-32 - 36) + 2(-128 + 72)$$

$$= -234 + 204 - 112 = -142$$

The numerator of y,

$$N_y = \begin{vmatrix} 1 & -13 & 2 \\ 2 & 32 & 3 \\ 3 & 12 & -1 \end{vmatrix} = 1(-32 - 36) - (-13)(-2 - 9) + 2(24 - 96)$$

$$= -68 - 143 - 144 = -355$$

The numerator of z,

$$N_z = \begin{vmatrix} 1 & 3 & -13 \\ 2 & -6 & 32 \\ 3 & -4 & 12 \end{vmatrix} = 1(-72 + 128) - 3(24 - 96) + (-13)(-8 + 18)$$

$$= 56 + 216 - 130 = 142$$

Then $\quad x = \dfrac{N_x}{D} = \dfrac{-142}{71} = -2, \qquad y = \dfrac{N_y}{D} = \dfrac{-355}{71} = -5, \qquad z = \dfrac{N_z}{D} = \dfrac{142}{71} = 2$

(See Problems 21.4–21.5.)

A DETERMINANT of order n consists of n^2 numbers called elements arranged in n rows and n columns, and enclosed by two vertical lines. For example,

$$D_1 = |a_1| \qquad D_2 = \begin{vmatrix} a_1 & b_1 \\ a_2 & b_2 \end{vmatrix} \qquad D_3 = \begin{vmatrix} a_1 & b_1 & c_1 \\ a_2 & b_2 & c_2 \\ a_3 & b_3 & c_3 \end{vmatrix} \qquad D_4 = \begin{vmatrix} a_1 & b_1 & c_1 & d_1 \\ a_2 & b_2 & c_2 & d_2 \\ a_3 & b_3 & c_3 & d_3 \\ a_4 & b_4 & c_4 & d_4 \end{vmatrix}$$

are determinants of orders one, two, three, and four, respectively. In this notation the letters designate columns and the subscripts designate rows. Thus, all elements with letter c are in the third column and all elements with subscript 2 are in the second row.

THE MINOR OF A GIVEN ELEMENT of a determinant is the determinant of the elements which remain after deleting the row and the column in which the given element stands. For example, the minor of a_1 in D_4 is

$$\begin{vmatrix} b_2 & c_2 & d_2 \\ b_3 & c_3 & d_3 \\ b_4 & c_4 & d_4 \end{vmatrix}$$

and the minor of b_3 is

$$\begin{vmatrix} a_1 & c_1 & d_1 \\ a_2 & c_2 & d_2 \\ a_4 & c_4 & d_4 \end{vmatrix}$$

Note that the minor of a given element contains no element having either the letter or the subscript of the given element. (See Problem 21.6.)

THE VALUE OF A DETERMINANT of order one is the single element of the determinant. A determinant of order $n > 1$ may be expressed as the sum of n products formed by multiplying each element of any chosen row (column) by its minor and prefixing a proper sign. The proper sign associated with each product is $(-1)^{i+j}$, where i is the number of the row and j is the number of the column in which the element stands. For example, for D_3 above,

$$D_3 = -a_2\begin{vmatrix} b_1 & c_1 \\ b_3 & c_3 \end{vmatrix} + b_2\begin{vmatrix} a_1 & c_1 \\ a_3 & c_3 \end{vmatrix} - c_2\begin{vmatrix} a_1 & b_1 \\ a_3 & b_3 \end{vmatrix}$$

is the expansion of D_3, along the second row. The sign given to the first product is $-$, since a_2 stands in the second row and first column, and $(-1)^{2+1} = -1$. In all, there are six expansions of D_3 along its rows and columns yielding identical results when the minors are evaluated.

There are eight expansions of D_4 along its rows and columns, of which

$$D_4 = +a_1\begin{vmatrix} b_2 & c_2 & d_2 \\ b_3 & c_3 & d_3 \\ b_4 & c_4 & d_4 \end{vmatrix} - b_1\begin{vmatrix} a_2 & c_2 & d_2 \\ a_3 & c_3 & d_3 \\ a_4 & c_4 & d_4 \end{vmatrix} + c_1\begin{vmatrix} a_2 & b_2 & d_2 \\ a_3 & b_3 & d_3 \\ a_4 & b_4 & d_4 \end{vmatrix} - d_1\begin{vmatrix} a_2 & b_2 & c_2 \\ a_3 & b_3 & c_3 \\ a_4 & b_4 & c_4 \end{vmatrix}$$

(along the first row)

$$D_4 = +a_1\begin{vmatrix} b_2 & c_2 & d_2 \\ b_3 & c_3 & d_3 \\ b_4 & c_4 & d_4 \end{vmatrix} - a_2\begin{vmatrix} b_1 & c_1 & d_1 \\ b_3 & c_3 & d_3 \\ b_4 & c_4 & d_4 \end{vmatrix} + a_3\begin{vmatrix} b_1 & c_1 & d_1 \\ b_2 & c_2 & d_2 \\ b_4 & c_4 & d_4 \end{vmatrix} - a_4\begin{vmatrix} b_1 & c_1 & d_1 \\ b_2 & c_2 & d_2 \\ b_3 & c_3 & d_3 \end{vmatrix}$$

(along the first column)

$$D_4 = -a_4\begin{vmatrix} b_1 & c_1 & d_1 \\ b_2 & c_2 & d_2 \\ b_3 & c_3 & d_3 \end{vmatrix} + b_4\begin{vmatrix} a_1 & c_1 & d_1 \\ a_2 & c_2 & d_2 \\ a_3 & c_3 & d_3 \end{vmatrix} - c_4\begin{vmatrix} a_1 & b_1 & d_1 \\ a_2 & b_2 & d_2 \\ a_3 & b_3 & d_3 \end{vmatrix} + d_4\begin{vmatrix} a_1 & b_1 & c_1 \\ a_2 & b_2 & c_2 \\ a_3 & b_3 & c_3 \end{vmatrix}$$

(along the fourth row)

$$D_4 = -b_1\begin{vmatrix} a_2 & c_2 & d_2 \\ a_3 & c_3 & d_3 \\ a_4 & c_4 & d_4 \end{vmatrix} + b_2\begin{vmatrix} a_1 & c_1 & d_1 \\ a_3 & c_3 & d_3 \\ a_4 & c_4 & d_4 \end{vmatrix} - b_3\begin{vmatrix} a_1 & c_1 & d_1 \\ a_2 & c_2 & d_2 \\ a_4 & c_4 & d_4 \end{vmatrix} + b_4\begin{vmatrix} a_1 & c_1 & d_1 \\ a_2 & c_2 & d_2 \\ a_3 & c_3 & d_3 \end{vmatrix}$$

(along the second column)

are examples. (See Problem 21.7.)

All computer mathematics packages of software give you the capability to easily evaluate determinants. The reader should try at least one of these (Maple, or others) to gain some familiarity with these kinds of computer computations.

THE COFACTOR OF AN ELEMENT of a determinant is the minor of that element together with the sign associated with the product of that element and its minor in the expansion of the determinant. The cofactors of the elements $a_1, a_2, b_1, b_3, c_1, \ldots$ will be denoted by $A_1, A_2, B_1, B_3, C_1, \ldots$. Thus, the cofactor of c_1 in D_3 is $C_1 = +\begin{vmatrix} a_2 & b_2 \\ a_3 & b_3 \end{vmatrix}$ and the cofactor of b_3 is $B_3 = -\begin{vmatrix} a_1 & c_1 \\ a_2 & c_2 \end{vmatrix}$.

When cofactors are used, the expansions of D_4 given above take the more compact form

$$\begin{aligned} D_4 &= a_1A_1 + b_1B_1 + c_1C_1 + d_1D_1 && \text{(along the first row)} \\ &= a_1A_1 + a_2A_2 + a_3A_3 + a_4A_4 && \text{(along the first column)} \\ &= a_4A_4 + b_4B_4 + c_4C_4 + d_4D_4 && \text{(along the fourth row)} \\ &= b_1B_1 + b_2B_2 + b_3B_3 + b_4B_4 && \text{(along the second column)} \end{aligned}$$

(See Problems 21.8–21.9.)

PROPERTIES OF DETERMINANTS. Subject always to our assumption of equivalent expansions of a determinant along any of its rows or columns, the following theorems may be proved by mathematical induction.

THEOREM I. If two rows (or two columns) of a determinant are identical, the value of the determinant is zero. For example,

$$\begin{vmatrix} 2 & 3 & 2 \\ 3 & 1 & 3 \\ 1 & 4 & 1 \end{vmatrix} = 0$$

COROLLARY I. If each of the elements of a row (or a column) of a determinant is multiplied by the cofactor of the corresponding element of another row (column), the sum of the products is zero.

THEOREM II. If the elements of a row (or a column) of a determinant are multiplied by any number m, the determinant is multiplied by m. For example,

$$5\begin{vmatrix} 2 & 3 & 4 \\ 3 & -1 & 2 \\ 1 & 4 & -3 \end{vmatrix} = \begin{vmatrix} 10 & 3 & 4 \\ 15 & -1 & 2 \\ 5 & 4 & -3 \end{vmatrix} = \begin{vmatrix} 2 & 3 & 4 \\ 15 & -5 & 10 \\ 1 & 4 & -3 \end{vmatrix}$$

THEOREM III. If each of the elements of a row (or a column) of a determinant is expressed as the sum of two or more numbers, the determinant may be written as the sum of two or more determinants. For example,

$$\begin{vmatrix} 2 & 5 & 4 \\ 4 & -2 & 3 \\ 1 & -4 & 3 \end{vmatrix} = \begin{vmatrix} -2+4 & 5 & 4 \\ 3+1 & -2 & 3 \\ 1+0 & -4 & 3 \end{vmatrix} = \begin{vmatrix} -2 & 5 & 4 \\ 3 & -2 & 3 \\ 1 & -4 & 3 \end{vmatrix} + \begin{vmatrix} 4 & 5 & 4 \\ 1 & -2 & 3 \\ 0 & -4 & 3 \end{vmatrix}$$

THEOREM IV. If to the elements of any row (or any column) of a determinant there is added m times the corresponding elements of another row (another column), the value of the determinants is unchanged. For example,

$$\begin{vmatrix} -2 & 5 & 4 \\ 3 & -2 & 2 \\ 1 & -4 & 3 \end{vmatrix} = \begin{vmatrix} -2 & 5+4(-2) & 4 \\ 3 & -2+4(3) & 2 \\ 1 & -4+4(1) & 3 \end{vmatrix} = \begin{vmatrix} -2 & -3 & 4 \\ 3 & 10 & 2 \\ 1 & 0 & 3 \end{vmatrix}$$

(See Problem 21.10.)

EVALUATION OF DETERMINANTS. A determinant of any order may be evaluated by expanding it and all subsequent determinants (minors) thus obtained along a row or column. This procedure may be greatly simplified by the use of Theorem IV. In Problem 21.11, (*a*) and (*b*), a row (column) containing an element $+1$ or -1 is used to obtain an equivalent determinant having an element 0 in another row (column). In (*c*) and (*d*), the same theorem has been used to obtain an element $+1$ or -1; this procedure is to be followed when the given determinant is lacking in these elements.

The revised procedure consists in first obtaining an equivalent determinant in which all the elements, save one, in some row (column) are zeros and then expanding along that row (column).

EXAMPLE 3. Evaluate $\begin{vmatrix} 1 & 4 & 3 & 1 \\ 2 & 8 & 2 & 5 \\ 4 & -4 & -1 & -3 \\ 2 & 5 & 3 & 3 \end{vmatrix}$

Using the first column since it contains the element 1 in the first row, we obtain an equivalent determinant all of whose elements, save the first, in the first row are zeros. We have

$$\begin{vmatrix} 1 & 4 & 3 & 1 \\ 2 & 8 & 2 & 5 \\ 4 & -4 & -1 & -3 \\ 2 & 5 & 3 & 3 \end{vmatrix} = \begin{vmatrix} 1 & 4+(-4)1 & 3+(-3)1 & 1+(-1)1 \\ 2 & 8+(-4)2 & 2+(-3)2 & 5+(-1)2 \\ 4 & -4+(-4)4 & -1+(-3)4 & -3+(-1)4 \\ 2 & 5+(-4)2 & 3+(-3)2 & 3+(-1)2 \end{vmatrix} = \begin{vmatrix} 1 & 0 & 0 & 0 \\ 2 & 0 & -4 & 3 \\ 4 & -20 & -13 & -7 \\ 2 & -3 & -3 & 1 \end{vmatrix}$$

$$= \begin{vmatrix} 0 & -4 & 3 \\ -20 & -13 & -7 \\ -3 & -3 & 1 \end{vmatrix} \quad \text{(by expanding along the first row)}$$

Expanding the resulting determinant along the first row to take full advantage of the element 0, we have

$$\begin{vmatrix} 0 & -4 & 3 \\ -20 & -13 & -7 \\ -3 & -3 & 1 \end{vmatrix} = 4(-20-21) + 3(60-39) = -101$$

SYSTEMS OF *n* LINEAR EQUATIONS IN *n* UNKNOWNS. Consider, for the sake of brevity, the system of four linear equations in four unknowns

$$\begin{cases} a_1x + b_1y + c_1z + d_1w = k_1 \\ a_2x + b_2y + c_2z + d_2w = k_2 \\ a_3x + b_3y + c_3z + d_3w = k_3 \\ a_4x + b_4y + c_4z + d_4w = k_4 \end{cases} \tag{21.3}$$

in which each equation is written with the unknowns x, y, z, w in that order on the left side and the constant term on the right side. Form

$$D = \begin{vmatrix} a_1 & b_1 & c_1 & d_1 \\ a_2 & b_2 & c_2 & d_2 \\ a_3 & b_3 & c_3 & d_3 \\ a_4 & b_4 & c_4 & d_4 \end{vmatrix}, \quad \text{the determinant of the coefficients of the unknowns,}$$

and from it the determinants

$$N_x = \begin{vmatrix} k_1 & b_1 & c_1 & d_1 \\ k_2 & b_2 & c_2 & d_2 \\ k_3 & b_3 & c_3 & d_3 \\ k_4 & b_4 & c_4 & d_4 \end{vmatrix}, \quad N_y = \begin{vmatrix} a_1 & k_1 & c_1 & d_1 \\ a_2 & k_2 & c_2 & d_2 \\ a_3 & k_3 & c_3 & d_3 \\ a_4 & k_4 & c_4 & d_4 \end{vmatrix}, \quad N_z = \begin{vmatrix} a_1 & b_1 & k_1 & d_1 \\ a_2 & b_2 & k_2 & d_2 \\ a_3 & b_3 & k_3 & d_3 \\ a_4 & b_4 & k_4 & d_4 \end{vmatrix}$$

$$N_w = \begin{vmatrix} a_1 & b_1 & c_1 & k_1 \\ a_2 & b_2 & c_2 & k_2 \\ a_3 & b_3 & c_3 & k_3 \\ a_4 & b_4 & c_4 & k_4 \end{vmatrix}$$

by replacing the column of coefficients of the indicated unknown by the column of constants.

CRAMER'S RULE STATES THAT:

(*a*) If $D \neq 0$, the system (*21.1*) has the unique solution

$$x = N_x/D, \quad y = N_y/D, \quad z = N_z/D, \quad w = N_w/D$$

(*b*) If $D = 0$ and at least one of $N_x, N_y, N_z, N_w \neq 0$, the system has no solution. For, if $D = 0$ and $N_x \neq 0$, then $x \cdot D = N_x$ leads to a contradiction. Such systems are called *inconsistent*. (See Problem 21.3.)

(*c*) If $D = 0$ and $N_x = N_y = N_z = N_w = 0$, the system may or may not have a solutions. A system having an infinite number of solutions is called *dependent*.

For systems of three or four equations, the simplest procedure is to evaluate D. If $D \neq 0$, proceed as in (*A*); if $D = 0$, proceed as in Chapter 5. (See Problems 21.13–21.16.)

SYSTEMS OF m LINEAR EQUATIONS IN $n > m$ UNKNOWNS. Ordinarily if there are fewer equations than unknowns, the system will have an infinite number of solutions.

To solve a consistent system of m equations, solve for m of the unknowns (in certain cases for $p < m$ of the unknowns) in terms of the others. (See Problem 21.17.)

SYSTEMS OF n EQUATIONS IN $m < n$ UNKNOWNS. Ordinarily if there are more equations than unknowns the system is inconsistent. However, if $p \leq m$ of the equations have a solution and if this solution satisfies each of the remaining equations, the system is consistent. (See Problem 21.18.)

A HOMOGENEOUS EQUATION is one in which all terms are of the same degree; otherwise, the equation is called *nonhomogeneous*. For example, the linear equation $2x + 3y - 4z = 5$ is nonhomogeneous, while $2x + 3y - 4z = 0$ is homogeneous. (The term "5" in the first equation has degree 0, while all other terms are of degree 1.)

Every system of homogeneous linear equations

$$\begin{gathered} a_1x + b_1y + c_1z + \cdots = 0 \\ a_2x + b_2y + c_2z + \cdots = 0 \\ \vdots \\ a_nx + b_ny + c_nz + \cdots = 0 \end{gathered}$$

always has the *trivial solution* $x = 0, y = 0, z = 0, \ldots$.

A system of n homogeneous linear equations in n unknowns has *only* the trivial solution if D, the determinant of the coefficients, is not equal to zero. If $D = 0$, the system has nontrivial solutions as well. (See Problem 21.19.)

ONE LINEAR AND ONE QUADRATIC EQUATION

Procedure: Solve the linear equation for one of the two unknowns (your choice) and substitute in the quadratic equation. Since this results in a quadratic equation in one unknown, the system can always be solved.

EXAMPLE 4. Solve the system $\begin{cases} 4x^2 + 3y^3 = 16 \\ 5x + y = 7 \end{cases}$

Solve the linear equation for y: $y = 7 - 5x$. Substitute in the quadratic equation:

$$4x^2 + 3(7 - 5x)^2 = 16$$

$$4x^2 + 3(49 - 70 + 25x^2) = 16$$

$$79x^2 - 210x + 131 = (x - 1)(79x - 131) = 0$$

and $x = 1, \frac{131}{79}$.

When $x = 1, y = 7 - 5x = 2$; when $x = \frac{131}{79}, y = -\frac{102}{79}$. The solutions are $x = 1, y = 2$ and $x = \frac{131}{79}, y = -\frac{102}{79}$.

The locus of the linear equation is the straight line and the locus of the quadratic equation is the ellipse in Fig. 21-1. (See Problems 21.20–21.21.)

TWO QUADRATIC EQUATIONS. In general, solving a system of two quadratic equations in two unknowns involves solving an equation of the fourth degree in one of the unknowns. Since the solution of the general equation of the fourth degree in one unknown is beyond the scope of this book, only those systems which require the solution of a quadratic equation in one unknown will be treated here.

TWO QUADRATIC EQUATIONS OF THE FORM $ax^2 + by^2 = c$

Procedure: Eliminate one of the unknowns by the method of addition for simultaneous equations in Chapter 5.

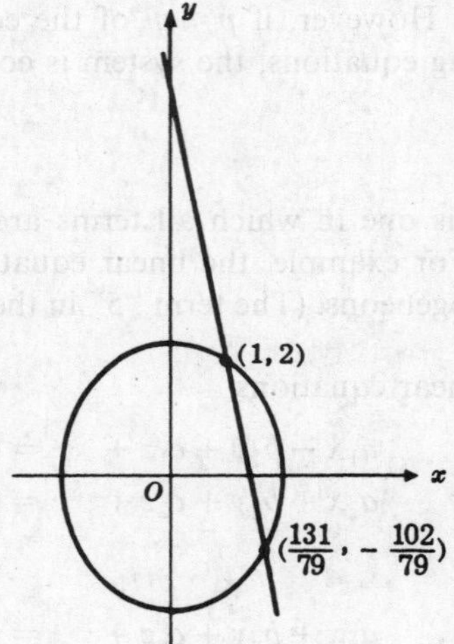

Fig. 21-1

EXAMPLE 5. Solve the system $\begin{cases} 4x^2 + 9y^2 = 72 & (21.4) \\ 3x^2 - 2y^2 = 19 & (21.5) \end{cases}$

$$\begin{aligned} \text{Multiply } (38.1) \text{ by 2:} \quad & 8x^2 + 18y^2 = 144 \\ \text{Multiply } (38.2) \text{ by 9:} \quad & 27x^2 - 18y^2 = 171 \\ \text{Add:} \quad & 35x^2 \qquad\quad = 315 \end{aligned}$$

Then $x^2 = 9$ and $x = \pm 3$.

When $x = 3$, (38.1) gives $9y^2 = 72 - 4x^2 = 72 - 36 = 36, y^2 = 4$, and $y = \pm 2$.

When $x = -3$, (38.1) gives $9y^2 = 72 - 36 = 36,\ y^2 = 4$, and $y = \pm 2$.

The four solutions $x = 3,\ y = 2; x = 3,\ y = -2;\ x = -3,\ y = 2;\ x = -3,\ y = -2$ may also be written as $x = \pm 3,\ y = \pm 2;\ x = \pm 3,\ y = \pm 2$. By convention, we read the two upper signs and the two lower signs in the latter form.

The ellipse and the hyperbola intersect in the points (3, 2), (3, –2), (–3, 2), (–3, –2). See Fig. 21-2. (See Problems 21.22–21.23.)

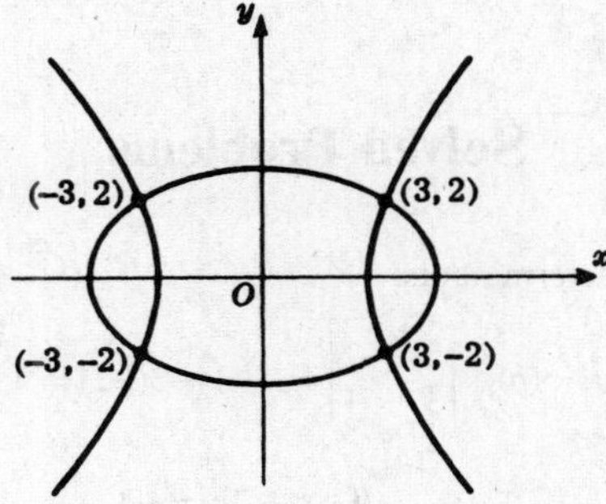

Fig. 21-2

TWO QUADRATIC EQUATIONS, ONE HOMOGENEOUS. An expression, as $2x^2 - 3xy + y^2$, whose terms are all of the same degree in the variables, is called *homogeneous*. A homogeneous expression equated to zero is called a *homogeneous equation*. A homogeneous quadratic equation in two unknowns can always be solved for one of the unknowns in terms of the other.

EXAMPLE 6. Solve the system $\begin{cases} x^2 - 3xy + 2y^2 = 0 & (21.6) \\ 2x^2 + 3xy - y^2 = 13 & (21.7) \end{cases}$

Solve (*21.6*) for x in terms of y: $(x - y)(x - 2y) = 0$ and $x = y,\ x = 2y$.
Solve the systems (see Example 1):

$$\begin{cases} 2x^2 + 3xy - y^2 = 13 \\ x = y \end{cases} \qquad \begin{cases} 2x^2 + 3xy - y^2 = 13 \\ x = 2y \end{cases}$$

$$2y^2 + 3y^2 - y^2 = 4y^2 = 13 \qquad 8y^2 + 6y^2 - y^2 = 13y^2 = 13$$

$$y^2 = \tfrac{13}{4},\ y = \pm\frac{\sqrt{13}}{2} \qquad y^2 = 1,\ y = \pm 1$$

Then $x = y = \pm\sqrt{13}/2$. Then $x = 2y = \pm 2$.

The solutions are $x = \sqrt{13}/2,\ y = \sqrt{13}/2;\ x = -\sqrt{13}/2,\ y = -\sqrt{13}/2;\ x = 2,\ y = 1;\ x = -2,\ y = -1$ or $x = \pm\sqrt{13}/2,\ y = \pm\sqrt{13}/2;\ x = \pm 2,\ y = \pm 1$. (See Problem 21.36.)

TWO QUADRATIC EQUATIONS OF THE FORM $ax^2 + bxy + cy^2 = d$

Procedure: Combine the two given equations to obtain a homogeneous equation. Solve, as in Example 3, the system consisting of this homogeneous equation and either of the given equations. (See Problems 21.37–21.38.)

TWO QUADRATIC EQUATIONS, EACH SYMMETRICAL IN x AND y. An equation, as $2x^2 - 3xy + 2y^2 + 5x + 5y = 1$, which is unchanged when the two unknowns are interchanged is called a *symmetrical equation*.

Procedure: Substitute $x = u + v$ and $y = u - v$ and then eliminate v^2 from the resulting equations. (See Problem 21.39.)

Frequently, a careful study of a given system will reveal some special device for solving it. (See Problems 21.28–21.32.)

Solved Problems

21.1 Evaluate each of the following determinants:

(a) $\begin{vmatrix} 2 & 3 \\ 4 & 5 \end{vmatrix} = 2 \cdot 5 - 4 \cdot 3 = -2$ (b) $\begin{vmatrix} 5 & -2 \\ 3 & 1 \end{vmatrix} = 5 \cdot 1 - 3(-2) = 11$

21.2 Solve for x and y, using determinants: (a) $\begin{cases} x + 2y = -4 \\ 5x + 3y = 1 \end{cases}$, (b) $\begin{cases} ax - 2by = c \\ 2ax - 3by = 4c \end{cases}$

(a) $D = \begin{vmatrix} 1 & 2 \\ 5 & 3 \end{vmatrix} = 3 - 10 = -7, \quad N_x = \begin{vmatrix} -4 & 2 \\ 1 & 3 \end{vmatrix} = -12 - 2 = -14, \quad N_y = \begin{vmatrix} 1 & -4 \\ 5 & 1 \end{vmatrix} = 1 + 20 = 21$

$$x = \frac{N_x}{D} = \frac{-14}{-7} = 2, \quad y = \frac{N_y}{D} = \frac{21}{-7} = -3$$

(b) $D = \begin{vmatrix} a & -2b \\ 2a & -3b \end{vmatrix} = ab, \quad N_x = \begin{vmatrix} c & -2b \\ 4c & -3b \end{vmatrix} = 5bc, \quad N_y = \begin{vmatrix} a & c \\ 2a & 4c \end{vmatrix} = 2ac$

$$x = \frac{N_x}{D} = \frac{5bc}{ab} = \frac{5c}{a}, \quad y = \frac{N_y}{D} = \frac{2ac}{ab} = \frac{2c}{b}$$

21.3 Evaluate the following determinants:

(a) $\begin{vmatrix} 2 & -2 & -1 \\ 6 & 1 & -1 \\ 4 & 3 & 5 \end{vmatrix} = 2\begin{vmatrix} 1 & -1 \\ 3 & 5 \end{vmatrix} - (-2)\begin{vmatrix} 6 & -1 \\ 4 & 5 \end{vmatrix} + (-1)\begin{vmatrix} 6 & 1 \\ 4 & 3 \end{vmatrix}$

$$= 2(5 + 3) + 2(30 + 4) - (18 - 4) = 2 \cdot 8 + 2 \cdot 34 - 14 = 70$$

(b) $\begin{vmatrix} 2 & 5 & 0 \\ 0 & 3 & 4 \\ -5 & 3 & 6 \end{vmatrix} = 2\begin{vmatrix} 3 & 4 \\ 3 & 6 \end{vmatrix} - 5\begin{vmatrix} 0 & 4 \\ -5 & 6 \end{vmatrix} + 0\begin{vmatrix} 0 & 3 \\ -5 & 3 \end{vmatrix} = 2(18 - 12) - 5(0 + 20) + 0 = -88$

(c) $\begin{vmatrix} 3 & 2 & 1 \\ 1 & -2 & 4 \\ 4 & 2 & 3 \end{vmatrix} = 3(-6 - 8) - 2(3 - 16) + 1(2 + 8) = -6$

(d) $\begin{vmatrix} 4 & -3 & 2 \\ 5 & 9 & -7 \\ 4 & -1 & 7 \end{vmatrix} = 4(36 - 7) + 3(20 + 28) + 2(-5 - 36) = 178$

21.4 Solve using determinants: $\begin{cases} 2x - 3y + 2z = 6 \\ x + 8y + 3z = -31 \\ 3x - 2y + z = -5 \end{cases}$

We evaluate $D = \begin{vmatrix} 2 & -3 & 2 \\ 1 & 8 & 3 \\ 3 & -2 & 1 \end{vmatrix} = 2(8+6) + 3(1-9) + 2(-2-24) = -48$

$$N_x = \begin{vmatrix} 6 & -3 & 2 \\ -31 & 8 & 3 \\ -5 & -2 & 1 \end{vmatrix} = 68(8+6) + 3(-31+15) + 2(62+40) = 240$$

$$N_y = \begin{vmatrix} 2 & 6 & 2 \\ 1 & -31 & 3 \\ 3 & -5 & 1 \end{vmatrix} = 2(-31+15) - 6(1-9) + 2(-5+93) = 192$$

$$N_z = \begin{vmatrix} 2 & -3 & 6 \\ 1 & 8 & -31 \\ 3 & -2 & -5 \end{vmatrix} = 2(-40-62) + 3(-5+93) + 6(-2-24) = -96$$

Then $x = \dfrac{N_x}{D} = \dfrac{240}{-48} = -5, \quad y = \dfrac{N_y}{D} = \dfrac{192}{-48} = -4, \quad z = \dfrac{N_z}{D} = \dfrac{-96}{-48} = -2$

21.5 Solve, using determinants: $\begin{cases} 2x + y = 2 \\ z - 4y = 0 \\ 4x + z = 6 \end{cases}$

$$D = \begin{vmatrix} 2 & 1 & 0 \\ 0 & -4 & 1 \\ 4 & 0 & 1 \end{vmatrix} = -4, \quad N_x = \begin{vmatrix} 2 & 1 & 0 \\ 0 & -4 & 1 \\ 6 & 0 & 1 \end{vmatrix} = -2, \quad N_y = \begin{vmatrix} 2 & 2 & 0 \\ 0 & 0 & 1 \\ 4 & 6 & 1 \end{vmatrix} = -4, \quad N_z = \begin{vmatrix} 2 & 1 & 2 \\ 0 & -4 & 0 \\ 4 & 0 & 6 \end{vmatrix} = -16$$

Then $x = \dfrac{-2}{-4} = \dfrac{1}{2}, \quad y = \dfrac{-4}{-4} = 1, \text{ and } z = \dfrac{-16}{-4} = 4.$

21.6 Write the minors of the elements a_1, b_3, c_2 of D_3.

The minor of a_1 is $\begin{vmatrix} b_2 & c_2 \\ b_3 & c_3 \end{vmatrix}$, the minor of b_3 is $\begin{vmatrix} a_1 & c_1 \\ a_2 & c_2 \end{vmatrix}$, the minor of c_1 is $\begin{vmatrix} a_2 & b_2 \\ a_3 & b_3 \end{vmatrix}$.

21.7 Evaluate: (*a*) $\begin{vmatrix} -1 & 3 & -4 \\ 0 & 2 & 0 \\ 2 & -3 & 5 \end{vmatrix}$ by expanding along the second row

(*b*) $\begin{vmatrix} 8 & 0 & 2 & 0 \\ 5 & 1 & -3 & 0 \\ -4 & 3 & 7 & -3 \\ 4 & 0 & 6 & 0 \end{vmatrix}$ by expanding along the fourth column

(*a*) $\begin{vmatrix} -1 & 3 & -4 \\ 0 & 2 & 0 \\ 2 & -3 & 5 \end{vmatrix} = -0\begin{vmatrix} 3 & -4 \\ -3 & 5 \end{vmatrix} + 2\begin{vmatrix} -1 & -4 \\ 2 & 5 \end{vmatrix} - 0\begin{vmatrix} -1 & 3 \\ 2 & -3 \end{vmatrix} = 2\begin{vmatrix} -1 & -4 \\ 2 & 5 \end{vmatrix} = 2 \cdot 3 = 6$

(*b*) $\begin{vmatrix} 8 & 0 & 2 & 0 \\ 5 & 1 & -3 & 0 \\ -4 & 3 & 7 & -3 \\ 4 & 0 & 6 & 0 \end{vmatrix} = -(-3)\begin{vmatrix} 8 & 0 & 2 \\ 5 & 1 & -3 \\ 4 & 0 & 6 \end{vmatrix} = 3\left\{+1\begin{vmatrix} 8 & 2 \\ 4 & 6 \end{vmatrix}\right\} = 3(48-8) = 120$

21.8 (*a*) Write the cofactors of the elements a_1, b_3, c_2, d_4 of D_4.

The cofactor of a_1 is $A_1 = +\begin{vmatrix} b_2 & c_2 & d_2 \\ b_3 & c_3 & d_3 \\ b_4 & c_4 & d_4 \end{vmatrix}$, of b_3 is $B_3 = -\begin{vmatrix} a_1 & c_1 & d_1 \\ a_2 & c_2 & d_2 \\ a_4 & c_4 & d_4 \end{vmatrix}$

of c_2 is $C_2 = -\begin{vmatrix} a_1 & b_1 & d_1 \\ a_3 & b_3 & d_3 \\ a_4 & b_4 & d_4 \end{vmatrix}$, of d_4 is $D_4 = +\begin{vmatrix} a_1 & b_1 & c_1 \\ a_2 & b_2 & c_2 \\ a_3 & b_3 & c_3 \end{vmatrix}$

(*b*) Write the expansion of D_4 along (1) the second row, (2) the third column, using cofactors.
(1) $D_4 = a_2A_2 + b_2B_2 + c_2C_2 + d_2D_2$
(2) $D_4 = c_1C_1 + c_2C_2 + c_3C_3 + c_4C_4$

21.9 Express $g_1C_1 + g_2C_2 + g_3C_3 + g_4C_4$, where the C_i are cofactors of the elements c_i of D_4, as a determinant.

Since the cofactors C_i contain no elements with basal letter c, we replace c_1, c_2, c_3, c_4 by g_1, g_2, g_3, g_4 respectively, in Problem 20.3(*b*) and obtain

$$g_1C_1 + g_2C_2 + g_3C_3 + g_4C_4 = \begin{vmatrix} a_1 & b_1 & g_1 & d_1 \\ a_2 & b_2 & g_2 & d_2 \\ a_3 & b_3 & g_3 & d_3 \\ a_4 & b_4 & g_4 & d_4 \end{vmatrix}$$

21.10 Prove by induction: If two rows (or two columns) of a determinant are identical, the value of the determinant is zero.

The theorem is true for determinants of order two since $\begin{vmatrix} a_1 & a_1 \\ a_2 & a_2 \end{vmatrix} = a_1a_2 - a_1a_2 = 0$.

Let us assume the theorem true for determinants of order k and consider a determinant D of order $(k+1)$ in which two columns are identical. When D is expanded along any column, other than the two with identical elements, each cofactor involved is a determinant of order k with two columns identical and, by assumption, is equal to zero. Thus, D is equal to zero and the theorem is proved by induction.

21.11 From the determinant $\begin{vmatrix} 1 & 4 & 3 & 1 \\ 2 & 8 & 2 & 5 \\ 4 & -4 & -1 & -3 \\ 2 & 5 & 3 & 3 \end{vmatrix}$ obtain an equivalent determinant.

(*a*) By adding -4 times the elements of the first column to the corresponding elements of the second column
(*b*) By adding 3 times the elements of the third row to the corresponding elements of the fourth row
(*c*) By adding -1 times the elements of the third column to the corresponding elements of the second column
(*d*) By adding -2 times the elements of the fourth row to the corresponding elements of the second row

(*a*)
$$\begin{vmatrix} 1 & 4 & 3 & 1 \\ 2 & 8 & 2 & 5 \\ 4 & -4 & -1 & -3 \\ 2 & 5 & 3 & 3 \end{vmatrix} = \begin{vmatrix} 1 & 4+(-4)1 & 3 & 1 \\ 2 & 8+(-4)2 & 2 & 5 \\ 4 & -4+(-4)4 & -1 & -3 \\ 2 & 5+(-4)2 & 3 & 3 \end{vmatrix} = \begin{vmatrix} 1 & 0 & 3 & 1 \\ 2 & 0 & 2 & 5 \\ 4 & -20 & -1 & -3 \\ 2 & -3 & 3 & 3 \end{vmatrix}$$

(*b*)
$$\begin{vmatrix} 1 & 4 & 3 & 1 \\ 2 & 8 & 2 & 5 \\ 4 & -4 & -1 & -3 \\ 2 & 5 & 3 & 3 \end{vmatrix} = \begin{vmatrix} 1 & 4 & 3 & 1 \\ 2 & 8 & 2 & 5 \\ 4 & -4 & -1 & -3 \\ 2+(3)4 & 5+(3)(-4) & 3+(3)(-1) & 3+(3)(-3) \end{vmatrix} = \begin{vmatrix} 1 & 4 & 3 & 1 \\ 2 & 8 & 2 & 5 \\ 4 & -4 & -1 & -3 \\ 14 & -7 & 0 & -6 \end{vmatrix}$$

(*c*)
$$\begin{vmatrix} 1 & 4 & 3 & 1 \\ 2 & 8 & 2 & 5 \\ 4 & -4 & -1 & -3 \\ 2 & 5 & 3 & 3 \end{vmatrix} = \begin{vmatrix} 1 & 4+(-1)3 & 3 & 1 \\ 2 & 8+(-1)2 & 2 & 5 \\ 4 & -4+(-1)(-1) & -1 & -3 \\ 2 & 5+(-1)3 & 3 & 3 \end{vmatrix} = \begin{vmatrix} 1 & 1 & 3 & 1 \\ 2 & 6 & 2 & 5 \\ 4 & -3 & -1 & -3 \\ 2 & 2 & 3 & 3 \end{vmatrix}$$

(*d*)
$$\begin{vmatrix} 1 & 4 & 3 & 1 \\ 2 & 8 & 2 & 5 \\ 4 & -4 & -1 & -3 \\ 2 & 5 & 3 & 3 \end{vmatrix} = \begin{vmatrix} 1 & 4 & 3 & 1 \\ 2+(-2)2 & 8+(-2)5 & 2+(-2)3 & 5+(-2)3 \\ 4 & -4 & -1 & -3 \\ 2 & 5 & 3 & 3 \end{vmatrix} = \begin{vmatrix} 1 & 4 & 3 & 1 \\ -2 & -2 & -4 & -1 \\ 4 & -4 & -1 & -3 \\ 2 & 5 & 3 & 3 \end{vmatrix}$$

21.12 Evaluate:

(*a*)
$$\begin{vmatrix} 50 & 2 & -9 \\ 250 & -10 & 45 \\ -150 & 6 & 27 \end{vmatrix} = 50\cdot 2\cdot 9\begin{vmatrix} 1 & 1 & -1 \\ 5 & -5 & 5 \\ -3 & 3 & 3 \end{vmatrix} = 900\cdot 5\cdot 3\begin{vmatrix} 1 & 1 & -1 \\ 1 & -1 & 1 \\ -1 & 1 & 1 \end{vmatrix} = 13\,500\begin{vmatrix} 2 & 0 & 0 \\ 1 & -1 & 1 \\ -1 & 1 & 1 \end{vmatrix}$$
$$= 27\,000\begin{vmatrix} -1 & 1 \\ 1 & 1 \end{vmatrix} = -54\,000$$

(*b*)
$$\begin{vmatrix} 3 & 2 & -3 & 1 \\ -1 & -3 & 5 & 2 \\ 2 & 1 & 6 & -3 \\ 5 & 4 & -3 & 4 \end{vmatrix} = \begin{vmatrix} 3+(-3)1 & 2+(-2)1 & -3+3(1) & 1 \\ -1+(-3)2 & -3+(-2)2 & 5+(3)2 & 2 \\ 2+(-3)(-3) & 1+(-2)(-3) & 6+(3)(-3) & -3 \\ 5+(-3)4 & 4+(-2)4 & -3+(3)4 & 4 \end{vmatrix} = \begin{vmatrix} 0 & 0 & 0 & 1 \\ -7 & -7 & 11 & 2 \\ 11 & 7 & -3 & -3 \\ -7 & -4 & 9 & 4 \end{vmatrix}$$
$$= -\begin{vmatrix} -7 & -7 & 11 \\ 11 & 7 & -3 \\ -7 & -4 & 9 \end{vmatrix} = -\begin{vmatrix} 0 & -7 & 11 \\ 4 & 7 & -3 \\ -3 & -4 & 9 \end{vmatrix} = -[7(36-9)+11(-16+21)] = -244$$

(*c*)
$$\begin{vmatrix} 2 & -1 & -2 & 3 \\ 3 & 2 & 4 & -1 \\ 2 & 4 & 1 & -5 \\ 4 & -3 & 2 & 1 \end{vmatrix} = \begin{vmatrix} 0 & -1 & 0 & 0 \\ 7 & 2 & 0 & 5 \\ 10 & 4 & -7 & 7 \\ -2 & -3 & 8 & -8 \end{vmatrix} = \begin{vmatrix} 7 & 0 & 5 \\ 10 & -7 & 7 \\ -2 & 8 & -8 \end{vmatrix} = 5\cdot 66 = 330$$

21.13 Solve the system
$$\begin{cases} 3x - 2y - z - 4w = 7 & (1) \\ x + 3z + 2w = -10 & (2) \\ x + 4y + 2z + w = 0 & (3) \\ 2x + 3y + 3w = 1 & (4) \end{cases}$$

We find

$$D = \begin{vmatrix} 3 & -2 & -1 & -4 \\ 1 & 0 & 3 & 2 \\ 1 & 4 & 2 & 1 \\ 2 & 3 & 0 & 3 \end{vmatrix} = \begin{vmatrix} 3 & -2 & -10 & -10 \\ 1 & 0 & 0 & 0 \\ 1 & 4 & -1 & -1 \\ 2 & 3 & -6 & -1 \end{vmatrix} = -\begin{vmatrix} -2 & -10 & -10 \\ 4 & -1 & -1 \\ 3 & -6 & -1 \end{vmatrix} = 2\begin{vmatrix} 1 & 5 & 5 \\ 4 & -1 & -1 \\ 3 & -6 & -1 \end{vmatrix} = -210$$

$$N_x = \begin{vmatrix} 7 & -2 & -1 & -4 \\ -10 & 0 & 3 & 2 \\ 0 & 4 & 2 & 1 \\ 1 & 3 & 0 & 3 \end{vmatrix} = \begin{vmatrix} 0 & -23 & -1 & -25 \\ 0 & 30 & 3 & 32 \\ 0 & 4 & 2 & 1 \\ 1 & 3 & 0 & 3 \end{vmatrix} = -\begin{vmatrix} -23 & -1 & -25 \\ 30 & 3 & 32 \\ 4 & 2 & 1 \end{vmatrix} = -\begin{vmatrix} -23 & -1 & -2 \\ 30 & 3 & 2 \\ 4 & 2 & -3 \end{vmatrix} = -105$$

$$N_y = \begin{vmatrix} 3 & 7 & -1 & -4 \\ 1 & -10 & 3 & 2 \\ 1 & 0 & 2 & 1 \\ 2 & 1 & 0 & 3 \end{vmatrix} = \begin{vmatrix} 3 & 7 & -7 & -7 \\ 1 & -10 & 1 & 1 \\ 1 & 0 & 0 & 0 \\ 2 & 1 & -4 & 1 \end{vmatrix} = \begin{vmatrix} 7 & -7 & -7 \\ -10 & 1 & 1 \\ 1 & -4 & 1 \end{vmatrix} = 7\begin{vmatrix} 1 & -1 & -1 \\ -10 & 1 & 1 \\ 1 & -4 & 1 \end{vmatrix} = -315$$

$$N_z = \begin{vmatrix} 3 & -2 & 7 & -4 \\ 1 & 0 & -10 & 2 \\ 1 & 4 & 0 & 1 \\ 2 & 3 & 1 & 3 \end{vmatrix} = \begin{vmatrix} 3 & -14 & 7 & -7 \\ 1 & -4 & -10 & 1 \\ 1 & 0 & 0 & 0 \\ 2 & -5 & 1 & 1 \end{vmatrix} = \begin{vmatrix} -14 & 7 & -7 \\ -4 & -10 & 1 \\ -5 & 1 & 1 \end{vmatrix} = -7\begin{vmatrix} 2 & -1 & 1 \\ -4 & -10 & 1 \\ -5 & 1 & 1 \end{vmatrix} = 525$$

$$N_w = \begin{vmatrix} 3 & -2 & -1 & 7 \\ 1 & 0 & 3 & -10 \\ 1 & 4 & 2 & 0 \\ 2 & 3 & 0 & 1 \end{vmatrix} = \begin{vmatrix} 3 & -14 & -7 & 7 \\ 1 & -4 & 1 & -10 \\ 1 & 0 & 0 & 0 \\ 2 & -5 & -4 & 1 \end{vmatrix} = \begin{vmatrix} -14 & -7 & 7 \\ -4 & 1 & -10 \\ -5 & -4 & 1 \end{vmatrix} = -7\begin{vmatrix} 2 & 1 & -1 \\ -4 & 1 & -10 \\ -5 & -4 & 1 \end{vmatrix} = 315$$

Then

$$x = \frac{N_x}{D} = \frac{-105}{-210} = \frac{1}{2},\quad y = \frac{N_y}{D} = \frac{-315}{-210} = \frac{3}{2},\quad z = \frac{N_z}{D} = \frac{525}{-210} = -\frac{5}{2},\quad w = \frac{N_w}{D} = \frac{315}{-210} = -\frac{3}{2}.$$

Check. Using (*1*), $3\left(\frac{1}{2}\right) - 2\left(\frac{3}{2}\right) - \left(-\frac{5}{2}\right) - 4\left(-\frac{3}{2}\right) = \dfrac{3-6+5+12}{2} = 7.$

[NOTE: The above system permits some variation in procedure. For example, having found $x = \frac{1}{2}$ and $y = \frac{3}{2}$ using determinants, the value of w may be obtained by substituting in (*4*)

$$2\left(\tfrac{1}{2}\right) + 3\left(\tfrac{3}{2}\right) + 3w = 1, \qquad 3w = -\tfrac{9}{2}, \qquad w = -\tfrac{3}{2}$$

and the value of z may then be obtained by substituting in (*2*)

$$\left(\tfrac{1}{2}\right) + 3z + 2\left(-\tfrac{3}{2}\right) = -10, \qquad 3z = -\tfrac{15}{2}, \qquad z = -\tfrac{5}{2}.$$

The solution may be checked by substituting in (*1*) or (*3*).]

21.14 Solve the system $\begin{cases} 2x + y + 5z + w = 5 & (1) \\ x + y - 3z - 4w = -1 & (2) \\ 3x + 6y - 2z + w = 8 & (3) \\ 2x + 2y + 2z - 3w = 2 & (4) \end{cases}$

We have $D = \begin{vmatrix} 2 & 1 & 5 & 1 \\ 1 & 1 & -3 & -4 \\ 3 & 6 & -2 & 1 \\ 2 & 2 & 2 & -3 \end{vmatrix} = -120, \qquad N_x = \begin{vmatrix} 5 & 1 & 5 & 1 \\ -1 & 1 & -3 & -4 \\ 8 & 6 & -2 & 1 \\ 2 & 2 & 2 & -3 \end{vmatrix} = -240,$

$$N_y = \begin{vmatrix} 2 & 5 & 5 & 1 \\ 1 & -1 & -3 & -4 \\ 3 & 8 & -2 & 1 \\ 2 & 2 & 2 & -3 \end{vmatrix} = -24, \qquad N_z = \begin{vmatrix} 2 & 1 & 5 & 1 \\ 1 & 1 & -1 & -4 \\ 3 & 6 & 8 & 1 \\ 2 & 2 & 2 & -3 \end{vmatrix} = 0.$$

Then $\quad x = \dfrac{N_x}{D} = \dfrac{-240}{-120} = 2, \qquad y = \dfrac{N_y}{D} = \dfrac{-24}{-120} = \dfrac{1}{5} \qquad$ and $\qquad z = \dfrac{Nz}{D} = \dfrac{0}{-120} = 0.$

Substituting in (*1*), $2(2) + \left(\frac{1}{5}\right) + 5(0) + w = 5$ and $w = \frac{4}{5}$.
Check. Using (*2*), $(2) + \left(\frac{1}{5}\right) - 3(0) - 4\left(\frac{4}{5}\right) = -1$.

21.15 Show that the system $\begin{cases} 2x - y + 5z + w = 2 \\ x + y - z - 4w = 1 \\ 3x + 6y + 8z + w = 3 \\ 2x + 2y + 2z - 3w = 1 \end{cases}$ is inconsistent.

Since $D = \begin{vmatrix} 2 & 1 & 5 & 1 \\ 1 & 1 & -1 & -4 \\ 3 & 6 & 8 & 1 \\ 2 & 2 & 2 & -3 \end{vmatrix} = 0$ while $N_x = \begin{vmatrix} 2 & 1 & 5 & 1 \\ 1 & 1 & -1 & -4 \\ 3 & 6 & 8 & 1 \\ 1 & 2 & 2 & -3 \end{vmatrix} = -80 \neq 0$, the system is inconsistent.

21.16 Solve when possible:

(*a*) $\begin{cases} 2x - 3y + z = 0 & (1) \\ x + 5y - 3z = 3 & (2) \\ 5x + 12y - 8z = 9 & (3) \end{cases}$ (*c*) $\begin{cases} 6x - 2y + z = 1 & (1) \\ x - 4y + 2z = 0 & (2) \\ 4x + 6y - 3z = 0 & (3) \end{cases}$

(*b*) $\begin{cases} x + 2y + 3z = 2 & (1) \\ 2x + 4y + z = -1 & (2) \\ 3x + 6y + 5z = 2 & (3) \end{cases}$ (*d*) $\begin{cases} x + 2y - 3z + 5w = 11 & (1) \\ 4x - y + z - 2w = 0 & (2) \\ 2x + 4y - 6z + 10w = 22 & (3) \\ 5x + y - 2z + 3w = 11 & (4) \end{cases}$

(*a*) Here $D = 0$; we shall eliminate the variable x.

$$(1) - 2(2): \quad -13y + 7z = -6$$
$$(3) - 5(2): \quad -13y + 7z = -6$$

Then $y = \dfrac{7z+6}{13}$ and from (2), $x = 3 - 5y + 3z = \dfrac{4z+9}{13}$.

The solutions may be written as $x = \dfrac{4a+9}{13}$, $y = \dfrac{7a+6}{13}$, $z = a$, where a is arbitrary.

(*b*) Here $D = 0$; we shall eliminate x.

$$(2) - 2(1): \quad -5z = -5$$
$$(3) - 3(1): \quad -4z = -4$$

Then $z = 1$ and each of the given equations reduces to $x + 2y = -1$. Note that the same situation arises when y is eliminated.

The solution may be written $x = -1 - 2y$, $z = 1$ or as $x = -1 - 2a$, $y = a$, $z = 1$, where a is arbitrary.

(*c*) Here $D = 0$; we shall eliminate z.

$$(2) - 2(1): \quad -11x = -2$$
$$(3) + 3(1): \quad 22x = 3$$

The system is inconsistent.

(*d*) Here $D = 0$; we shall eliminate x.

$$(2) - 4(1): \quad -9y + 13z - 22w = -44$$
$$(3) - 2(1): \quad 0 = 0$$
$$(4) - 5(1): \quad -9y + 13z - 22w = -44$$

Then $y = \dfrac{44 + 13z - 22w}{9}$ and, from (1), $x = \dfrac{11 + z - w}{9}$. The solutions are $x = \dfrac{11 + a - b}{9}$, $y = \dfrac{44 + 13a - 22b}{9}$, $z = a$, $w = b$, where a and b are arbitrary.

21.17 (*a*) The system of two equations in four unknowns

$$\begin{cases} x + 2y + 3z - 4w = 5 \\ 3x - y - 5z - 5w = 1 \end{cases}$$

may be solved for any two of the unknowns in terms of the others; for example, $x = 1 + z + 2w$, $y = 2 - 2z + w$.

(*b*) The system of three equations in four unknowns

$$\begin{cases} x + 2y + 3z - 4w = 5 \\ 3x - y - 5z - 5w = 1 \\ 2x + 3y + z - w = 8 \end{cases}$$

may be solved for any three of the unknowns in terms of the fourth; for example, $x = 1 + 4w$, $y = 2 - 3w$, $z = 2w$.

(*c*) The system of three equations in four unknowns

$$\begin{cases} x + 2y + 3z - 4w = 5 \\ 3x - y - 5z - 5w = 1 \\ 2x + 3y + z - w = 8 \end{cases}$$

may be solved for any two of the unknowns in terms of the others. Note that the third equation is the same as the second minus the first. We solve any two of these equations, say the first and second, and obtain the solution given in (*a*) above.

21.18 Solve when possible.

(a) $\begin{cases} 3x - 2y = 1 \\ 4x + 3y = 41 \\ 6x + 2y = 23 \end{cases}$ (b) $\begin{cases} 3x + y = 1 \\ 5x - 2y = 20 \\ 4x + 5y = -17 \end{cases}$ (c) $\begin{cases} x + y + z = 2 \\ 4x + 5y - 3z = -15 \\ 5x - 3y + 4z = 23 \\ 7x - y + 6z = 27 \end{cases}$ (d) $\begin{cases} x + y = 5 \\ y + z = 8 \\ x + z = 7 \\ 5x - 5y + z = 1 \end{cases}$

(a) The system $\begin{cases} 3x - 2y = 1 \\ 4x + 3y = 41 \end{cases}$ has solution $x = 5$, $y = 7$.

Since $6x + 2y = 6(5) + 2(7) \neq 23$, the given system is inconsistent.

(b) The system $\begin{cases} 3x + y = 1 \\ 5x - 2y = 20 \end{cases}$ has solution $x = 2$, $y = -5$.

Since $4x + 5y = 4(2) + 5(-5) = -17$, the given system is consistent with solution $x = 2$, $y = -5$.

(c) The system $\begin{cases} x + y + z = 2 \\ 4x + 5y - 3z = -15 \\ 5x - 3y + 4z = 23 \end{cases}$ has solution $x = 1$, $y = -2$, $z = 3$.

Since $7x - y + 6z = 7(1) - (-2) + 6(3) = 27$, the given system is consistent with solution $x = 1$, $y = -2$, $z = 3$.

(d) The system $\begin{cases} x + y = 5 \\ y + z = 8 \\ x + z = 7 \end{cases}$ has solution $x = 2$, $y = 3$, $z = 5$.

Since $5x - 5y + z = 5(2) - 5(3) + (5) \neq 1$, the given system is inconsistent.

(NOTE: If the constant of the fourth equation of the system were changed from 1 to 0, the resulting system would be consistent.)

21.19 Examine the following systems for nontrivial solutions:

(a) $\begin{cases} 2x - 3y + 3z = 0 \\ 3x - 4y + 5z = 0 \\ 5x + y + 2z = 0 \end{cases}$ (b) $\begin{cases} 4x + y - 2z = 0 \\ x - 2y + z = 0 \\ 11x - 4y - z = 0 \end{cases}$ (c) $\begin{cases} x + 2y + z = 0 \\ 3x + 6y + 3z = 0 \\ 5x + 10y + 5z = 0 \end{cases}$ (d) $\begin{cases} 2x + y = 0 \\ 3y - 2z = 0 \\ 2y + w = 0 \\ 4x - w = 0 \end{cases}$

(a) Since $D = \begin{vmatrix} 2 & -3 & 3 \\ 3 & -4 & 5 \\ 5 & 1 & 2 \end{vmatrix} \neq 0$, the system has only the trivial solution.

(b) Since $D = \begin{vmatrix} 4 & 1 & -2 \\ 1 & -2 & 1 \\ 11 & -4 & -1 \end{vmatrix} = 0$, there are nontrivial solutions.

The system $\begin{cases} 4x + y = 2z \\ x - 2y = -z \end{cases}$, for which $D = \begin{vmatrix} 4 & 1 \\ 1 & -2 \end{vmatrix} \neq 0$, has the solution $x = \frac{z}{3}$, $y = \frac{2z}{3}$. This solution may be written as $x = \frac{a}{3}$, $y = \frac{2a}{3}$, $z = a$ or $x = a$, $y = 2a$, $z = 3a$, where a is arbitrary, or as $x:y:z = 1:2:3$.

(c) Here $D = \begin{vmatrix} 1 & 2 & 1 \\ 3 & 6 & 3 \\ 5 & 10 & 5 \end{vmatrix} = 0$ and there are nontrivial solutions.

Since the minor of every element of D is zero, we cannot proceed as in (b). We solve the first equation for $x = -2y - z$ and write the solution as $x = -2a - b$, $y = a$, $z = b$, where a and b are arbitrary.

(*d*) Here $D = \begin{vmatrix} 2 & 1 & 0 & 0 \\ 0 & 3 & -2 & 0 \\ 0 & 2 & 0 & 1 \\ 4 & 0 & 0 & -1 \end{vmatrix} = 0$ and there are nontrivial solutions.

Take $x = a$, where a is arbitrary. From the first equation, $y = -2a$; from the second, $2z = 3y = -6a$ and $z = -3a$; and from the fourth equation, $w = 4a$.

Thus the solution is $x = a$, $y = -2a$, $z = -3a$, $w = 4a$, or $x:y:z:w = 1:-2:-3:4$.

21.20 Solve the system $\begin{cases} 2y^2 - 3x = 0 & (1) \\ 4y - x = 6 & (2) \end{cases}$

Solve (2) for $x: x = 4y - 6$. Substitute in (1):

$$2y^2 - 3(4y - 6) = 2(y - 3)^2 = 0 \qquad y = 3, 3$$

When $y = 3$: $x = 4y - 6 = 12 - 6 = 6$. The solutions are $x = 6, y = 3; x = 6, y = 3$. The straight line is tangent to the parabola (see Fig. 21-3) at (6, 3).

21.21 Solve the system $\begin{cases} y^2 - 4y - 3x + 1 = 0 & (1) \\ 3y - 4x = 7 & (2) \end{cases}$

Solve (2) for $x: x = \frac{1}{4}(3y - 7)$. Substitute in (1):

$$y^2 - 4y - \tfrac{3}{4}(3y - 7) + 1 = 0$$

$$4y^2 - 16y - 9y + 21 + 4 = 4y^2 - 25y + 25 = (y - 5)(4y - 5) = 0 \quad \text{or} \quad y = 5 \quad \text{and} \quad y = \tfrac{5}{4}$$

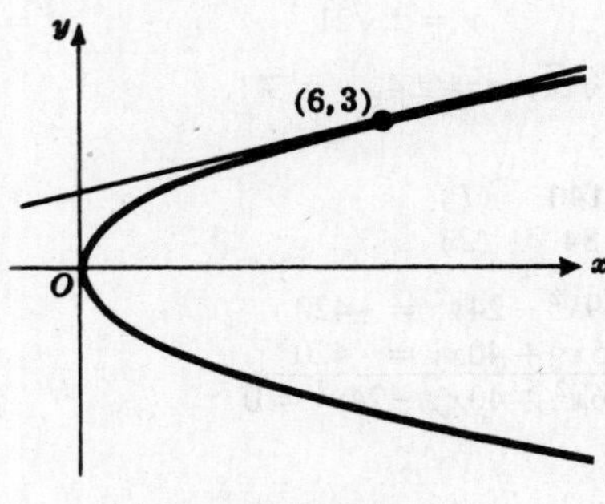

Fig. 21-3

When $y = 5, x = \frac{1}{4}(3y - 7) = 2$; when $y = \frac{5}{4}, x = \frac{1}{4}(3y - 7) = -\frac{13}{16}$. The solutions are $x = 2, y = 5$; $x = -\frac{13}{16}$, $y = \frac{5}{4}$. The straight line intersects the parabola in the points (2, 5) and $(-\frac{13}{16}, \frac{5}{4})$.

21.22 Solve the system $\begin{cases} 3x^2 - y^2 = 27 & (1) \\ x^2 - y^2 = -45 & (2) \end{cases}$

Subtract: $2x^2 = 72, x^2 = 36$, and $x = \pm 6$.

When $x = 6, y^2 = x^2 + 45 = 36 + 45 = 81$, and $y = \pm 9$.

When $x = -6$, $y^2 = x^2 + 45 = 36 + 45 = 81$, and $y = \pm 9$. The solutions are $x = \pm 6$, $y = \pm 9$; $x = \pm 6$, $y = \pm 9$. The two hyperbolas intersect in the points (6, 9), (−6, 9), (−6, −9) and (6, −9).

21.23 Solve the system $\begin{cases} 5x^2 + 3y^2 = 92 & (1) \\ 2x^2 + 5y^2 = 52 & (2) \end{cases}$

$$\begin{array}{rrl} \text{Multiply } (1) \text{ by } 5: & 25x^2 + 15y^2 = & 460 \\ \text{Multiply } (2) \text{ by } -3: & -6x^2 - 15y^2 = & -156 \\ \hline \text{Add:} & 19x^2 \qquad = & 304 \qquad x^2 = 16 \quad \text{and} \quad x = \pm 4 \end{array}$$

When $x = \pm 4$: $3y^2 = 92 - 5x^2 = 92 - 80 = 12$; $y^2 = 4$ and $y = \pm 2$. The solutions are $x = \pm 4$, $y = \pm 2$; $x = \pm 4$, $y = \pm 2$. See Fig. 21-4.

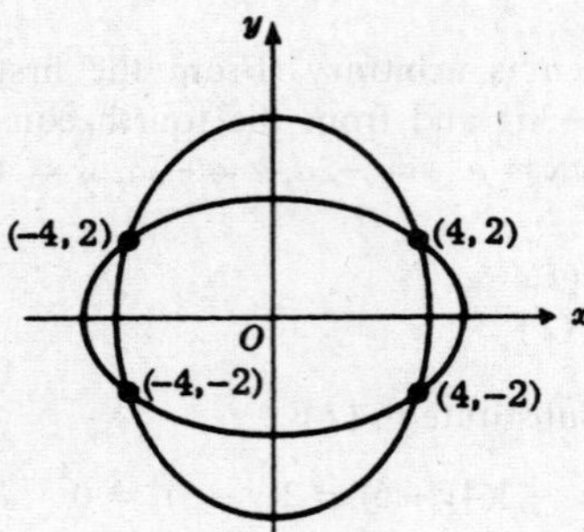

Fig. 21-4

21.24 Solve the system $\begin{cases} x^2 + 4xy = 0 & (1) \\ x^2 - xy + y^2 = 21 & (2) \end{cases}$

Solve (*1*) for x: $x(x + 4y) = 0$ and $x = 0, x = -4y$.
Solve the systems

$$\begin{cases} x^2 - xy + y^2 = 21 \\ x = 0 \end{cases} \qquad \begin{cases} x^2 - xy + y^2 = 21 \\ x = -4y \end{cases}$$

$$y^2 = 21, \quad y = \pm\sqrt{21} \qquad y^2 = 1, \quad y = \pm 1; \quad x = -4y \mp 4$$

The solutions are $x = 0$, $y = \pm\sqrt{21}$; $x = \pm 4$, $y = \mp 1$.

21.25 Solve the system $\begin{cases} 3x^2 + 8y^2 = 140 & (1) \\ 5x^2 + 8xy = 84 & (2) \end{cases}$

$$\begin{array}{rl} \text{Multiply (1) by } -3: & -9x^2 - 24y^2 = -420 \\ \text{Multiply (2) by 5:} & 25x^2 + 40xy = 420 \\ \hline \text{Add:} & 16x^2 + 40xy - 24y^2 = 0 \end{array}$$

Then

$$8(2x^2 + 5xy - 3y^2) = 8(2x - y)(x + 3y) = 0 \quad \text{and} \quad x = \tfrac{1}{2}y, \; x = -3y.$$

Solve the systems

$$\begin{cases} 3x^2 + 8y^2 = 140 \\ x = \frac{1}{2}y \end{cases} \qquad \begin{cases} 3x^2 + 8y^2 = 140 \\ x = -3y \end{cases}$$

$$\tfrac{3}{4}y^2 + 8y^2 = \tfrac{35}{4}y^2 = 140 \qquad 27y^2 + 8y^2 = 35y^2 = 140$$

$$y^2 = 16, \quad y = \pm 4, \quad x = \tfrac{1}{2}y = \pm 2 \qquad y^2 = 4, \quad y = \pm 2; \quad x = -3y = \mp 6$$

The solutions are $x = \pm 2$, $y = \pm 4$; $x = \mp 6$, $y = \pm 2$.

21.26 Solve the system $\begin{cases} x^2 - 3xy + 2y^2 = 15 & (1) \\ 2x^2 + y^2 = 6 & (2) \end{cases}$

$$\begin{array}{rl} \text{Multiply (1) by } -2: & -2x^2 + 6xy - 4y^2 = -30 \\ \text{Multiply (2) by 5:} & 10x^2 \qquad\quad + 5y^2 = 30 \\ \hline \text{Add:} & 8x^2 + 6xy + y^2 = (4x + y)(2x + y) = 0. \end{array}$$

Then $y = -4x$ and $y = -2x$.

Solve the systems

$$\begin{cases} 2x^2 + y^2 = 6 \\ \quad y = -4x \end{cases} \qquad \begin{cases} 2x^2 + y^2 = 6 \\ \quad y = -2x \end{cases}$$

$$2x^2 + 16x^2 = 18x^2 = 6, x^2 = 1/3; \qquad 2x^2 + 4x^2 = 6x^2 = 6, x^2 = 1;$$

$$x = \pm\sqrt{3}/3 \text{ and } y = -4x = \mp 4\sqrt{3}/3 \qquad x = \pm 1 \text{ and } y = -2x = \mp 2$$

The solutions are $x = \pm\sqrt{3}/3, y = \mp 4\sqrt{3}/3; x \pm 1, y = \mp 2$.

21.27 Solve the system $\begin{cases} x^2 + y^2 + 3x + 3y = 8 \\ xy + 4x + 4y = 2 \end{cases}$

Substitute $x = u + v, y = u - v$ in the given system:

$$(u+v)^2 + (u-v)^2 + 3(u+v) + 3(u-v) = 2u^2 + 2v^2 + 6u = 8 \quad (1)$$

$$(u+v)(u-v) + 4(u+v) + 4(u-v) = u^2 - v^2 + 8u = 2 \quad (2)$$

Add (*1*) and 2(*2*):

$$4u^2 + 22u - 12 = 2(2u-1)(u+6) = 0; \quad u = \tfrac{1}{2}, -6.$$

For $u = \frac{1}{2}$, (*2*) yields $v^2 = u^2 + 8u - 2 = \frac{1}{4} + 4 - 2 = \frac{9}{4}; v = \pm\frac{3}{2}$.

When $u = \frac{1}{2}, v = \frac{3}{2} : x = u + v = 2, y = u - v = -1$.
When $u = \frac{1}{2}, v = -\frac{3}{2} : x = u + v = -1, y = u - v = 2$.

For $u = -6$, (*2*) yields $v^2 = u^2 + 8u - 2 = 36 - 48 - 2 = -14; v = \pm i\sqrt{14}$.

When $u = -6, v = i\sqrt{14} : x = u + v = -6 + i\sqrt{14}, y = u - v = -6 - i\sqrt{14}$.
When $u = -6, v = -i\sqrt{14} : x = u + v = -6 - i\sqrt{14}, y = u - v = -6 + i\sqrt{14}$.

The solutions are $x = 2, y = -1; x = -1, y = 2; x = -6 \pm i\sqrt{14}, y = -6 \mp i\sqrt{14}$.

21.28 Solve the system $\begin{cases} x^2 + y^2 = 25 & (1) \\ \quad xy = 12 & (2) \end{cases}$

Multiply (*2*) by 2 and add to (*1*): $x^2 + 2xy + y^2 = 49$ or $x + y = \pm 7$.
Multiply (*2*) by -2 and add to (*1*): $x^2 - 2xy + y^2 = 1$ or $x - y = \pm 1$.
Solve the systems

$$\begin{cases} x + y = 7 \\ x - y = 1 \end{cases} \qquad \begin{cases} x + y = 7 \\ x - y = -1 \end{cases}$$

$$2x = 8; \quad x = 4 \qquad 2x = 6; \quad x = 3$$

$$y = 7 - x = 3 \qquad y = 7 - x = 4$$

$$\begin{cases} x + y = -7 \\ x - y = \ \ 1 \end{cases} \qquad \begin{cases} x + y = -7 \\ x - y = -1 \end{cases}$$

$$2x = -6; \quad x = -3 \qquad 2x = -8; \quad x = -4$$

$$y = -7 - x = -4 \qquad y = -7 - x = -3$$

The solutions are $x = \pm 4, y = \pm 3; x = \pm 3, y = \pm 4$. See Fig. 21-5.

Alternate Solutions. Solve (*2*) for $y = 12/x$ and substitute (*1*). The resulting quartic $x^4 - 25x^2 + 144 = 0$ can be factored readily.

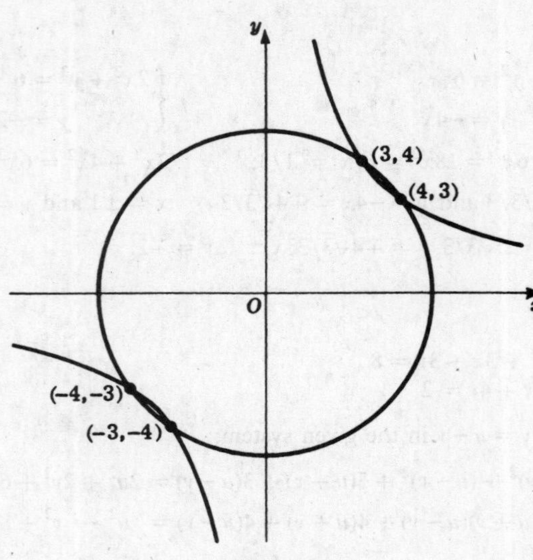

Fig. 21-5

21.29 Solve the system $\begin{cases} x^2 - xy - 12y^2 = 8 & (1) \\ x^2 + xy - 10y^2 = 20 & (2) \end{cases}$

This system may be solved by the procedure used in Problems 21.25 and 21.26. Here we give an alternate solution.

Procedure: Substitute $y = mx$ in the given equations to obtain a system in the unknowns m and x: then eliminate x to obtain a quadratic in m.

Put $y = mx$ in (1) and (2):

$$x^2 - mx^2 - 12m^2x^2 = x^2(1 - m - 12m^2) = 8$$
$$x^2 + mx^2 - 10m^2x^2 = x^2(1 + m - 10m^2) = 20$$

Now

$$x^2 = \frac{8}{1 - m - 12m^2} \quad \text{and} \quad x^2 = \frac{20}{1 + m - 10m^2}$$

so that

$$\frac{8}{1 - m - 12m^2} = \frac{20}{1 + m - 10m^2}$$
$$8 + 8m - 80m^2 = 20 - 20m - 240m^2$$
$$160m^2 + 28m - 12 = 4(5m - 1)(8m + 3) = 0 \quad \text{and} \quad m = \tfrac{1}{5}, m = -\tfrac{3}{8}$$

then $m = \frac{1}{5}$: $x^2 = \dfrac{8}{1 - m - 12m^2} = 25; x = \pm 5, y = mx - \dfrac{1}{5}(\pm 5) = \pm 1.$

then $m = -\frac{3}{8}$:

$$x^2 = \frac{8}{1 - m - 12m^2} = -\frac{128}{5}; \quad x = \pm\frac{8i\sqrt{10}}{5}, \quad y = mx = \mp\frac{3i\sqrt{10}}{5}.$$

The solutions are $x = \pm 5, y = \pm 1; x = \pm\dfrac{8i\sqrt{10}}{5}, y = \mp\dfrac{3i\sqrt{10}}{5}$.

21.30 Solve the system $\begin{cases} x^3 - y^3 = 19 & (1) \\ x^2 + xy + y^2 = 19 & (2) \end{cases}$

Divide (1) by (2): $x - y = 1$.

Solve the system $\begin{cases} x^2 + xy + y^2 = 19 & (2) \\ x - y = 1 & (3) \end{cases}$

Solve (*3*) for $x: x = y + 1$.
Substitute in (*2*): $(y + 1)^2 + (y + 1)y + y^2 = 3y^2 + 3y + 1 = 19$.
Then $3y^2 + 3y - 18 = 3(y + 3)(y - 2) = 0$ and $y = -3, 2$.
When $y = -3, x = y + 1 = -2$; when $y = 2, x = y + 1 = 3$.

The solutions are $x = -2, y = -3; x = 3, y = 2$.

21.31 Solve the system $\begin{cases} (2x - y)^2 - 4(2x - y) = 5 & (1) \\ x^2 - y^2 = 3 & (2) \end{cases}$

Factor (*1*): $(2x - y)^2 - 4(2x - y) - 5 = (2x - y - 5)(2x - y + 1) = 0$. Then $2x - y = 5$ and $2x - y = -1$.
Solve the systems

$$\begin{cases} x^2 - y^2 = 3 \\ 2x - y = 5 \end{cases} \qquad \begin{cases} x^2 - y^2 = 3 \\ 2x - y = -1 \end{cases}$$

$$y = 2x - 5 \qquad y = 2x + 1$$
$$x^2 - (2x - 5)^2 = 3 \qquad x^2 - (2x + 1)^2 = 3$$
$$3x^2 - 20x + 28 = (x - 2)(3x - 14) = 0 \qquad 3x^2 + 4x + 4 = 0$$
$$x = 2, \tfrac{14}{3} \qquad x = \frac{-4 \pm \sqrt{16 - 48}}{6} = \frac{-2 \pm 2i\sqrt{2}}{3}$$

When $x = 2, y = 2x - 5 = -1$. $\qquad y = 2x + 1 = \dfrac{-1 \pm 4i\sqrt{2}}{3}$

When $x = \frac{14}{3}, y = 2x - 5 = \frac{13}{3}$.

The solutions are $x = 2, y = -1; x = \frac{14}{3}, y = \frac{13}{3}; x = \dfrac{-2 \pm 2i\sqrt{2}}{3}, y = \dfrac{-1 \pm 4i\sqrt{2}}{3}$.

21.32 Solve the system $\begin{cases} 5/x^2 + 3/y^2 = 32 & (1) \\ 4xy = 1 & (2) \end{cases}$

Write (*1*) as $3x^2 + 5y^2 = 32x^2y^2 = 2(4xy)^2$. Substitute (2): $3^2 + 5y^2 = 2(1)^2 = 2$ (*3*)
Subtract 2(*2*) from (*3*): $3x^2 - 8xy + 5y^2 = 0$. Then $(x - y)(3x - 5y) = 0$ and $x = y, x = 5y/3$.
Solve the systems

$$\begin{cases} 4xy = 1 \\ x = y \end{cases} \qquad \begin{cases} 4xy = 1 \\ x = 5y/3 \end{cases}$$

$$4y^2 = 1; \quad y = \pm\tfrac{1}{2} \qquad \frac{20}{3}y^2 = 1, \quad y^2 = \frac{3}{20} = \frac{15}{100};$$

and $x = y = \pm\frac{1}{2}$.

$$y = \pm\frac{\sqrt{15}}{10} \quad \text{and} \quad x = \frac{5}{3}y = \pm\frac{\sqrt{15}}{6}.$$

The solutions are $x = \pm\frac{1}{2}, y = \pm\frac{1}{2}; x = \pm\dfrac{\sqrt{15}}{6}, y = \pm\dfrac{\sqrt{15}}{10}$

Supplementary Problems

21.33 Evaluate. (*a*) $\begin{vmatrix} 1 & 2 \\ -3 & 1 \end{vmatrix}$ (*b*) $\begin{vmatrix} 2 & 0 \\ -5 & 1 \end{vmatrix}$ (*c*) $\begin{vmatrix} 3 & 2 \\ 1 & -4 \end{vmatrix}$ (*d*) $\begin{vmatrix} 2 & -10 \\ 3 & -15 \end{vmatrix}$

Ans. (*a*) 7 (*b*) 2 (*c*) −14 (*d*) 0

21.34 Solve, using determinants. (*a*) $\begin{cases} 2x + y = 4 \\ 3x + 4y = 1 \end{cases}$ (*b*) $\begin{cases} 5x + 2y = 2 \\ 3x - 5y = 26 \end{cases}$ (*c*) $\begin{cases} 5x + 3y = -6 \\ 3x + 5y = -18 \end{cases}$

Ans. (*a*) $x = 3, y = -2$ (*b*) $x = 2, y = -4$ (*c*) $x = \frac{3}{2}, y = -\frac{9}{2}$

21.35 Evaluate. (*a*) $\begin{vmatrix} 1 & 2 & 3 \\ 2 & -3 & 4 \\ -3 & 4 & 5 \end{vmatrix}$ (*b*) $\begin{vmatrix} 2 & -1 & 3 \\ 2 & 1 & 5 \\ -2 & 3 & 2 \end{vmatrix}$ (*c*) $\begin{vmatrix} -2 & 3 & 1 \\ 3 & 5 & 1 \\ 4 & -2 & 1 \end{vmatrix}$ (*d*) $\begin{vmatrix} 2 & 6 & 1 \\ 4 & -4 & 1 \\ 3 & 1 & 1 \end{vmatrix}$

Ans. (*a*) -78 (*b*) -4 (*c*) -37 (*d*) 0

21.36 Solve, using determinants.

(*a*) $\begin{cases} x+2y+2z=4 \\ 3x-y+4z=25 \\ 3x+2y-z=-4 \end{cases}$ (*b*) $\begin{cases} 2x-3y+5z=4 \\ 3x-2y+2z=3 \\ 4x+y-4z=-6 \end{cases}$ (*c*) $\begin{cases} \dfrac{1}{x}+\dfrac{2}{y}+\dfrac{1}{z}=2 \\ \dfrac{3}{x}-\dfrac{4}{y}-\dfrac{2}{z}=1 \\ \dfrac{2}{x}+\dfrac{5}{y}-\dfrac{2}{z}=3 \end{cases}$

Hint: In (*c*) solve first for $1/x$, $1/y$, $1/z$.

Ans. (*a*) $x=2,\ y=-3,\ z=4$ (*b*) $x=-\frac{1}{3},\ y=\frac{2}{3},\ z=\frac{4}{3}$ (*c*) $x=1,\ y=z=3$

21.37 Verify, by evaluating the determinants.

(*a*) $\begin{vmatrix} 1 & 2 & 3 \\ a & 2a & 3a \\ 8 & 9 & 10 \end{vmatrix}=0$ (*b*) $\begin{vmatrix} 2 & 1 & -1 \\ 3 & 4 & 2 \\ -2 & -5 & 3 \end{vmatrix}=\begin{vmatrix} 0 & 1 & 0 \\ -5 & 4 & 6 \\ 8 & -5 & -2 \end{vmatrix}$

(*c*) $\begin{vmatrix} 3 & 4 & 5 \\ 7 & -2 & 3 \\ 2 & 5 & -1 \end{vmatrix}=-\begin{vmatrix} 4 & -2 & 5 \\ 3 & 7 & 2 \\ 5 & 3 & -1 \end{vmatrix}$ (*d*) $\begin{vmatrix} 4 & 2 & 5 \\ -7 & -3 & 1 \\ 9 & 4 & 8 \end{vmatrix}+\begin{vmatrix} -1 & 2 & 5 \\ 5 & -3 & 1 \\ -3 & 4 & 8 \end{vmatrix}=\begin{vmatrix} 3 & 2 & 5 \\ -2 & -3 & 1 \\ 6 & 4 & 8 \end{vmatrix}$

21.38 Solve, using determinants. $\begin{vmatrix} x & y & 1 \\ 1 & -1 & 1 \\ 13 & 2 & 1 \end{vmatrix}=0,\quad \begin{vmatrix} x & y & 1 \\ 3 & 2 & 1 \\ -6 & -4 & 1 \end{vmatrix}=0.$

Ans. $x=-3,\ y=-2$

21.39 Verify: The value of a determinant is unchanged if the rows are written as columns or if its columns are written as rows.

Hint: Show that $\begin{vmatrix} a_1 & b_1 & c_1 \\ a_2 & b_2 & c_2 \\ a_3 & b_3 & c_3 \end{vmatrix}=\begin{vmatrix} a_1 & a_2 & a_3 \\ b_1 & b_2 & b_3 \\ c_1 & c_2 & c_3 \end{vmatrix}$.

21.40 Prove by induction:

(*a*) The expansion of a determinant of order *n* contains *n*! terms.
(*b*) If two rows (or two columns) of a determinant are interchanged, the sign of the determinant is changed.

Hint: Show that $\begin{vmatrix} a_1 & b_1 \\ a_2 & b_2 \end{vmatrix}=-\begin{vmatrix} b_1 & a_1 \\ b_2 & a_2 \end{vmatrix}$ and proceed as in Problem 21.10.

21.41 Show, without expanding the determinants, that

(*a*) $\begin{vmatrix} 3 & 2 & 1 & 4 \\ -1 & 5 & 2 & 6 \\ 2 & -4 & 7 & -5 \\ -2 & 1 & 3 & 5 \end{vmatrix}=-\begin{vmatrix} 3 & 2 & 1 & 4 \\ 2 & -4 & 7 & -5 \\ -1 & 5 & 2 & 6 \\ -2 & 1 & 3 & 5 \end{vmatrix}=-\begin{vmatrix} 3 & 2 & 1 & 4 \\ -2 & 1 & 3 & 5 \\ 2 & -4 & 7 & -5 \\ -1 & 5 & 2 & 6 \end{vmatrix}=-\begin{vmatrix} 3 & -2 & 2 & -1 \\ 2 & 1 & -4 & 5 \\ 1 & 3 & 7 & 2 \\ 4 & 5 & -5 & 6 \end{vmatrix}$

(b) $\begin{vmatrix} 10 & 0 & -2 \\ -10 & 3 & -4 \\ -5 & -2 & 3 \end{vmatrix} = -5\begin{vmatrix} -2 & 0 & -2 \\ 2 & 3 & -4 \\ 1 & -2 & 3 \end{vmatrix} = 10\begin{vmatrix} 1 & 0 & 1 \\ 2 & 3 & -4 \\ 1 & -2 & 3 \end{vmatrix}$

(c) $\begin{vmatrix} -3 & -1 & 2 \\ 1 & -3 & 3 \\ 4 & -2 & 1 \end{vmatrix} = 0$ **Hint**: Subtract the first row from the second.

(d) $\begin{vmatrix} a-3b & a+b & a+5b & e \\ a-2b & a+2b & a+6b & f \\ a-b & a+3b & a+7b & g \\ a & a+4b & a+8b & h \end{vmatrix} = 0$ (NOTE: The elements of the first three columns form an arithmetic progression.)

(e) $\begin{vmatrix} 0 & a_1 & a_2 \\ -a_1 & 0 & a_3 \\ -a_2 & -a_3 & 0 \end{vmatrix} = 0$ **Hint**: Write the rows as columns and factor −1 from each row.

21.42 Verify: If the corresponding elements of two rows (or two columns) of a determinant are proportional, the value of the determinant is zero.

21.43 Evaluate each of the following determinants:

(a) $\begin{vmatrix} -3 & 6 & -1 & 1 \\ -4 & -3 & -2 & 4 \\ 5 & -4 & 1 & 3 \\ -1 & -5 & 0 & -1 \end{vmatrix}$ (b) $\begin{vmatrix} 1 & -1 & 1 & 2 \\ 3 & -2 & 4 & -3 \\ 5 & 4 & 1 & 2 \\ -3 & 0 & 3 & 1 \end{vmatrix}$ (c) $\begin{vmatrix} 3 & -1 & -1 & -4 \\ 2 & 3 & -6 & 1 \\ 4 & -1 & 3 & 1 \\ 3 & -1 & 5 & 2 \end{vmatrix}$

(d) $\begin{vmatrix} 2 & 4 & 4 & -3 & 4 \\ 1 & 3 & 1 & 0 & -1 \\ -1 & 3 & 1 & 2 & -1 \\ 4 & 8 & 11 & -10 & 9 \\ 2 & 6 & 9 & -12 & 5 \end{vmatrix}$

Ans. (a) −50 (b) −397 (c) −78 (d) −316

21.44 Show that $\begin{vmatrix} 1 & 1 & 1 & 1 \\ a & b & c & d \\ a^2 & b^2 & c^2 & d^2 \\ a^3 & b^3 & c^3 & d^3 \end{vmatrix} = (b-a)(c-a)(d-a)(c-b)(d-b)(d-c)$

21.45 Write a determinant that is equal to $x^2 - 1$. (See Problem 21.44.)

21.46 Repeat Problem 20.7 above using a computer software package to evaluate the given determinants.

21.47 Solve, using determinants.

(a) $\begin{cases} x+y+z=6 \\ y+z+w=9 \\ z+w+x=8 \\ w+x+y=7 \end{cases}$ (b) $\begin{cases} 3x-2y+2z+w=5 \\ 2x+4y-z-2w=3 \\ 3x+7y-z+3w=23 \\ x-3y+2z-3w=-12 \end{cases}$ (c) $\begin{cases} x+y+z+w=2 \\ 2x+3y-2z-w=5 \\ 3x-2y+z+3w=4 \\ 5x+2y+3z-2w=-4 \end{cases}$

Ans. (a) $x=1, y=2, z=3, w=4$ (b) $x=2, y=1, z=-1, w=3$
(c) $x=\frac{1}{2}, y=1, z=-\frac{3}{2}, w=2$

21.48 Test for consistency, and solve when possible.

(a) $\begin{cases} 2x + 3y - 4z = 1 \\ 3x - y + 2z = -2 \\ 5x - 9y + 14z = 3 \end{cases}$ (b) $\begin{cases} x + 7y + 5z = -22 \\ x - 9y - 11z = 26 \\ x - y - 3z = -22 \end{cases}$ (c) $\begin{cases} x + y + z = 4 \\ 2x - 4y + 11z = -7 \\ 4x + 6y + z = 21 \end{cases}$

Ans. (a) Inconsistent (b) $x = 2z - 1$, $y = -z - 3$ (c) $x = \frac{1}{2}(3 - 5z)$, $y = \frac{1}{2}(5 + 3z)$

21.49 Solve, when possible.

(a) $\begin{cases} x - 3y + 11 = 0 \\ 3x + 2y - 33 = 0 \\ 2x - 3y + 4 = 0 \end{cases}$ (b) $\begin{cases} x - 2y - 8 = 0 \\ 3x + y - 3 = 0 \\ x - 10y + 32 = 0 \end{cases}$ (c) $\begin{cases} 2x - 3y - 7 = 0 \\ 5x + 4y + 17 = 0 \\ 4x - y + 1 = 0 \end{cases}$

Ans. (a) $x = 7$, $y = 6$ (b) No solution (c) $x = -1$, $y = -3$

21.50 Solve the systems.

(a) $\begin{cases} 2x - y + 3z = 8 \\ x + 3y - 2z = -3 \end{cases}$ (b) $\begin{cases} 4x + 2y + z = 13 \\ 2x + y - 2z = -6 \end{cases}$ (c) $\begin{cases} 4x + 2y - 6z + w = 10 \\ 3x - y - 9z - w = 7 \\ 7x + y - 11z - w = 13 \end{cases}$

Ans. (a) $x = 3 - z$, $y = -2 + z$ (b) $y = 4 - 2x$, $z = 5$
(c) $x = 7w/10$, $y = 2 - 23w/20$, $z = -1 + w/4$

21.51 Examine for nontrivial solutions.

(a) $\begin{cases} 3x + y - 9z = 0 \\ 4x - 3y + z = 0 \\ 6x - 11y + 21z = 0 \end{cases}$ (b) $\begin{cases} 2x - 3y - 5z = 0 \\ x + 2y - 13z = 0 \\ 9x - 10y - 30z = 0 \end{cases}$ (c) $\begin{cases} 2x - 3y + 2z - 9w = 0 \\ x + 4y - z + 3w = 0 \\ 3x - 2y - 2z - 6w = 0 \\ 7x + 11y + 3z - 6w = 0 \end{cases}$

Ans. (a) $x = 2z$, $y = 3z$ (b) $x = y = z = 0$ (c) $x = 2w$, $y = -w$, $z = w$

21.52 Is it the case that the converse of the statement "A system of n homogeneous linear equations in n unknowns has only the trivial solution if D, the determinant of the coefficients, is not equal to zero is true?" In other words, is this an "if and only if" statement?

21.53 Solve Problem 21.51 using a computer software package.

Solve.

21.54 $\begin{cases} xy + y^2 = 5 \\ 2x + 3y = 7 \end{cases}$ *Ans.* $x = -4, y = 5$; $x = \frac{1}{2}, y = 2$

21.55 $\begin{cases} y = x^2 - x - 1 \\ y = 2x + 3 \end{cases}$ *Ans.* $x = -1, y = 1$; $x = 4, y = 11$

21.56 $\begin{cases} 3x^3 - 7y^2 = 12 \\ x - 3y = -2 \end{cases}$ *Ans.* $x = -2, y = 0$; $x = \frac{17}{5}, y = \frac{9}{5}$

21.57 $\begin{cases} x^2 + 3y^2 = 43 \\ 3x^2 + y^2 = 57 \end{cases}$ *Ans.* $x = \pm 4, y = \pm 3$; $x = \pm 4, y = \mp 3$

21.58 $\begin{cases} 9x^2 + y^2 = 90 \\ x^2 + 9y^2 = 90 \end{cases}$ *Ans.* $x = \pm 3, y = \pm 3$; $x = \pm 3, y = \mp 3$

21.59 $\begin{cases} 2/x^2 - 3/y^2 = 5 \\ 1/x^2 + 2/y^2 = 6 \end{cases}$ *Ans.* $x = \pm\frac{1}{2}, y = \pm 1$; $x = \pm\frac{1}{2}, y = \mp 1$

21.60 $\begin{cases} x^2 - xy^2 + y^2 = 28 \\ 2x^2 + 3xy - 2y^2 = 0 \end{cases}$ *Ans.* $x = \pm 4, y = \mp 2$; $x = \pm\sqrt{21}/3, y = \pm 4\sqrt{21}/3$

21.61 $\begin{cases} x^2 - xy^2 - 12y^2 = 0 \\ x^2 + xy - 10y^2 = 20 \end{cases}$ *Ans.* $x = \pm 4\sqrt{2}, y = \pm\sqrt{2}$; $x = \mp 3i\sqrt{5}, y = \pm i\sqrt{5}$

21.62 $\begin{cases} 6x^2 + 3xy^2 + 2y^2 = 24 \\ 3x^2 + 2xy + 2y^2 = 18 \end{cases}$ *Ans.* $x = \pm 2, y = \mp 3$; $x = \pm\sqrt{30}/5, y = \pm 2\sqrt{30}/5$

21.63 $\begin{cases} y^2 = 4x - 8 \\ y^2 = -6x + 32 \end{cases}$ *Ans.* $x = 4, y = \pm 2\sqrt{2}$

21.64 $\begin{cases} x^2 - y^2 = 16 \\ y^2 = 2x - 1 \end{cases}$ *Ans.* $x = 5, y = \pm 3$; $x = -3, y = \pm i\sqrt{7}$

21.65 $\begin{cases} 2x^2 + y^2 = 6 \\ x^2 + y^2 + 2x = 3 \end{cases}$ *Ans.* $x = -1, y = \pm 2$; $x = 3, y = \pm 2i\sqrt{3}$

21.66 $\begin{cases} x^2 + y^2 - 2x - 2y = 12 \\ xy = 6 \end{cases}$ *Ans.* $x = 3 \pm \sqrt{3}, y = 3 \mp \sqrt{3}$; $x = -2 \pm i\sqrt{2}, y = -2 \mp i\sqrt{2}$

21.67 $\begin{cases} x^3 - y^3 = 28 \\ x - y = 4 \end{cases}$ *Ans.* $x = 1, y = -3$; $x = 3, y = -1$

21.68 $\begin{cases} x + y + 3\sqrt{x+y} = 18 \\ x - y - 2\sqrt{x-y} = 15 \end{cases}$ **Hint:** Let $\sqrt{x+y} = u$, $\sqrt{x-y} = v$. *Ans.* $x = 17,\ \ y = -18$

21.69 Two numbers differ by 2 and their squares differ by 48. Find the numbers. *Ans.* 11, 13

21.70 The sum of the circumference of two circles is 88 cm and the sum of their areas is $\frac{2200}{7}\text{cm}^2$, when $\frac{22}{7}$ is used for π. Find the radius of each circle. *Ans.* 6 cm, 8 cm

21.71 A party costing \$30 is planned. It is found that by adding three more to the group, the cost per person would be reduced by 50 cents. For how many was the party originally planned? *Ans.* 12

Vector Calculus: Grad, Div, and Curl

INTRODUCTION. In the previous chapters, we have explored a number of topics. These topics include scalar functions that associate a single quantity (such as temperature or density) with a particular point in space (a space that may consist of one or more dimensions). We have also considered vector quantities, such as the velocity of a baseball or the force on a door, which include both a magnitude and a direction. We extended the idea of an object that acts like a vector to a continuous distribution, or vector field, in which there is a magnitude and a position associated with every point in space. The electric field, the gravitational field, and the distribution of molecular velocities (if we ignore the gaps between the molecules) all provide examples of such fields. For both the scalar and the vector functions, we considered both functions of a single variable and functions of multiple variables.

We applied the ideas and techniques of both differential and integral calculus to each of these cases and are now ready to combine some of these ideas in order to solve more sophisticated problems in both mathematics and physics. Calculus is the mathematics of change, and so we now in a position to consider more subtle and sophisticated kinds of changes in physical quantities than we have done so far. When coupled with the experiments and observations that form the basis of physical law, the discussion that follows will allow for the analysis of a wide array of physical situations, including the direction of greatest increase of the temperature in a room, sources and "sinks" of electric charge, and the turbulent flow of liquids.

THE GRADIENT AND DEL OPERATOR. We begin by considering the density distribution of matter within a solid. As before, we can define an average density $\bar{\rho}$ throughout the solid, where $\bar{\rho}$ is equal to the total mass divided by the total volume of the solid. Of course, there will in general be variations in density within the solid, with some regions having a greater mass per unit volume than others. Within a tiny volume centered at a point (x,y,z), the *local density* $\rho(x,y,z)$ is equal to the mass within that (tiny) volume divided by the tiny volume itself. Hence, the mass of this volume is $\Delta m = \rho(x,y,z)\,\Delta V$. In the differential limit, this becomes $\mathrm{d}m = \rho(x,y,z)\mathrm{d}V$.

Suppose we want to know how quickly the density $\rho(x,y,z)$ is changing as a function of position. We might want to know in what directions it increases most rapidly, decreases most rapidly, and stays unchanged. There are seemingly an infinite number of different directions from a particular point from which the density changes and it is unclear that we will want to examine all of them!

How, then, do we calculate the rate of change of the density in an arbitrary direction? Fortunately, we are aided by a theorem that relates differential changes in a function to its partial derivatives:

$$d\rho = \frac{\partial \rho}{\partial x}dx + \frac{\partial \rho}{\partial y}dy + \frac{\partial \rho}{\partial z}dz \tag{22.1}$$

For example, if we know the function $\rho = \rho(x,y,z)$, then $\partial\rho/\partial x$ gives the change in density with respect to the x direction. Similarly, we can calculate $\partial\rho/\partial y$ and $\partial\rho/\partial z$ to find the density variations with respect to the y and z axes. We can then use Eq. (22.1) total change in density for simultaneous variations in all three directions from the point (x,y,z) that we are considering.

We can express this relationship as a kind of dot product:

$$\mathrm{d}\rho = \vec{\nabla}\rho \cdot \mathrm{d}\vec{l} \tag{22.2}$$

where we define the vector

$$\vec{\nabla}\rho = \left(\hat{x}\frac{\partial \rho}{\partial x} + \hat{y}\frac{\partial \rho}{\partial y} + \hat{z}\frac{\partial \rho}{\partial z}\right) \tag{22.3}$$

as the gradient of ρ and the differential length dl is defined by

$$\mathrm{d}\vec{l} = \hat{x}\mathrm{d}x + \hat{y}\mathrm{d}y + \hat{z}\mathrm{d}z^{*} \tag{22.4}$$

By writing things in this form, we are able to address our original challenge of determining the rate of change of a scalar function (in this case $\rho(x,y,z)$) in the vicinity of a point (x,y,z). We do so by noting that, according to (*22.2*)

$$|\mathrm{d}\rho| = \left|\vec{\nabla}\rho\right|\left|\mathrm{d}\vec{l}\right| \cos\theta \tag{22.5}$$

where θ is the angle between the vectors $\vec{\nabla}\rho$ and $\mathrm{d}\vec{l}$. For a particular point (x,y,z), the value of $|\mathrm{d}\rho|$ will increase most rapidly when $\vec{\nabla}\rho$ and $\mathrm{d}\vec{l}$ are aligned[†] (since $\theta = 0$ corresponds to the maximum value of cos $\theta(= 1)$). Hence, $\vec{\nabla}\rho$ points in the direction of maximum increase of ρ. This is a very useful property that allows one to find the maximum increase of any scalar function at any point.

Similarly, when for changes that are perpendicular to the direction of the gradient, $\theta = 90°$ and $\cos\theta = 0$, leading to $\mathrm{d}\rho = 0$—that is, no change in the density. In other words, *movement along the paths that are perpendicular to the gradient keeps the value of* ρ *unchanged.*

Tracing out the path of the curve defined by a lack of change in the gradient then corresponds to a contour line along on which every point has the same value of the density. Doing so for several different density values, then results in the kind of contour map shown in Fig. 22-1, in which a function $z = f(x,y)$ is shown along with the contour lines and the perpendicular lines to them that point along the direction of the gradient—that is, along the direction of greatest ascent (descent). Fig. 22-1 indicates the function plotted as a three-dimensional figure, while Fig. 22-2 shows the two-dimensional projection of the contour lines. Of course, while we have used density as a physical example, the same discussion could be applied to a number of different physical quantities that vary, more or less continuously, with position. One could then picture similar contour maps showing changes (and constant values) of altitude as a function of position on a two-dimensional grid (the kind of contour map with which hikers and backpackers are familiar). In addition, electric fields, gravitational fields, temperatures, and many other physical quantities can be represented in a similar fashion.

---

[*] Note that the factors on the right-hand side of equation (*22.2*) give nine terms involving the products of unit vectors. Recall that, because the unit vectors are mutually perpendicular only terms with factors, such as $\hat{x}\cdot\hat{x}$, $\hat{y}\cdot\hat{y}$ and $\hat{z}\cdot\hat{z}$ appear above—the others vanish because, when a and b are mutually perpendicular, $\vec{a}\cdot\vec{b} = |a||b| \sin(\theta) = |a||b| \sin(90) = 0$.

[†] When the two vectors are aligned, the angle θ between them is 0, and $\mathrm{d}\rho$ reaches its maximum value according to equation (*22.5*).

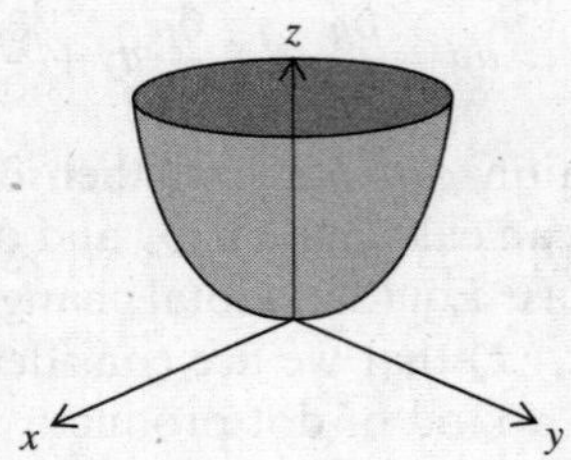

Fig. 22-1

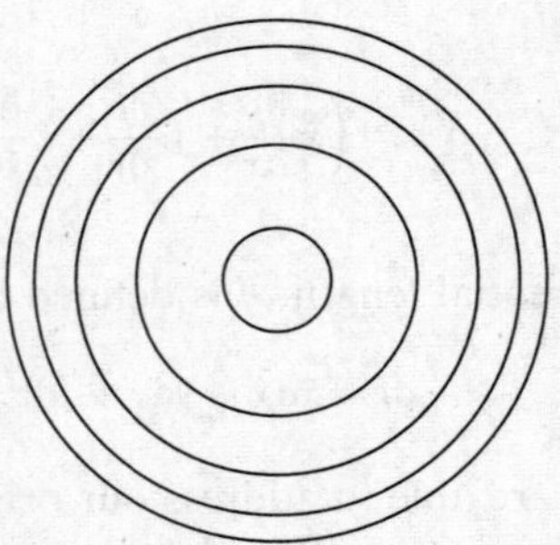

Fig. 22-2

EXAMPLE 1. Find the direction of maximum increase and the contour lines of the function $\rho = x^2 + y^2$.

We use a 2D density function here because it is easier to visualize. (Think of a sprinkling of sand in which the grain density increases with distance according to the above formula.)*

According to our above discussion, the direction of maximum increase is given by $\vec{\nabla}\rho$, which we calculate as:

$$\vec{\nabla}\rho = \left(\hat{x}\frac{\partial \rho}{\partial x} + \hat{y}\frac{\partial \rho}{\partial y}\right) = \left(\hat{x}\frac{\partial(x^2+y^2)}{\partial x} + \hat{y}\frac{\partial(x^2+y^2)}{\partial y}\right) = 2x\hat{x} + 2y\hat{y}$$

Hence, along the horizontal axis ($y = 0$), the direction of maximum increase is in the vertical direction and along the vertical axis, it is in the horizontal direction. Because the slope of the gradient vector in this example is y/x, the direction of maximal increase is in the radial direction. The contours that are perpendicular to these vectors form circles around the origin. In general, the contour lines will not be so simple, as indicated in the previous figure.

Recall that the gradient of a function $f(x,y,z)$ is defined by

$$\vec{\nabla}f = \left(\hat{x}\frac{\partial f}{\partial x} + \hat{y}\frac{\partial f}{\partial y} + \hat{z}\frac{\partial f}{\partial z}\right) \tag{22.6}$$

The definition makes it clear that the gradient $\vec{\nabla}f$ (often pronounced "grad f") is a vector function (with both magnitude and direction) that is "built from" the scalar function $f(x,y,z)$ (with only magnitude). In the above expression, we can consider ∇ (pronounced "del") to be a separate entity (and a rather odd-looking one) defined by:

$$\vec{\nabla} = \left(\hat{x}\frac{\partial}{\partial x} + \hat{y}\frac{\partial}{\partial y} + \hat{z}\frac{\partial}{\partial z}\right) \tag{22.7}$$

The gradient $\vec{\nabla}f$ then comes about from $\vec{\nabla}$ "operating" on $f(x,y,z)$ to produce the expression shown in (*22.6*). $\vec{\nabla}$ is known as a mathematical operator, operating on the scalar function $f(x,y,z)$ to produce the vector function $\vec{\nabla}f$. It should not be construed as multiplying the function. Note that in order to have meaning, the operator needs

* Note that, since $r = \sqrt{x^2 + y^2}$ describes the radius of a circle, this function describes a density distribution that is increasing according to the square of the distance from the center.

to have a function on which to operate. $\vec{\nabla}$ is just one example of an array of mathematical operators that act upon various functions; we shall consider them at great length in a later chapter.

THE DIVERGENCE. We considered in the previous section the use of the gradient in helping us to determine, for a particular function, the direction of maximum increase of a (scalar) function (we used the density function as an example) and also the construction of contour lines on which the function assumed a constant value. In doing so, we used the gradient—a vector function—to help us determining the properties of a scalar function. We now reverse the situation: we construct a scalar quantity that helps us analyze a vector function.

The particular property among vector functions that we want to consider is their tendency to either "diverge from" or "converge toward" a particular point. We illustrate what we are talking about in Fig. 22-3, using, as an example, a vector function representing. Note that, in the vicinity of the indicated point (x,y,z), the vectors are pointing outward, and we say that the function at this point has a large positive divergence. In Fig. 22-4, we see that the vectors are pointing inward and we say that the function has a large negative divergence. In Fig. 22-5, we see that both the vectors are pointing into and away from the point and so the divergence is somewhere in between.

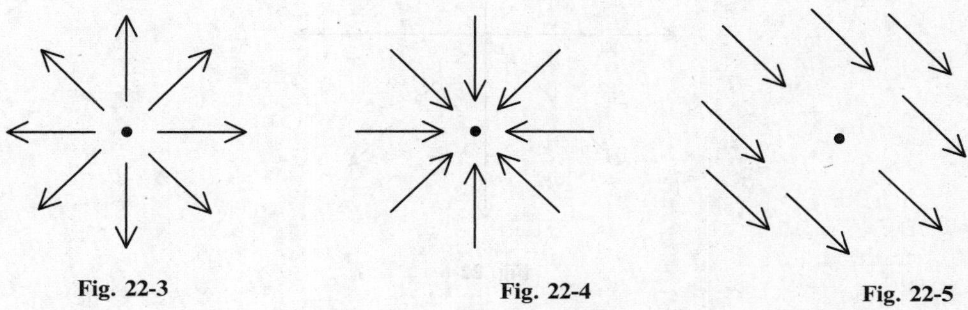

Fig. 22-3 Fig. 22-4 Fig. 22-5

DIVERGENCE AND SOURCES. Consider, for example, a garden hose with a nozzle that sprays water in all directions. If we consider the droplets to be spread continuously throughout space, then we can consider the velocity distribution of the droplets to be a vector function that is a continuous function of space. If we idealize the place where the water leaves the nozzle to be a point, then we can say that, in a tiny volume surrounding this point, the velocity vectors, pointing outward from the point, have a *positive divergence*—loosely speaking, there is more water flowing outward from this point than is flowing in. This is equivalent to saying that this point is a source of water, or more generally a *source*. As we shall see, *vector function sources are equivalent to points of positive divergence*.

DIVERGENCE AND SINKS. Conversely, we can imagine that this is a drainpipe somewhere along the ground where this water is spraying. At the drainpipe, water is gathering together and moving through the pipe. If one examines a small volume surrounding this point, the velocity vectors are converging toward the drainpipe (which we again idealize as a single point in space). We say that this point is a sink (rather than a source) of water and that it has a negative divergence.

What about those points in between that are neither a source nor a sink of water? Within small volumes surrounding such points, water is merely flowing through, with some velocity vectors pointing in toward and others pointing outward from the volume. What is the divergence of points within such volumes? You might think that the answer depends upon the particular location or the details of the velocity distribution of the water, but this turns out not to be the case. As we shall see below, the divergence of such points, which act as neither a source nor a sink, is *zero*.

Of course, those are qualitative descriptions of divergence, and we need to find a working mathematical definition that will ultimately allow us to solve problems. We shall define the divergence first and then demonstrate that it has conforms to our qualitative thinking. Finally, we will try to give an intuitive sense for why the definition works.

First, the definition: the divergence of a function E is defined by $\nabla \cdot E$, where ∇ is the del operator defined above and the dot product corresponds to the usual definition. In other words,

$$\vec{\nabla} \cdot \vec{E} = \nabla = \left(\hat{x}\frac{\partial}{\partial x} + \hat{y}\frac{\partial}{\partial y} + \hat{z}\frac{\partial}{\partial z}\right) \cdot (E_x\hat{x} + E_y\hat{y} + E_z\hat{z})$$

Therefore, $\vec{\nabla} \cdot \vec{E} = \dfrac{\partial E_x}{\partial x} + \dfrac{\partial E_y}{\partial y} + \dfrac{\partial E_z}{\partial z}$, which we define as the divergence of a vector function $\vec{E}(x, y, z)$

EXAMPLE 2. Compute the divergence of the vector field $\vec{F}(x, y, z) = x\hat{x} + y\hat{y}$, which is sketched below.

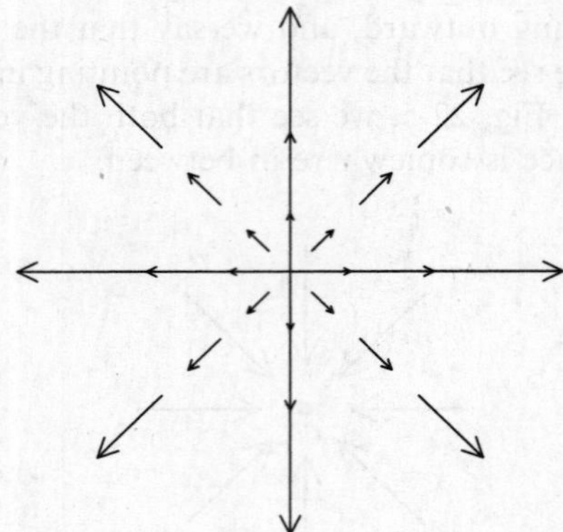

Fig. 22-6

The divergence is $\vec{\nabla} \cdot \vec{F} = \dfrac{\partial F_x}{\partial x} + \dfrac{\partial F_y}{\partial y} + \dfrac{\partial F_z}{\partial z} = \dfrac{\partial x}{\partial x} + \dfrac{\partial y}{\partial y} + \dfrac{\partial (0)}{\partial z} = 1 + 1 = 2.$

Therefore, we say that this function has a positive divergence of 2 at all points—that it is "spreading out" at each point. Loosely, this means that the amount of "stuff" coming out of each point is greater than the amount of stuff going in. We will clarify these ideas in the following chapter.

Although in the preceding example, the divergence happens to have a constant value for all points, in general, this will not be the case. For example, if the vector field is instead $G(x, y, z) = x^2\hat{x} + y^2 x\hat{y} + z^2\hat{z}$, then the divergence of G is

$$\vec{\nabla} \cdot \vec{G} = \frac{\partial G_x}{\partial x} + \frac{\partial G_y}{\partial y} + \frac{\partial G_z}{\partial z} = \frac{\partial (x^2)}{\partial x} + \frac{\partial (y^2)}{\partial y} + \frac{\partial (z^2)}{\partial z} = 2x + 2y + 2z,$$

which clearly varies depending upon the specific coordinates (x,y,z).

Note that ∇ acts as if it were a vector in many ways: we form the dot product between it and a vector in the same manner as we do between two ordinary vectors. For example, the following two identities can be shown to hold:

$$\vec{\nabla}(ab) = a\vec{\nabla}b + b\vec{\nabla}a$$

$$\vec{\nabla}(\vec{A} \cdot \vec{B}) = \vec{A} \times (\vec{\nabla} \times \vec{B}) + \vec{B} \times (\vec{\nabla} \times \vec{A}) + (\vec{A} \cdot \vec{\nabla})\vec{B}$$

It is important to keep in mind, however, that $\vec{\nabla}$ is not a vector but an *operator*, despite the temptation to think otherwise.

THE DIVERGENCE AND THE EQUATION OF CONTINUITY. Let us consider the velocity distribution $\vec{F}(x, y, z)$ of a fluid (liquid or gas) in the vicinity of an infinitesimal rectangular solid that has a corner at (x,y,z) and with sides equal to dx, dy, and dz. This velocity distribution is a vector with x, y, and z components. The volume might contain sources, sinks, both or neither. What we do know is that

The net amount flowing from sources (the sources of minus the sinks) equals to the net amount flowing across walls (the amount flowing out minus the amount flowing in).

The amount flowing from any sources in this cube is just equal to the volume of water flowing per second:

$$\int \rho \, dV$$

This quantity—the volume of water per second emerging from the cube—must equal to the volume of water flowing across the cube faces. Let us calculate the total flux flowing across the six faces of the solid shown in Fig. 22-7.

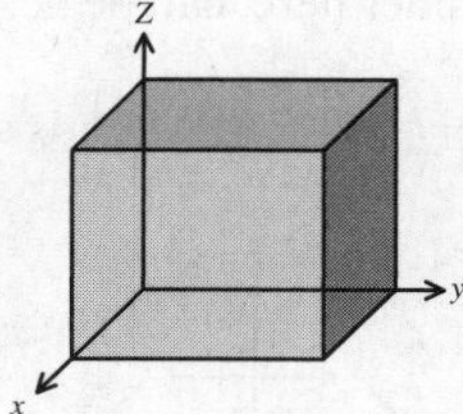

Fig. 22-7

We begin by considering

$$\begin{aligned}
&[\vec{F}(x+dx,y,z)-\vec{F}(x,y,z)]\cdot\hat{x}dydz \\
&= \delta F_x dydz \\
&= \frac{\partial F_x}{\partial x} dxdydz
\end{aligned}$$

Extending this argument to the other four sides of the cube gives us the total flow across the faces:

$$\begin{aligned}
\int \vec{F}\cdot d\vec{A} &= \int\left(\frac{\partial F_x}{\partial x}dxdydz + \frac{\partial F_y}{\partial y}dxdydz + \frac{\partial F_z}{\partial z}dxdydz\right) \\
&= \int\left(\frac{\partial F_x}{\partial x} + \frac{\partial F_y}{\partial y} + \frac{\partial F_z}{\partial z}\right)dxdydz \\
&= \int\left(\frac{\partial F_x}{\partial x} + \frac{\partial F_y}{\partial y} + \frac{\partial F_z}{\partial z}\right)dV
\end{aligned}$$

Since these two things are equal, we apparently have

$$\int \rho dV = \int\left(\frac{\partial F_x}{\partial x} + \frac{\partial F_y}{\partial y} + \frac{\partial F_z}{\partial z}\right)dV$$

$$\therefore \int \rho dV = \int \vec{\nabla}\cdot\vec{F}dV$$

Since this equality holds true for all points in space, we can equate the integrands:

$$\vec{\nabla}\cdot\vec{F} = \rho$$

This is one form—the differential form—of the *Equation of Continuity*, which relates the strength of a field source (whether positive, negative, or zero) to the convergence or spreading out of a vector field at that point. It recurs throughout fluid mechanics, electromagnetism, and other areas.

THE CURL—AND STOKES' THEOREM. The curl of a vector, evaluated at a particular point, is a vector that indicates the tendency of a vector field to rotate around that point. A whirlpool is an example of a point of large curl. In general, a point of nonzero curl will rotate a paddlewheel placed at that point.

The magnitude of the curl at a point gives a measure of the tendency to rotate around that point and its direction indicates the direction of rotation (clockwise or counterclockwise).

Let us define the curl and then show that this definition is in accord with the preceding description. Using the del operator defined above and the usual vector cross product, the curl of vector $\vec{A}$ is defined as:

$$\vec{\nabla}\times\vec{A}=\left(\frac{\partial A_z}{\partial y}-\frac{\partial A_y}{\partial z}\right)\hat{x}+\left(\frac{\partial A_x}{\partial z}-\frac{\partial A_z}{\partial x}\right)\hat{y}+\left(\frac{\partial A_y}{\partial x}-\frac{\partial A_x}{\partial y}\right)\hat{z}$$

This is potentially a confusing array of partial derivatives that can be challenging to remember. The use of determinants, which we do not fully consider here, can be helpful in evaluating expressions of this sort.*.

EXAMPLE 3. Consider the function $\vec{A}(x, y) = -y\hat{x} + x\hat{y}$, which is sketched in Fig. 22-8.

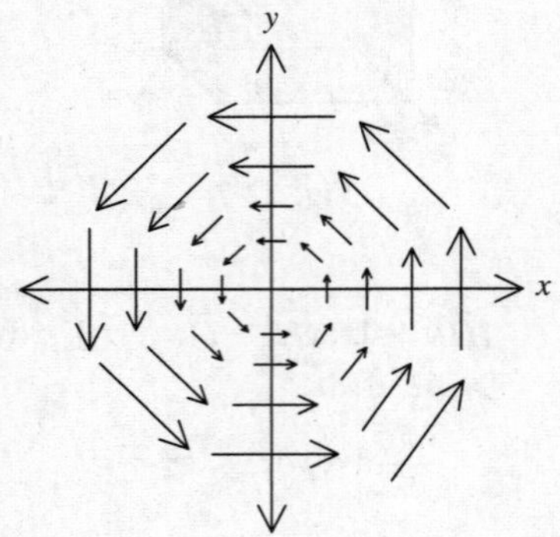

Fig. 22-8

According to the definition above, the curl of $\vec{A}$ is given by

$$\vec{\nabla}\times\vec{A}=\begin{vmatrix}\hat{x} & \hat{y} & \hat{z}\\ \frac{\partial}{\partial x} & \frac{\partial}{\partial y} & \frac{\partial}{\partial z}\\ A_x & A_y & A_z\end{vmatrix}=\left(\frac{\partial A_z}{\partial y}-\frac{\partial A_y}{\partial z}\right)\hat{x}+\left(\frac{\partial A_x}{\partial z}-\frac{\partial A_z}{\partial x}\right)\hat{y}+\left(\frac{\partial A_y}{\partial x}-\frac{\partial A_x}{\partial y}\right)\hat{z}$$

Substituting $A_x = -y$, $A_y = x$, and $A_z = 0$, we immediately get $\vec{\nabla}\times\vec{A}=\left(\frac{\partial A_y}{\partial x}-\frac{\partial A_x}{\partial y}\right)\hat{z}=(1+1)\hat{z}=2\hat{z}$ (with all other partial derivatives vanishing).

Note that the curl of a vector is itself a vector—in the case of this example, one that points along the z axis. A paddlewheel placed at any point in Fig. 22-8 will indeed rotate—most notably, perhaps, at the center but also at any other point, where the "higher velocity" of the "water" further from the origin will dominate the rotation.

For simplicity, we have couched the above discussion of the divergence, the gradient, and the curl in rectangular coordinates. However, one can imagine that one might want to know, for example, the divergence of a function that is expressed in spherical or cylindrical coordinates. To do so, one needs to know the form of the grad operator, divergence, gradient, and curl in these other coordinate systems. We demonstrate how these can be obtained in some of the problems below. Evaluation of expressions in spherical and cylindrical coordinates is usually straightforward (if occasionally tedious).

---

* The expression can be easily obtained using the determinant method of cross product evaluation:

$$\vec{\nabla}\times\vec{A}=\begin{vmatrix}\hat{x} & \hat{y} & \hat{z}\\ \frac{\partial}{\partial x} & \frac{\partial}{\partial y} & \frac{\partial}{\partial z}\\ A_x & A_y & A_z\end{vmatrix}=\left(\frac{\partial A_z}{\partial y}-\frac{\partial A_y}{\partial z}\right)\hat{x}+\left(\frac{\partial A_x}{\partial z}-\frac{\partial A_z}{\partial x}\right)\hat{y}+\left(\frac{\partial A_y}{\partial x}-\frac{\partial A_x}{\partial y}\right)\hat{z}$$

where, as before, we are treating the $\vec{\nabla}$ operator as an ordinary vector in the evaluation of the cross product.

DIVERGENCE, GRADIENT, AND CURL IN SPHERICAL COORDINATES.

$$\text{divergence}: \quad \vec{\nabla}\cdot\vec{F} = \frac{1}{r^2}\frac{\partial}{\partial r}(r^2 F_r) + \frac{1}{r\sin\theta}\frac{\partial}{\partial\theta}(\sin\theta F_\theta) + \frac{1}{r\sin\theta}\frac{\partial F_\phi}{\partial\phi}$$

$$\text{gradient}: \quad \vec{\nabla}f = \frac{\partial f}{\partial r}\hat{r} + \frac{1}{r}\frac{\partial f}{\partial\theta}\hat{\theta} + \frac{\partial f}{\partial\phi}\hat{\phi}$$

$$\text{curl}: \quad \vec{\nabla}\times\vec{F} = \frac{1}{r\sin\theta}\left[\frac{\partial}{\partial\theta}(\sin\theta F_\theta) - \frac{\partial F_\theta}{\partial\phi}\right]\hat{r}$$

DIVERGENCE, GRADIENT, AND CURL IN CYLINDRICAL COORDINATES.

$$\text{divergence}: \quad \vec{\nabla}\cdot\vec{F} = \frac{1}{r}\frac{\partial}{\partial r}(rF_r) + \frac{1}{r}\frac{\partial F_\phi}{\partial\phi} + \frac{\partial F_z}{\partial z}$$

$$\text{gradient}: \quad \vec{\nabla}f = \frac{\partial f}{\partial r}\hat{r} + \frac{1}{r}\frac{\partial f}{\partial\phi}\hat{\phi} + \frac{\partial f}{\partial z}\hat{z}$$

$$\text{curl}: \quad \vec{\nabla}\times\vec{F} = \left[\frac{1}{r}\frac{\partial F_z}{\partial\phi} - \frac{\partial F_\phi}{\partial z}\right]\hat{r} + \left[\frac{\partial F_r}{\partial z} - \frac{\partial F_z}{\partial r}\right]\hat{\phi} + \frac{1}{r}\left[\frac{\partial}{\partial r}(rF_\phi) - \frac{\partial F_r}{\partial\phi}\right]\hat{z}$$

EXAMPLE 4. Evaluate, in spherical coordinates, the divergence of $\vec{G} = re^{-2\pi\theta}\hat{r} + 2r^2\phi\hat{\theta} + 3r^3\theta\hat{\phi}$

$$\vec{\nabla}\cdot\vec{G} = \frac{1}{r^2}\frac{\partial}{\partial r}(r^2 G_r) + \frac{1}{r\sin\theta}\frac{\partial}{\partial\theta}(\sin\theta G_\theta) + \frac{1}{r\sin\theta}\frac{\partial G_\phi}{\partial\phi}$$

$$= \frac{1}{r^2}\frac{\partial}{\partial r}(r^2(re^{-2\pi\theta})) + \frac{1}{r\sin\theta}\frac{\partial}{\partial\theta}(\sin\theta(2r^2\phi)) + \frac{1}{r\sin\theta}\frac{\partial(3r^3\theta)}{\partial\phi}$$

$$= \frac{1}{r^2}(3r^2e^{-2\pi\theta}) + \frac{1}{r\sin\theta}(\cos\theta(2r^2\phi)) + 0$$

$$= 3e^{-2\pi\theta} + \frac{2r\phi}{\tan\theta}$$

EXAMPLE 5. Evaluate the curl of $\vec{A} = r^2\hat{r} + 3r\phi\hat{\phi} + 12z^3\hat{z}$ in cylindrical coordinates

Solution: The curl of $\vec{A}$ is

$$\vec{\nabla}\times\vec{A} = \left[\frac{1}{r}\frac{\partial A_z}{\partial\phi} - \frac{\partial A_\phi}{\partial z}\right]\hat{r} + \left[\frac{\partial A_r}{\partial z} - \frac{\partial A_z}{\partial r}\right]\hat{\phi} + \frac{1}{r}\left[\frac{\partial}{\partial r}(rA_\phi) - \frac{\partial A_r}{\partial\phi}\right]\hat{z}$$

$$= \left[\frac{1}{r}\frac{\partial(12z^3)}{\partial\phi} - \frac{\partial A_\phi}{\partial z}\right]\hat{r} + \left[\frac{\partial(r^2)}{\partial z} - \frac{\partial(12z^3)}{\partial r}\right]\hat{\phi} + \frac{1}{r}\left[\frac{\partial}{\partial r}(r(3r\phi)) - \frac{\partial(r^2)}{\partial\phi}\right]\hat{z}$$

$$= [0-0]\hat{r} + [0-0]\hat{\phi} + \frac{1}{r}[6r-0]\hat{z}$$

$$= 6\hat{z}$$

Solved Problems

22.1 Find the gradient of the following functions: (*a*) $f(x,y,z) = x^2y + yx^3 + 6z$; (*b*) $g(x,y,z) = 4x^3 + 7y^2 + 8xyz$; and (*c*) $h(x,y,z) = Ayz^4e^{2x+3} + By^3\sin(kx+\phi)$, where A, B, k and ϕ are constants.

(*a*) $\vec{\nabla}f = \hat{x}\frac{\partial f}{\partial x} + \hat{y}\frac{\partial f}{\partial y} + \hat{z}\frac{\partial f}{\partial z} = \hat{x}(2xy + 3x^2y) + \hat{y}(x^2 + x^3)$

(*b*) $\vec{\nabla}g = \hat{x}\frac{\partial g}{\partial x} + \hat{y}\frac{\partial g}{\partial y} + \hat{z}\frac{\partial g}{\partial z} = \hat{x}(2xy + 3x^2y) + \hat{y}(x^2 + x^3)$

(*c*) $\vec{\nabla}h = \hat{x}\frac{\partial h}{\partial x} + \hat{y}\frac{\partial h}{\partial y} + \hat{z}\frac{\partial h}{\partial z} = (2Ayz^4e^{2x+3} + Bky^3\cos(kx+\phi))\hat{x} + (Az^4e^{2x+3} + 3By^2\sin(kx+\phi))\hat{y} + (4Ayz^3e^{2x+3})\hat{z}$

22.2 Find the divergence of the following functions: (*a*) $\vec{F} = 3x^2y\hat{x} - 6xz\hat{y} + 4e^{3z}\hat{z}$; (*b*) $\vec{T} = (x-5)^2y\hat{x} + 4\sin(3y)\hat{y} - 2z^2x\hat{z}$; (*c*) $\vec{E} = e^{2x}\cos(3y)\hat{x} + 4z^2x\hat{y}$

(*a*) $\vec{\nabla}\cdot\vec{F} = \frac{\partial F_x}{\partial x} + \frac{\partial F_y}{\partial y} + \frac{\partial F_z}{\partial z} = 6xy + 12e^{3z}$

(*b*) $\vec{\nabla}\cdot\vec{T} = \frac{\partial T_x}{\partial x} + \frac{\partial T_y}{\partial y} + \frac{\partial T_z}{\partial z} = 2(x-5)y + 12\cos(3y) - 4zx$

(*c*) $\vec{\nabla}\cdot\vec{E} = \frac{\partial E_x}{\partial x} + \frac{\partial E_y}{\partial y} + \frac{\partial E_z}{\partial z} = 2e^{2x}\cos(3y)$

22.3 Find the curl of the following vector functions: (*a*) $\vec{J} = 3x^2yz\hat{x} + 4z^5y\hat{y} - 12xy\hat{z}$; (*b*) $\vec{B} = x^2\hat{x} + zy^3\hat{y} - 2xy\hat{z}$; (*c*) $\vec{S} = 4x^2z\hat{x} - 6e^z\hat{y} + 4\sin^2(3x)\hat{z}$

(*a*) $\vec{\nabla}\times\vec{J} = \left(\frac{\partial J_z}{\partial y} - \frac{\partial J_y}{\partial z}\right)\hat{x} + \left(\frac{\partial J_x}{\partial z} - \frac{\partial J_z}{\partial x}\right)\hat{y} + \left(\frac{\partial J_y}{\partial x} - \frac{\partial J_x}{\partial y}\right)\hat{z}$
$= (-12x - 20z^4y)\hat{x} + (3x^2y)\hat{y} + (-3x^2z)\hat{z}$

(*b*) $\vec{\nabla}\times\vec{B} = \left(\frac{\partial B_z}{\partial y} - \frac{\partial B_y}{\partial z}\right)\hat{x} + \left(\frac{\partial B_x}{\partial z} - \frac{\partial B_z}{\partial x}\right)\hat{y} + \left(\frac{\partial B_y}{\partial x} - \frac{\partial B_x}{\partial y}\right)\hat{z}$
$= (-2x - y^3)\hat{x} + (2y)\hat{y}$

(*c*) $\vec{\nabla}\times\vec{S} = \left(\frac{\partial S_z}{\partial y} - \frac{\partial S_y}{\partial z}\right)\hat{x} + \left(\frac{\partial S_x}{\partial z} - \frac{\partial S_z}{\partial x}\right)\hat{y} + \left(\frac{\partial S_y}{\partial x} - \frac{\partial S_x}{\partial y}\right)\hat{z}$
$= (4x^2 - 12\sin^2(3x))\hat{y}$

22.4 The height of the terrain surrounding a country village varies according to the formula $h(x, y) = k(-5x^2 + 6x + 30xy - 4y^2 + 200y)$, where $h(x,y)$ is measured in meters and where x and y are the respective positions of east and north of a marker at the village center, respectively (measured in kilometers).

(*a*) What is the location of the extreme value (top or bottom) of the terrain? (*b*) What is the height at this location?
$\vec{\nabla}h = 0 \Rightarrow \hat{x}\frac{\partial h}{\partial x} + \hat{y}\frac{\partial h}{\partial y} = \hat{x}k(-10x + 6 + 30y) + k(30x - 8y + 200)\hat{y} = 0$
Setting each of the components equal to zero, we then have:

$$-10x + 30y = -6$$
$$30x - 8y = -200$$

from which it follows that $x = -7.4$ and $y = -2.7$.
The extreme value of the height is therefore located 2.66 km south and 7.38 km west of the town marker.

(*b*) The height is given by

$$\begin{aligned} h(x, y) &= -5x^2 + 6x + 30xy - 4y^2 + 200y \\ &= -5(-7.38)^2 + 6(-7.38) + 30(-7.38)(-2.66) - 4(-2.66)^2 + 200(-2.66) \\ &= -288\text{ m} \end{aligned}$$

This is apparently a low point (rather than the top of a hill) located 288 m below the elevation of the town marker.

(*c*) The slope along the direction of steepest ascent at this location is given simply by the gradient at this point:

22.5 Show that the equation of continuity:

$$\vec{\nabla}\cdot\vec{J} + \frac{\partial\rho}{\partial t} = 0$$

is satisfied for an electrical current of current density J (that is, current per unit cross sectional area) emanating from a differential volume dV containing a source of charge with charge density ρ and bounded by a surface S.

For such a surface, we know that the flux of fluid flowing across all surfaces is

$$I = \int \vec{J}\cdot d\vec{A}$$

From the discussion above, we know that $\int \vec{J}\cdot d\vec{A} = \int \vec{\nabla}\cdot\vec{J}dV$.

The fluid flowing across the boundary, each second must match the change per second in the total fluid in the volume:

$$I = \frac{dM}{dt} = \frac{d[\int \rho dV]}{dt} = \int \frac{d\rho}{dt} dV$$

Equating the two expressions, we get:

$$\int \vec{\nabla} \cdot \vec{J} dV = \int \frac{d\rho}{dt} dV$$

Since this expression is true for all volumes, we may equate the integrands, yielding:

$$\vec{\nabla} \cdot \vec{J} + \frac{\partial \rho}{\partial t} = 0,$$

which is the equation of continuity.

22.6 What is the divergence of the electric field of a charge-free region of space? The divergence of the electric field and the charge density are related through:

$$\vec{\nabla} \cdot \vec{E} = \frac{\rho}{\varepsilon_0}$$

Because there is no source of charge at this point, the divergence is 0.

22.7 Find the curl of a vector field $\vec{F} = 3x^2y\hat{x} + (3y^2 - 4z)\hat{y} + 4z^2x\hat{z}$ at the point $(x,y,z) = (3,4,6)$.

The curl of $\vec{F}$ is:

$$\vec{\nabla} \times \vec{F} = \left(\frac{\partial F_z}{\partial y} - \frac{\partial F_y}{\partial z}\right)\hat{x} + \left(\frac{\partial F_x}{\partial z} - \frac{\partial F_z}{\partial x}\right)\hat{y} + \left(\frac{\partial F_y}{\partial x} - \frac{\partial F_x}{\partial y}\right)\hat{z}$$

Evaluating the partial derivatives at the chosen point, we get:

$$\frac{\partial F_z}{\partial y} = \frac{\partial(4z^2x)}{\partial y} = 0$$

$$\frac{\partial F_y}{\partial z} = \frac{\partial(3y^2 - 4z)}{\partial y} = 6y = 6(4) = 24$$

and so on for the remaining terms, yielding

$$\vec{\nabla} \times \vec{F} = -24\hat{x} - 144\hat{y} - 27\hat{z}$$

Supplementary Problems

22.8 Compute the gradient of the following functions:

(*a*) $3x^2y + 4z^2 + 12yx$

(*b*) $3xyz - 2yz^4 + 12x \sin x$

(*c*) xyz

(*d*) $x + 3y + 4ze^{-x}$

(*e*) $xe^{-xy} + y^3z$

(*f*) $yxz \sin(5\pi x)$

Ans. (*a*) $(6xy + 12y)\hat{x} + (3x^2 + 12x)\hat{y} + 8z\hat{z}$

(*b*) $(3yz + 12x\cos x + 12\sin x)\hat{x} + (3xz - 2z^4)\hat{y} + (3xy - 8yz^3)\hat{z}$

(*c*) $yz\hat{x} + xz\hat{y} + xy\hat{z}$

(*d*) $(1 - 4ze^{-x})\hat{x} + 3\hat{y} - 4e^{-x}\hat{z}$

(*e*) $\hat{x}(-xye^{-xy} + e^{-xy}) + \hat{y}(-x^2e^{-xy} + 3y^2z) + y^3\hat{z}$.

22.9 Compute the divergence of the following functions:

(*a*) $4y^2\hat{\imath} + 6zx\hat{\jmath} + 2yz^3\hat{k}$

(*b*) $zx^2\hat{\imath} + x\sin y\hat{\jmath} + 2e^{-5y}\hat{k}$

(*c*) $4ye^{-5z}\hat{\imath} + 15xyz^2\hat{\jmath} + 2y^3z^3\hat{k}$

(*d*) $3z^5y^2\hat{\imath} + 4zx\hat{\jmath} + 2y\tan(yx)\hat{k}$

(*e*) $40zx^2\hat{\imath} + 6zx\hat{\jmath} + 2yz^3\hat{k}$

(*f*) $4y^2x\hat{\imath} + 30x\hat{\jmath} + 2y\hat{k}$

Ans. (*a*) $6yz^2$ (*b*) $2zx + x\cos y$ (*c*) $15xz^2 + 2y^3$ (*d*) 0 (*e*) $80zx + 6yz^2$ (*f*) $4y^2$.

22.10 Compute the curl of the following functions:

(*a*) $x^4\hat{\imath} + 2z^4x\hat{\jmath} + yz^2\hat{k}$

(*b*) $x^3\hat{\imath} + 5x\cos y\hat{\jmath} + 2(x^2 - 1)\hat{k}$

(*c*) $8x\hat{\imath} + z^2\hat{\jmath} + 4y^3\hat{k}$

(*d*) $y\hat{\imath} + z^4\hat{\jmath} + 2yz\hat{k}$

(*e*) $xy\hat{\imath} + (6zx + 4)\hat{\jmath} + 2yz^3\hat{k}$

(*g*) $y^2zx\hat{\imath} + 2x\hat{\jmath} + 4y\hat{k}$

Ans. (*a*) $(z^2 - 8z^3x)\hat{\imath} + 2z^4\hat{k}$;

(*b*) $4x\hat{\jmath} + 5\cos y\hat{k}$;

(*c*) $(12y^2 - 2z)\hat{\imath}$;

(*d*) $(2z - 4z^3)\hat{\imath} - \hat{k}$;

(*e*) $(2z^3 - 6x)\hat{\imath} + (6z - x)\hat{k}$

(*f*) $4\hat{\imath} + y^2x\hat{\jmath} + (2 - 2x)\hat{k}$

22.11. Compute the expressions below for these functions:

$f = 3xyz$

$\vec{G} = 3xy^2\hat{\imath} + 6zx\hat{\jmath} + 2yz^3\hat{k}$

(*a*) $\vec{\nabla} \times (\vec{\nabla} f)$ (*b*) $G \times \vec{\nabla} f$ (*c*) $\vec{\nabla} \times \vec{G}$

Ans. (*a*) 0 (*b*) $(6zx - 3xy)\hat{x} + (6y^2z^4 - 9x^2y^3)\hat{y} + (9x^2y^2z - 18xyz^2)\hat{z}$ (*c*) $(2z^3 - 6x)\hat{\imath} + 6z\hat{k}$

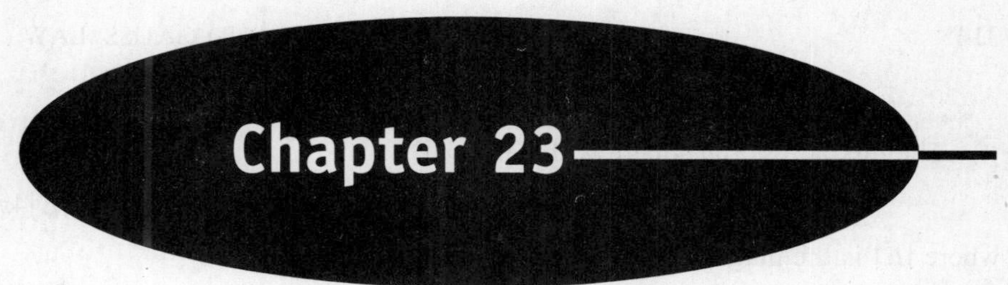

Vector Calculus: Flux and Gauss' Law

FLUX. Think of some kind of "stuff" that flows through a cross-sectional area, as indicated in Fig. 23-1. It could, for example, be a river flowing across the cross section of a riverbank, with the water's velocity vectors as indicated. It could also be an electric or gravitational field with field lines "piercing" the boundary as shown. We can describe this in general as a vector field that crosses a surface. The amount of "stuff" that flows perpendicular to the surface is the *flux* of the material (or of a vector field) across the surface.

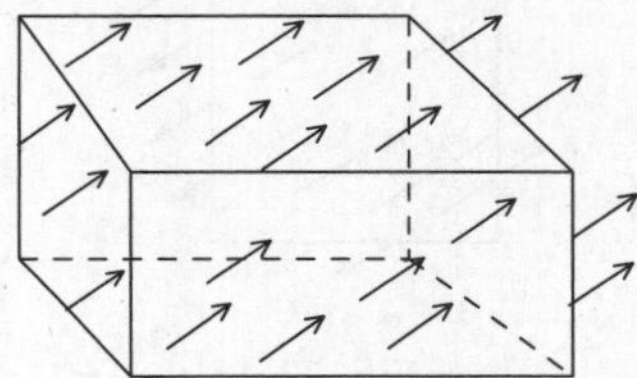

Fig. 23-1

A simple example is shown in Fig. 23-1. Assume that the electric field shown has a constant magnitude throughout the area of the cross-sectional boundary and that its angle with respect to the surface also remains constant. The flux Φ of the electric field E across the surface A is

$$\Phi = |\vec{E}| \cos(\theta) A \tag{23.1}$$

If the electric field emanates radially from a point charge (see Figs. 23-2 and 23-3 for the cases of positive and negative charges), then the electric field is everywhere perpendicular to the surface of an imaginary surrounding sphere.

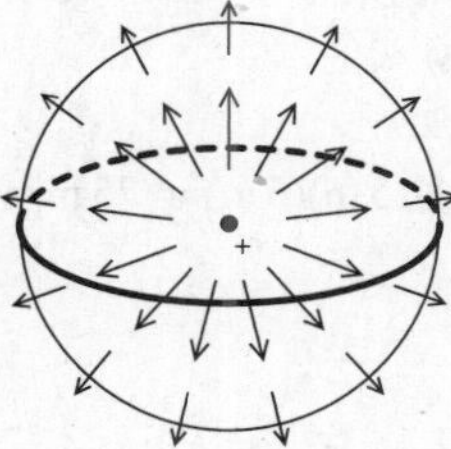

Fig. 23-2

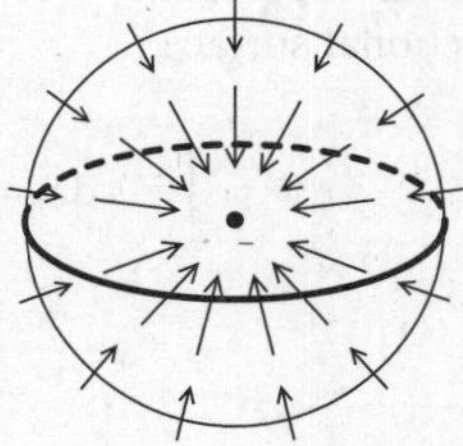

Fig. 23-3

The flux across the sphere is therefore

$$\Phi = E_{\perp} A = |\vec{E}| \cos\theta A = |\vec{E}| \cos(0) 4\pi r^2 = |\vec{E}| 4\pi r^2 \qquad (23.2)$$

where $|\vec{E}|$ is the magnitude of the electric field at the surface.*

FLUX—INTEGRATION OVER DIFFERENTIAL ELEMENTS. When the vector field varies in magnitude and/or direction over the cross-sectional boundary, then the total flux is equal to the integral over differential flux contributions. For example, consider the vector field $\vec{F}$ and the area element $d\vec{S}$ as shown in Fig. 23-3. The differential flux of $\vec{F}$ across $d\vec{S}$ is

$$d\Phi = \vec{F} \cdot d\vec{S}.$$

This is just the product of the area element and the component of the field perpendicular to that element. If the field makes an angle θ with the normal to the area, as indicated in Fig. 23-4, then the flux

$$d\Phi = \vec{F} \cdot d\vec{S} = |\vec{F}||d\vec{S}| \cos\theta \qquad (23.3)$$

where θ is the angle between the field direction and the normal to the area element. If the field is perpendicular to the surface, then the flux

$$d\Phi = F dS \qquad (23.4)$$

If the field is parallel to the surface, then there is no component of the field normal to the surface and the flux is zero.

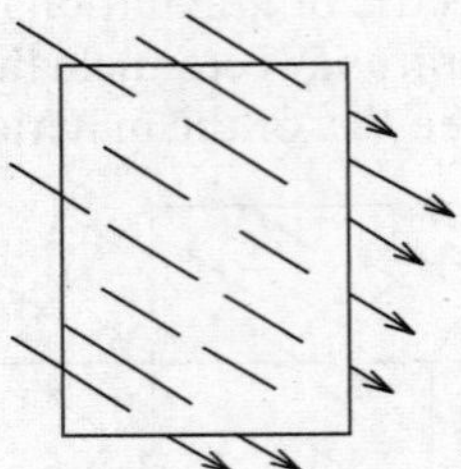

Fig. 23-4

The total flux across an area A due to a field F is simply the sum of all of the differential flux contributions

$$\Phi = \int \vec{F} \cdot d\vec{S} \qquad (23.5)$$

EXAMPLE 1. Find the volume of water flowing across the cross section of a river, assuming that the water flows perpendicular to the cross section and has a constant velocity of 1 m/s. The river at this cross section is 15 m across and 5 m deep.

Using equation (*23.5*) with the velocity vector field $\vec{v}$, the flux $\Phi = \int \vec{v} \cdot d\vec{S}$. Because $\vec{v}$ is constant and perpendicular to the cross-sectional surface,

$$\Phi = \int \vec{v} \cdot d\vec{S} = vS = (1\,\text{m/s})(15\,\text{m})(5\,\text{m}) = 75\,\text{m/s}$$

* It is interesting to note that, for the point charge mentioned here, $|\vec{E}| \propto \frac{1}{r^2}$ from Coulomb's Law, so that the total flux through a sphere is independent of the radius.

EXAMPLE 2. Find the flux of the magnetic field $\vec{B}$ across a surface if the magnetic field is everywhere parallel to the surface.

The flux is zero because there is no component of the field moving perpendicular to the surface. From the definition above, only such components contribute to the flux.

Note that the contribution to the flux is negative when $\cos\theta < 0$, i.e., when the angle between the vector field and the unit normal to the surface exceeds 90°.

EXAMPLE 3. Find the flux due to a vector field $\vec{F}(x, y) = 3x^2y\hat{x} + 2yz\hat{y} + 4x^3y^2\hat{z}$ incident on the rectangle at $z = 1$ bounded by $x = 1$, $x = 3$, $y = 2$, and $y = 5$.

$$\Phi = \int \vec{F} \cdot d\vec{S} = \int \vec{F}(x, y, z) \cdot d\vec{S} = \int (3x^2y\hat{x} + 2yz\hat{y} + 4x^3y^2\hat{z}) \cdot (dx\,dy\,\hat{z})$$

$$= \int 4x^3y^2 dx\,dy = 4\int_1^3 x^3 dx \int_2^5 y^2 dy = 4\frac{x^4}{4}\Big|_1^3 \frac{y^3}{3}\Big|_2^5 = \frac{4}{3}(3^4 - 1^4)(5^3 - 2^3) = \frac{4}{3}(80)(117) = 12,480$$

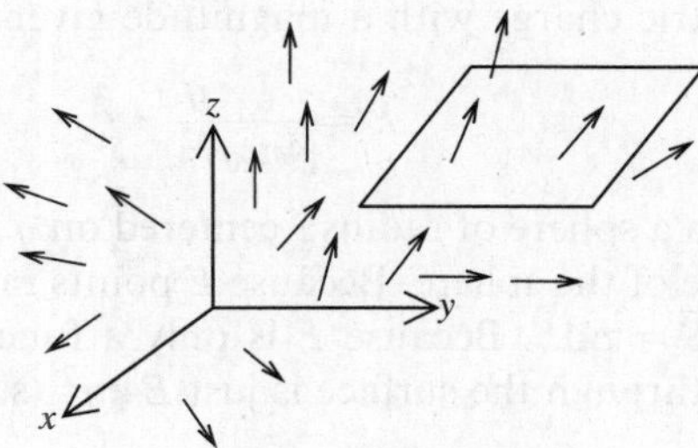

Fig. 23-5

EXAMPLE 4. Compute the flux due to an electric field $\vec{E}$ across an area S that is 10 m by 10 m in each of the following situations, which are illustrated in the figures below:

(*a*) a constant electric field of 10 N/C perpendicular to an area S

(*b*) constant electric field at an angle of 35° relative to the normal to S

(*a*) the flux is $\int \vec{E} \cdot d\vec{S} = ES = (10\,\text{N/C})(100\,\text{m}^2) = 1,000\,\frac{\text{N m}^2}{\text{C}}$

(*b*) the flux is $\int \vec{E} \cdot d\vec{S} = ES\cos\theta = 10(100)\cos(35°) = 819\,\frac{\text{N m}^2}{\text{C}}$

FLUX THROUGH A CLOSED SURFACE. So far we have defined the flux across a "patch" of surface which intercepts some portion of the vector field. Now let us consider a *closed surface* which completely encloses some volume. The differential area vector $d\vec{A}$ points from the inside to the outside of the surface. The *net flux* is equal to the difference between the flux pointing out of the volume and that pointing inward.

FLUX AND SOURCES. As we mentioned at the outset of this chapter, flux describes the amount of "flow" of a field across a surface.* In the case of fluid flow, the fluid volume crossing a boundary per second depends upon the product of its velocity and the area of the boundary.† In the physical world, there must be a source of this field—a mountain lake, perhaps, in the case of the river flow, and electric charge in the case of the electric field. Not surprisingly, the amount of flow depends upon the "strength" of the field source.

* "Flux" is derived from the Greek word for flow.

† Note that the dimensions of volume per unit time (m^3/s) is equal to the product of the dimensions of velocity (m/s) and area (m^2).

SOURCES AND SINKS. This same net flux flowing through this closed surface is related to *the sources and sinks of water within the volume.*[*] A source is a place from which water appears (e.g., a reservoir spouting water into the river) and a sink is where it disappears (e.g., another reservoir sucking water from the river).

Another example is the electric field, for which electric charges serve as sources and sinks. As shown in the previous chapter, electric field lines diverge from positive charges and converge toward negative charges. Hence, positive (negative) charges serve as sources (sinks) of the electric field. Again, the net flux of the electric field across the closed volume is related to the sources and sinks of electric field (i.e., the positive and negative charges) within the volume.

Then for both the river example and the example of the electric field, the net flux will be positive (negative) when there is a source (sink).

GAUSS' LAW. The relationship between one vector field—the electric field—and its source (the electric charge) can be derived from Coulomb's Law and is known as Gauss' Law. Suppose an electric field $\vec{E}$ radiates out from a point-like electric charge with a magnitude given by

$$E = \frac{1}{4\pi\varepsilon_0}\frac{q}{r^2}$$

The flux from this field through a sphere of radius r centered on q is given by $\int \vec{E} \cdot \mathrm{d}\vec{S}$, where $\mathrm{d}S$ is a small element of area on the surface of the sphere. Because E points radially outward, it is perpendicular to this area element and so $\vec{E} \cdot \mathrm{d}\vec{S} = E\mathrm{d}S$. Because E is only a function of distance r (and not of its angular orientation), the flux of E through the surface is just $E4\pi r^2$ (since $E4\pi r^2$ is the total surface area of a sphere of radius r). Hence,

$$\text{flux} = \int \vec{E} \cdot \mathrm{d}\vec{S} = E4\pi r^2$$

We now use the following relationship, derived from Coulomb's Law, for the electric field

$$E = \frac{1}{4\pi\varepsilon_0}\frac{q}{r^2}$$

$$\text{flux} = \int \vec{E} \cdot \mathrm{d}\vec{S} = E4\pi r^2 = \frac{1}{4\pi\varepsilon_0}\frac{q}{r^2}(4\pi r^2) = \frac{q}{\varepsilon_0}$$

Hence, we see that the flux due to the electric field and its source—the electric charge—are related through

$$\int \vec{E} \cdot \mathrm{d}\vec{S} = \frac{q}{\varepsilon_0}$$

We have derived this result—known as *Gauss' Law* for electrostatics—for a sphere but *it can be shown to be true for any closed surface containing q*. Note that the net flux in a region of space only depends upon the electric charge in that region. In particular, there is no net flux associated with a charge-free region of space. Gauss' Law is useful for determining the electric field in situations involving a high degree of symmetry, as demonstrated in some of the solved problems that follow.

---

[*] Note that, for sources or sinks *outside* of the volume, there is no contribution to the total flux. For closed volumes, the unit normal is defined as pointing outward. A source (sink) outside of the volume contributes negative flux upon entering the volume and then contributes positively as it leaves, and one can show that these two contributions exactly cancel each other. This is reasonable—the water flowing into a portion of a river ends up leaving it, regardless of the details of the shape of the river.

Solved Problems

23.1 Consider an electric field of magnitude 2×10^4 N/C directed along the y axis. What is the flux due to the electric field through a circular plane that is 1 m in if the plane is (*a*) parallel to the xz plane, (*b*) perpendicular to the xz plane, and (*c*) at 45° with respect to the plane?

(*a*) The flux is $\Phi = \int \vec{E} \cdot d\vec{S} = \int |\vec{E}||d\vec{S}| \cos 0 = |\vec{E}| \int |d\vec{S}| = EA = (2 \times 10^4 \text{ N/C})\pi m^2 = 6.28 \times 10^4 \text{ Nm}^2/\text{C}$

(*b*) The flux is $\Phi = \int \vec{E} \cdot d\vec{S} = \int |\vec{E}||d\vec{S}| \cos 90° = 0$

(*c*) The flux is $\Phi = \int \vec{E} \cdot d\vec{S} = \int |\vec{E}||d\vec{S}| \cos 45° = |\vec{E}|(A)\frac{1}{\sqrt{2}} = (2 \times 10^4 \text{ N/C})(\pi m^2)(.707) = 4.4 \frac{\text{Nm}^2}{\text{C}}$

23.2 A magnetic field of strength 10 T is incident perpendicularly upon a 5 m·3 m rectangular loop. What is the magnetic flux through the loop?

The flux is $\Phi = \int \vec{B} \cdot d\vec{S}$. Because the magnetic field is constant over the face of the loop,

$$\Phi = \int \vec{B} \cdot d\vec{S} = BA = (10\,\text{T})(5\,\text{m})(3\,\text{m}) = 150\,\text{T}\,\text{m}^2$$

23.3 A hollow plastic cube with tiny holes in the sides is dropped into a river. What is the net flux of water through it?

Since there is no source of water inside the cube (equal amount of water flows in and out), the net flux is simply zero.

(This is a much easier approach than trying to calculate the amount of water flowing across each of the faces of the cube—especially, since we are not told the size of the cube or the velocity distribution of the river water where it is dropped! Instead, we need only to recall that the net flux emerging from an object is proportional to the strength of the source inside. Since the cube is neither a source nor a sink of water, its source strength is zero and so is the next flux.)

23.4 A line charge of density λ per unit length lies along an infinite wire as shown in Fig. 23-6. Using Gauss' Law and symmetry arguments, determine the shape of the electric field.

The line charge is shown in Fig. 23-6, surrounded by an imaginary cylinder of length L and radius r.

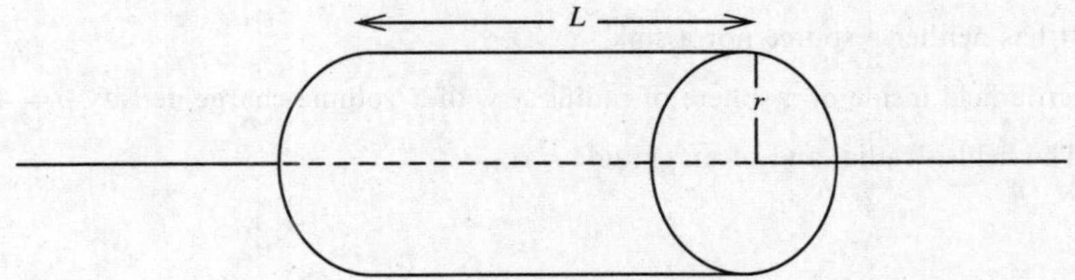

Fig. 23-6

Because the charge density is the charge per unit length, the total charge q in the cylinder is ρL.

From Gauss' Law, we know that $\varepsilon_0 \int \vec{E} \cdot d\vec{S} = q$

Therefore, $\varepsilon_0 \int \vec{E} \cdot d\vec{S} = \lambda L$

To evaluate the flux through the cylinder—including across its barrel and its ends—we invoke a symmetry argument. We know that electric field lines flow out from positive charges, so they must be emerging from the line charge. Because the line charge is of infinite length, symmetry demands that any electric field line that is not perpendicular to the line charge must be cancelled by a corresponding contribution (see figure). Hence, all electric field lines are perpendicular to the line charge. Because of the rotational symmetry, the magnitude of the electric field $|\vec{E}|$ will be a constant at any particular radius, regardless of the angle.

Hence, $\varepsilon_0 \int \vec{E} \cdot d\vec{S} = \varepsilon_0 |\vec{E}| A = \varepsilon_0 |\vec{E}| 2\pi r L = \lambda L$

where we have used the area of the cylinder barrel, recognizing that, because E is parallel to the plane of the barrel ends, the flux through those ends is zero.

Thus, $\varepsilon_0|\vec{E}|2\pi r = \lambda \Rightarrow |\vec{E}|A = \frac{\lambda}{2\pi\varepsilon_0 r}$

So the electric field decreases inversely with the distance from the line of charge.

23.5 A point charge of $+q$ is placed at the center of an equilateral pyramid. What is the electric flux through each face?

From symmetry, the flux through each of the four faces must be equal. From Gauss' Law, the total flux is equal to

$$\int \vec{E} \cdot d\vec{S} = \frac{q}{\varepsilon_0}$$

Therefore, the flux through each face is simply $\frac{q}{\varepsilon_0}$.

Supplementary Problems

23.6 Find the flux of the vector $\vec{f}(x, y, z) = 3y^2\hat{x} + zx^2\hat{y} + 3y\hat{z}$ across a rectangle in the xy plane occupying $0 \leq x \leq 3; 2 \leq y \leq 6$.

Ans. 144

23.7 Find the electric field due to an infinite plane that has a surface charge density σ, where the surface charge density is measured in C/m^2.

Ans. $\sigma/2\varepsilon$

23.8 Consider a current of water that flows with a velocity

$$v = 5\hat{x} + z\hat{y} - y\hat{z}$$

Does this current have a source or sink at the origin?

Ans. It has neither a source nor a sink.

23.9 Find the electric field inside of a sphere of radius R with a volume charge density $\rho = 4r$.

Ans. The field is radial and of magnitude r^2/ε_0.

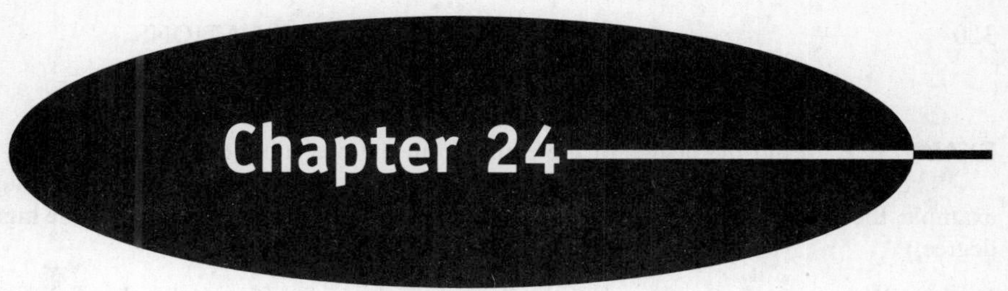

Chapter 24

Differential Equations

INTRODUCTION. Many relationships in physics are expressed in terms of the relationships among quantities and their rates of change. An equation that includes a function and any of its first and/or higher order derivatives as variables is called a *differential equation*. Differential equations are used throughout the natural and social sciences and are of central importance in the modeling of scientific data. For example, the movement of an object acted upon by forces external to it is expressed by Newton's second law, $\vec{F} = m\frac{d^2\vec{x}}{dt^2}$, where $\vec{F}$ is the external force acting upon an object and $\frac{d^2\vec{x}}{dt^2}$ is the acceleration of the object. Depending upon the particular situation, $\vec{F}$ may be constant, may be a function of position, velocity, time and/or other quantities—or may be some complicated combination of these quantities as well as their first or higher derivatives. Hence, Newton's second law—and any number of other relations throughout physics—constitutes a differential equation that relates these quantities. Often we want to know x, or its derivative, as a function of time, and so we need to know how to solve such equations.

Things can be relatively simple or they can be so complicated as to defy an analytical solution. While there is no all-purpose technique for solving any arbitrary differential equation, there are a number of techniques for solving the ones that arise most frequently in physics, particularly at the undergraduate level. We will introduce several such techniques in the context of a number of different physics problems. This chapter serves as an introduction to differential equations and focuses on a few specific situations that recur throughout physics and that are relatively straightforward to handle. The equations that we consider are restricted to *linear differential equations*—ones that do not involve higher powers of a function or any of its derivatives, viz:

$$a_0 x + a_1\frac{dx}{dt} + a_2\frac{d^2x}{dt^2} + \ldots + a_n\frac{d^nx}{dt^n} = c$$

where a_n and c are real constants.

The *order* of a differential equation is the order of the highest derivative that it includes. For example, a second-order differential equation would contain a second derivative and no higher derivatives. An *ordinary differential equation* is an equation that contains only a function of one variable along with its derivatives. A *partial differential equation* is an equation that contains a function of more than one variable along with its partial derivatives. A *homogeneous* equation is one in which, as written above, the right-hand side is zero; otherwise, it is said to be *non-homogeneous*.

The order of a differential equation is the highest derivative that appears in the equation.

EXAMPLE 1. Find the order of $3\left(\frac{d^2f}{dx^2}\right)^4 + \left(\frac{df}{dx}\right)^7 + 4x^5 = 6$.

The order of this equation is 2, because the highest derivative that appears is a second derivative.

The *degree* of a differential equation is the power of the highest derivative term in the equation.

EXAMPLE 2. Find the degree of the equation in the preceding example.

The degree of the equation is 4, because the highest derivative is raised to the fourth power. (Note that, in this example, the first derivative is raised to a higher power but only the exponent of the highest derivative determines the degree.)

In this chapter, we provide solutions for a number of different kinds of differential equations, typically highlighting their use within different physical contexts. We caution the reader that an understanding of the mathematics is not a substitute for an understanding of the physical principles. While we touch briefly upon the physical laws used (e.g., Newton's second law, Kirchoff's laws, etc.), the focus is on the equations rather than on the physical principles. The reader is encouraged to explore the physical concepts behind these equations in order to gain a deeper understanding.

EXAMPLE 3. Write differential equations that describe the motion of an object along the x axis that is subject to:

(*a*) a constant force
(*b*) a force proportional to its velocity

For (*a*): Using Newton's law, $F_0 = m\frac{d^2x}{dt^2}$, where F_0 is the magnitude of the constant force.

For (*b*): $m\frac{d^2x}{dt^2} = b\frac{dx}{dt}$, where $\frac{dx}{dt}$ is the speed of the object and b is the constant of proportionality.

VERIFYING A SOLUTION. Because the solution of many differential equations is tedious (if not downright difficult), textbooks often leave it to the reader to demonstrate, through direct substitution into the differential equation, that a proposed solution actually satisfies the equation.

EXAMPLE 4. Show that the function $g(x) = 3x^2 + 6x + 3$ satisfies the equation

$$g(x) - \frac{dg}{dx} + \frac{d^2g}{dx^2} - 3x^2 - 3 = 0.$$

Substituting the proposed function into the equation, we get:

$$g(x) - \frac{dg}{dx} + \frac{d^2g}{dx^2} - 3x^2 - 3 = (3x^2 + 6x + 3) - (6x + 6) + 6 - 3x^2 - 3 = 0$$

thereby satisfying the equation.

BOUNDARY CONDITIONS. The solution to a differential equation generally includes one or more arbitrary constants, the value(s) of which must be determined from the specifics of the physical problem. For example, a vibrating string that is fixed at two ends will constrain the displacement of the string at those positions, and the solution to the corresponding differential equation will need to reflect this. As another example, the differential equation expressing falling objects near the Earth's surface does not include either the initial position or the initial velocity of an object. Such boundary conditions are called *initial conditions* and will constrain the range of possible solutions of the differential equation. We shall see several examples of the application of boundary conditions to differential equations in this chapter, but the full treatment of boundary conditions is beyond the scope of this book.

EXAMPLE 5. A billiard ball moving in a straight line at 1 m/s at $t = 0$ then undergoes a steady deceleration of $a = -.05\,\text{m/s}^2$. What is it speed as a function of time?

We know that

$$\frac{dv}{dt} = a \tag{24.1}$$

so that

$$dv = a\,dt$$

$$v = \left(\int_{t_0}^{t} a\,dt\right) + v_0 = \left(a\int_{t_0}^{t} dt\right) + v_0 = a(t - t_0) + v_0$$

Substituting $t_0 = 0$ and the given value of a, we conclude that in SI units,

$$v = v_0 - .05t$$

where we use the symbol v_0 to represent the constant that arises in the integration. v_0 is apparently the initial velocity since from the last equation, $v = v_0$ at the initial time $t = t_i$. The main point here is that, while the form of the solution is a consequence of the differential equation (*24.1*), the value of the constant v_0 is not specified by the equation and needs to be obtained elsewhere. In this case, then, one needs additional information—for example, that the velocity of the ball at $t = 0$ is $v_0 = 1.5$ m/s.

EXAMPLE 6. Find the equation describing the position of an automobile as a function of time assuming that the car accelerates at 3 m/s^2 and has a velocity of 13 m/s at $t = 3$.

$$a = 3\,\mathrm{m/s^2}$$

$$\frac{\mathrm{d}v}{\mathrm{d}t} = 3$$

$$\mathrm{d}v = 3\,\mathrm{d}t$$

$$\int \mathrm{d}v = \int 3\,\mathrm{d}t$$

$$v = 3t + v_0$$

$$\frac{\mathrm{d}x}{\mathrm{d}t} = 3t + v_0$$

$$\mathrm{d}x = 3t\,\mathrm{d}t + v_0\mathrm{d}t$$

$$\int \mathrm{d}x = 3\int t\,\mathrm{d}t + v_0 \int \mathrm{d}t$$

$$x = \frac{3}{2}t^2 + v_0 t + x_0$$

where x_0 and v_0 are constants the values of which are not determined by the differential equation (*24.1*). Instead, they are determined from the specified initial conditions that the position of the car at $x(t = 2)$ is 5 and that $v(t = 3) = 13$.

Then,

$$5 = \frac{3}{2}(4) + 2v_0 + x_0 \rightarrow -1 = 2v_0 + x_0 \tag{24.2}$$

$$13 = 3(3) + v_0 \rightarrow v_0 = 4 \tag{24.3}$$

Substituting (*24.3*) into (*24.2*) yields $x_0 = -9$.
Hence, the specific solution to this problem is

$$x = \frac{3}{2}t^2 + 4t - 9$$

SOLVING DIFFERENTIAL EQUATIONS. Because of the large variety of differential equations that occur throughout physics, there is no general prescription for solving them. Instead, we restrict ourselves in this chapter to some of the more common kinds of differential equations found within the undergraduate physics and engineering curriculum. These cases include both homogeneous (i.e., lacking a constant term in the equation) and non-homogeneous situations involving:

(*1*) Equations with the derivative of a function and a polynomial in the independent variable.
(*2*) Equations linear in a function and its first derivative.
(*3*) Equations linear in a function and its second derivative.
(*4*) Equations linear in a function, its first derivative and its second derivative.

We examine both homogeneous and non-homogeneous versions.

Despite its brevity, the form of the equations described recur frequently throughout a diverse range of physics topics.

CASE 1: AN EQUATION WITH THE DERIVATIVE OF A FUNCTION AND A POLYNOMIAL. An example of such an equation is

$$\frac{dy}{dx} = c$$

The equation is easily solved using the techniques described in Chapters 16 and 17. Hence, because $\frac{dy}{dx} = c$, then $dy = c\,dx$ and $y = cx + d$. If y itself is the derivative of another function of interest, then it can be determined through application of the polynomial integration law $\int x^n\,dx = \frac{x^{n+1}}{n+1}$. This is demonstrated in Example 7 below.

EXAMPLE 7. Find the displacement y and the speed v_y of King Kong as a function of time as he falls from the top of the Empire State Building, assuming that he falls from the 102nd floor (height = 1250 ft). Calculate as well the total time involved in the fall. Ignore the effects of air resistance.

We know that the acceleration of an object in free fall is a constant, which we denote as g and which is known to be approximately equal to 32 ft/s^2 (or 9.80 m/s^2). Because the acceleration is the time derivative of velocity and the second derivative of displacement with respect to time, we have:

$$a = \frac{d^2y}{dt^2} = g$$

Let us choose the initial vertical position to be $y_0 = 0$ and choose the vertical axis to extend downward. Then, the ground corresponds to $y = 1250$ ft. Since the velocity $v = \frac{dy}{dt}$, we therefore have:

$$\frac{d^2y}{dt^2} = \frac{dv}{dt} = g$$

Solving for v, we get

$$dv = g\,dt$$
$$v = gt + C$$

If King Kong drops from rest, we know that $v = 0$ at $t = 0$, and so C must equal 0. Hence, the vertical speed as a function of time is simply

$$v = gt$$

Substituting dy/dt for v in the above equation, we then have:

$$\frac{dy}{dt} = gt$$
$$dy = gt\,dt$$
$$y = \int gt\,dt = g\int t\,dt = \frac{1}{2}gt^2 + y_0$$

where y_0 is the integration constant and represents the initial vertical position of the ape at $t = 0$.

When the initial velocity is not zero, then the displacement can be written as:

$$y = y_0 + v_0t + \frac{1}{2}gt^2$$

which is one of the kinematic equations of motion.

Substituting in the value of $y = 1250$ ft and $y_0 = 0$, we can then find the time at which King Kong reaches street level ($y = 0$):

$$y - y_0 = \frac{1}{2}gt^2$$

Rearranging terms to solve for the elapsed time for the ape to reach ground level and substituting the above values, we get

$$t = \sqrt{\frac{2(y - y_0)}{g}} = \sqrt{\frac{2(1,250 - 0)}{32}} = 8.8\text{ s}$$

CASE 2A: AN EQUATION WITH A FUNCTION AND ITS DERIVATIVE (BUT NO CONSTANT TERM). An example of such a function is the equation

$$a_0 x + a_1 \frac{dx}{dt} = 0 \tag{24.4}$$

where the dependent variable x is a function of the independent variable t (i.e., $x = x(t)$) and the coefficients a_0 and a_1 represent constants. A *general solution* to this equation is the exponential function

$$x(t) = Ae^{-\frac{a_0}{a_1}t} \tag{24.5}$$

where A is a constant. We demonstrate the technique for finding this solution in the examples that follow. For now, the reader can verify that *(24.5)* is indeed a solution to *(24.4)* by direct substitution into that equation.

EXAMPLE 8. An object is launched horizontally into a medium which exerts a resistive force $F = -bv$, where F is the force, v is the velocity of the object, and b is a constant. If the object is initially launched at 50 m/s, how fast will it be going after 10 s (assume $b = 0.2$ N s/m and ignore the influence of gravity).

From Newton's second law, we know that $F = m\frac{dv}{dt}$.

$$F = -bv$$
$$F = m\frac{dv}{dt}$$
$$-bv = m\frac{dv}{dt}$$
$$\frac{dv}{v} = -\frac{b}{m}dt$$
$$\int \frac{dv}{v} = -\frac{b}{m}\int dt$$
$$\ln(v) = -\frac{b}{m}t + C$$
$$v = e^{-\frac{b}{m}t+C} = e^{-\frac{b}{m}t}e^{C}$$

If we call the initial velocity (when $t = 0$) v_0, then, from the preceding equation, e^C must equal v_0. We therefore have:

$$v = v_0 e^{-\frac{b}{m}t}$$

A sketch of this curve is shown in Fig. 24-1 for various values of the ratio b/m. Note that relatively high values of b/m correspond to a relatively high frictional force.

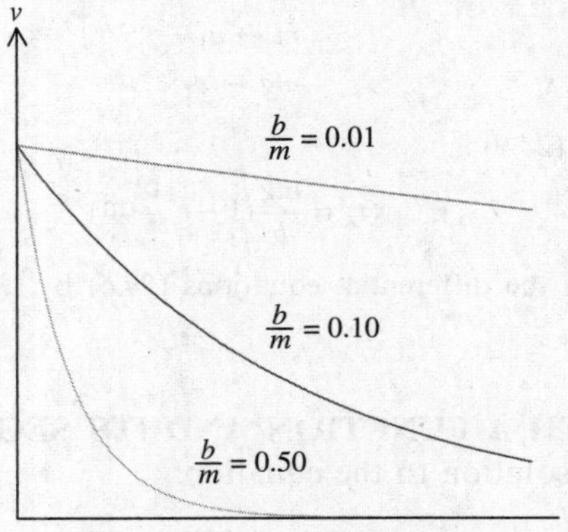

Fig. 24-1

CASE 2B: AN EQUATION WITH A FUNCTION, ITS DERIVATIVE, AND A CONSTANT TERM. We seek a solution to the equation

$$a_0 x + a_1 \frac{dx}{dt} = c_1 \tag{24.6}$$

where x is a function of t and a_0, a_1, and c_1 are constants. The solution to this kind of equation, which we explore in the example and solved problems that follow, is:

$$x(t) = \frac{c_1}{a_0}(1 - e^{-a_0 t/a_1}) \tag{24.7}$$

EXAMPLE 9. We consider *a falling object subject to air resistance near the Earth's surface.*

We begin with Newton's second law:

$$F = ma$$

$$F_{\text{net}} = m\frac{dv}{dt}$$

The net force is due to two forces: the downward force of gravity and the upward force of air resistance.

$$m\frac{dv}{dt} = F_{\text{gravity}} + F_{\text{air}}$$

Near the surface of the Earth, the force of gravity is simply equal to mg, where m is the mass of the object and g is the acceleration due to gravity (approximately 9.8 m/s^2). The force due to air resistance is assumed, as is often the case in such situations, to be proportional to the object's velocity and in the direction that opposes the downward acceleration (because the faster it moves, the more air molecules will collide with it per second, thereby opposing the downward accelerating force due to gravity). That is, the force of air resistance is represented as

$$F_{\text{air}} = -bv$$

where v is the speed of the object and b is a constant called the *damping coefficient* and we are observing the object in a reference frame in which the surrounding air is at rest. Objects which are streamlined to minimize air resistance, like sports cars, have low damping coefficients—and therefore do not generate much air resistance—while other objects, like falling paper plates, have high damping coefficients and therefore high air resistance.

The net force on the object is the sum of the gravitational force and the force of air resistance:

$$m\frac{dv}{dt} = mg - bv \tag{24.8}$$

Rearranging the terms of (*24.8*), we get

$$bv + m\frac{dv}{dt} = mg \tag{24.9}$$

This is a linear differential equation that includes a function ($v(t)$), its derivative (dv/dt) and a constant ($-bv$). It therefore has the same form as (*24.6*) and aside from the names of the constants, the same solution. The variable $v(t)$ corresponds to the variable $x(t)$ in (*24.7*). Comparing the respective coefficients we see that:

$$b \leftrightarrow a_0$$
$$m \leftrightarrow a_1$$
$$mg \leftrightarrow c_1$$

Apparently, then, the solution to (*24.9*) is

$$v(t) = \frac{mg}{b}(1 - e^{-bt/m}) \tag{24.10}$$

In Problem 24.27 the solution of the differential equations (*24.6*) by (*24.7*) is shown explicitly; we content ourselves here with only the results.

CASE 3A: AN EQUATION WITH A FUNCTION AND ITS SECOND DERIVATIVE (BUT NO CONSTANT TERM). We seek a solution to the equation:

$$a_0 x + a_2 \frac{d^2x}{dt^2} = 0$$

We need to distinguish two separate situations* when $\frac{a_0}{a_2} > 0$ and $\frac{a_0}{a_2} < 0$

Situation 1: $a_0/a_2 > 0$. The solution to this equation is a periodic one:

$$x = A\sin(\omega t + \delta), \text{ where } \omega = \sqrt{\frac{a_0}{a_2}} \tag{24.11}$$

and A and δ are arbitrary constants representing, respectively, the amplitude and phase shift of the sine wave (Fig. 24-2). The amplitude A and the phase shift ϕ are not determined from the differential equation above but rather from the specifics of a particular problem, as is demonstrated below.

The solution indicated above can be expressed in other forms. For example, because the sine and cosine differ only by a phase shift of 90° (i.e., $\sin(x + 90°) = \cos(x)$, the solution could equally well have been represented as:

$$x = A\cos(\omega t + \delta_2) \text{ or}$$
$$x = B_1\cos(\omega t) + B_2\sin(\omega t)$$

where A and B are arbitrary constants determined by boundary conditions (often the value of an initial displacement and/or velocity)

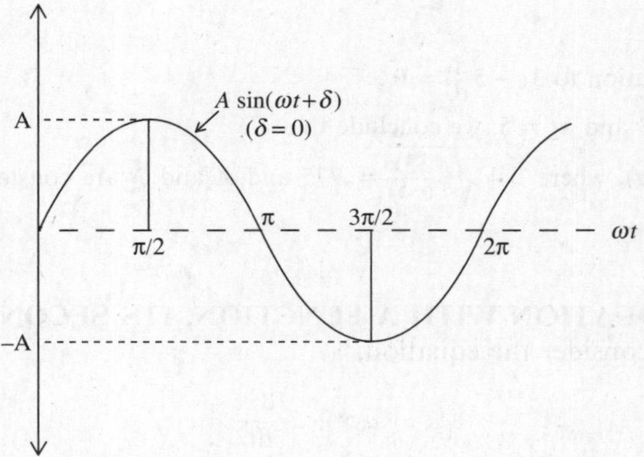

Fig. 24-2

EXAMPLE 10. (*a*) Find the natural frequency ω of a simple harmonic oscillator consisting of a mass m attached to a spring with a spring constant of k. (*b*) Find the period of oscillation assuming a mass of 1 kg and $k = 3$ N/m.

(*a*) Most springs obey, at least approximately, Hooke's law, which states that, for small displacements, the force that tends to restore the spring to its equilibrium position is proportional to the displacement of the spring from that position. In other words,

$$F = -kx, \text{ where } k \text{ is the spring constant.}$$

Combining this with Newton's second law ($F = ma = m\frac{d^2x}{dt^2}$), we get:

$$F = -kx$$

$$F = m\frac{d^2x}{dt^2}$$

* The remaining situation—when $a_0 = 0$—is easily solved, corresponding to $\frac{d^2x}{dt^2} = 0$ or $x = bt + c$ for some constants b and c.

$$m\frac{d^2x}{dt^2} = -kx$$

$$m\frac{d^2x}{dt^2} + kx = 0$$

$$\frac{d^2x}{dt^2} + \frac{k}{m}x = 0$$

Using (*24.11*), we see that $x = A\sin(\omega t + \delta)$, where $\omega = \sqrt{\frac{k}{m}}$

(*b*) $\omega = \sqrt{\frac{k}{m}} = \sqrt{\frac{3N/m}{1\,\text{kg}}} = 1.732\,\text{s}^{-1}$

Situation 2: $a_0/a_2 < 0$. We consider again the equation $a_0 x + a_2\frac{d^2x}{dt^2} = 0$, now considering the case when $a_0/a_2 < 0$. The solution to this equation is:

$$x = A\cosh(\alpha t) + B\sinh(\alpha t) \qquad (24.12)$$

where $\cosh(\alpha t) = \frac{e^{\alpha t} + e^{-\alpha t}}{2}$, $\sinh(\alpha t) = \frac{e^{\alpha t} - e^{-\alpha t}}{2}$, and $\alpha = \sqrt{-\frac{a_0}{a_2}}$.

EXAMPLE 11. Find the solution to $3y - 5\frac{dy}{dz} = 0$

From (*24.12*), with $a_0 = 3$ and $a_2 = 5$, we conclude that

$y = A\cosh(\alpha z) + B\sinh(\alpha z)$, where $\alpha = \sqrt{-\frac{(3)}{(-5)}} = .775$ and A and B are constants that are determined from initial conditions.

CASE 3B: A LINEAR EQUATION WITH A FUNCTION, ITS SECOND DERIVATIVE, AND A CONSTANT TERM. We consider the equation:

$$a_0 x + a_2\frac{d^2x}{dt^2} = c_1 \qquad (24.13)$$

which is identical to that of Case 3A except for the addition of the constant term c_1 on the right-hand side of (*24.13*). (We consider here only the case $^{a_0}/_{a_1} > 0$ as the one arising more frequently at the undergraduate level.) The solution to this equation is identical to that of the homogeneous equation (*24.11*) except for the addition of a constant term:

$$x = A\sin(\omega t + \delta) + \frac{c_1}{a_0} \qquad (24.14)$$

where, again,

$$\omega = \sqrt{\frac{a_0}{a_2}}.$$

The solution to this equation is identical to that of the homogeneous equation (*24.11*) except for the addition of the constant term on the right-hand side of (*24.14*).

EXAMPLE 12. A simple harmonic oscillator, consisting of a mass at the end of a (massless) spring, is held vertically in a uniform gravitational field. Analyze its equilibrium position and subsequent motion when displaced from rest.

At rest, the mass will be in an equilibrium position that will be displaced from the position it would occupy in the absence of gravity. Because equilibrium is defined as that point at which there is no net force, the downward force of gravity must equal the upward restoring force ($= -kx$) of the oscillator at this point. If we call $x = 0$ the equilibrium point in the absence of gravity and $x' = 0$ the equilibrium point in the presence of gravity, then

$$m\frac{d^2x}{dt^2} = mg - kx$$

$$m\frac{d^2x}{dt^2} + kx = mg \qquad (24.15)$$

At equilibrium, $mg = kx \rightarrow x = \frac{mg}{k}$

If we now let $x' = 0$ be the location of the equilibrium point in the presence of gravity and x' the vertical (downward) displacement from that point (Fig. 24-1), then

$$x = x' + \frac{mg}{k} \qquad (24.16)$$

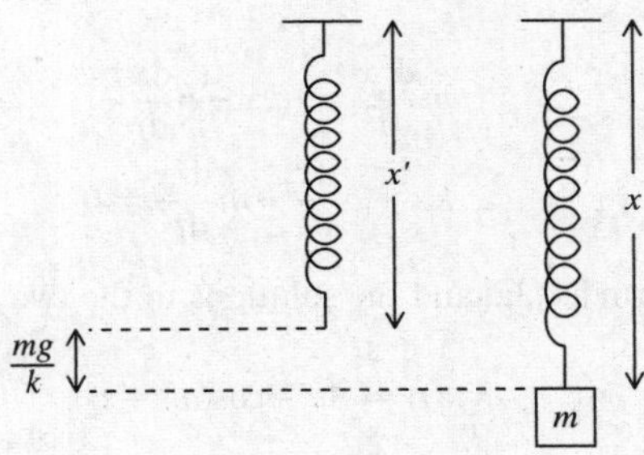

Fig. 24-3

Substituting this into the previous expression, we get:

$$m\frac{d^2\left(x' + \frac{mg}{k}\right)}{dt^2} + k\left(x' + \frac{mg}{k}\right) = mg$$

Since the derivative of the constant term mg/k is zero, we get:

$$m\frac{d^2x'}{dt^2} + kx' = 0$$

This equation has the same form as that discussed in Case 3A above, except that x is now replaced by x'. Hence, the solution (in terms of x') has the same form as in that example:

$$x' = A\sin(\omega t + \delta), \text{ where } \omega = \sqrt{\frac{k}{m}}$$

Because x' and x are related by the equation (*24.16*), we conclude that:

$$x = A\sin(\omega t + \delta) + \frac{mg}{k}, \text{ where } \omega = \sqrt{\frac{k}{m}}$$

Note that the same result could have been obtained immediately in comparing (*24.15*) with (*24.13*) but without the same level of physical description.

CASE 4A: A LINEAR EQUATION WITH A FUNCTION, ITS DERIVATIVE, AND ITS SECOND DERIVATIVE (BUT NO CONSTANT TERM)

The following equation includes a function along with its first and second derivatives:

$$a_0x + a_1\frac{dx}{dt} + a_2\frac{d^2x}{dt^2} = 0$$

We do not present a general solution to this equation but rather one for which the coefficients obey the relation $a_1^2 < 4a_0a_2$. Under these circumstances, the solution is given by:

$$x(t) = Ae^{\frac{a_1t}{2a_2}}\cos(\omega t + \phi)$$

where

$$\omega = \sqrt{\frac{a_0}{a_2} - \left(\frac{a_1}{2a_2}\right)^2}.$$

EXAMPLE 13. A simple harmonic oscillator with spring constant k moves horizontally subject to a frictional force $\vec{F} = -b\vec{v}$, where $\vec{v}$ is the velocity and b is the damping constant. Find the motion of the oscillator as a function of time, assuming it is released 10 cm from its equilibrium point at $t = 0$.

From Newton's second law, we know that $\vec{F}_{TOT} = m\frac{d^2\vec{x}}{dt^2}$.

In this case, F_{TOT} is the sum of the separate forces acting upon the mass:

$F_{spring} = -kx$, where k is the spring constant and x is the displacement from the equilibrium position

$F_{friction} = -bv$, where v is the velocity of the mass and b is the damping constant

The equation of motion is then:

$$m\frac{d^2x}{dt^2} = -kx - b\frac{dx}{dt}$$
$$kx + b\frac{dx}{dt} + m\frac{d^2x}{dt^2} = 0 \tag{24.17}$$

Equation (*24.17*) is of the same form as [3a] and the solutions to the two equations will therefore also have the same form.

$$x(t) = Ae^{-\frac{bt}{2m}}\cos(\omega t + \phi) \tag{24.18}$$

has the behavior described and is in fact a solution of (*24.17*), provided that

$$\omega = \sqrt{\omega_0^2 - \left(\frac{b}{2m}\right)^2},$$

where $\omega_0 = \sqrt{k/m}$ is the frequency of the undamped oscillator.

We can also appeal to our physical intuition for a solution in the case of the damped spring. We know that, in the presence of a frictional force, a spring will typically diminish in amplitude but continue to oscillate. Hence, we expect that $x(t)$ will be a periodic function with an ever-diminishing amplitude—exactly the behavior of (*24.18*).

If the damping constant b is not too large—i.e., if the maximum damping force is small compared to the maximum spring force—then these damped oscillations will occur, and we say that the oscillator is *underdamped.* When b is equal to $2m\omega_0$, then the mass approaches equilibrium as quickly as possible without successive oscillation, and the oscillator is said to be *critically damped.* The damper on a door or other device may be critically damped so that it will return to its equilibrium position as quickly as possible without "overshooting". When b exceeds this value, then the approach to equilibrium occurs more slowly, and the oscillator is said to be *overdamped.* Each of these situations is shown in Fig. 24-4.

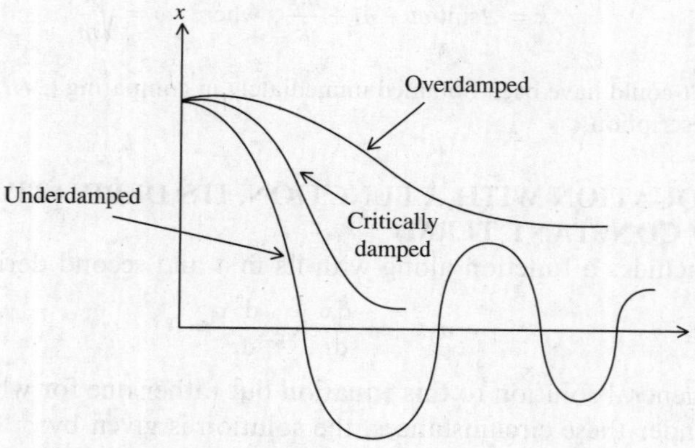

Fig. 24-4

CASE 4B: A LINEAR EQUATION WITH A FUNCTION, ITS DERIVATIVE, ITS SECOND DERIVATIVE, AND A DRIVING TERM

We finally examine an equation that includes a function, its derivative and second derivative, and a periodic function of the independent variable (in this case, t):

$$a_0 x + a_1 \frac{dx}{dt} + a_2 \frac{d^2x}{dt^2} = F_0 \sin(\omega t + \delta) \qquad (24.19)$$

The function is identical to that of Case 4A except for the term on the right-hand side of the equation.

This kind of equation typically arises in mechanical or electrical systems that possess some kind of natural frequency (e.g., the frequency of a pendulum swing or of the charge oscillating between circuit elements). When such systems are subject to an external stimulus (e.g., a mechanical force or an voltage source), they respond. When the external stimulus is of a periodic nature (as represented by the right-hand side of (*24.19*)), then the response of the system depends upon how closely the frequency of the external source matches the natural frequency of the system. When the two are closely matched, then the response is maximal. An example of resonance is the response of a swing when a child extends her legs at a frequency that matches the swing's natural frequency. Other examples in mechanics, electrical systems, and atomic systems abound.

We consider here only the behavior of the system after a sufficiently long time has elapsed to allow the system to reach a steady state. Under those circumstances, the solution to (*24.19*) is a periodic solution:

$$x = A\cos(\omega t + \delta) \qquad (24.20)$$

where ω is the frequency of the external force. The amplitude A is given by:

$$A = \frac{1}{a_2} \frac{F_0}{\sqrt{(\omega^2 - \omega_0^2)^2 + \left(\frac{a_1 \omega}{a_2}\right)^2}} \qquad (24.21)$$

where, as in the solution to the homogeneous form of this equation, $\omega_0 = \sqrt{\frac{a_0}{a_2} - \left(\frac{a_1}{2a_2}\right)^2}$

EXAMPLE 14. Find the steady-state solution to a mechanical oscillator with mass m, spring constant k, and damping constant b driven by a force $F = F_0 \sin(\omega t + \delta)$.

According to Newton's second law, we have:

$m\frac{d^2x}{dt^2} = -kx - b\frac{dx}{dt} + F_0 \sin(\omega t + \delta)$, or, after rearranging terms:

$$kx + b\frac{dx}{dt} + m\frac{d^2x}{dt^2} = F_0 \sin(\omega t + \delta) \qquad (24.22)$$

This corresponds to the form of (*24.19*) with $a_0 = k$, $a_1 = b$, and $a_2 = m$.

Substituting these values into (*24.22*) and (*24.21*), we get

$$x = A\cos(\omega t + \delta) \qquad (24.23a)$$

where

$$A = \frac{1}{m} \frac{F_0}{\sqrt{(\omega^2 - \omega_0^2)^2 + \left(\frac{b\omega}{m}\right)^2}} \qquad (24.23b)$$

where ω_0 is the damped frequency,

$$\omega_0 = \sqrt{\frac{k}{m} - \left(\frac{b}{2m}\right)^2} \qquad (24.23c)$$

From a physical standpoint, the external force is pouring energy into the system while the damping force absorbs it. As time progresses, the energy coming in and the energy dissipated become equal. When the applied frequency ω with which energy is applied nearly matches the natural frequency ω_0 of the system, then one can show that energy is transmitted into the system with high efficiency. This is reflected by the behavior of the denominator of

A as ω approaches ω_0. This results in a large amplitude—i.e., relatively large oscillations—which is termed *resonance*. A graph of A as a function of ω is shown in Fig. 24-5.

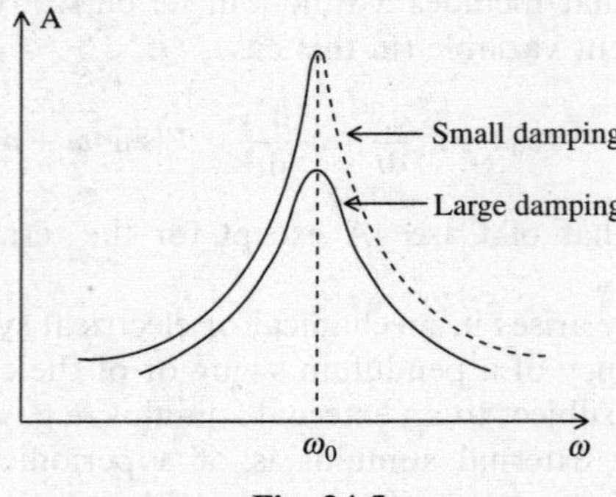

Fig. 24-5

CORRESPONDENCE BETWEEN MECHANICAL AND ELECTRICAL SYSTEMS. In discussing differential equations, we have so far limited ourselves to mechanical systems involving forces and motion. We have not yet discussed electrical circuits and the differential equations that determine the flow of charge in such circuits. There is an interesting connection between the equations that govern mechanical systems and those that govern electrical circuits.

We first summarize some essential elements of circuit theory:

Table 24.1

In this chapter, a number of examples and problems involve electrical circuits. For such circuits, *Kirchoff's law* is widely used:

The sum of the potential differences (or "voltage drops") around any closed loop of a circuit is equal to zero).

In addition, the following relations involving the potential difference across individual circuit elements are also widely used:

Resistor: $V_R = IR = R\frac{dq}{dt}$, where R is the resistance and $I = \frac{dq}{dt}$ is the current flowing through the resistor.

Inductor: $V_L = LI = L\frac{di}{dt} = L\frac{d^2q}{dt^2}$, where L is the inductance and $\frac{di}{d_t}\left(=\frac{d^2q}{dt^2}\right)$ is the time rate of change of the current.

Capacitor: $V_C = \frac{q}{C}$, where q is the charge on either plate of the capacitor and C is the capacitance.

For mechanical systems containing a simple harmonic oscillator, a velocity-dependent damping force and an external driving term, we have, using Newton's second law:

$$m\frac{d^2x}{dt^2} = -kx - b\frac{dx}{dt} + F_{\text{driving}} \rightarrow kx + b\frac{dx}{dt} + m\frac{d^2x}{dt^2} = F_{\text{driving}}$$

For electrical circuits, on the other hand, Kirchoff's law says that the sum of the voltage drops around a closed circuit—consisting, potentially, of resistors, inductors, capacitors, and voltage sources—is zero:

$$L\frac{d^2q}{dt^2} = -\frac{q}{C} - R\frac{dq}{dt} - v_{\text{source}} = 0 \rightarrow L\frac{d^2q}{dt^2} + R\frac{dq}{dt} + \frac{q}{C} = v_{\text{source}}$$

Comparing the two equations above, we see that both the mechanical and electrical systems can be described by non-homogeneous linear differential equations in displacement and charge and their respective first and second derivatives. Each version of the differential equation has the same solution, once we know which values of L, R, and C to substitute for the corresponding values of m, b, and k.

In fact, comparing directly the above equations, it seems that:

springs behave *like capacitors*, where the spring constant k corresponds to $1/C$
frictional forces behave like *resistors*, where the frictional constant b corresponds to R
inductors behave like *masses*, where the mass m corresponds to L

These correspondences have been made on the basis of comparing the charge $q(t)$ in electrical systems with the displacement $x(t)$ in mechanical ones and do not necessarily hold when comparing other quantities. Aside from this caveat, however, we see that if we construct analogous mechanical and electrical circuits and find the solution to one system, the substitutions listed above will provide a solution to the other system.

EXAMPLE 15. Find the charge on a capacitor in a series LC circuit as a function of time by analyzing the mechanical analogy. (Assume that a charge q_0 resides on the plates of the capacitor, with the switch closed at $t = 0$.)

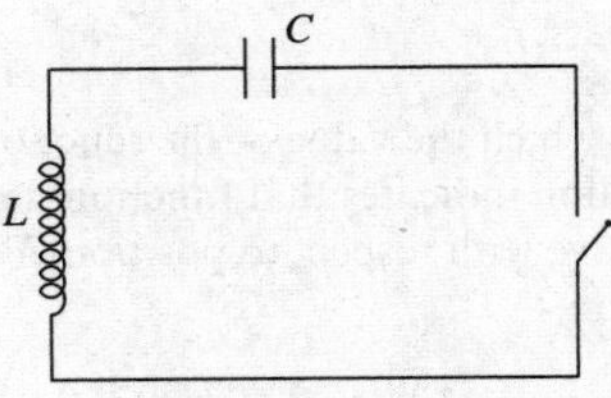

Fig. 24-6

For the LC circuit (a circuit including a capacitor and an inductor), we have, using Kirchoff's laws:

$$-\frac{q}{C} - L\frac{\mathrm{d}i}{\mathrm{d}t} = 0$$

Since $i = \frac{\mathrm{d}q}{\mathrm{d}t}, \frac{\mathrm{d}i}{\mathrm{d}t} = \frac{\mathrm{d}}{\mathrm{d}t}\left(\frac{\mathrm{d}q}{\mathrm{d}t}\right) = \frac{\mathrm{d}^2q}{\mathrm{d}t^2}$. Substituting this into the previous equation, we get:

$$\frac{q}{C} + L\frac{\mathrm{d}^2q}{\mathrm{d}t^2} = 0$$

Identifying $1/C$ with a spring constant k and L with a mass, this is apparently the electrical analog of a mass at the end of a simple harmonic oscillator, for which we know the solution *(24.11)* is:

$$x = A\sin(\omega t + \delta),$$

where

$$\omega = \sqrt{\frac{k}{m}}.$$

Making the identifications above, and substituting the charge q for the displacement x we then have:

$$q = q_o\sin(\omega t + \delta),$$

where $\omega = \sqrt{\frac{1}{LC}}$ and q_0 is an arbitrary constant to be determined by the initial conditions (the amount of charge that happens to be on the capacitor at time $t = 0$).

PARTIAL DIFFERENTIAL EQUATIONS. An equation that includes a function of more than one variable along with one or more of its partial derivatives is called a partial differential equation. The solution of such equations can be notoriously difficult, with many amenable only to numerical computation. An extensive treatment of such equations is beyond the scope of this book. However, because of its frequent use and its simplicity, one particularly useful tool in the solution of such equations—the technique of the *separation of variables*—is presented.

THE SEPARATION OF VARIABLES METHOD. The idea behind this method is that the solution of a partial differential equation may be a multivariable function that can be decomposed into a combination (often a product) of functions of a single variable. If successful, then substitution of a "trial function" into the original partial differential equation yields two or more differential equations, each dependent on only one variable (i.e., "ordinary" differential equations). These equations then may be solvable by the kinds of methods described earlier in this chapter.

The technique is illustrated in the following example.

EXAMPLE 16. The *Schrödinger equation* for a particle in one dimension x describes the evolution over time of the quantum mechanical wave function $\Psi(x, t)$ of that particle. (We do not discuss the physical significance of this function beyond noting that $|\Psi(x, t)|^2$ is a measure of the probability of finding the particle in the vicinity of a particular location at a particular time.) If we restrict ourselves to potential functions V that depend only upon the position x (and not on the time), then the equation is given by

$$-\frac{\hbar^2}{2m}\frac{\partial^2\Psi(x,t)}{\partial x^2} + V(x)\Psi(x,t) = i\hbar\frac{\partial\Psi}{\partial t} \tag{24.24}$$

where $\hbar = h/2\pi$ is Planck's constant.

This is a partial differential equation in which the value of the function $\Psi(x, t)$ is a unknown function of both its position x and its time t. As the above equation indicates that function depends upon the first partial derivative with respect to time and its second partial derivative with respect to position. To solve this equation, we use the method of separation of variables to guess at a solution:

$$\Psi(x,t) = \psi(x)T(t) \tag{24.25}$$

where $\psi(x)$ is a function only of x and $T(t)$ is a function only of t.

Substituting this into the previous equation, we get:

$$-\frac{\hbar^2}{2m}\frac{\partial^2[\psi(x)T(t)]}{\partial x^2} + V(x)[\psi(x)T(t)] = i\hbar\frac{\partial[\psi(x)T(t)]}{\partial t}$$

Because $T(t)$ does not depend on x and $\psi(x)$ does not depend upon t, we can rewrite this as:

$$-\frac{\hbar^2}{2m}T(t)\frac{d^2\psi(x)}{dx^2} + V(x)\psi(x)T(t) = i\hbar\psi(x)\frac{dT(t)}{dt},$$

where $\frac{d^2\psi(x)}{dx^2}$ and $\frac{dT(t)}{dt}$ are ordinary derivatives (i.e., dependent on only one variable).

Dividing both sides by $\Psi(x, t) = \psi(x)T(t)$, we get:

$$-\frac{\hbar^2}{2m}\frac{1}{\psi(x)}\frac{d^2\psi(x)}{dx^2} + V(x) = i\hbar\frac{1}{T(t)}\frac{dT(t)}{dt}$$

What is remarkable about this equation is that the left-hand side depends only upon the position (and not upon the time) and the right-hand side depends only upon the time (and not upon the position). This appears at first to be contradictory: as time progresses, for example, the value of the left-hand side will not change but the value of the right-hand side generally will. The only way to avoid this apparent contradiction is if both sides are equal to a constant value. In the above case, if we call this constant E, then we can separately set both sides of this equation to E, yielding:

$$-\frac{\hbar^2}{2m}\frac{1}{\psi(x)}\frac{d^2\psi(x)}{dx^2} + V(x) = E$$

and

$$i\hbar\frac{1}{T(t)}\frac{dT(t)}{dt} = E$$

As a result of the substitution (*24.25*), then, the partial differential equation (*24.24*) has been decomposed into the two ordinary differential equations above. It can be shown that for the particular case $V(x) = 0$, the first equation has the solution

$\psi(x) = Ae^{ikx}$, where A is an undetermined constant and $k = \frac{\sqrt{2mE}}{\hbar}$
and the second equation has the solution
$T(t) = Be^{i\omega t}$, where $\omega = \frac{E}{\hbar}$.
We can then patch a solution together using (*24.23*) to find the solution to (*24.22*):
$\Psi(x, t) = \psi(x)T(t) = Ce^{i(kx-\omega t)}$ with k and ω defined above. (C is a constant to be determined from boundary conditions, which we do not consider here.)

Solved Problems

24.1 (*a*) Show that a series LRC circuit driven by a sinusoidally varying voltage $V(t)$ is described by the same form of differential equation as a damped driven mechanical oscillator. (*b*) Find the steady-state solution for an LRC circuit using by considering the solution of the mechanical oscillator.

(*a*) Using Kirchoff's laws, we know that

$$V_0\cos(\omega t) - \frac{q}{C} - R\frac{dq}{dt} - L\frac{d^2q}{dt^2} = 0 \rightarrow \frac{q}{C} + R\frac{dq}{dt} + L\frac{d^2q}{dt^2} = V_0\cos(\omega t)$$

The damped driven harmonic oscillator is described using Newton's laws by:

$$m\frac{d^2x}{dt^2} = -kx - b\frac{dx}{dt} + F_0\cos\omega t \rightarrow kx + b\frac{dx}{dt} + m\frac{d^2x}{dt^2} = F_0\cos\omega t$$

The steady-state solution to the damped driven harmonic oscillator is given by (*24.23*(*a–c*)):

$$x = A\cos(\omega t + \delta)$$

where ω is the frequency of the external force and the amplitude A is given by:

$$A = \frac{1}{m}\frac{F_0}{\sqrt{(\omega^2 - \omega_0^2)^2 + \left(\frac{b\omega}{m}\right)^2}}$$

where ω_0 is the natural frequency of the system.
Making the identification of Table 24.1, where $k \leftrightarrow 1/C$, $m \leftrightarrow L$, and $b \leftrightarrow R$, we immediately conclude that, in the case of the LCR circuit,

$$q = q_0\cos(\omega t + \delta)$$

where ω is the frequency of the external force and the amplitude A is given by:

$$A = \frac{1}{m}\frac{V_0}{\sqrt{(\omega^2 - \omega_0^2)^2 + \left(\frac{R\omega}{L}\right)^2}}$$

where

$$\omega_0 = \sqrt{\frac{1}{LC} - \left(\frac{R}{2L}\right)^2}.$$

24.2 (*a*) Write down the differential equation that describes the motion of the damped harmonic oscillator in which a mass m at the end of a spring with spring constant k is subject to a damping force of $-bv$. (*b*) Show by direct substitution that $x(t) = Ae^{-bt/2m}\cos(\omega t + \delta)$ is a solution of this equation. (*c*) Find the time required for a mass to reach $1/e$ of its initial amplitude if the mass $m = 1$ kg and $b = 1.2$ kg/s

(*a*) Using Newton's second law,

$$F = m\frac{d^2x}{dt^2} = -kx - b\frac{dx}{dt}$$

Therefore,

$$kx + b\frac{dx}{dt} + m\frac{d^2x}{dt^2} = 0 \tag{24.26}$$

The proposed solution is $x(t) = Ae^{-bt/2m}\cos(\omega t + \delta)$ (24.27a)

where

$$\omega = \sqrt{\omega_0^2 - \left(\frac{b}{2m}\right)^2} \tag{24.27b}$$

and $\omega_0 = \sqrt{k/m}$. We differentiate this expression and its derivative in order to evaluate each of the terms in (24.26):

$$kx = Am\omega_0^2 e^{-bt/2m}\cos(\omega t + \delta)$$

(where we have used the relation $\omega_0^2 = k/m$ to replace k)

$$b\frac{dx}{dt} = -b\omega Ae^{-bt/2m}\sin(\omega t + \delta) - \frac{b^2}{2m}Ae^{-bt/2m}\cos(\omega t + \delta)$$

$$m\frac{d^2x}{dt^2} = m\omega\frac{b}{2m}Ae^{-bt/2m}\sin(\omega t + \delta) - m\omega^2 Ae^{-bt/2m}\cos(\omega t + \delta)$$

$$+ m\left(\frac{b}{2m}\right)^2 Ae^{-bt/2m}\cos(\omega t + \delta) + mA\left(\frac{b}{2m}\right)\omega e^{-bt/2m}\sin(\omega t + \delta)$$

Adding these three terms together, we get:

$$kx + b\frac{dx}{dt} + m\frac{d^2x}{dt^2} = Am\omega_0^2 e^{-bt/2m}\cos(\omega t + \delta) - b\omega Ae^{-bt/2m}\sin(\omega t + \delta) - \frac{b^2}{2m}Ae^{-bt/2m}\cos(\omega t + \delta)$$

$$+ m\omega\frac{b}{2m}Ae^{-bt/2m}\sin(\omega t + \delta) - m\omega^2 Ae^{-bt/2m}\cos(\omega t + \delta)$$

$$+ m\left(\frac{b}{2m}\right)^2 Ae^{-bt/2m}\cos(\omega t + \delta) + mA\left(\frac{b}{2m}\right)\omega e^{-bt/2m}\sin(\omega t + \delta)$$

Upon rearrangement, this becomes:

$$kx + b\frac{dx}{dt} + m\frac{d^2x}{dt^2} = Ae^{-bt/2m}\left(m\omega_0^2 - \frac{b^2}{2m} - m\omega^2 + m\left(\frac{b}{2m}\right)^2\right)\cos(\omega t + \delta)$$

$$\left(-b\omega + \omega\frac{b}{2} + \left(\frac{b}{2}\right)\omega\right)Ae^{-bt/2m}\sin(\omega t + \delta)$$

The coefficients of the second term add to zero and we are left with:

$$kx + b\frac{dx}{dt} + m\frac{d^2x}{dt^2} = \left(m\omega_0^2 - m\omega^2 - \frac{b^2}{4m}\right)Ae^{-bt/2m}\cos(\omega t + \delta) \tag{24.28}$$

But since, from above we know that $\omega = \sqrt{\omega_0^2 - \left(\frac{b}{2m}\right)^2}$, then $m\omega^2 = m\omega_0^2 - \frac{b^2}{4m}$ and the right-hand side of (24.26) equals zero, thereby establishing that (24.27(a, b)) is a solution of (24.26)

24.3 (a) Find the voltage as a function of time across the capacitor in a driven LRC circuit consisting of an inductor, a resistor, a capacitor, and an oscillating voltage source $V = V_0 \sin(\omega t + \delta)$ in series. (b) What is the voltage across the inductor as a function of time? (c) Determine the voltages across each of these circuit elements at $t = 0.05$ s. (d) Determine the voltages across each of these circuit elements at $t = 5$ s.

$$V_C = \frac{q}{C}$$

$$V_R = R\frac{dq}{dt}$$

$$V_L = -L\frac{d^2q}{dt^2}$$

If the circuit is discharging, then $V_0 = V_C + V_R + V_C$

$$V_0 = V_R + V_C + V_L V_0$$

$$V_0 - R\frac{dq}{dt} - \frac{q}{C} - R\frac{d^2q}{dt^2} = 0$$

$$L\frac{d^2q}{dt^2} + R\frac{dq}{dt} + \frac{q}{C} = V_0$$

Identifying this equation as an example of Case 4A (*24.17.0*), we identify $1/C$, R, and L with the coefficients of (*24.17.0*) to obtain:

$$q = q_0 e^{-bt}\cos(\omega t + \phi)$$

where, using the equation following (24.17.1), $\omega = \sqrt{\frac{1}{LC} - \left(\frac{R}{2L}\right)^2}$ (In this expression, we have used the expression following (*24.18*) for the analogous mechanical system and made the identifications $k \leftrightarrow \frac{1}{C}$, $b \leftrightarrow R$, and $m \leftrightarrow L$).

24.4 A mass of 2 kg is suspended from a spring with spring constant $k = 5$ N/m in a container immersed in oil. The frictional force between the mass and the oil is equal to $-bv$, where v is the velocity of the mass and b has the value of 5 N s/m. If the mass is initial stretched 10 cm from the equilibrium point of the spring, then what is the:

(*a*) undamped frequency?
(*b*) damped frequency?
(*c*) amplitude as a function of time?
(*d*) time when the amplitude reaches 1/e of its original height?

$$m\frac{d^2x}{dt^2} = -kx - b\frac{dx}{dt}$$

$$\frac{d^2x}{dt^2} = -\frac{k}{m}x - \frac{b}{m}\frac{dx}{dt}$$

(*a*) The undamped frequency $f_{\text{undamped}} = \frac{\omega}{2\pi} = \frac{1}{2\pi}\sqrt{\frac{k}{m}} = \frac{1}{2\pi}\sqrt{\frac{k}{m}} = \frac{1}{2(3.14)}\sqrt{\frac{5\text{N/m}}{2\text{kg}}} = .25/\text{s}$

(*b*) The damped frequency $f_{\text{damped}} = \frac{\omega'}{2\pi} = \frac{1}{2\pi}\sqrt{\frac{k}{m} - \frac{b^2}{4m^2}} = \frac{1}{2(3.14)}\sqrt{\frac{(5\text{N/m})}{(2\text{ kg})} - \frac{(5\text{N} - \text{s/m})^2}{4(2\text{ kg})^2}} = .15\text{ s}^{-1}$

(c) $x = A_0 e^{-\frac{bt}{2m}}\sin(\omega' t + \delta)$

$$A(t) = A_0 e^{-\frac{bt}{2m}} = (10\text{ cm})e^{-\frac{(5)t}{2(2)}} = 10e^{-1.25t}\text{cm}$$

(*d*) $\frac{bt}{2m} = 1 \to t = \frac{2m}{b} = \frac{2(2\text{kg})}{5\text{N} - \text{s/m}} = .8\text{ s}$

24.5 What is the value of a function that is:

(*a*) equal everywhere to its derivative?
(*b*) equal everywhere to twice its derivative?
(*c*) equal everywhere to twice its second derivative?

(*a*) $y = \frac{dy}{dx} \rightarrow \frac{dy}{y} = dx \rightarrow \ln y = x + c \rightarrow y = De^x$, where D is an arbitrary constant

(*b*) $y = 2\frac{dy}{dx} \rightarrow \frac{dy}{y} = \frac{dx}{2} \rightarrow \int \frac{dy}{y} = \int \frac{dx}{2} \rightarrow \ln y = \frac{x}{2} + c \rightarrow y = e^{\frac{x}{2}+c} = De^{\frac{x}{2}}$

(*c*) $y = \frac{1}{2}\frac{d^2y}{dx^2} \rightarrow y - \frac{1}{2}\frac{d^2y}{dx^2} = 0$

This equation is of the form of Case 3A (where $a_0/a_2 < 0$), and the answer is

$$y = A\cosh(\sqrt{2}x) + B\sinh(\sqrt{2}x)$$

where $\cosh(\sqrt{2}x) = \frac{e^{\sqrt{2}t} + e^{-\sqrt{2}t}}{2}$ and $\sinh(\sqrt{2}x) = \frac{e^{\sqrt{2}t} - e^{-\sqrt{2}t}}{2}$

24.6 $f(x) = f'(x) + f''(x)$. Find $f(x)$.

$$(x) - f'(x) + f''(x) = 0$$

This is of the form of Case 4A—a linear, homogeneous differential equation of order 2, where $a_0 = 1$, $a_1 = -1$, and $a_2 = 1$. Using (24.17.0) and (24.17.1), we immediately conclude that:

$$f(x) = Ae^{-t/2}\cos(\omega t + \delta),$$

where

$$\omega = \sqrt{1 - \left(\frac{1}{2}\right)^{-2}} = .87\,\text{Hz}$$

24.7 An RC circuit consists of a resistor and a capacitor in series, usually with an additional voltage source. Assume that an RC circuit has $R = 50$ K, $C = 1\,\mu$F and a constant voltage source.

(*a*) Describe both the quantitative and the qualitative behavior of this circuit after the switch closes the circuit, allowing charge to flow; (*b*) What is the charging time to $1/e$ its peak value?; (*c*) What is the discharge time to $1/e$ its peak value?

(*a*) Using Kirchoff's law, we have:

$$\frac{q}{C} + R\frac{dq}{dt} = V_0$$

This is of the form of Case 2B, consisting of a linear first-order differential equation with a constant term (i.e., a non-homogeneous equation). Making the identification of the terms of this equation with (*24.6*), we substitute into (*24.7*) to get:

$$q(t) = CV_0(1 - e^{-t/RC})$$

The voltage across the capacitor is given by $V_C = \frac{q}{C}$.
Therefore, $V_C = V_0(1 - e^{-t/RC})$

Qualitatively, when we close the switch, charge will flow as a result of the "electrical pressure" of the voltage source, causing current to flow through the resistor and charge to build up on opposite sides of the capacitor. As the voltage difference across the capacitor increases, the current through the circuit will slow as the capacitor voltage approaches the voltage source. The above equation describes this behavior quantitatively, building up from $V_C = 0$ at time $= 0$ and approaching $V_C = V_0$ for large values of the time t.

(*b*) The charging time for the capacitor to reach $(1 - 1/e)$ of its peak value is a typical measure of the "rise time" of the capacitor in this circuit. Since $V_C/V_0 = (1 - e^{-t/RC})$, the circuit will reach $(1 - 1/e)$ of its peak value when $t = RC$. Therefore, the charging time is $t = RC = (50\,\text{k}\Omega)(10^{-6}\,F) = 50 \times 10^{-3}\,\text{s} = 50\,\text{ms}$.

(*c*) To discharge the circuit, the switch is closed so that the resistor and the capacitor, after having been charged as described in (*a*), are the only elements in the closed circuit. Under these circumstances, current will flow through the resistor due to the potential difference across the capacitor. This will occur according to Kirchoff's laws:

$$\frac{q}{C}+R\frac{dq}{dt}=0$$

This is of the form of Case 2A—a linear, first-order homogeneous differential equation. According to (*24.6*) and (*24.7*), it has the solution:

$$q(t)=q_0e^{-\frac{t}{RC}},$$

where q_0 is the charge on one of the capacitor plates at $t=0$.

Since $V_C=\frac{q}{C}$, $V_C=V_0e^{-t/RC}$. The reduction in the voltage across the capacitor from $V_c=V_0$ at $t=0$ to V_c approaching 0 as t becomes very large is reflected in this equation.

24.8 An *LR* circuit consists of a voltage source ε, a resistor *R*, and an inductor *L* connected in series. (*a*) Using Kirchoff's law, find the differential equation that describes current flow in this circuit. (*b*) Show through direct substitution into this equation that the current is described by $I=\frac{\varepsilon}{R}(1-e^{-t/\tau})$, where $\tau={}^L/_R$.

(*a*) From Kirchoff's laws, we know that the sum of the voltage drops around a closed circuit is zero. The voltage drop across a voltage source (ε), across a resistor ($=IR$) and across an inductor $\left(=L\frac{dI}{dt}\right)$ therefore add to zero:

$$\varepsilon-IR-L\frac{dI}{dt}=0 \tag{24.29}$$

(*b*) Substituting the proposed solution $I=\frac{\varepsilon}{R}(1-e^{-t/\tau})$ into the left-hand side of the previous equation, we get:

$$\begin{aligned}
&\varepsilon-\left(\frac{\varepsilon}{R}(1-e^{-t/\tau})\right)R-L\frac{d\left(\frac{\varepsilon}{R}(1-e^{-t/\tau})\right)}{dt}\\
&=\varepsilon-\varepsilon+\varepsilon e^{-t/\tau}-L\frac{d(\frac{\varepsilon}{R}(1-e^{-t/\tau}))}{dt}\\
&=\varepsilon e^{-t/\tau}+L\frac{\varepsilon}{R}\frac{d(e^{-t/\tau}))}{dt}\\
&=\varepsilon e^{-t/\tau}-\frac{L}{\tau}\frac{\varepsilon}{R}e^{-t/\tau}
\end{aligned}$$

Substituting $\tau={}^L/_R$ into the above equation

$$\begin{aligned}
&=\varepsilon e^{-t/\tau}-\frac{L}{{}^L/_R}\frac{\varepsilon}{R}e^{-t/\tau}\\
&=\varepsilon e^{-t/\tau}-\varepsilon e^{-t/\tau}\\
&=0
\end{aligned}$$

which matches the right-hand side of (*24.27*), demonstrating that this solution does indeed satisfy the equation.

24.9 An *RL* circuit with $R=150\ \Omega$ and $L=100$ H is being discharged (cf. Fishbane 33–10a). $E=10$ V, $R=150\ \Omega$, $L=100$ H. Using the equation $I=\frac{\varepsilon}{R}(1-e^{-t/\tau})$,where $\tau=L/R$, find the current flowing through the circuit as a function of time during the first (*a*) 10–6 s; (*b*) 10–3 s; (*c*) 1 s; (*d*) 10 s.

(*a*) $I=\frac{\varepsilon}{R}(1-e^{-t/\tau})=\frac{10\ V}{(150\ \Omega)}(1-e^{-(10^{-6}s)/(100\ H/150\ \Omega)})=1.00\times10^{-6}\text{ amp}=1.00\ \mu A$

Through substitution of the indicated times in (*b*) through (*d*) into the same equation, we get:

(*b*) For $t=10^{-3}$ s: $I=1.00$ mA
(*c*) For $t=1$ s: $I=670$ mA
(*d*) For $t=10$ s: $I=670$ mA

24.10 Consider an object falling through the atmosphere subject to a resistive force $-bv$ that falls from rest at a height y_0 at the time $t = 0$. Determine the velocity of this object as a function of time.

We begin with Newton's second law:

$$F = m\frac{dv}{dt}$$

Choosing the downward direction as that of positive displacement

$$mg - bv = m\frac{dv}{dt}$$

Aside from the different names given to the variables, this is of exactly the same form as in the previous problem. Using the same reasoning, we guess at a solution that is an exponential plus a constant. Hence

$$v = v_0 e^{-Dt} + F,$$

where D and F represent constants to be determined.

Then, $mg = bv_0 e^{-Dt} + bF - Dmv_0 e^{-Dt}$

Equating separately the quantities that are constant and those that depend upon the time, we get:

$$mg = bF \text{ and } 1 = FD$$

Therefore, $F = mg/b$ and $D = b/F_{mg}$

Therefore, $v = v_0 e^{-bt/mg} + mg/b$

If we know further that the object is at rest at $t = 0$, then, substituting in $v = 0$ in the above equation, we get $0 = v_0 e^0 + mg/b$, so that $v_0 = -mg/b$.

Hence, the general solution is $v = -mg/be^{-bt/mg} + mg/b$, or

$$v = \frac{mg}{b}(1 - e^{-bt/mg}) \qquad (24.30)$$

Note that this solution is identical in form to the one immediately preceding Eq. 24.29. In fact, if, in that equation, we replace V_0 with mg, replace $R\frac{dq}{dt}$ with b, and replace L with m, then *(24.30)* immediately follows.

Note that the velocity approaches the equilibrium value of mg/b. This makes sense since, in equilibrium the force of gravity mg must equal the opposing force bv. How rapidly the approach to equilibrium takes place depends upon the ratio of b/mg in the exponential function. Since bv determines the air resistance, this again makes sense: at any particular velocity v, the deceleration is higher for larger values of b.

24.11 A battery, a resistor, and a capacitor are connected in series. At time $t = 0$, a switch closes the circuit. Find the voltage across the capacitor as a function of time (*a*) using a previously determined solution to the appropriate differential equation; (*b*) using a trial function for $q(t)$ that is the sum of an exponential in time and a constant. (*c*) What are the steady-state value of the capacitor charge q and its potential difference V_C? (*d*) What fraction of the steady-state values will be reached after a time $t = RC$ (the "time constant" of the circuit)?

Once the switch is closed, current will flow through the resistor and, under the influence of the battery, one capacitor plate will become negatively charged while the other becomes positively charged. As a result, the potential difference between the capacitor plates will increase until it ultimately approaches that across the battery terminals. During this time, the potential difference across the resistor approaches zero. Hence, while current initially flows through the resistor, it drops off over time, approaching zero. At all times, the potential difference across the battery is equal to the sum of the potential drops across the resistor and the capacitor.

We know that, because the potential difference across the battery is equal to the sum of those across the resistor and the capacitor at the time the switch is closed,

$$V_0 = V_C + V_R$$

According to Ohm's law, $V_R = IR = R\frac{dq}{dt}$, where the current I through the resistor is equal to $\frac{dq}{dt}$. From the definition of capacitance, we know that $V_C = \frac{q}{C}$, where q is the charge on one of the capacitor plates.

(From the law of the conservation of total charge, we know that the rate of change of the capacitor charge $\frac{dq}{dt}$ matches the rate of charge flow through the resistor.)

Therefore,

$$V_0 = \frac{q}{C} + R\frac{dq}{dt} \tag{24.31}$$

Rearranging this, we get:

$\frac{q}{C} + R\frac{dq}{dt} = V_0$, which is of the form (*24.6*). According to (*24.7*), its solution is:

$$q = CV_0(1 - e^{-t/RC})$$

(*b*) Derive the same result by substituting a trial function for $q(t)$ that is the sum of an exponential in time and a constant.

A function that satisfies the specified form is:

$$q = Ae^{bt} + D,$$

where we can determine the constants A, b, and D through substitution in the above equation.

Using $\frac{dq}{dt} = Abe^{bt}$ and substituting q and $\frac{dq}{dt}$ into (*24.31*), we then have:

$$V_0 = \frac{A}{C}e^{bt} + \frac{D}{C} + RAbe^{bt}$$

In order for this equation to be true at ALL times, we must equate the terms that depend upon time and those that do not. Hence,

$$V_0 = D/C,$$

so that $D = CV_0$ and $\frac{A}{C}e^{bt} = RAbe^{bt}$. This implies that $\frac{A}{C} = RAb$, so that $b = 1/RC$ (we ignore the alternative possibility, that $A = 0$, which corresponds to a static charge ($q = D$), rather than the time-varying situation considered here.)

So $q = Ae^{-t/RC} + CV_0$

Since the charge q on the capacitor plates at $t = 0$ is zero, then

$$q(0) = A + CV_0 = 0,$$

so that

$$A = -CV_0$$

Therefore,

$$q = -CV_0e^{-t/RC} + CV_0 = CV_0(1 - e^{-t/RC})$$

Therefore,

$$q = CV_0(1 - e^{-t/RC})$$

Finally, since

$$V_C = q/C(24.34PRE)$$

we have

$$V_C = V_0(1 - e^{-t/RC}) \tag{24.34}$$

which is equation (*24.33*) above.

(*c*) What are the steady-state values of q and V?

Note that the charge on the plates ultimately (as t approaches infinity) approaches

$$q = CV_0(1 - e^{-\infty}) = CV_0(1 - 0) = CV_0$$

Similarly, $V_C = V_0(1 - e^{-t/RC})$ will approach V_0 as t approaches infinity.

(*d*) What fraction of the steady-state values will be reached after a time $t = RC$ (the "time constant" of the circuit)?

Using (24.34) and the equation immediately preceding it, we see that, after a time $t = RC$ has elapsed, both the capacitor charge and the potential difference across the plates will have built up to $(1 - e^{-1}) \approx 63\%$ of their final values.

24.12 The Laplace equation, applied to a charge-free region of space, takes the following form:

$$\nabla^2 V = 0 \tag{24.35}$$

Show, in rectangular coordinates, that this partial differential equation can be decomposed into three ordinary differential equations using the method of separation of variables.

$$\nabla^2 V = \frac{\partial^2 V}{\partial x^2} + \frac{\partial^2 V}{\partial y^2} + \frac{\partial^2 V}{\partial z^2}$$

$$\frac{\partial^2 V}{\partial x^2} + \frac{\partial^2 V}{\partial y^2} + \frac{\partial^2 V}{\partial z^2} = 0$$

If $V(x,y,z) = X(x)Y(y)Z(z)$, then

$$\frac{\partial^2[X(x)Y(y)Z(z)]}{\partial x^2} + \frac{\partial^2[X(x)Y(y)Z(z)]}{\partial y^2} + \frac{\partial^2[X(x)Y(y)Z(z)]}{\partial z^2} = 0$$

Taking terms outside of the brackets that are constant with respect to the partial differentiations, we have:

$$Y(y)Z(z)\frac{\mathrm{d}^2 X(x)}{\mathrm{d}x^2} + X(x)Z(z)\frac{\mathrm{d}^2 Y(y)}{\mathrm{d}y^2} + X(x)Y(y)\frac{\mathrm{d}^2 Z(z)}{\mathrm{d}z^2} = 0$$

Dividing each term by $X(x)Y(y)Z(z)$ then yields:

$$\frac{1}{X(x)}\frac{\mathrm{d}^2 X(x)}{\mathrm{d}x^2} + \frac{1}{Y(y)}\frac{\mathrm{d}^2 Y(y)}{\mathrm{d}y^2} + \frac{1}{Z(z)}\frac{\mathrm{d}^2 Z(z)}{\mathrm{d}z^2} = 0$$

Since each of these terms depends upon a different independent variable, each of the terms must be a constant in order for the equation to be valid—otherwise a change in, for example, x would alter the value of the first term without changing the value of the second or third terms (both of which are independent of changes in x), thereby altering the zero sum of the terms. Hence, each of the terms must be a constant (although not necessarily the same constant).

$$\frac{1}{X(x)}\frac{\mathrm{d}^2 X(x)}{\mathrm{d}x^2} = \alpha \tag{24.36a}$$

$$\frac{1}{Y(y)}\frac{\mathrm{d}^2 Y(y)}{\mathrm{d}y^2} = \beta \tag{24.36b}$$

$$\frac{1}{Z(z)}\frac{\mathrm{d}^2 Z(z)}{\mathrm{d}z^2} = \gamma \tag{24.36c}$$

Note that the original partial differential equation (*24.35*) has been decomposed into three ordinary differential equations. Each of those equations is a linear homogeneous differential equation involving a function and its second derivative. Additional information in the form of boundary conditions is needed in order to determine a solution for each of these equations, which—depending on the signs of α, β, and γ—will be either of the form (*24.11*) or (*24.12*). We have seen that the solutions to these equations are, respectively, either trigonometric or hyperbolic trigonometric functions. Note that the decomposition from the partial differential equation above into the three ordinary differential equations (*24.36(a–c)*) depends upon the original potential $V(x,y,z)$ being a product of three independent functions $X(x)Y(y)Z(z)$ and will not work in the most general case.

24.13 The displacement $f(x,t)$ of a string that is fixed at two ends $x = 0$ and $x = a$ is described by the wave equation

$$\frac{1}{c^2}\frac{\partial^2 f}{\partial t^2} - \frac{\partial^2 f}{\partial x^2} = 0 \qquad (24.37)$$

(*a*) Applying to find two independent equations in x and t starting with the and applying the technique of the separation of variables.

Let us assume that $f(x,t) = X(x)T(t)$. Substituting into the above equation, we get

$$\frac{1}{c^2}\frac{\partial^2[X(x)T(t)]}{\partial t^2} - \frac{\partial^2[X(x)T(t)]}{\partial x^2} = 0$$

Factoring out the variables not depending upon the partial differentiations, we have:

$$\frac{1}{c^2}X(x)\frac{d^2T(t)}{dt^2} - T(t)\frac{d^2X(x)}{dx^2} = 0.$$

Dividing both sides by $X(x)Y(y)$ and rearranging terms, we get:

$$\frac{1}{X(x)}\frac{d^2X(x)}{dx^2} = \frac{1}{c^2}\frac{1}{T(t)}\frac{d^2T(t)}{dt^2} \qquad (24.38)$$

Because the left-hand side of (*24.38*) does not depend upon x and the right-hand side does not depend upon t, both must be equal to the same constant:

$$\frac{1}{X(x)}\frac{d^2X(x)}{dx^2} = \alpha$$

$$\frac{1}{c^2}\frac{1}{T(t)}\frac{d^2T(t)}{dt^2} = \alpha$$

Upon rearrangement, these become:

$$\alpha X(x) - \frac{d^2X(x)}{dx^2} = 0$$

$$\alpha c^2 T(t) - \frac{d^2T(t)}{dt^2} = 0$$

24.14 Find the electrical analog to a mass falling under constant gravity subject to free fall and find its solution.

For example, consider the circuit consisting of a voltage source, a resistor, and a capacitor as shown in Fig. 24-7. Using Kirchoff's laws, we know that the sum of the voltage drops around the closed loop shown is zero:

$$v - R\frac{dq}{dt} - \frac{q}{C} = 0$$

which, upon rearrangement, becomes

$$\frac{q}{C} + R\frac{dq}{dt} = v$$

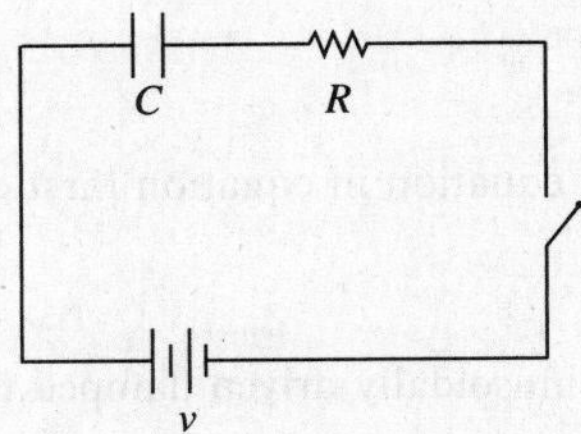

Fig. 24-7 (*RC* circuit with voltage source)

This is a linear equation that includes a function ($q(t)$), its derivative (dq/dt), and a constant (V)—an equation that matches the form of (*24.6*). $a_0x + a_1\frac{dx}{dt} = c_1$, where $a_0 = \frac{1}{C}$, $a_1 = R$, and $c_1 = v$.

Substituting these values into the solution for (*24.6*), equation (*24.7*), we get:

$$q(t) = Cv(1 - e^{-t/RC})$$

If we compare this to the example from that section—that of a body subject to a constant gravitational force and a air drag proportional to velocity—then we see that the equations are of the same form:

$$bv + m\frac{dv}{dt} = mg \quad \text{(mass with gravity and frictional drag)}$$

$$\frac{q}{C} + R\frac{dq}{dt} = v \quad \text{(capacitor with constant voltage source and resistor)}$$

and the solutions are also of the same form:

$$v(t) = \frac{mg}{b}(1 - e^{-bt/m}) \quad \text{(solution for mechanical situation)}$$

$$q(t) = Cv(1 - e^{-t/RC}) \quad \text{(solution for electrical situation)}$$

There is nothing particularly surprising here—we are simply noting that very different physical situations that are subject to the same form of differential equations then have the same form of solution, despite the very different physical interpretation of the corresponding constants and variables in each equation.

Supplementary Problems

24.15 Find the time required for an *RC* circuit to charge up to 90% of its steady-state value. Assume that $R = 50\ \text{k}\,\Omega$ and $C = 1$ pF.

Ans. 2.30 ns

24.16 Consider a capacitor and an inductor in series in a closed circuit. The voltages across each of these circuit elements will oscillate over time with a particular frequency. What is the frequency of a 50 mH inductor and a 10 μF capacitor in such a configuration?

Ans. 225 Hz

24.17 A function $f(x)$ is equal to twice its second derivative. Find the function.

Ans. $f(x) = \frac{1}{2}Ae^{\frac{1}{2}x}$, where A is an undetermined constant

24.18 Show through direct substitution that $x(t) = Ae^{-\frac{a_0}{a_1}t}$ (*24.5*) is a solution to $a_0x + a_1\frac{dx}{dt} = 0$ (*24.4*)

24.19 What is the period of oscillation of a 3 kg mass attached to the end of a spring with spring constant $k = 0.75$ N/m?

Ans. 13 s

24.20 What is the order of the differential equation in equation [first equation in chapter]?

Ans. It is of order *n*.

24.21 What is the electrical analogy to a sinusoidally driven damped harmonic oscillator?

Ans. An LRC circuit driven by a sinusoidal voltage source

24.22 An object is dropped from rest at $t = 0$ from 1000 m above the surface of the Earth. In the absence of air resistance, in how many seconds will it strike the ground?

Ans. 14.2 s

24.23 Find the solution to the equation $5y + y' - 3 = 0$

Ans. $y = \frac{3}{5}(1 - e^{-5t})$

24.24 Find the solution to $10f(x) + 3f''(x) = 0$

Ans. $f(x) = A\sin(\sqrt{\frac{10}{3}}t + \delta)$, where δ is an undetermined constant.

24.25 Find the solution to $x + 3\frac{dx}{dt} + 5\frac{d^2x}{dt^2} = 4\sin(3t + 6)$

Ans. $x = A\cos(3t + \delta)$, where $A = .23$ and δ is an undetermined constant.

24.26 Show by direct substitution that $x(t) = Ae^{-\frac{a_0}{a_1}t}$ is a solution of $a_0x + a_1\frac{dx}{dt} = 0$.

24.27 Show by direct substitution that $x(t) = \frac{c_1}{a_0}(1 - e^{-a_0t/a_1})$ is a solution of $a_0x + a_1\frac{dx}{dt} = c_1$

24.28 Show by direct substitution that $x = A\sin(\omega t + \delta)$ is a solution of $a_0x + a_2\frac{d^2x}{dt^2} = 0$, where $\omega = \sqrt{\frac{a_0}{a_2}}$ and $a_0/a_2 > 0$

24.29 Show by direct substitution that Case 4 $x = A\cosh(\alpha t) + B\sinh(\alpha t)$ is a solution of

$$a_0x + a_2\frac{d^2x}{dt^2} = 0$$

where $a_0/a_2 < 0$ and the hyperbolic cosine and hyperbolic sine functions are defined, respectively, by $\cosh(\alpha t) = \frac{e^{\alpha t} + e^{-\alpha t}}{2}$ and $\sinh(\alpha t) = \frac{e^{\alpha t} - e^{-\alpha t}}{2}$.

24.30 Find the order and degree of each of the following equations:

(*a*) $y' + 2y'' + y^3 = 6$

(*b*) $x^3 - \frac{dx}{dt} = 0$

(*c*) $x^3 = \left(\frac{d^2x}{dt^2}\right)^4$

(*d*) $e^{-x} + x^3 = \left(\frac{dx}{dt}\right)^4$

(*e*) $4y^5 + 12\frac{d^{(6)}y}{dx^{(6)}} = 0$

Ans. (*a*) Second order, first degree; (*b*) first order, first degree; (*c*) second order, fourth degree; (*d*) first order, fourth degree; (*e*) sixth order, first degree

24.31 Classify each of the following situations according to whether it corresponds to an underdamped, overdamped, or a critically damped oscillator.

(*a*) $3x + 2\frac{dx}{dt} + \frac{d^2x}{dt^2} = 0$

(*b*) $\sqrt{3}x + \frac{1}{2}\frac{dx}{dt} + \frac{d^2x}{dt^2} = 0$

(*c*) $8x + 9\frac{dx}{dt} + \frac{7}{3}\frac{d^2x}{dt^2} = 0$

(*d*) $\frac{100}{9}x + \sqrt{7}\frac{dx}{dt} + 45\frac{d^2x}{dt^2} = 0$

(*e*) $2x + 10\frac{dx}{dt} + 8\frac{d^2x}{dt^2} = 0$

(*f*) $2x + 8\frac{dx}{dt} + 8\frac{d^2x}{dt^2} = 0$

Ans. (*a*) overdamped; (*b*) overdamped; (*c*) underdamed; (*d*) overdamped; (*e*) underdamped; (*f*) critically damped

24.32 Classify each of the following situations according to whether it corresponds to an underdamped, overdamped or a critically damped oscillator.

(*g*) $k = \sqrt{3}; m = 10; b = .05$
(*h*) $k = 9.7; m = 1.12; b = .07$
(*i*) $k = 1; m = 25; b = 10$
(*j*) $k = 4.3; m = 6; b = .3$
(*k*) $k = 8.25; m = 6; b = 14$

Ans. (*g*) underdamped; (*h*) underdamped; (*i*) critically damped; (*j*) underdamped; (*k*) overdamped

24.33 (*a*) Show that (*24.11*) is a solution to $a_0x + a_2\frac{d^2x}{dt^2} = 0$, when $a_0/a_2 > 0$. (*b*) Show that (*24.12*) is a solution to this same equation, when $a_0/a_2 < 0$.

24.34 Show through direct substitution that for the case of a falling object subject to air resistance, equation (*24.10*) is a solution to (*24.8*).

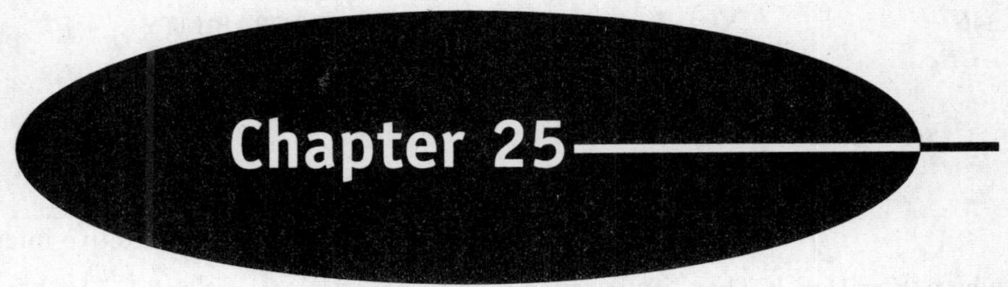

Elementary Probability

ANY ARRANGEMENT OF A SET OF OBJECTS in a definite order is called a *permutation* of the set taken all at a time. For example, *abcd, acbd, bdca* are permutations of a set of letters, a, b, c, d taken all at a time.

If a set contains n objects, any ordered arrangement of any $r \leq n$ of the objects is called a permutation of the n object taken r at a time. For example, *ab, ba, ca, db* are permutations of the $n = 4$ letters a, b, c, d taken $r = 2$ at a time, while *abc, adb, bad, cad* are permutations of the $n = 4$ letters taken $r = 3$ at a time. The number of permutations of n objects taken r at a time is denoted by nPr, where $r \leq n$.

THE NUMBER OF PERMUTATIONS which may be formed in each situation can be found by means of the

FUNDAMENTAL PRINCIPLE: If one thing can be done in u different ways, if after it has been done in any one of these, a second thing can be done in v different ways, if after it has been done in any one of these, a third thing can be done in w different ways, . . . , the several things can be done in the order stated in $u \cdot v \cdot w \cdot \cdot \cdot$ different ways.

EXAMPLE 1. In how many ways can 6 students be assigned to (*a*) row of 6 seats, (*b*) a row of 8 seats?

(*a*) Let the seats be denoted xxxxxx. The seat on the left may be assigned to any one of the 6 students; that is, it may be assigned in 6 different ways. After the assignment has been made, the next seat may be assigned to any one of the 5 remaining students. After the assignment has been made, the next seat may be assigned to any one of the 4 remaining students, and so on. Placing the number of ways in which each seat may be assigned under the x marking the seat, we have

$$\begin{matrix} \text{x} & \text{x} & \text{x} & \text{x} & \text{x} & \text{x} \\ 6 & 5 & 4 & 3 & 2 & 1 \end{matrix}$$

By the fundamental principle, the seats may be assigned in

$$6 \cdot 5 \cdot 4 \cdot 3 \cdot 2 \cdot 1 = 720 \text{ ways}$$

The reader should assure him- or herself that the seats might have been assigned to the students with the same result.

(*b*) Here each student must be assigned a seat. The first student may be assigned any one of the 8 seats, the second student any one of the 7 remaining seats, and so on. Letting x represent a student, we have

$$\begin{matrix} \text{x} & \text{x} & \text{x} & \text{x} & \text{x} & \text{x} \\ 8 & 7 & 6 & 5 & 4 & 3 \end{matrix}$$

and the assignment may be made in $8 \cdot 7 \cdot 6 \cdot 5 \cdot 4 \cdot 3 = 20\,160$ ways.

(See Problem 25.1–25.5.)

Define $n!$ (n factorial) to be

$$n \cdot (n-1) \cdot (n-2) \cdot \cdots \cdot (2)(1) \text{ for positive integers } n,$$

where $0! = 1! = 1$. Then, if we define $\binom{n}{k}$ to be $\frac{n!}{k!(n-k)!}$, we call $\binom{n}{k}$ a binomial coefficient and note that

$$(a+b)^n = \sum_{k=0}^{n} \binom{n}{k} a^{n-k} b^k, \qquad n \geq 1.$$

(See Problems 25.6–25.7.)

PERMUTATIONS OF OBJECTS NOT ALL DIFFERENT. If there are n objects of which k are alike while the remaining $(n-k)$ objects are different from them and from each other, it is clear that the number of different permutations of the n objects taken all together is not $n!$.

EXAMPLE 2. How many different permutations of four letters can be formed using the letters of the word *bass*?

For the moment, think of the given letters as b, a, s_1, s_2 so that they are all different. Then

$$bas_1s_2 \quad as_1bs_2 \quad s_2s_1ba \quad s_1as_2b \quad bs_1s_2a$$

$$bas_2s_1 \quad as_2bs_1 \quad s_1s_2ba \quad s_2as_1b \quad bs_2s_1a$$

are 10 of the 24 permutations of the four letters taken all together. However, when the subscripts are removed, it is seen that the two permutations in each column are alike.

Thus, there are $\frac{1 \cdot 2 \cdot 3 \cdot 4}{1 \cdot 2} = 12$ different permutations.

In general, given n objects of which k_1 are one sort, k_2 of another, k_3 of another,..., then the number of different permutations that can be made from the n objects taken all together is

$$\frac{n!}{k_1!\,k_2!\,k_3!\ldots}$$

(See Problem 25.8.)

IN GENERAL: The number of permutations, nPr, of different objects taken $r < n$ at a time is $\frac{n!}{(n-r)!}$. The number of permutation of n objects taken n at a time, nPn, is $n!$.

THE COMBINATIONS of n objects taken r at a time consist of all possible sets of r of the objects, without regard to the order of arrangement. The number of combinations of n objects taken r at a time will be denoted by ${}_nC_r$.

For example, the combinations of the $n = 4$ letters a, b, c, d taken $r = 3$ at a time, are

$$abc, \qquad abd, \qquad acd, \qquad bcd$$

Thus, ${}_4C_3 = 4$. When the letters of each combination are rearranged (in 3! ways), we obtain the ${}_4P_3$ permutations of the 4 letters taken 3 at a time. Hence, ${}_4P_3 = 3!({}_4C_3)$ and ${}_4C_3 = {}_4P_3/3!$.

The number of combinations of n different objects taken r at a time is equal to the number of permutations of the n objects taken r at a time divided by factorial r, or

$${}_nC_r = \frac{{}_nP_r}{r!} = \frac{n(n-1)\cdots(n-r+1)}{1 \cdot 2 \cdots r}$$

(For a proof, see Problem 25.9.)

EXAMPLE 3. From a shelf containing 12 different toys, a child is permitted to select 3. In how many ways can this be done?

The required number is

$$_{12}C_3 = \frac{_{12}P_3}{3!} = \frac{12 \cdot 11 \cdot 10}{1 \cdot 2 \cdot 3} = 220.$$

Notice that $_nC_r$ is the rth term's coefficient in the binomial theorem.

IN ESTIMATING THE PROBABILITY that a given event will or will not happen, we may, as in the case of drawing a face card from an ordinary deck, count the different ways in which this event may or may not happen. On the other hand, in the case of estimating the probability that a person who is now 25 years old will live to receive a bequest at age 30, we are forced to depend upon such knowledge of what has happened on similar occasions in the past as is available. In the first case, the result in called *mathematical* or *theoretical probability*; in the latter case, the result is called *statistical* or *empirical probability*.

MATHEMATICAL PROBABILITY. If an event must result in some one of n, $(n \neq 0)$ different but *equally likely ways* and if a certain s of these ways are considered successes and the other $f = n - s$ ways are considered failures, then the probability of success in a given trial is $p = s/n$ and the probability of failure is $q = f/n$. Since $p + q = \dfrac{s+f}{n} = \dfrac{n}{n} = 1$, $p = 1 - q$ and $q = 1 - p$.

EXAMPLE 4. One card is drawn from an ordinary deck. What is the probability (*a*) that it is a red card, (*b*) that it is a spade, (*c*) that it is a king, (*d*) that it is not the ace of hearts?

One card can be drawn from the desk in $n = 52$ different ways.

(*a*) A red card can be drawn from the deck in $s = 26$ different ways. Thus, the probability of drawing a red card is $s/n = \frac{26}{52} = \frac{1}{2}$.

(*b*) A spade can be drawn from the deck in 13 different ways. The probability of drawing a spade is $\frac{13}{52} = \frac{1}{4}$.

(*c*) A king can be drawn in 4 ways. The required probability is $\frac{4}{52} = \frac{1}{13}$.

(*d*) The ace of hearts can be drawn in 1 way; the probability of drawing the ace of hearts is $\frac{1}{52}$. Thus, the probability of *not* drawing the ace of hearts is $1 - \frac{1}{52} = \frac{51}{52}$.

(See Problems 25.19–25.22.)

[NOTE: We write $P(A)$ to denote the "probability that A occurs."]

Two or more events are called *mutually exclusive* if not more than one of them can occur in a single trial. Thus, the drawing of a jack and the drawing of a queen on a single draw from an ordinary deck are mutually exclusive events; however, the drawing of a jack and the drawing of a spade are not mutually exclusive.

EXAMPLE 5. Find the probability of drawing a jack or a queen from an ordinary deck of cards.

Since there are four jacks and four queens, $s = 8$ and $p = \frac{8}{52} = \frac{2}{13}$. Now the probability of drawing a jack is $\frac{1}{13}$, the probability of drawing a queen is $\frac{1}{13}$, and the required probability is $\frac{2}{13} = \frac{1}{13} + \frac{1}{13}$.

We have verified

THEOREM ***A***. The probability that some one of a set of mutually exclusive events will happen at a single trial is the sum of their separate probabilities of happening.

THEOREM ***A'***. The probability that A will occur or B will occur, $P(A \cup B) = P(A) + P(B) - P(A \cap B)$. (NOTE: "$\cup$" refers to set union and "$\cap$" to set intersection.)

$A \cup B$ is the set of all elements belonging either to A or to B or to both

$= \{x | x \text{ is in } A \text{ or } B \text{ or both}\}$.

$A \cap B$ is the set of all elements common to A and B

$= \{x | x \text{ is in } A \text{ and } x \text{ is in } B\}$.

(See Fig. 25-1.)

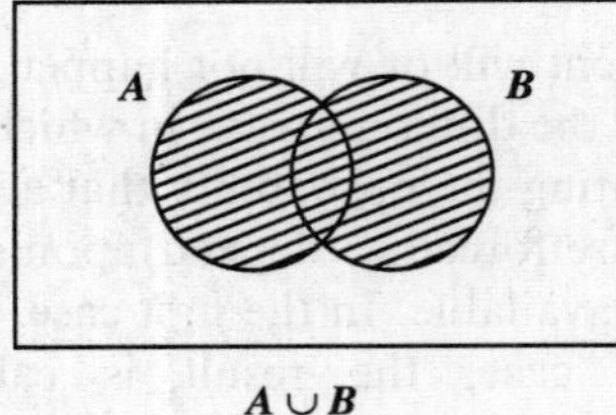

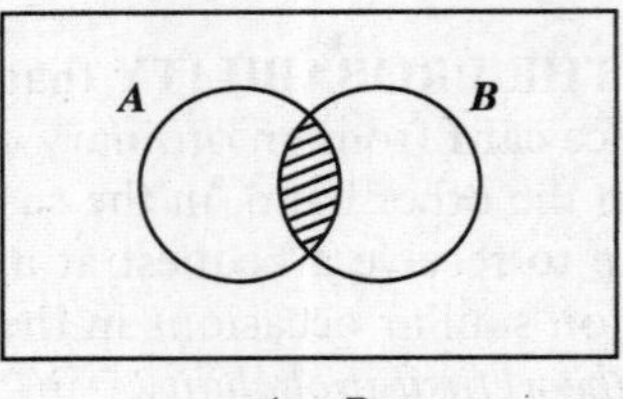

Fig. 25-1

For example, if $A = \{1, 2, 3\}$ and $B = \{2, 3, 7\}$, then $A \cup B = \{1, 2, 3, 7\}$ and $A \cap B = \{2, 3\}$.

Two events A and B are called *independent* if the happening of one does not affect the happening of the other. Thus, in a toss of two dice, the fall of either does not affect the fall of the other. However, in drawing two cards from a deck, the probability of obtaining a red card on the second draw depends upon whether or not a red card was obtained on the draw of the first card. Two such events are called *dependent*. More explicitly, if $P(A \cap B) = P(A) \cdot P(B)$, then A and B are independent.

EXAMPLE 6. One bag contains 4 white and 4 black balls, a second bag contains 3 white and 6 black balls, and a third contains 1 white and 5 black balls. If one ball is drawn from each bag, find the probability that all are white.

A ball can be drawn from the first bag in any one of 8 ways, from the second in any one of 9 ways, and from the third in any one of 6 ways; hence, three balls can be drawn from each bag in $8 \cdot 9 \cdot 6$ ways. A white ball can be drawn from the first bag in 4 ways, from the second in 3 ways, and from the third in 1 way; hence, three white balls can be drawn one from each bag in $4 \cdot 3 \cdot 1$ ways. Thus the required probability is

$$\frac{4 \cdot 3 \cdot 1}{8 \cdot 9 \cdot 6} = \frac{1}{36}$$

Now drawing a white ball from one bag does not affect the drawing of a white ball from another, so that here we are concerned with three independent events. The probability of drawing a white ball from the first bag is $\frac{4}{5}$, from the second is $\frac{1}{9}$, and from the third bag is $\frac{1}{6}$.

Since the probability of drawing three white balls, one from each bag, is $\frac{4}{8} \cdot \frac{3}{9} \cdot \frac{1}{6}$, we have verified

THEOREM *B*. The probability that all of a set of independent events will happen in a single trial is the product of their separate probabilities.

THEOREM *C*. (concerning dependent events). If the probability that an event will happen is p_1, and if after it has happened the probability that a second event will happen is p_2, the probability that the two events will happen in that order is $p_1 p_2$.

EXAMPLE 7. Two cards are drawn from an ordinary deck. Find the probability that both are face cards (king, queen, jack) if (*a*) the first card drawn is replaced before the second is drawn, (*b*) the first card drawn is not replaced before the second is drawn.

(*a*) Since each drawing is made from a complete deck, we have the case of two independent events. The probability of drawing a face card in a single draw is $\frac{12}{52}$; thus the probability of drawing two face cards, under the conditions imposed, is $\frac{12}{52} \cdot \frac{12}{52} = \frac{9}{169}$.

(*b*) Here the two events are dependent. The probability that the first drawing results in a face card is $\frac{12}{52}$. Now, of the 51 cards remaining in the deck, there are 11 face cards; the probability that the second drawing results in a face card is $\frac{11}{51}$. Hence, the probability of drawing two face cards is $\frac{12}{52} \cdot \frac{11}{51} = \frac{11}{221}$.

(See Problems 25.23–25.28.)

EXAMPLE 8. Two dice are tossed six times. Find the probability (*a*) that 7 will show on the first four tosses and will not show on the other two, (*b*) that 7 will show on exactly four of the tosses.

The probability that 7 will show on a single toss is $p = \frac{1}{6}$ and the probability that 7 will not show is $q = 1 - p = \frac{5}{6}$.

(*a*) The probability that 7 will show on the first four tosses and will not show on the other two is

$$\frac{1}{6} \cdot \frac{1}{6} \cdot \frac{1}{6} \cdot \frac{1}{6} \cdot \frac{5}{6} \cdot \frac{5}{6} = \frac{25}{46\,656}$$

(*b*) The four tosses on which 7 is to show may be selected in ${}_6C_4 = 15$ ways. Since these 15 ways constitue mutually exclusive events and the probability of any one of them is $\left(\frac{1}{6}\right)^4\left(\frac{5}{6}\right)^2$, the probability that 7 will show exactly four times in six tosses is ${}_6C_4\left(\frac{1}{6}\right)^4\left(\frac{5}{6}\right)^2 = \frac{125}{15\,552}$.

We have verified

THEOREM *D*. If p is the probability that an event will happen and q is the probability that it will fail to happen at a given trial, the probability that it will happen exactly r times in n trials is ${}_nC_r p^r q^{n-r}$. (See Problems 25.29–25.30.)

EMPIRICAL PROBABILITY. If an event has been observed to happen s times in n trials, the ratio $p = s/n$ is defined as the *empirical probability* that the event will happen at any future trial. The confidence which can be placed in such probabilities depends in a large measure on the number of observations used. Life insurance companies, for example, base their preminum rate on empirical probabilities. For this purpose they use a mortality table based on an enormous number of observations over the years.

The American Experience Table of Mortality begins with 100 000 persons all of age 10 years and indicates the number of the group who die each year thereafter. In using this table, it will be assumed that the laws stated above for mathematical probability hold also for empirical probability.

EXAMPLE 9. Find the probability that a person 20 years old (*a*) will die during the year, (*b*) will die during the next 10 years, (*c*) will reach age 75.

(*a*) Of the 100 000 persons alive at age 10 years, 92 637 are alive at age 20 years. Of these 92 637 a total of 723 will die during the year. The probability that a person 20 years of age will die during the year is $\frac{723}{92\,637} = 0.0078$.

(*b*) Of the 92 637 who reach age 20 years, 85 441 reach age 30 years; thus, $92\,637 - 85\,441 = 7196$ die during the 10-year period. The required probability is $\frac{7196}{92\,637} = 0.0777$.

(*c*) Of the 92 637 alive at age 20 years, 26 237 will reach age 75 years. The required probability is $\frac{26\,237}{92\,637} = 0.2832$.

(See Problem 25.68.)

Solved Problems

25.1 Using the letters of the word MARKING and calling any arrangement a word, (*a*) how many different 7-letter words can be formed, (*b*) how many different 3-letter words can be formed?

(*a*) We must fill each of the positions xxxxxxx with a different letter. The first position may be filled in 7 ways, the second in 6 ways, and so on.

Thus, we have $\underset{7}{\text{x}}\ \underset{6}{\text{x}}\ \underset{5}{\text{x}}\ \underset{4}{\text{x}}\ \underset{3}{\text{x}}\ \underset{2}{\text{x}}\ \underset{1}{\text{x}}$ and there are $7 \cdot 6 \cdot 5 \cdot 4 \cdot 3 \cdot 2 \cdot 1 = 5040$ words.

(*b*) We must fill of the positions xxx with a different letter. The first position can be filled in 7 ways, the second in 6 ways, and the third in 5 ways. Thus, there are $7 \cdot 6 \cdot 5 = 210$ words.

25.2 In forming 5-letter words using the letters of the word EQUATIONS, (*a*) how many consist only of vowels, (*b*) how many contain all of the consonants, (*c*) how many begin with E and end in S, (*d*) how many begin with a consonant, (*e*) how many contain N, (*f*) how many in which the vowels and consonants alternate, (*g*) how many in which Q is immediately followed by U?

There are 9 letters, consisting of 5 vowels and 4 consonants.

(*a*) There are five places to be filled and 5 vowels at our disposal. Hence, we can form $5 \cdot 4 \cdot 3 \cdot 2 \cdot 1 = 120$ words.

(*b*) Each word is to contain the 4 consonants and one of the 5 vowels. There are now six things to do: first pick the vowel to be used (in 5 ways), next place the vowel (in 5 ways), and fill the remaining four positions with consonants.

We have $\underset{5}{\text{x}}\ \underset{5}{\text{x}}\ \underset{4}{\text{x}}\ \underset{3}{\text{x}}\ \underset{2}{\text{x}}\ \underset{1}{\text{x}}$; hence there are $5 \cdot 5 \cdot 4 \cdot 3 \cdot 2 \cdot 1 = 600$ words.

(*c*) Indicate the fact that the position of certain letters is fixed writing $\overset{\text{E}}{\text{x}}\ \text{x}\ \text{x}\ \text{x}\ \overset{\text{S}}{\text{x}}$.

Now there are just three positions to be filled and 7 letters at our disposal. Thus, there are $7 \cdot 6 \cdot 5 = 210$ words.

(*d*) Here we have $\underset{4}{\overset{\text{c}}{\text{x}}}\ \underset{8}{\text{x}}\ \underset{7}{\text{x}}\ \underset{6}{\text{x}}\ \underset{5}{\text{x}}$ since, after filling the first position with any one of the 4 consonants, there are 8 letters remaining. Hence, there are $4 \cdot 8 \cdot 7 \cdot 6 \cdot 5 = 6720$ words.

(*e*) There are five things to do: first, place the letter N in any one of the five positions and then fill the other four positions from among the 8 letters remaining.

We have $\underset{5}{\text{x}}\ \underset{8}{\text{x}}\ \underset{7}{\text{x}}\ \underset{6}{\text{x}}\ \underset{5}{\text{x}}$; hence, there are $5 \cdot 8 \cdot 7 \cdot 6 \cdot 5 = 8400$ words.

(*f*) We may have $\underset{5}{\overset{\text{v}}{\text{x}}}\ \underset{4}{\overset{\text{c}}{\text{x}}}\ \underset{4}{\overset{\text{v}}{\text{x}}}\ \underset{3}{\overset{\text{c}}{\text{x}}}\ \underset{3}{\overset{\text{v}}{\text{x}}}$ or $\underset{4}{\overset{\text{c}}{\text{x}}}\ \underset{5}{\overset{\text{v}}{\text{x}}}\ \underset{3}{\overset{\text{c}}{\text{x}}}\ \underset{4}{\overset{\text{v}}{\text{x}}}\ \underset{2}{\overset{\text{c}}{\text{x}}}$. Hence, there are $5 \cdot 4 \cdot 4 \cdot 3 \cdot 3 + 4 \cdot 5 \cdot 3 \cdot 4 \cdot 2 =$ 1200 words.

(*g*) First we place Q so that U may follow it (Q may occupy any of the first four positions but not the last), next we place U (in only 1 way), and then we fill the three other positions from among the 7 letters remaining.

Thus, we have $\underset{4}{\text{x}}\ \underset{1}{\text{x}}\ \underset{7}{\text{x}}\ \underset{6}{\text{x}}\ \underset{5}{\text{x}}$ and there $4 \cdot 1 \cdot 7 \cdot 6 \cdot 5 = 840$ words.

25.3 If repetitions are not allowed, (*a*) how many three-digit numbers can be formed with the digits 0, 1, 2, 3, 4, 5, 6, 7, 8, 9? (*b*) How many of these are odd numbers? (*c*) How many are even numbers? (*d*) How many are divisble by 5? (*e*) How many are greater than 600?

In each case we have $\overset{\neq 0}{\text{x}}\ \text{x}\ \text{x}$.

(*a*) The position on the left can be filled in 9 ways (0 cannot be used), the middle position can be filled in 9 ways (0 can be used), and the position on the right can be filled in 8 ways. Thus, there are $9 \cdot 9 \cdot 8 = 648$ numbers.

(*b*) We have $\overset{\neq 0}{\text{x}}\ \text{x}\ \overset{\text{odd}}{\text{x}}$. Care must be exercised here in choosing the order in which to fill the positions. If the position on the left is filled first (in 9 ways), we cannot determine the number of ways in which the position on the right can be filled since, if the former is filled with an odd digit there are 4 ways of filling the latter but if the former is filled with an even digit there are 5 ways of filling the latter.

We fill first the position on the right (in 5 ways), then the position on the left (in 8 ways, since one odd digit and 0 are excluded), and the middle position (in 8 ways, since two digits are now excluded). Thus, there are $8 \cdot 8 \cdot 5 = 320$ numbers.

(*c*) We have $\underset{\text{x}}{\neq 0}\quad \text{x}\quad \underset{\text{x}}{\text{even}}$. We note that the argument above excludes the possibility of first filling the position on the left. But if we fill the first position on the right (in 5 ways), were unable to determine the number of ways the position on the left can be filled (9 ways if 0 was used on the right, 8 ways if 2, 4, 6, or 8 was used). Thus, we must separate the two cases.

First, we form all numbers ending in 0; there are $9 \cdot 8 \cdot 1$ of them. Next, we form all numbers ending in 2, 4, 6, or 8; there $8 \cdot 8 \cdot 4$ of them. Thus, in all, there are $9 \cdot 8 \cdot 1 + 8 \cdot 8 \cdot 4 = 328$ numbers.

As a check, we have 320 odd and 328 even numbers for a total of 648 as found in (*a*) above.

(*d*) A number is divisible by 5 if and only if it ends in 0 or 5. There are $9 \cdot 8 \cdot 1$ numbers ending in 0 and $8 \cdot 8 \cdot 1$ numbers ending in 5. Hence, in all, there are $9 \cdot 8 \cdot 1 + 8 \cdot 8 \cdot 1 = 136$ numbers divisible by 5.

(*e*) The position on the left can be filled in 4 ways (with 6, 7, 8, or 9) and the remaining positions in $9 \cdot 8$ ways. Thus,there are $4 \cdot 9 \cdot 8 = 288$ numbers.

25.4 Solve Problem 16.3 (*a*), (*b*), (*c*), (*d*) if any digit may be used once, twice, or three times in forming the three-digit number.

(*a*) The position on the left can be filled in 9 ways and each of other position can be filled in 10 ways. Thus, there are $9 \cdot 10 \cdot 10 = 900$ numbers.

(*b*) The position on the right can be filled in 5 ways, the middle position in 10 ways, and the position on the left in 9 ways. Thus, there are $9 \cdot 10 \cdot 5 = 450$ numbers.

(*c*) There are $9 \cdot 10 \cdot 5 = 450$ even numbers.

(*d*) There are $9 \cdot 10 \cdot 1 = 90$ numbers ending in 0 and the same number ending in 5. Thus, there are 180 numbers divisible by 5.

25.5 In how many ways can 10 boys be arranged (*a*) in a straight line, (*b*) in a circle?

(*a*) The boys may be arranged in a straight line in $10 \cdot 9 \cdot 8 \cdot 7 \cdot 6 \cdot 5 \cdot 4 \cdot 3 \cdot 2 \cdot 1$ ways.
(*b*) We first place a boy at any point on the circle. The other 9 boys may then be arranged in $9 \cdot 8 \cdot 7 \cdot 6 \cdot 5 \cdot 4 \cdot 3 \cdot 2 \cdot 1$ ways.

This is an example of a *circular permutation*. In general, n objects may be arranged in a circle in $(n-1)(n-2) \cdot \cdot \cdot 2 \cdot 1$ ways.

25.6 Evaluate.

(*a*) $\dfrac{8!}{3!} = \dfrac{1 \cdot 2 \cdot 3 \cdot 4 \cdot 5 \cdot 6 \cdot 7 \cdot 8}{1 \cdot 2 \cdot 3} = 6720$ (*c*) $\dfrac{10!}{3!3!4!} = \dfrac{1 \cdot 2 \cdot 3 \cdot 4 \cdot 5 \cdot 6 \cdot 7 \cdot 8 \cdot 9 \cdot 10}{1 \cdot 2 \cdot 3 \cdot 1 \cdot 2 \cdot 3 \cdot 1 \cdot 2 \cdot 3 \cdot 4} = 4200$

(*b*) $\dfrac{7!}{6!} = \dfrac{1 \cdot 2 \cdot 3 \cdot 4 \cdot 5 \cdot 6 \cdot 7}{1 \cdot 2 \cdot 3 \cdot 4 \cdot 5 \cdot 6} = 7$ (*d*) $\dfrac{(r+1)!}{(r-1)!} = \dfrac{1 \cdot 2 \cdot \cdot \cdot (r-1)r(r+1)}{1 \cdot 2 \cdot \cdot \cdot (r-1)} = r(r+1)$

25.7 Solve for n, given (*a*) ${}_nP_2 = 110$, (*b*) ${}_nP_4 = 30\,{}_nP_2$.

(*a*) ${}_nP_2 = n(n-1) = n^2 - n - 110$. Then $n^2 - n = 110 = (n-11)(n+10) = 0$ and, since n is positive, $n = 11$.

(*b*) We have $n(n-1)(n-2)(n-3) = 30n(n-1)$ or $n(n-1)(n-2)(n-3) - 30n(n-1) = 0$.

Then $n(n-1)[(n-2)(n-3) - 30] = n(n-1)(n^2 - 5n - 24) = n(n-1)(n-8)(n+3) = 0$.

Since $n \geq 4$, the required solution is $n = 8$.

25.8 (*a*) How many permutations can be made of the letters, taken all together, of the "word" MASSESS?
(*b*) In how many ways will the four S's be together? (*c*) How many will end in SS?

(*a*) There are seven letters of which four are S's. The number of permutations is $7!/4! = 210$.

(*b*) First, permute the non-S's in $1 \cdot 2 \cdot 3 = 6$ ways and then place the four S's at the ends or between any two letters in each of the six permutations. Thus, there will be $4 \cdot 6 = 24$ permutations.

(*c*) After filling the last two places with S, we have to fill five places with 5 letters of which 2 are S's. Thus, there are $5!/2! = 60$ permutations.

25.9 Derive the formula ${}_nC_r = \dfrac{{}_nP_r}{r!}$.

From each of the ${}_nC_r$ combinations of n objects taken r at a time, $r!$ permutations can be formed. Since two combinations differ at least in one element, the ${}_nC_r \cdot r!$ permutations thus formed are precisely the number of permutations ${}_nP_r$ of n objects taken r at a time. Thus,

$${}_nC_r \cdot r! = {}_nP_r \quad \text{and} \quad {}_nC_r = \frac{nP_r}{r!}$$

25.10 Show (*a*) ${}_nC_r = \dfrac{n!}{r!(n-r)!}$, (*b*) ${}_nC_r = {}_nC_{n-r}$.

(*a*) $${}_nC_r = \frac{n(n-1)\cdots(n-r+1)}{1\cdot 2\cdots r} = \frac{n(n-1)\cdots(n-r+1)}{1\cdot 2\cdots r}\cdot\frac{(n-r)\cdots 2\cdot 1}{1\cdot 2\cdots(n-r)} = \frac{n!}{r!(n-r)!}$$

(*b*) $${}_nC_r = \frac{n!}{r!(n-r)!} = \frac{n(n-1)\cdots(r+1)r(r-1)\cdots 2\cdot 1}{1\cdot 2\cdots r\cdot(n-r)!} = \frac{n(n-1)\cdots(r+1)}{(n-r)!} = \frac{{}_nP_{n-r}}{(n-r)!} = {}_nC_{n-r}$$

25.11 Compute (*a*) ${}_{10}C_2$, (*b*) ${}_{12}C_5$, (*c*) ${}_{15}C_{12}$, (*d*) ${}_{25}C_{21}$.

(*a*) ${}_{10}C_2 = \dfrac{10\cdot 9}{1\cdot 2} = 45$ (*c*) ${}_{15}C_{12} = {}_{15}C_3 = \dfrac{15\cdot 14\cdot 13}{1\cdot 2\cdot 3} = 455$

(*b*) ${}_{12}C_5 = \dfrac{12\cdot 11\cdot 10\cdot 9\cdot 8}{1\cdot 2\cdot 3\cdot 4\cdot 5} = 792$ (*d*) ${}_{25}C_{21} = {}_{25}C_4 = \dfrac{25\cdot 24\cdot 23\cdot 22}{1\cdot 2\cdot 3\cdot 4} = 12\,650$

25.12 A lady gives a dinner party for six guests. (*a*) In how many ways may they be selected from among 10 friends? (*b*) In how many ways if two of the friends will not attend the party together?

(*a*) The six guests may be selected in ${}_{10}C_6 = {}_{10}C_4 = \dfrac{10\cdot 9\cdot 8\cdot 7}{1\cdot 2\cdot 3\cdot 4} = 210$ ways.

(*b*) Let A and B denote the two who will not attend together. If neither A nor B is included, the guests may be selected in ${}_8C_6 = {}_8C_2 = \dfrac{8\cdot 7}{1\cdot 2} = 28$ ways. If one of A and B is included, the guests may be selected in $2\cdot{}_8C_5 = 2\cdot{}_8C_3 = 2\left(\dfrac{8\cdot 7\cdot 6}{1\cdot 2\cdot 3}\right) = 112$ ways. Thus, the six guests may be selected in $28 + 112 = 140$ ways.

25.13 A committee of 5 is to be selected from 12 seniors and 8 juniors. In how many ways can this be done (*a*) if the committee is to consist of 3 seniors and 2 juniors, (*b*) if the committee is to contain at least 3 seniors and 1 junior?

(*a*) With each of the ${}_{12}C_3$ selections of 3 seniors, we may associate any one of the ${}_8C_2$ selections of 2 juniors. Thus, a committee can be selected in ${}_{12}C_3 \cdot {}_8C_2 = \dfrac{12\cdot 11\cdot 10}{1\cdot 2\cdot 3}\cdot\dfrac{8\cdot 7}{1\cdot 2} = 6160$ ways.

(*b*) The committee may consist of 3 seniors and 2 juniors or of 4 seniors and 1 junior. A committee of 3 seniors and 2 juniors can be selected in 6160 ways, and a committee of 4 seniors and 1 junior can be selected in ${}_{12}C_4 \cdot {}_8C_1 = 3960$ ways. In all, a committee may be selected in $6160 + 3960 = 10\,120$ ways.

25.14 There are ten points $A, B, \ldots$, in a plane, no three on the same straight line. (*a*) How many lines are determined by the points? (*b*) How many of the lines pass through A? (*c*) How many triangles are determined by the points? (*d*) How many of the triangles have A as a vertex? (*e*) How many of the triangles have AB as a side?

(*a*) Since any two points determine a line, there are ${}_{10}C_2 = 45$ lines.

(*b*) To determine a line through A, one other point must be selected. Thus, there are nine lines through A.

(*c*) Since any three of the points determine a triangle, there are ${}_{10}C_3 = 120$ triangles.

(*d*) Two additional points are needed to form a triangle. These points may be selected from the nine points in ${}_9C_2 = 36$ ways.

(*e*) One additional point is needed; there are eight triangles having AB as a side.

25.15 In how many ways may 12 persons be divided into three groups (*a*) of 2, 4, and 6 persons, (*b*) of 4 persons each?

(*a*) The groups of two can be selected in ${}_{12}C_2$ ways, then the group of four in ${}_{10}C_4$ ways, and the group of six in ${}_6C_6 = 1$ way. Thus, the division may be made in ${}_{12}C_2 \cdot {}_{10}C_4 \cdot 1 = 13\,860$ ways.

(*b*) One group of four can be selected in ${}_{12}C_4$ ways, then another in ${}_8C_4$ ways, and the third in 1 way. Since the order in which the groups are formed is now immaterial, the division may be made in ${}_{12}C_4 \cdot {}_8C_4 \cdot 1 \div 3! = 5775$ ways.

25.16 The English alphabet consists of 21 consonants and 5 vowels.

(*a*) In how many ways can 4 consonants and 2 vowels be selected?
(*b*) How many words consisting of 4 consonants and 2 vowels can be formed?
(*c*) How many of the words in (*b*) begin with R?
(*d*) How many of the words in (*c*) contain E?

(*a*) The 4 consonants can be selected in ${}_{21}C_4$ ways and the 2 vowels can be selected in ${}_5C_2$ ways. Thus, the selections may be made in ${}_{21}C_4 \cdot {}_5C_2 = 59\ 850$ ways.

(*b*) From each of the selections in (*a*), 6! words may be formed by permuting the letters. Therefore, $59850 \cdot 6! = 43\,092\,000$ words can be formed.

(*c*) Since the position of the consonant R is fixed, we must select 3 other consonants (in ${}_{20}C_3$ ways) and 2 vowels (in ${}_5C_2$ ways), and arrange each selection of 5 letters in all possible ways. Thus, there are ${}_{20}C_3 \cdot {}_5C_2 \cdot 5! = 1\ 368\,000$ words.

(*d*) Since the position of the consonant R is fixed but the position of the vowel E is not, we must select 3 other consonants (in ${}_{20}C_3$ ways) and 1 other vowel (in four ways), and arrange each set of 5 letters in all possible ways. Thus, there are ${}_{20}C_3 \cdot 4 \cdot 5! = 547\,200$ words.

25.17 From an ordinary deck of playing cards, in how many different ways can five cards be dealt (*a*) consisting of spades only, (*b*) consisting of black cards only, (*c*) containing the four aces, (*d*) consisting of three cards of one suit and two of another, (*e*) consisting of three kings and a pair, (*f*) consisting of three of one kind and two of another?

(*a*) From the 13 spades, 5 can be selected in ${}_{13}C_5 = 1287$ ways.

(*b*) From the 26 black cards, 5 can be selected in ${}_{26}C_5 = 65\ 780$ ways.

(*c*) One card must be selected from the 48 remaining cards. This can be done in 48 different ways.

(*d*) A suit can be selected in four ways and three cards from the suit can be selected in ${}_{13}C_3$ ways; a second suit can now be selected in three ways and two cards of this suit in ${}_{13}C_2$ ways. Thus, three cards of one suit and two of another can be selected in $4 \cdot {}_{13}C_3 \cdot 3 \cdot {}_{13}C_2 = 267\,696$ ways.

(*e*) Three kings can be selected from the four kings in ${}_4C_3$ ways, another kind can be selected in 12 ways, and two cards of this kind can be selected in ${}_4C_2$ ways. Thus, three kings and another pair can be dealt in ${}_4C_3 \cdot 12 \cdot {}_4C_2 = 288$ ways.

(*f*) A kind can be selected in 13 ways and three of this kind in ${}_4C_3$ ways; another kind can be selected in 12 ways and two of this kind can be selected in ${}_4C_2$ ways. Thus, 3 of one kind and 2 of another can be dealt in $13 \cdot {}_4C_3 \cdot 12 \cdot {}_4C_2 = 3744$ ways.

25.18 (*a*) Prove: The total number of combinations of n objects taken successively $1, 2, 3, \ldots, n$ at a time is $2^n - 1$.
(*b*) In how many different ways can one invite one or more of five friends to the movies?

(*a*) The total number of combinations is

$${}_nC_1 + {}_nC_2 + {}_nC_3 + \cdots + {}_nC_n = 2^n - 1$$

Use the fact that

$${}_nC_0 + {}_nC_1 + {}_nC_2 + \cdots + {}_nC_n = 2^n$$

(*b*) The number is ${}_5C_1 + {}_5C_2 + {}_5C_3 + {}_5C_4 + {}_5C_5 = 2^5 - 1 = 31$.

25.19 One ball is drawn from a bag containing 3 white, 4 red, and 5 black balls. What is the probability (*a*) that it is white, (*b*) that it is white or red, (*c*) that it is not red?

A ball can be drawn from the bag in $n = 3 + 4 + 5 = 12$ different ways.

(*a*) A white ball can be drawn in $s = 3$ different ways. Thus, the probability of drawing a white ball is $p = s/n = \frac{3}{12} = \frac{1}{4}$.

(*b*) Here success consists in drawing either a white or a red ball; hence, $s = 3 + 4 = 7$ and the required probability is $\frac{7}{12}$.

(*c*) The probability of drawing a red ball is $p = \frac{4}{12} = \frac{1}{3}$. The probability that the ball drawn is *not* red is $1 - p = 1 - \frac{1}{3} = \frac{2}{3}$. This problem may also be solved as in (*b*).

25.20 If two dice are tossed, what is the probability (*a*) of throwing a total of 7, (*b*) of throwing a total of 8, (*c*) of throwing a total of 10 or more, (*d*) of both dice showing the same number?

Two dice may turn up in $6 \times 6 = 36$ ways.

(*a*) A total of 7 may result in 6 ways. (1, 6; 6, 1; 2, 5; 5, 2; 3, 4; 4, 3). The probability of a throw of 7 is then $\frac{6}{36} = \frac{1}{6}$.

(*b*) A total of 8 may result in 5 ways (2, 6; 6, 2; 3, 5; 5, 3; 4, 4). The probability of a throw of 8 is $\frac{5}{36}$.

(*c*) Here, success consists in throwing a total of 10, 11, or 12. Since a total of 10 may result in 3 ways, a total of 11 in 2 ways, and a total of 12 in 1 way, the probability is $(3 + 2 + 1)/36 = \frac{1}{6}$.

(*d*) The probability that the second die will show the same number as the first is $\frac{1}{6}$. This problem may also be solved by counting the number of successes 1, 1; 2, 2; etc.

25.21 If five coins are tossed, what is the probability (*a*) that all will show heads, (*b*) that exactly three will show heads, (*c*) that at least three will show heads?

Each coin can turn up in 2 ways; hence, the five coins can turn up in $2^5 = 32$ ways. (The assumption here is that HHHTT and THHHT are different results.)

(*a*) Five heads can turn up in only 1 way; hence, the probability of a toss of five heads is $\frac{1}{32}$.

(*b*) Exactly three heads can turn up in ${}_5C_3 = 10$ ways; thus, the probability of a toss of exactly three heads is $\frac{10}{32} = \frac{5}{16}$.

(*c*) Success here consists of a throw of exactly three, exactly four, or all heads. Exactly three heads can turn up in 10 ways, exactly four heads in 5 ways, and all heads in 1 way. Thus, the probability of throwing at least three heads is $(10 + 5 + 1)/32 = \frac{1}{2}$.

25.22 If three cards are drawn from an ordinary deck, find the probability (*a*) that all are red cards, (*b*) that all are of the same suit, (*c*) that all are aces.

Three cards may be drawn from a deck in ${}_{52}C_3$ ways.

(a) Three red cards may be drawn from the 26 red cards in ${}_{26}C_3$ ways. Hence, the probability of drawing three red cards in a single draw is $\frac{{}_{26}C_3}{{}_{52}C_3} = \frac{26 \cdot 25 \cdot 24}{1 \cdot 2 \cdot 3} \cdot \frac{1 \cdot 2 \cdot 3}{52 \cdot 51 \cdot 50} = \frac{2}{17}$.

(b) There are 4 ways of choosing a suit and ${}_{13}C_3$ ways of selecting three cards of that suit. Thus, the probability of drawing three cards of the same suit is $4 \cdot {}_{13}C_3/{}_{52}C_3 = \frac{22}{425}$.

This problem may also be solved as follows: The first card drawn determines a suit. The deck now contains 51 cards of which 12 are of that suit; hence, the probability that the next two cards drawn will be of that suit is ${}_{12}C_2/{}_{51}C_2 = \frac{22}{425}$, as before.

(c) Three aces may be selected from the four aces in 4 ways; hence, the required probability is $\frac{4}{{}_{52}C_3} = \frac{1}{5525}$.

25.23 One bag contains 8 black balls and a second bag contains 1 white and 6 black balls. One of the bags is selected and then a ball is drawn from that bag. What is the probability that it is the white ball?

The probability that the second bag is chosen is $\frac{1}{2}$ and the probability that the white ball is drawn from this bag is $\frac{1}{7}$. Thus, the required probability is $\frac{1}{2}(\frac{1}{7}) = \frac{1}{14}$.

25.24 Two cards are drawn in succession from an ordinary deck. What is the probability (a) that the first will be the jack of diamonds and the second will be the queen of spades, (b) that the first will be a diamond and the second a spade, (c) that both cards are diamonds or both are spades?

(a) The probability that the first card is the jack of diamonds is $\frac{1}{52}$ and the probability that the second is the queen of spades is $\frac{1}{51}$. The required probability is $(\frac{1}{52})(\frac{1}{51}) = \frac{1}{2652}$.

(b) The probability that the first card is a diamond is $\frac{13}{52}$ and the probability that the second card is a spade is $\frac{13}{51}$. The probability of the required draw is $\frac{13}{52} \cdot \frac{13}{51} = \frac{13}{204}$.

(c) The probability that both cards are of a specified suit is $\frac{13}{52} \cdot \frac{12}{51}$. Thus, the probability that both are diamonds or both are spades is $\frac{13}{52} \cdot \frac{12}{51} + \frac{13}{52} \cdot \frac{12}{51} = \frac{2}{17}$.

25.25 A, B, and C work independently on a problem. If the respective probabilities that they will solve it are $\frac{1}{2}, \frac{1}{3}, \frac{2}{5}$, find the probability that the problem will be solved.

The problem will be solved unless all three fail; the probability that this will happen is $\frac{1}{2} \cdot \frac{2}{3} \cdot \frac{3}{5} = \frac{1}{5}$. Thus, the probability that the problem will be solved is $1 - \frac{1}{5} = \frac{4}{5}$.

25.26 A tosses a coin and if a head appears he wins the game; if a tail appears, B tosses the coin under the same conditions, and so on. If the stakes are \$15, find the expectation of each.

We first compute the probability that A will win. The probability that he will win on the first toss is $\frac{1}{2}$; the probability that he will win on his second toss (that is, that A first tosses a tail, B tosses a tail, and A then tosses a head) is $\frac{1}{2} \cdot \frac{1}{2} \cdot \frac{1}{2} = (\frac{1}{2})^3$; the probability that he will win on his third toss (that is, that A first tosses a tail, B tosses a tail, A tosses a tail, B tosses a tail, and A then tosses a head) is $\frac{1}{2} \cdot \frac{1}{2} \cdot \frac{1}{2} \cdot \frac{1}{2} \cdot \frac{1}{2} = (\frac{1}{2})^5$, and so on. Thus, the probability that A will win is

$$\frac{1}{2} + \frac{1}{2^3} + \frac{1}{2^5} + \cdots = \frac{\frac{1}{2}}{1 - 1/2^2} = \frac{2}{3}$$

and his expectation is $\frac{2}{3}(\$15) = \$10$. Then B's expectation is \$5.

25.27 On a toss of two dice, X throws a total of 5. Find the probability that he will throw another 5 before he throws 7.

X will succeed should he throw a total of 5 on the next toss, or should he not throw 5 or 7 on this toss but throw 5 on the next, or should he not throw 5 or 7 on either of these tosses but throw 5 on the next, and so on. The respective probabilities are $\frac{4}{36}, \frac{26}{36}\cdot\frac{4}{36}, \frac{26}{36}\cdot\frac{26}{36}\cdot\frac{4}{36}$, and so on. Thus, the probability that he throws 5 before 7 is

$$\frac{4}{36}+\frac{26}{36}\cdot\frac{4}{36}+\frac{26}{36}\cdot\frac{26}{36}\cdot\frac{4}{36}+\cdots=\frac{\frac{4}{36}}{1-\frac{26}{36}}=\frac{2}{5}$$

25.28 A bag contains 2 white and 3 black balls. A ball is drawn 5 times, each being replaced before another is drawn. Find the probability that (*a*) the first 4 balls drawn are white and the last is black, (*b*) exactly 4 of the balls drawn are white, (*c*) at least 4 of the balls drawn are white, (*d*) at least 1 ball is white.

The probability of drawing a white ball is $p=\frac{2}{5}$ and the probability of drawing a black ball is $\frac{3}{5}$.

(*a*) The probability that the first 4 are white and the last black is $\frac{2}{5}\cdot\frac{2}{5}\cdot\frac{2}{5}\cdot\frac{2}{5}\cdot\frac{3}{5}=\frac{48}{3125}$.

(*b*) Here $n=5, r=4$; the probability that exactly 4 of the balls drawn are white is

$$nC_r p^r q^{n-r} = {}_5C_4(\tfrac{2}{5})^4(\tfrac{3}{5}) = \tfrac{48}{625}$$

(*c*) Since success consists of drawing either 4 white and 1 black ball or 5 white balls, the probability is

$${}_5C_4(\tfrac{2}{5})^4(\tfrac{3}{5}) + {}_5C_5(\tfrac{2}{5})^5 = \tfrac{272}{3125}$$

(*d*) Here failure consists of drawing 5 black balls. Since the probability of failure is $(\frac{3}{5})^5=\frac{243}{3125}$, the probability of success is $1-\frac{243}{3125}=\frac{2882}{3125}$. The problem may also be solved as in (*c*).

25.29 One bag contains 2 white balls and 2 black balls, and another contains 3 white balls and 5 black balls. At five different trials, a bag is chosen at random and 1 ball is drawn from that bag and replaced. Find the probability (*a*) that exactly 3 white balls are drawn, (*b*) that at least 3 white balls are drawn.

At any trial the probability that a white ball is drawn from the first bag is $\frac{1}{2}\cdot\frac{1}{2}$ and the probability that a white ball is drawn from the second bag is $\frac{1}{2}(\frac{3}{8})$. Thus, the probability that a white ball is drawn at any trial is $p=\frac{1}{2}\cdot\frac{1}{2}+\frac{1}{2}(\frac{3}{8})=\frac{7}{16}$, and the probability that a black ball is drawn is $q=\frac{9}{16}$.

(*a*) The probability of drawing exactly 3 white balls in 5 trials is ${}_5C_3\left(\frac{7}{16}\right)^3\left(\frac{9}{16}\right)^2=\frac{138\,915}{524\,288}$.

(*b*) The probability of drawing at least 3 white balls in 5 trials is

$${}_5C_3\left(\frac{7}{16}\right)^3\left(\frac{9}{16}\right)^2+{}_5C_4\left(\frac{7}{16}\right)^4\left(\frac{9}{16}\right)+{}_5C_5\left(\frac{7}{16}\right)^5=\frac{201\,341}{524\,288}$$

25.30 A husband is 35 years old and his wife is 28. Find the probability that at the end of 20 years (*a*) the husband will be alive, (*b*) the wife will be alive, (*c*) both will be alive, (*d*) both will be dead, (*e*) the wife will be alive and the husband will not, (*f*) one will be alive but not the other.

(*a*) Of the 81 822 alive at age 35, 64 563 will reach age 55. The probability that the husband will be alive at the end of 20 years is $\dfrac{64\,563}{81\,822}=0.7890$.

(*b*) The probability that the wife will be alive at the end of 20 years is $\dfrac{71\,627}{86\,878}=0.8245$.

(*c*) Since the survival of the husband and of the wife are independent events, the probability that both are alive after 20 years is $\dfrac{64\,563}{81\,822}\cdot\dfrac{71\,627}{86\,878}=0.6506$.

(*d*) From (*a*) 17 259 of the 81 822 alive at age 35 will not reach age 55; thus, the probability that the husband will not live for 20 years is $\dfrac{17\,259}{81\,822}$. Similarly, the probability that the wife will not live for 20 years is $\dfrac{15\,251}{86\,878}$. Hence, the probability that after 20 years both will be dead is $\dfrac{17\,259}{81\,822}\cdot\dfrac{15\,251}{86\,878}=0.0370$.

(*e*) The probability that the husband will be dead and the wife will be alive after 20 years is $\frac{17\,259}{81\,822} \cdot \frac{71\,627}{86\,878} = 0.1739$.

(*f*) The probability that the wife will survive but the husband will not is found in (*e*). The probability that the husband will survive but the wife will not is $\frac{64\,563}{81\,822} \cdot \frac{15\,251}{86\,878}$. Thus, the probability that just one will survive is $\frac{17\,259}{81\,822} \cdot \frac{71\,627}{86\,878} + \frac{64\,563}{81\,822} \cdot \frac{15\,251}{86\,878} = 0.3116$.

Supplementary Problems

25.31 In how many different ways can 5 persons be seated on a bench?

Ans. 120

25.32 In how many ways can the offices of chairman, vice-chairman, secretary, and treasurer be filled from a committee of seven?

Ans. 840

25.33 How many 3-digit numbers can be formed with the digits $1, 2, \ldots, 9$, if no digit is repeated in any number?

Ans. 504

25.34 How many 3-digit odd numbers can be formed with the digits $1, 2, 3, \ldots, 9$, if no digit is repeated in any number?

Ans. 280

25.35 How many 3-digit number > 300 can be formed with the digits 1, 2, 3, 4, 5, 6, if no digit is repeated in any number?

Ans. 80

25.36 How many 4-digit numbers > 3000 can be formed with the digits 2, 3, 4, 5 if repetitions of digits (*a*) are not allowed, (*b*) are allowed?

Ans. (*a*) 18 (*b*) 192

25.37 In how many ways can 3 girls and 3 boys be seated in a row if boys and girls alternate?

Ans. 72

25.38 In how many ways can 2 letters be mailed if 5 letter boxes are available?

Ans. 25

25.39 Seven-letter words are formed using the letters of the word BLACKER. (*a*) How many can be formed? (*b*) How many which end in R? (*c*) How many in which E immediately follows K? (*d*) How many do not begin with B? (*e*) How many in which the vowels are separated by exactly two letters? (*f*) How many in which the vowels are separated by two or more letters?

Ans. (*a*) 5040 (*b*) 720 (*c*) 720 (*d*) 4320 (*e*) 960 (*f*) 2400

25.40 Eight books are to be arranged on a shelf. (*a*) In how many ways can this be done? (*b*) In how many ways if two of the books are to be placed together? (*c*) In how many ways if five of the books have red binding and three have blue binding, and the books of the same color are to be kept together? (*d*) In how many ways if four of the books belong to a numbered set and are to be kept together and in order?

Ans. (*a*) 40 320 (*b*) 10 080 (*c*) 1440 (*d*) 120

25.41 How many six-letter words can be formed using the letters of the word ASSIST (*a*) in which the S's alternate with other letters? (*b*) in which the three S's are together? (*c*) which begin and end with S? (*d*) which neither begin nor end with S?

Ans. (*a*) 12 (*b*) 24 (*c*) 24 (*d*) 24

25.42 (*a*) In how many ways can 8 persons be seated about a round table? (*b*) With 8 beads of different colors, how many bracelets can be formed by stringing them all together?

Ans. (*a*) 5040 (*b*) 2520

25.43 How many signals can be made with 3 white, 3 green, and 2 blue flags by arranging them on a mast (*a*) all at a time? (*b*) three at a time? (*c*) five at a time?

Ans. (*a*) 560 (*b*) 26 (*c*) 170

25.44 A car will hold 2 in the front seat and 1 in the rear seat. If among 6 persons only 2 can drive, in how many ways can the car be filled?

Ans. 40

25.45 A chorus consists of 6 boys and 6 girls. How many arrangements of them can be made (*a*) in a row, facing front, the boys and girls alternating? (*b*) in two rows, facing front, with a boy behind each girl? (*c*) in a ring with the boys facing the center and the girls facing away from the center? (*d*) in two concentric rings, both facing the center, with a boy behind each girl?

Ans. (*a*) 1 036 800 (*b*) 518 400 (*c*) 39 916 800 (*d*) 86 400

25.46 (*a*) In how many ways can 10 boys take positions in a straight line if two particular boys must not stand side by side? (*b*) In how many ways can 10 boys take positions about a round table if two particular boys must not be seated side by side?

Ans. (*a*) $8 \cdot 9!$ (*b*) $7 \cdot 8!$

25.47 A man has 5 large books, 7 medium-sized books, and 3 small books. In how many different ways can they be arranged on a shelf if all books of the same size are to be kept together?

Ans. 21 772 800

25.48 (*a*) How many words can be made from the letters of the word MASSACHUSETTS taken all together? (*b*) Of the words in (*a*), how many begin and end with SS? (*c*) Of the words in (*a*), how many begin and end with S? (*d*) Show that there are as many words having H as middle letter as there are circular permutations, using all letters.

Ans. (*a*) 64 864 800 (*b*) 90 720 (*c*) 4 989 600

25.49 Evaluate (*a*) ${}_6C_2$ (*b*) ${}_8C_6$ (*c*) ${}_nC_3$

Ans. (*a*) 15 (*b*) 28 (*c*) $\dfrac{n(n-1)(n-2)}{1 \cdot 2 \cdot 3}$

25.50 Find n if (*a*) $_nC_2 = 55$, (*b*) $_nC_3 = 84$, (*c*) $_{2n}C_3 = 11 \cdot {}_nC_3$

Ans. (*a*) 11 (*b*) 9 (*c*) 6b=3pt>

25.51 Two dice can be tossed in 36 ways. In how many of these is the sum equal to (*a*) 4, (*b*) 7, (*c*) 11?

Ans. (*a*) 3 (*b*) 6 (*c*) 2

25.52 Four delegates are to be chosen from eight members of a club. (*a*) How many choices are possible? (*b*) How many contain member A? (*c*) How many contain A or B but not both?

Ans. (*a*) 70 (*b*) 35 (*c*) 40

25.53 A party of 8 boys and 8 girls are going on a picnic. Six of the party go in one automobile, four go in another, and the rest walk. (*a*) In how many ways can the party be distributed for the trip? (*b*) In how many ways if no girl walks?

Ans. (*a*) 1 681 680 (*b*) 5880

25.54 Solve Problem 17.15 if the owner of each car (a boy) drives his own car.

Ans. (*a*) 168 168 (*b*) 56

25.55 How many selection of five letters each can be made from the letters of the word CANADIANS?

Ans. 41

25.56 A bag contains nine balls numbered $1, 2, \ldots, 9$. In how many ways can two balls be drawn so that (*a*) both are odd? (*b*) their sum is odd?

Ans. (*a*) 10 (*b*) 20

25.57 How many diagonals has (*a*) a hexagon, (*b*) an octagon, (*c*) an n-gon?

Ans. (*a*) 9 (*b*) 20 (*c*) $\frac{1}{2}n(n-3)$

25.58 (*a*) How many words consisting of 3 consonants and 2 vowels can be formed from 10 consonants and 5 vowels? (*b*) In how many of these will the consonants occupy the odd places?

Ans. (*a*) 144 000 (*b*) 14 400

25.59 Three balls are drawn from a bag containing five red, four white, and three black balls. In how many ways can this can be done if (*a*) each is of a different color? (*b*) they are of the same color? (*c*) exacly two are red? (*d*) at least two are red?

Ans. (*a*) 60 (*b*) 15 (*c*) 70 (*d*) 80

25.60 A squad is made up of 10 privates and 5 privates first class. (*a*) In how many ways can a detail of 4 privates and 2 privates first class be formed? (*b*) On how many of these details will private X serve? (*c*) On how many will private X but not private first class Y serve?

Ans. (*a*) 2100 (*b*) 840 (*c*) 504

25.61 A civic club has 60 members including 2 bankers, 4 lawyers, and 5 doctors. In how many ways can a committee of 10 be formed to contain 1 banker, 2 lawyers, and 2 doctors?

Ans. 228 826 080

25.62 How many committees of two or more can be selected from 10 people?

Ans. $2^{10} - 11$

25.63 Hands consisting of three cards are dealt from an ordinary deck. Show that a hand consisting of three different kinds should show 352 times as often as a hand consisting of three cards of the same kind.

25.64 Prove (*a*) ${}_nC_r + {}_nC_{r+1} = {}_{n+1}C_{r+1}$ (*b*) ${}_{2n}C_n = 2 \cdot {}_{2n-1}C_{n-1}$

25.65 Prove that ${}_nC_n = 1$.

25.66 Prove that ${}_{n+1}C_n = 2(n + 1)$.

25.67 Derive a formula for ${}_{n+2}C_n$.

25.68 One ball is drawn from a bag containing 4 white and 6 black balls. Find the probability that it is (*a*) white, (*b*) black.

Ans. (*a*) $\frac{2}{5}$ (*b*) $\frac{3}{5}$

25.69 Three balls are drawn together from a bag containing 8 white and 12 black balls. Find the probability that (*a*) all are white, (*b*) just two are white, (*c*) just one is white, (*d*) all are black.

Ans. (*a*) $\frac{14}{285}$ (*b*) $\frac{28}{95}$ (*c*) $\frac{44}{95}$ (*d*) $\frac{11}{57}$

25.70 Ten students are seated at random in a row. Find the probability that two particular students are not seated side by side.

Ans. $\frac{4}{5}$

25.71 If a die is cast three times, find the probability (*a*) that an even number will be thrown each time, (*b*) that an odd number will appear just once, (*c*) that the sum of the three numbers will be even.

Ans. (*a*) $\frac{1}{8}$ (*b*) $\frac{3}{8}$ (*c*) $\frac{1}{2}$

25.72 From a box containing 10 cards numbered 1, 2, 3, ..., 10, four cards are drawn. Find the probability that their sum will be even (*a*) if the cards are drawn together, (*b*) if each card drawn is replaced before the next is drawn.

Ans. (*a*) $\frac{11}{21}$ (*b*) $\frac{1}{2}$

25.73 *A* and *B*, having equal skill, are playing a game of three points. After *A* has won 2 points and *B* has won 1 point, what is the probability that *A* will win the game?

Ans. $\frac{3}{4}$

25.74 One bag contains 3 white and 2 black balls, and another contains 2 white and 3 black balls. A ball is drawn from the second bag and placed in the first; then a ball is drawn from the first bag and placed in the second. When the pair of operations is repeated, what is the probability that the first bag will contain 5 white balls?

Ans. $\frac{1}{225}$

25.75 Three bags contain respectively 2 white and 1 back ball, 3 white and 3 black balls, 6 white and 2 black balls. Two bags are selected and a ball is drawn from each. Find the probability (*a*) that both balls are white, (*b*) that both balls are of the same color.

Ans. (*a*) $\frac{29}{72}$ (*b*) $\frac{19}{36}$

25.76 If four trials are made in Problem 18.20, find the probability (*a*) that the first two will result in pairs of white balls and the other two in pairs of black balls, (*b*) that a pair of black balls will be obtained at least three times.

Ans. (*a*) $\frac{841}{331\,776}$ (*b*) $\frac{125}{36\,864}$

25.77 Five cards numbered 1, 2, 3, 4, 5 respectively are placed in a revolving box. If the cards are drawn one at a time from the box, what is the probability that they will be drwn in their natural order?

Ans. $\frac{1}{120}$

25.78 Brown, Jones, and Smith shoot at a target in alphabetical order with probabilities $\frac{1}{4}, \frac{1}{3}, \frac{1}{2}$ respectively of hitting the bull's eye. (*a*) Find the probability that on his first shot each will be the first to hit the bull's-eye. (*b*) Find the probability that the bull's-eye is not hit on the first round. (*c*) Find the probability that the first to hit the bull's-eye is Jones on his second shot.

Ans. (*a*) $\frac{1}{4}, \frac{1}{4}, \frac{1}{4}$ (*b*) $\frac{1}{4}$ (*c*) $\frac{1}{16}$

25.79 The probability that X will win a game of checkers is $\frac{2}{5}$. In a five-game match, what is the probability (*a*) that X will win the first, third, and fifth games, and lose the others? (*b*) that he will win exactly three games? (*c*) that he will win at least three games?

Ans. (*a*) $\frac{72}{3125}$ (*b*) $\frac{144}{625}$ (*c*) $\frac{992}{3125}$

25.80 Three pennies are tossed at the same time. Find the probability that two are heads and one is a tail.

Ans. $\frac{3}{8}$

Chapter 26

Infinite Series

GENERAL TERM OF A SEQUENCE. Frequently the law of formation of a given sequence may be stated by giving a representative or *general term* of the sequence. This general term is a function of n, where n is the number of the term in the sequence. For this reason, it is also called the nth term of the sequence.

When the general term is given, it is a simple matter to write as many terms of the sequence as desired.

EXAMPLE 1

(*a*) Write the first four terms and the tenth term of the sequence whose general term is $1/n$.

The first term ($n = 1$) is $\frac{1}{1} = 1$, the second term ($n = 2$) is $\frac{1}{2}$, and so on.The first four terms are $1, \frac{1}{2}, \frac{1}{3}, \frac{1}{4}$ and the tenth term is $\frac{1}{10}$.

(*b*) Write the first terms and the ninth term of the sequence whose general term is $(-1)^{n-1}\dfrac{2n}{n^2+1}$.

The first term ($n = 1$) is $(-1)^{1-1}\dfrac{2\cdot 1}{1^2+1} = 1$, the second term ($n = 2$) is $(-1)^1\dfrac{2\cdot 2}{2^2+1} = -\dfrac{4}{5}$, and so on.

The first four terms are $1, -\frac{4}{5}, \frac{3}{5}, -\frac{8}{17}$ and the ninth term is $(-1)^8\dfrac{2\cdot 9}{9^2+1} = \dfrac{9}{41}$.

Note that the effect of the factor $(-1)^{n-1}$ is to produce a sequence whose terms have alternate signs, the sign of the first term being positive. The same pattern of signs is also produced by the factor $(-1)^{n+1}$. In order to produce a sequence whose terms alternate in sign, the first term being negative, the factor $(-1)^n$ is used.

When the first few terms of a sequence are given and they match an obvious pattern, the general term is obtained by inspection.

EXAMPLE 2. Obtain the general term for each of the sequences:

(*a*) $1, 4, 9, 16, 25, \ldots$.

The terms of the sequence are the squares of the positive integers; the general term is n^2.

(*b*) $3, 7, 11, 15, 19, 23, \ldots$.

This is an arithmetic progression having $a = 3$ and $d = 4$. The general term is $a + (n-1)d = 4n - 1$. Note, however, that the general term can be obtained about as easily by inspection.

(See Problems 26.1–26.3.)

LIMIT OF AN INFINITE SEQUENCE. From Example 2 of Chapter 8, the line $y = 2$ is a horizontal asymptote of $xy - 2x - 1 = 0$. To show this, let $P(x, y)$ move along the curve so that its abscissa takes on the values $10, 10^2, 10^3, \ldots, 10^n, \ldots$. Then the corresponding values of y are

$$2.1, 2.01, 2.001, \ldots, 2 + \frac{1}{10^n}, \ldots \tag{26.1}$$

and we infer that, by proceeding far enough along in this sequence, the difference between the terms of the sequence and 2 may be made as small as we please. This is equivalent to the following: Let ε denote a positive number, as small as we please; then there is a term of the sequence such that the difference between it and 2 is less than ε, and the same is true for all subsequent terms of the sequence. For example, let $\varepsilon = 1/10^{25}$ then the difference between the term $2 + 1/10^{26}$ and 2, $2 + 1/10^{26} - 2 = 1/10^{26}$, is less then $\varepsilon = 1/10^{25}$ and the same is true for the terms $2 + 1/10^{27}$, $2 + 1/10^{28}$, and so on.

The behavior of the terms of the sequence (*26.1*) discussed above is indicated by the statement: *The limit of the sequence* (*26.1*) *is 2.* In general, if, for an infinite sequence

$$s_1, s_2, s_3, \ldots, s_n, \ldots \tag{26.2}$$

and a positive number ε, however small, there exists a number s and a positive integer m such that for all $n > m$

$$|s - s_n| < \varepsilon,$$

then the limit of the sequence is s.

EXAMPLE 3. Show, using the above definition, that the limit of sequence (*26.1*) is 2.

Take $\varepsilon = 1/10^p$, where p is a positive integer as large as we please; thus, ε is a positive number as small as we please. We must produce a positive integer m (in other words, a term s_m) such that for $n > m$ (that is, for all subsequent terms) $|s - s_n| < \varepsilon$. Now

$$\left|2 - \left(2 + \frac{1}{10^n}\right)\right| < \frac{1}{10^p} \quad \text{or} \quad \frac{1}{10^n} < \frac{1}{10^p}$$

requires $n > p$. Thus, $m = p$ is the required value of m.

The statement that the limit of the sequence (*26.2*) is s describes the behavior of s_n as n increases without bound over the positive integers. Since we shall repeatedly be using the phrase "as n increases without bound" or the phrase "as n becomes infinite," which we shall take to be equivalent to the former phrase, we shall introduce the notation $n \to \infty$ for it. Thus the behavior of s_n may be described briefly by

$$\lim_{n\to\infty} s_n = s$$

(read: the limit of s_n, as n becomes infinite, is s).

We state, without proof, the following theorem:

If $\lim_{n\to\infty} s_n = s$ and $\lim_{n\to\infty} t_n = t$, then

$$(A) \quad \lim_{n\to\infty} (s_n \pm t_n) = \lim_{n\to\infty} s_n \pm \lim_{n\to\infty} t_n = s \pm t$$

$$(B) \quad \lim_{n\to\infty} (s_n \cdot t_n) = \lim_{n\to\infty} s_n \cdot \lim_{n\to\infty} t_n = s \cdot t$$

$$(C) \quad \lim_{n\to\infty} \frac{s_n}{t_n} = \frac{\lim_{n\to\infty} s_n}{\lim_{n\to\infty} t_n} = \frac{s}{t}, \quad \text{provided} \quad t \neq 0$$

or, in words, if each of two sequences approaches a limit, then the limits of the sum, difference, product, and quotient of the two sequences are equal, respectively, to the sum, difference, product, and quotient of their limits provided only that, in the case of the quotient, the limit of the denominator is not zero.

This theorem makes it possible to find the limit of a sequence directly from its general term. In this connection, we shall need

$$\lim_{n\to\infty} a = a, \qquad \text{where } a \text{ is any constant} \tag{26.3}$$

$$\lim_{n\to\infty} \frac{1}{n^k} = 0, \qquad k > 0 \text{ (See Problem 26.4.)} \tag{26.4}$$

$$\lim_{n\to\infty} \frac{1}{b^n} = 0, \qquad \text{where } b \text{ is a constant} > 1 \text{ (See Problem 26.5)} \tag{26.5}$$

(See Problem 26.6.)

THE FOLLOWING THEOREMS are useful in establishing whether or not certain sequences have a limit.

I. Suppose M is a fixed number, such that for all values of n,

$$s_n \le s_{n+1} \qquad \text{and} \qquad s_n \le M;$$

then $\lim_{n\to\infty} s_n$ exists and is $\le M$.

If, however, s_n eventually exceeds M, no matter how large M may be, $\lim_{n\to\infty} s_n$ does not exist.

II. Suppose M is a fixed number such that, for all values of n,

$$s_n \ge s_{n+1} \qquad \text{and} \qquad s_n \ge M;$$

then $\lim_{n\to\infty} s_n$ exists and is $\ge M$.

If, however, s_n is eventually smaller than M, no matter how small M may be, $\lim_{n\to\infty} s_n$ does not exist.

EXAMPLE 4

(*a*) For the sequence $\frac{5}{2}, 3, \frac{19}{6}, \frac{13}{4}, \ldots, \left(\frac{7}{2} - \frac{1}{n}\right), \ldots$, $s_n < s_{n+1}$ and $s_n < 4$, for all values of n; the sequence has a limit ≤ 4. In fact, $\lim_{n\to\infty} s_n = \frac{7}{2}$.

(*b*) For the sequence $3, 5, 7, 9, \ldots, 2n+1, \ldots$, $s_n < s_{n+1}$ but s_n will eventually exceed any chosen M, however large (if $M = 2^{1000} + 1$, then $2n + 1 > M$ for $n > 2^{999}$), and the sequences does not have a limit.

(See Problems 26.7–26.9.)

RECURSIVELY DEFINED SEQUENCES. Sequences can be defined recursively. For example, suppose that $a_1 = 1$ and $a_{n+1} = 2a_n$ for every natural number n. Then,

$$a_1 = 1, \qquad a_2 = a_{1+1} = 2a_1 = 2, \qquad a_3 = a_{2+1} = 2a_2 = 4, \qquad \text{etc.}$$

Thus, the sequence is $1, 2, 4, 8, \ldots$.

One famous such sequence is the Fibonacci sequence:

$$a_1 = 1, \qquad a_2 = 1, \qquad a_{n+2} = a_{n+1} + a_n.$$

The sequence is $1, 1, 2, 3, 5, 8, 13, \ldots$.

THE INDICATED SUM of the terms of an infinite sequence is called an *infinite series*. Let

$$s_1 + s_2 + s_3 + \cdots + s_n + \cdots \tag{26.6}$$

be such a series and define the sequence of *partial sums*

$$S_1 = s_1, \quad S_2 = s_1 + s_2, \quad \ldots, \quad S_n = s_1, + s_2 + \cdots + s_n, \quad \ldots$$

If $\lim_{n\to\infty} S_n$ exists, the series (*26.6*) is called *convergent*; if $\lim_{n\to\infty} S_n = S$, the series is said to converge to S. If $\lim_{n\to\infty} S_n$ does not exist, the series is called *divergent*.

EXAMPLE 5

(*a*) Every infinite geometric series

$$a + ar + ar^2 + \cdots + ar^{n-1} + \cdots$$

is convergent if $|r| < 1$ and is divergent if $|r| \geq 1$. (See Problem 26.10)

(*b*) The *harmonic series* $1 + \frac{1}{2} + \frac{1}{2} + \cdots + 1/n + \cdots$ is divergent. (See Problem 26.9 and Problem 26.11.)

A NECESSARY CONDITION THAT (*26.10*) BE CONVERGENT is $\lim_{n\to\infty} s_n = 0$; that is, if (*26.6*) is convergent then $\lim_{n\to\infty} s_n = 0$. However, the condition is *not sufficient* since the harmonic series is divergent although $\lim_{n\to\infty} s_n = \lim_{n\to\infty}(1/n) = 0$.

A SUFFICIENT CONDITION THAT (*26.10*) BE DIVERGENT is $\lim_{n\to\infty} s_n \neq 0$; that is, if $\lim_{n\to\infty} s_n$ exists and is different from 0, or if $\lim_{n\to\infty} s_n$ does not exist, the series is divergent. This, in turn, is not a necessary condition since the harmonic series is divergent although $\lim_{n\to\infty} s_n = 0$. (See Problem 26.12).

SERIES OF POSITIVE TERMS

COMPARISON TEST FOR CONVERGENCE of a series of positive terms.

I. If every term of a given series of positive terms is less than or equal to the corresponding term of a known convergent series from some point on in the series, the given series is convergent.

II. If every term of a given series of positive terms is equal to or greater than the corresponding term of a known divergent series from some point on in the series, the given series is divergent.

The following series will be found useful in making comparison tests:

(*a*) The geometric series $a + ar + ar^2 + \cdots + ar^n + \cdots$ which converges when $|r| < 1$ and diverges when $|r| \geq 1$

(*b*) The p series $1 + \dfrac{1}{2^p} + \dfrac{1}{3^p} + \cdots + \dfrac{1}{n^p} + \cdots$ which converges for $p > 1$ and diverges for $p \leq 1$

(*c*) Each new series tested

In comparing two series, it is not sufficient to examine the first few terms of each series. The *general terms must be compared*, since the comparison must be shown from some point on. (See Problems 26.13–26.15.)

THE RATIO TEST FOR CONVERGENCE. If, in a series of positive terms, the *test ratio*

$$r_n = \frac{s_{n+1}}{s_n}$$

approaches a limit R as $n \to \infty$, the series is convergent if $R < 1$ and is divergent if $R > 1$. If $R = 1$, the test fails to indicate convergency or divergency. (See Problem 26.16.)

SERIES WITH NEGATIVE TERMS

A SERIES WITH ALL ITS TERMS NEGATIVE may be treated as the negative of a series with all of its terms positive.

ALTERNATING SERIES. A series whose terms are alternately positive and negative, as

$$s_1 - s_2 + s_3 - \cdots + (-1)^{n-1} s_n + \cdots \tag{26.7}$$

where each s is positive, is called an *alternating series*.

An alternating series (*26.11*) is convergent provided $s_n \geq s_{n+1}$, for every value of n, and $\lim_{n \to \infty} s_n = 0$. (See Problem 26.17)

ABSOLUTELY CONVERGENT SERIES. A series (*26.6*), $s_1 + s_2 + s_3 + \cdots + s_n + \cdots$ in which some of the terms are positive and some are negative is called *absolutely convergent* if the series of absolute values of the terms

$$|s_1| + |s_2| + |s_3| + \cdots + |s_n| + \cdots \tag{26.8}$$

is convergent.

CONDITIONALLY CONVERGENT SERIES. A series (*26.6*), where some of the terms are positive and some are negative, is called *conditionally convergent* if it is convergent but the series of absolute values of its terms is divergent.

EXAMPLE 6. The series $1 - \frac{1}{2} + \frac{1}{3} - \frac{1}{4} + \cdots$ is convergent, but the series of absolute values of its terms $1 + \frac{1}{2} + \frac{1}{3} + \frac{1}{4} + \cdots$ is divergent. Thus, the given series is conditionally convergent.

THE GENERALIZED RATIO TEST. Let (*26.10*) $s_1 + s_2 + s_3 + \cdots + s_n + \cdots$ be a series some of whose terms are positive and some are negative. Let

$$\lim_{n \to \infty} \frac{|s_{n+1}|}{|s_n|} = R$$

The series (*26.10*) is absolutely convergent if $R < 1$ and is divergent if $R > 1$. If $R = 1$, the test fails. (See Problem 26.17.)

INFINITE SERIES OF THE FORM

$$c_0 + c_1 x + c_2 x^2 + \cdots + c_{n-1}(x)^{n-1} + \cdots \tag{26.9}$$

and

$$c_0 + c_1(x - a) + c_2(x - a)^2 + \cdots + c_{n-1}(x - a)^{n-1} + \cdots, \tag{26.10}$$

where a, c_0, c_1, $c_2, \ldots$ are constants, are called *power series*. The first is called a power series in x and the second a power series in $(x - a)$.

The power series (*26.9*) converges for $x = 0$ and (*26.10*) converges for $x = a$. Both series may converge for other values of x but not necessarily for every finite value of x. Our problem is to find for a given power series all values of x for which the series converges. In finding this set of values, called the *interval of convergence* of the series, the generalized ratio test of Chapter 26 will be used.

EXAMPLE 7. Find the interval of convergence of the series

$$x + x^2/2 + x^3/3 + \cdots$$

Since

$$|s_n| = \left|\frac{x^n}{n}\right| \quad \text{and} \quad |s_{n+1}| = \left|\frac{x^{n+1}}{n+1}\right|,$$

$$R = \lim_{n \to \infty} \frac{|s_{n+1}|}{|s_n|} = \lim_{n \to \infty} \left|\frac{x^{n+1}}{n+1} \cdot \frac{n}{x^n}\right| = \lim_{n \to \infty} \left|\frac{n}{n+1} x\right| = |x|.$$

Then, by the ratio test, the given series is convergent for all values of x such that $|x| < 1$, that is, for $-1 < x < 1$; the series is divergent for all values of x such that $|x| > 1$, that is, for $x < -1$ and $x > 1$; and the test fails for $x = \pm 1$.

But, when $x = 1$ the series is $1 + \frac{1}{2} + \frac{1}{3} + \frac{1}{4} + \cdots$ and is divergent, and when $x = -1$ the series is $-1 + \frac{1}{2} - \frac{1}{3} + \frac{1}{4} - \cdots$ and is convergent.

Thus, the series converges on the interval $-1 \le x < 1$. This interval may be represented graphically as in Fig. 26-1. The solid line represents the interval on which the series converges, the thin lines the intervals on which the series diverges. The solid circle represents the end point for which the series converges, the open circle the end point at which the series diverges. (See Problems 26.19–26.26.)

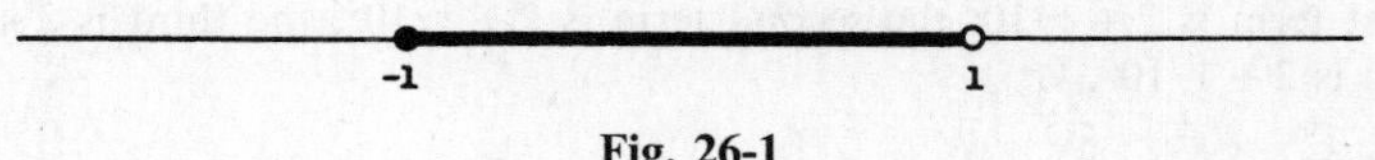

Fig. 26-1

Solved Problems

26.1 Write the first five terms and the tenth term of the sequence whose general term is

(*a*) $4n - 1$.

The first term is $4 \cdot 1 - 1 = 3$, the second term is $4 \cdot 2 - 1 = 7$, the third term is $4 \cdot 3 - 1 = 11$, the fourth term is $4 \cdot 4 - 1 = 15$, the fifth term is $4 \cdot 5 - 1 = 19$; the tenth term is $4 \cdot 10 - 1 = 39$.

(*b*) 2^{n-1}.

The first term is $2^{1-1} = 2^0 = 1$, the second term is $2^{2-1} = 2$, the third is $2^{3-1} = 2^2 = 4$, the fourth is $2^3 = 8$, the fifth is $2^4 = 16$; the tenth is $2^9 = 512$.

(*c*) $\dfrac{(-1)^{n-1}}{n+1}$.

The first term is $\dfrac{(-1)^{1-1}}{1+1} = \dfrac{(-1)^0}{2} = \dfrac{1}{2}$, the second is $\dfrac{(-1)^{2-1}}{2+1} = -\dfrac{1}{3}$, the third is $\dfrac{(-1)^2}{3+1} = \dfrac{1}{4}$, the fourth is $-\frac{1}{5}$, the fifth is $\frac{1}{6}$; the tenth is $-\frac{1}{11}$.

26.2 Write the first four terms of the sequences whose general term is

(*a*) $\dfrac{n+1}{n!}$.

The terms are $\dfrac{1+1}{1!}, \dfrac{2+1}{2!}, \dfrac{3+1}{3!}, \dfrac{4+1}{4!}$ or $2, \dfrac{3}{2}, \dfrac{2}{3}, \dfrac{5}{24}$.

(*b*) $\dfrac{x^{2n-1}}{(2n+1)!}$.

The required terms are $\dfrac{x}{3!}, \dfrac{x^3}{5!}, \dfrac{x^5}{7!}, \dfrac{x^7}{9!}$.

(*c*) $(-1)^{n-1}\dfrac{x^{2n-2}}{(n-1)!}$.

The terms are $\dfrac{x^0}{0!}, -\dfrac{x^2}{1!}, \dfrac{x^4}{2!}, -\dfrac{x^6}{3!}$ or $1, -x^2, \dfrac{x^4}{2}, \dfrac{x^6}{6}$.

26.3 Write the general term for each of the following sequences:

(*a*) $2, 4, 6, 8, 10, 12, \ldots$.

The first term is $2 \cdot 1$, the second is $2 \cdot 2$, the third is $2 \cdot 3$, etc.; the general term is $2n$.

(*b*) 1, 3, 5, 7, 9, 11, ...

Each term of the given sequence is 1 less than the corresponding term of the sequences in (*a*); the general term is $2n-1$.

(*c*) 2, 5, 8, 11, 14, ...

The first term is $3\cdot 1-1$, the second term is $3\cdot 2-1$, the third term is $3\cdot 3-1$, and so on; the general term is $3n-1$.

(*d*) 2, −5, 8, −11, 14, ...

This sequence may be obtained from that in (*c*) by changing the signs of alternate terms beginning with the second; the general term is $(-1)^{n-1}(3n-1)$.

(*e*) 2.1, 2.001, 2.001, 2.0001, ...

The first term is $2+1/10$, the second term is $2+1/10^2$, the third is $2+1/10^3$, and so on; the general term is $2+1/10^n$.

(*f*) $\frac{1}{8}, -\frac{1}{27}, \frac{1}{64}, -\frac{1}{125}, \ldots$

The successive denominators are the cubes of 2, 3, 4, 5... or of $1+1, 2+1, 3+1, 4+1, \ldots$; the general term is $\dfrac{(-1)^{n-1}}{(n+1)^3}$.

(*g*) $\dfrac{3}{1}, \dfrac{4}{1\cdot 2}, \dfrac{5}{1\cdot 2\cdot 3}, \ldots$

Rewriting the sequence as $\dfrac{1+2}{1!}, \dfrac{2+2}{2!}, \dfrac{3+2}{3!}, \ldots$, the general term is $\dfrac{n+2}{n!}$.

(*h*) $\dfrac{x}{2}, \dfrac{x^2}{6}, \dfrac{x^3}{24}, \dfrac{x^4}{120}, \ldots$

The denominators are $2!, 3!, \ldots, (n+1)!, \ldots$; the general term is $\dfrac{x^n}{(n+1)!}$.

(*i*) $x, \dfrac{-x^3}{3!}, \dfrac{x^5}{5!}, \dfrac{-x^7}{7!}, \ldots$

The exponents of x are $2\cdot 1-1, 2\cdot 2-1, 2\cdot 3-1, \ldots$; the general term is $(-1)^{n-1}\dfrac{x^{2n-1}}{(2n-1)!}$.

(*j*) $3, 4, \frac{5}{2}, 1, \frac{7}{24}, \ldots$

Rewrite the sequence as $\dfrac{3}{0!}, \dfrac{4}{1!}, \dfrac{5}{2!}, \dfrac{6}{3!}, \ldots$; the general term is $\dfrac{n+2}{(n-1)!}$.

26.4 Show that $\lim\limits_{n\to\infty} \dfrac{1}{n^k} = 0$ when $k > 0$.

Take $\varepsilon = 1/p^k$, where p is a positive integer as large as we please. We seek a positive number m such that for $n > m$, $|0 - 1/n^k| = 1/n^k < 1/p^k$. Since this inequality is satisfied when $n > p$, it is sufficient to take for m any number equal to or greater than p.

26.5 Show that $\lim\limits_{n\to\infty} \dfrac{1}{b^n} = 0$ when $b > 1$.

Take $\varepsilon = 1/b^p$, where p is a positive integer. Since $b > 1$, $b^p > 1$ and $\varepsilon = 1/b^p < 1$. Thus, ε may be made as small as we please by taking p sufficiently large. We seek a positive number m such that for $n > m$, $|0 - 1/b^n| = 1/b^n < 1/b^p$. Since $n > p$ satisfies the inequality, it is sufficient to take for m any number equal to or greater than p.

26.6 Evaluate each of the following:

(*a*) $\lim\limits_{n\to\infty}\left(\dfrac{3}{n}+\dfrac{5}{n^2}\right) = \lim\limits_{n\to\infty}\dfrac{3}{n} + \lim\limits_{n\to\infty}\dfrac{5}{n^2} = 0+0=0$

(*b*) $\lim\limits_{n\to\infty}\dfrac{n}{n+1} = \lim\limits_{n\to\infty}\dfrac{n/n}{\dfrac{n+1}{n}} = \lim\limits_{n\to\infty}\dfrac{1}{1+1/n} = \dfrac{\lim\limits_{n\to\infty} 1}{\lim\limits_{n\to\infty}(1+1/n)} = \dfrac{1}{1+0} = 1$

(*c*) $\lim\limits_{n\to\infty}\dfrac{n^2+2}{2n^2-3n} = \lim\limits_{n\to\infty}\dfrac{1+2/n^2}{2-3/n} = \dfrac{1+0}{2-0} = \dfrac{1}{2}$

(*d*) $\lim_{n\to\infty}\left(4-\frac{2^n-1}{2^{n+1}}\right)=4-\lim_{n\to\infty}\frac{2^n-1}{2^{n+1}}=4-\lim_{n\to\infty}\left(\frac{1}{2}-\frac{1}{2^{n+1}}\right)$

$$=4\frac{1}{2}-\lim_{n\to\infty}\left(-\frac{1}{2^{n+1}}\right)=4-\frac{1}{2}-0=3.5$$

26.7 Show that every infinite arithmetic sequence fails to have a limit except when $d=0$.

(*a*) If $d>0$, then $s_n=a+(n-1)d<s_{n+1}=a+nd$; but s_n eventually exceeds any previously selected M, however large. Thus, the sequence has no limit.

(*b*) If $d<0$, then $s_n>s_{n+1}$; but s_n eventually becomes smaller than any previously selected M, however small. Thus, the sequence has no limit.

(*c*) If $d=0$, the sequence is $a, a, a, \ldots, a, \ldots$ with limit a.

26.8 Show that the infinite geometric sequence $3, 6, 12, \ldots, 3\cdot 2^{n-1}, \ldots$, does not have a limit.

Here $s_n<s_{n+1}$; but $3\cdot 2^{n-1}$ may be made to exceed any previously selected M, however, large. The sequence has no limit.

26.9 Show that the following sequence does not have a limit:

$$1,\ 1+\frac{1}{2},\ 1+\frac{1}{2}+\frac{1}{3},\ 1+\frac{1}{2}+\frac{1}{3}+\frac{1}{4}+\cdots,\quad 1+\frac{1}{2}+\frac{1}{3}+\frac{1}{4}+\cdots+\frac{1}{n},\quad \ldots$$

Here $s_n<s_{n+1}$. Let M, as large as we please, be chosen, Now

$$1+\tfrac{1}{2}+\tfrac{1}{3}+\tfrac{1}{4}+\cdots=1+\tfrac{1}{2}+(\tfrac{1}{3}+\tfrac{1}{4})+(\tfrac{1}{5}+\tfrac{1}{6}+\tfrac{1}{7}+\tfrac{1}{8})+(\tfrac{1}{9}+\cdots+\tfrac{1}{16})+(\tfrac{1}{17}+\cdots+\tfrac{1}{32})+\cdots$$

and $\qquad \tfrac{1}{3}+\tfrac{1}{4}>\tfrac{1}{2},\quad \tfrac{1}{5}+\tfrac{1}{6}+\tfrac{1}{7}+\tfrac{1}{8}>\tfrac{1}{2},\quad \tfrac{1}{9}+\cdots+\tfrac{1}{16}>\tfrac{1}{2},$ and so on.

Since the sum of each group exceeds $\frac{1}{2}$ and we may add as many groups as we please, we can eventually obtain a sum of groups which exceeds M. Thus, the sequence has no limit.

26.10 Examine the infinite geometric series $a+ar+ar^2+\cdots+ar^n+\cdots$ for convergence and divergence.

The sum of the first n terms is $S_n=\dfrac{a(1-r^n)}{1-r}=\dfrac{a}{1-r}(1-r^n)$.

If $|r|<1$, $\lim_{n\to\infty} r^n=0$; then $\lim_{n\to\infty} S_n=\dfrac{a}{1-r}$ and the series is convergent.

If $|r|>1$, $\lim_{n\to\infty} r^n$ does not exist and $\lim_{n\to\infty} S_n$ does not exist; the series is divergent.

If $r=1$, the series is $a+a+a+\cdots+a+\cdots$; then $\lim_{n\to\infty} S_n=\lim_{n\to\infty} na$ does not exist.

If $r=-1$, the series is $a-a+a-a+\cdots$; then $S_n=a$ or 0 according as n is odd or even, and $\lim_{n\to\infty} S_n$ does not exist.

Thus, the infinite geometric series convergence to $\dfrac{a}{1-r}$ when $|r|<1$, and diverges when $|r|\geq 1$.

26.11 Show that the following series are convergent:

(*a*) $1+\dfrac{2}{3}+\dfrac{4}{9}+\dfrac{8}{27}+\cdots+\dfrac{2^{n-1}}{3^{n-1}}+\cdots$.

This is a geometric series with ratio $r=\frac{2}{3}$; then $|r|<1$ and the series is convergent.

(*b*) $2-\frac{3}{2}+\frac{9}{8}-\frac{27}{32}+\cdots+(-1)^{n-1}2(\frac{3}{4})^{n-1}+\cdots$.

This is a geometric series with ratio $r=-\frac{3}{4}$; then $|r|<1$ and the series is convergent.

(*c*) $1+\dfrac{1}{2^p}+\dfrac{1}{2^p}+\dfrac{1}{4^p}+\dfrac{1}{4^p}+\dfrac{1}{4^p}+\dfrac{1}{4^p}+\dfrac{1}{8^p}+\cdots, p>1$.

This series may be rewritten as $1+\dfrac{2}{2^p}+\dfrac{4}{4^p}+\dfrac{8}{8^p}+\cdots$, a geometric series with ratio $r=2/2^p$. Since $|r|<1$, when $p>1$, the series is convergent.

26.12 Show that the following series are divergent:

(*a*) $2+\frac{3}{2}+\frac{4}{3}+\cdots+\frac{n+1}{n}+\cdots$.

Since $\lim_{n\to\infty} s_n = \lim_{n\to\infty}\frac{n+1}{n} = \lim_{n\to\infty}\left(1+\frac{1}{n}\right) = 1 \neq 0$, the series is divergent.

(*b*) $\frac{1}{2}+\frac{3}{8}+\frac{5}{16}+\frac{9}{32}+\cdots+\frac{2^{n-1}+1}{4\cdot 2^{n-1}}+\cdots$.

Since $\lim_{n\to\infty} s_n = \lim_{n\to\infty}+\frac{2^{n-1}+1}{4\cdot 2^{n-1}} = \lim_{n\to\infty}\frac{1+1/2^{n-1}}{4} = \frac{1}{4} \neq 0$, the series is divergent.

26.13 Show that $1+\frac{1}{2^p}+\frac{1}{3^p}+\frac{1}{4^p}+\cdots+\frac{1}{n^p}+\cdots$ is divergent for $p \le 1$ and convergent for $p > 1$.

For $p = 1$, the series is the harmonic series and is divergent.

For $p < 1$, including negative values, $\frac{1}{n^p} \ge \frac{1}{n}$, for every n. Since every term of the given series is equal to or greater than the corresponding term of the harmonic series, the given series is divergent.

For $p > 1$, compare the series with the convergent series

$$1+\frac{1}{2^p}+\frac{1}{2^p}+\frac{1}{4^p}+\frac{1}{4^p}+\frac{1}{4^p}+\frac{1}{4^p}+\frac{1}{8^p}+\cdots \tag{1}$$

of Problem 26.11 (*c*). Since each term of the given series is less than or equal to the corresponding term of series (*1*), the given series is convergent.

26.14 Use Problem 26.13 to determine whether the following series are convergent or divergent:

(*a*) $1+\frac{1}{2\sqrt{2}}+\frac{1}{3\sqrt{3}}+\frac{1}{4\sqrt{4}}+\cdots$. The general term is $\frac{1}{n\sqrt{n}} = \frac{1}{n^{3/2}}$.

This is a p series with $p = \frac{3}{2} > 1$; the series is convergent.

(*b*) $1+4+9+16+\cdots$. The general term is $n^2 = \frac{1}{n^{-2}}$.

This is a p series with $p = -2 < 1$; the series is divergent.

(*c*) $1+\frac{\sqrt[4]{2}}{4}+\frac{\sqrt[4]{3}}{9}+\frac{\sqrt[4]{4}}{16}+\cdots$. The general term is $\frac{\sqrt[4]{n}}{n^2} = \frac{1}{n^{7/4}}$.

The series is convergent since $p = \frac{7}{4} > 1$.

26.15 Use the comparison test to determine whether each of the following is convergent or divergent:

(*a*) $1+\frac{1}{2!}+\frac{1}{3!}+\frac{1}{4!}+\cdots$.

The general term $\frac{1}{n!} \le \frac{1}{n^2}$. Thus, the terms of the given series are less than or equal to the corresponding terms of the p series with $p = 2$. The series is convergent.

(*b*) $\frac{1}{2}+\frac{1}{3}+\frac{1}{5}+\frac{1}{9}+\cdots$.

The general term $\frac{1}{1+2^{n-1}} \le \frac{1}{2^{n-1}}$. Thus, the terms of the given series are less than or equal to the corresponding terms of the geometric series with $a = 1$ and $r = \frac{1}{2}$. The series is convergent.

(*c*) $\frac{2}{1}+\frac{3}{4}+\frac{4}{9}+\frac{5}{16}+\cdots$.

The general term $\dfrac{n+1}{n^2}=\dfrac{1}{n}+\dfrac{1}{n^2}>\dfrac{1}{n}$. Thus, the terms of the given series are equal to or greater than the corresponding terms of the harmonic series. The series is divergent.

(*d*) $\frac{1}{3}+\frac{1}{12}+\frac{1}{27}+\frac{1}{48}+\cdots$.

The general term $\dfrac{1}{3\cdot n^2}\le\dfrac{1}{n^2}$. Thus, the terms of the given series are less than or equal to the corresponding terms of the p series with $p=2$. The series is convergent.

(*e*) $1+\dfrac{1}{2}+\dfrac{1}{3^2}+\dfrac{1}{4^3}+\dfrac{1}{5^4}+\cdots$.

The general term $\dfrac{1}{n^{n-1}}\le\dfrac{1}{n^2}$ for $n\ge 3$. Thus, neglecting the first two terms, the given series is term by term less than or equal to the corresponding terms of the p series with $p=2$. The given series is convergent.

26.16 Apply the ratio test to each of the following. If it fails, use some other method to determine convergency or divergency.

(*a*) $\dfrac{1}{2}+\dfrac{1}{2}+\dfrac{3}{8}+\dfrac{1}{4}+\cdots$ or $\dfrac{1}{2}+\dfrac{2}{2^2}+\dfrac{3}{2^3}+\dfrac{4}{2^4}+\cdots$.

For this series $s_n=\dfrac{n}{2^n}$, $s_{n+1}=\dfrac{n+1}{2^{n+1}}$, and $r_n=\dfrac{s_{n+1}}{s_n}=\dfrac{n+1}{2^{n+1}}\cdot\dfrac{2^n}{n}=\dfrac{n+1}{2n}$. Then $R=\lim\limits_{n\to\infty} r_n=\lim\limits_{n\to\infty}\dfrac{n+1}{2n}=\lim\limits_{n\to\infty}\dfrac{1+1/n}{2}=\dfrac{1}{2}<1$ and the series is convergent.

(*b*) $3+\dfrac{9}{2}+\dfrac{9}{2}+\dfrac{27}{8}+\dfrac{81}{40}+\cdots$ or $\dfrac{3}{1!}+\dfrac{3^2}{2!}+\dfrac{3^3}{3!}+\cdots$.

Here $s_n=\dfrac{3^n}{n!}$, $s_{n+1}=\dfrac{3^{n+1}}{(n+1)!}$, and $r_n=\dfrac{3^{n+1}}{(n+1)!}\cdot\dfrac{n!}{3^n}=\dfrac{3}{n+1}$. Then $R=\lim\limits_{n\to\infty}\dfrac{3}{n+1}=0$ and the series is convergent.

(*c*) $\dfrac{1}{1\cdot 1}+\dfrac{1}{2\cdot 3}+\dfrac{1}{3\cdot 5}+\dfrac{1}{4\cdot 7}+\cdots$.

Here $s_n=\dfrac{1}{n(2n-1)}$, $s_{n+1}=\dfrac{1}{(n+1)(2n+1)}$, and $r_n=\dfrac{n(2n-1)}{(n+1)(2n+1)}$.

Then $R=\lim\limits_{n\to\infty}\dfrac{n(2n-1)}{(n+1)(2n+1)}=\lim\limits_{n\to\infty}\dfrac{2-1/n}{(1+1/n)(2+1/n)}=1$ and the test fails.

Since $\dfrac{1}{n(2n-1)}\le\dfrac{1}{n^2}$, the given series is term by term less than or equal to the convergent p series, with $p=2$. The given series is convergent.

(*d*) $\dfrac{2}{1^2+1}+\dfrac{2^3}{2^2+2}+\dfrac{2^5}{3^2+3}+\cdots$.

Here $s_n=\dfrac{2^{2n-1}}{n^2+n}$, $s_{n+1}=\dfrac{2^{2n+1}}{(n+1)^2+(n+1)}$, and $r_n=\dfrac{2^{2n+1}}{(n+1)(n+2)}\cdot\dfrac{n(n+1)}{2^{2n-1}}=\dfrac{4n}{n+2}$. Then $R=\lim\limits_{n\to\infty}\dfrac{4}{1+2/n}=4$ and the series is divergent.

(*e*) $\frac{1}{5}+\frac{2}{25}+\frac{6}{125}+\frac{24}{625}+\cdots$.

In this series $s_n=\dfrac{n!}{5^n}$, $s_{n+1}=\dfrac{(n+1)!}{5^{n+1}}$, and $r_n=\dfrac{(n+1)!}{5^{n+1}}\cdot\dfrac{5^n}{n!}=\dfrac{n+1}{5}$. Now $\lim\limits_{n\to\infty} r_n$ does not exist. However, since $s_n\to\infty$ as $n\to\infty$, the series is divergent.

26.17 Test the following alternating series for convergence:

(*a*) $1 - \frac{1}{3} + \frac{1}{5} - \frac{1}{7} + \cdots$.

$s_n > s_{n+1}$, for all values of n, and $\lim_{n\to\infty} s_n = \lim_{n\to\infty} \frac{1}{2n-1} = 0$. The series is convergent.

(*b*) $\frac{1}{2^3} - \frac{2}{3^3} + \frac{3}{4^3} - \frac{4}{5^3} + \cdots$.

$s_n > s_{n+1}$, for all values of n, and $\lim_{n\to\infty} s_n = \lim_{n\to\infty} \frac{n}{(n+1)^3} = 0$. The series is convergent.

26.18 Investigate the following for absolute convergence, conditional convergence, or divergence:

(*a*) $1 - \frac{1}{2} + \frac{1}{4} - \frac{1}{8} + \cdots$.

Here $|s_n| = \frac{1}{2^{n-1}}$, $|s_{n+1}| = \frac{1}{2^n}$, and $R = \lim_{n\to\infty} \frac{|s_{n+1}|}{|s_n|} = \lim_{n\to\infty} \frac{2^{n-1}}{2^n} = \frac{1}{2} < 1$. The series is absolutely convergent.

(*b*) $1 - \frac{4}{1!} + \frac{4^2}{2!} - \frac{4^3}{3!} + \cdots$.

Here $|s_n| = \frac{4^{n-1}}{(n-1)!}$, $|s_{n+1}| = \frac{4^n}{n!}$, and $R = \lim_{n\to\infty} \frac{4}{n} = 0$. The series is absolutely convergent.

(*c*) $\frac{1}{2-\sqrt{2}} - \frac{1}{3-\sqrt{3}} + \frac{1}{4-\sqrt{4}} - \frac{1}{5-\sqrt{5}} + \cdots$.

The ratio test fails here.

Since $\frac{1}{n+1-\sqrt{n+1}} > \frac{1}{n+2-\sqrt{n+2}}$ and $\lim_{n\to\infty} \frac{1}{n+1-\sqrt{n+1}} = 0$, the series is convergent.

Since $\frac{1}{n+1-\sqrt{n+1}} > \frac{1}{n+1}$ for all values of n, the series of absolute values is term by term greater than the harmonic series, and thus is divergent. The given series is conditionally convergent.

In Problems 26.19–26.24, find the interval of convergence *including* the end points.

26.19 $1 + \frac{x}{1!} + \frac{x^2}{2!} + \frac{x^3}{3!} + \cdots$.

For this series

$$|s_n| = \left|\frac{x^{n-1}}{(n-1)!}\right|, \qquad |s_{n+1}| = \left|\frac{x^n}{n!}\right|, \qquad \text{and} \qquad R = \lim_{n\to\infty} \frac{|s_{n+1}|}{|s_n|} = \lim_{n\to\infty} \left|\frac{x^n}{n!} \cdot \frac{(n-1)!}{x^{n-1}}\right| = \lim_{n\to\infty} \left|\frac{x}{n}\right| = 0.$$

The series is *everywhere convergent*; that is, it is convergent for all finite values of x.

26.20 $1 + x + 2x^2 + 3x^3 + \cdots$.

Here $|s_n| = |(n-1)x^{n-1}|$, $|s_{n+1}| = |nx^n|$, and $R = \lim_{n\to\infty} \left|\frac{nx^n}{(n-1)x^{n-1}}\right| = \lim_{n\to\infty} \left|\frac{n}{n-1} x\right| = |x|$.

The series converges on the interval $-1 < x < 1$ and diverges on the intervals $x < -1$ and $x > 1$.

When $x = 1$, the series is $1 + 1 + 2 + 3 + \cdots$ and is divergent.

When $x = -1$, the series is $1 - 1 + 2 - 3 + \cdots$ and is divergent.

The interval of convergence $-1 < x < 1$ is indicated in Fig. 26-2.

-1 1

Fig. 26-2

26.21 $1+\frac{x}{2}+\frac{x^2}{4}+\frac{x^3}{8}+\cdots$.

Here $|s_n|=\left|\frac{x^{n-1}}{2^{n-1}}\right|$, $|s_{n+1}|=\left|\frac{x^n}{2^n}\right|$, and $R=\lim_{n\to\infty}\left|\frac{x^n}{2^n}\cdot\frac{2^{n-1}}{x^{n-1}}\right|=\lim_{n\to\infty}\left|\frac{x}{2}\right|=\frac{1}{2}|x|$.

The series converges for all values of x such that $\frac{1}{2}|x|<1$, that is, for $-2<x<2$, and diverges for $x<-2$ and $x>2$.

For $x=2$, the series is $1+1+1+1+\cdots$ and is divergent.
For $x=-2$, the series is $1-1+1-1+\cdots$ and is divergent.

The interval of convergence $-2<x<2$ is indicated in Fig. 26-3.

Fig. 26-3

26.22 $\frac{1!}{x+1}-\frac{2!}{(x+1)^2}+\frac{3!}{(x+1)^3}+\cdots$.

For this series $|s_n|=\left|\frac{n!}{(x+1)^n}\right|$, $|s_{n+1}|=\left|\frac{(n+1)!}{(x+1)^{n+1}}\right|$, and $R=\lim_{n\to\infty}\left|\frac{(n+1)!}{(x+1)^{n+1}}\cdot\frac{(x+1)^n}{n!}\right|=\lim_{n\to\infty}\left|\frac{n+1}{x+1}\right|$ does not exist.

Thus, the series diverges for every value of x.

26.23 $\frac{x+3}{1\cdot 4}+\frac{(x+3)^2}{2\cdot 4^2}+\frac{(x+3)^3}{3\cdot 4^3}+\cdots$.

For this series $|s_n|=\left|\frac{(x+3)^n}{n\cdot 4^n}\right|$, $|s_{n+1}|=\left|\frac{(x+3)^{n+1}}{(n+1)\cdot 4^{n+1}}\right|$, and

$$R=\lim_{n\to\infty}\left|\frac{(x+3)^{n+1}}{(n+1)4^{n+1}}\cdot\frac{n4^n}{(x+3)^n}\right|=\lim_{n\to\infty}\left|\frac{n}{n+1}\cdot\frac{x+3}{4}\right|=\frac{1}{4}|x+3|.$$

The series converges for all values of x such that $\frac{1}{4}|x+3|<1$, that is, for $-4<x+3<4$ or $-7<x<1$, and diverges for $x<-7$ and $x>1$.

For $x=-7$, the series is $-1+\frac{1}{2}-\frac{1}{3}+\cdots$ and is convergent.
For $x=1$, the series is $1+\frac{1}{2}+\frac{1}{3}+\cdots$ and is divergent.

The interval of convergence $-7\le x<1$ is indicated in Fig. 26-4.

Fig. 26-4

26.24 $\frac{1}{1\cdot 2\cdot 3}-\frac{(x+2)^2}{2\cdot 3\cdot 4}+\frac{(x+2)^4}{3\cdot 4\cdot 5}-\cdots$.

Here $|s_n|=\left|\frac{(x+2)^{2n-2}}{n(n+1)(n+2)}\right|$, $|s_{n+1}|=\left|\frac{(x+2)^{2n}}{(n+1)(n+2)(n+3)}\right|$, and

$$R=\lim_{n\to\infty}\left|\frac{(x+2)^{2n}}{(n+1)(n+2)(n+3)}\cdot\frac{n(n+1)(n+2)}{(x+2)^{2n-2}}\right|=\lim_{n\to\infty}\left|\frac{n}{n+3}(x+2)^2\right|=(x+2)^2.$$

The series converges for all values of x such that $(x+2)^2<1$, that is, for $-3<x<-1$, and diverges for $x<-3$ and $x>-1$.

For $x = -3$ and $x = -1$ the series is $\frac{1}{1\cdot 2\cdot 3} - \frac{1}{2\cdot 3\cdot 4} + \frac{1}{3\cdot 4\cdot 5} - \cdots$ and is convergent.

The interval of convergence $-3 \le x \le -1$ is indicated in Fig. 26-5.

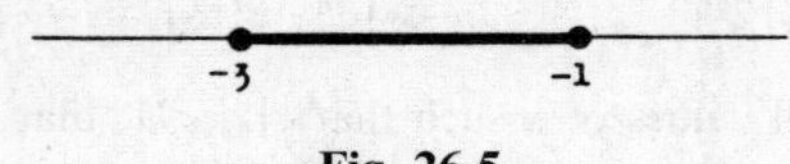

Fig. 26-5

26.25 Expand $(1+x)^{-1}$ as a power series in x and examine for convergence.

By division, $\frac{1}{x+1} = 1 - x + x^2 - x^3 + x^4 - \cdots$. Then $R = \lim_{n\to\infty} \left|\frac{x^{n+1}}{x^n}\right| = |x|$.

The series converges for $-1 < x < 1$, and diverges for $x < -1$ and $x > 1$.

For $x = 1$, the series is $1 - 1 + 1 - 1 + \cdots$ and is divergent.

For $x = -1$, the series is $1 + 1 + 1 + 1 + \cdots$ and is divergent.

The interval of convergence $-1 < x < 1$ is indicated in Fig. 26-6.

Fig. 26-6

Thus, the series $1 - x + x^2 - x^3 + x^4 - \cdots$ *represents* the function $f(x) = (1+x)^{-1}$ for all x such that $|x| < 1$. It does not represent the function for, say, $x = -4$. Note that $f(-4) = -\frac{1}{3}$, while for $x = -4$ the series is $1 + 4 + 16 + 64 + \cdots$.

26.26 Expand $(1+x)^{1/2}$ in a power series in x and examine for convergence.

By the binomial theorem

$$(1+x)^{1/2} = 1 + \frac{1}{2}x + \frac{(\frac{1}{2})(-\frac{1}{2})}{1\cdot 2}x^2 + \frac{(\frac{1}{2})(-\frac{1}{2})(-\frac{3}{2})}{1\cdot 2\cdot 3}x^3 + \frac{(\frac{1}{2})(-\frac{1}{2})(-\frac{3}{2})(-\frac{5}{2})}{1\cdot 2\cdot 3\cdot 4}x^4 + \cdots$$

Except for $n = 1$,

$$|s_n| = \left|\frac{(\frac{1}{2})(-\frac{1}{2})(-\frac{3}{2})\cdots(-2n+5)/2}{(n-1)!}x^{n-1}\right|,$$

$$|s_{n+1}| = \left|\frac{(\frac{1}{2})(-\frac{1}{2})(-\frac{3}{2})\cdots(-2n+3)/2}{n!}x^n\right|, \quad \text{and} \quad R = \lim_{n\to\infty}\left|\frac{-2n+3}{2n}x\right| = |x|.$$

The series is convergent for $-1 < x < 1$, and divergent for $x < -1$ and $x > 1$. An investigation at the end points is beyond the scope of this book.

Supplementary Problems

26.27 Write the first terms of the sequence whose general term is

(*a*) $\frac{1}{1+n}$ (*c*) $\frac{1}{3^n}$ (*e*) $\frac{n^2}{3n-2}$ (*g*) $(-1)^{n+1}\frac{1}{n!}$

(*b*) $\frac{1}{n+n\sqrt{n}}$ (*d*) $\frac{2n-1}{2n+3}$ (*f*) $(-1)^{n+1}\frac{2n+1}{2^{n+1}}$ (*h*) $\frac{n^2}{(2n)!}$

Ans. (*a*) $\frac{1}{2}, \frac{1}{3}, \frac{1}{4}, \frac{1}{5}$ (*c*) $\frac{1}{3}, \frac{1}{9}, \frac{1}{27}, \frac{1}{81}$ (*f*) $\frac{3}{4}, -\frac{5}{8}, \frac{7}{16}, -\frac{9}{32}$

(*b*) $\frac{1}{2}, \frac{1}{1+2\sqrt{2}}, \frac{1}{1+3\sqrt{3}}, \frac{1}{9}$ (*d*) $\frac{1}{5}, \frac{3}{7}, \frac{5}{9}, \frac{7}{11}$ (*g*) $1, -\frac{1}{2}, \frac{1}{6}, -\frac{1}{24}$

(*e*) $1, 1, \frac{9}{7}, \frac{8}{5}$ (*h*) $\frac{1}{2}, \frac{1}{6}, \frac{1}{80}, \frac{1}{2520}$

26.28 Write the general term of each sequence.

(*a*) $1, \frac{1}{3}, \frac{1}{5}, \frac{1}{7}, \ldots$ (*c*) $1, \frac{3}{5}, \frac{2}{5}, \frac{5}{17}, \frac{3}{13}, \ldots$ (*f*) $\frac{1}{3}, -\frac{1}{15}, \frac{1}{35}, -\frac{1}{63}, \ldots$

(*b*) $\frac{4}{1\cdot 3}, \frac{5}{2\cdot 4}, \frac{6}{3\cdot 5}, \frac{7}{4\cdot 6}, \ldots$ (*d*) $\frac{1}{2}, \frac{3}{8}, \frac{7}{24}, \frac{15}{64}, \ldots$ (*g*) $\frac{1}{2}, \frac{3}{8}, \frac{5}{16}, \frac{9}{32}, \ldots$

(*e*) $2, 1, \frac{8}{9}, 1, \frac{32}{25}, \frac{16}{9}, \ldots$ (*h*) $\frac{1}{2}, -x^2/4, x^4/6, -x^6/8, \ldots$

Ans. (*a*) $\frac{1}{2n-1}$ (*c*) $\frac{n+1}{n^2+1}$ (*e*) $\frac{2^n}{n^2}$ (*g*) $\frac{1+2^{n-1}}{2^{n+1}}$

(*b*) $\frac{n+3}{n(n+2)}$ (*d*) $\frac{2^n-1}{n\cdot 2^n}$ (*f*) $\frac{(-1)^{n+1}}{(2n-1)(2n+1)}$ (*h*) $(-1)^{n+1}\frac{x^{2n-2}}{2n}$

26.29 Evaluate. (*a*) $\lim_{n\to\infty}\left(2-\frac{1}{n}\right)$ (*d*) $\lim_{n\to\infty}\frac{2n^2+5n-6}{n^2+n-1}$ (*g*) $\lim_{n\to\infty}\frac{1}{2^n+1}$

(*b*) $\lim_{n\to\infty}\frac{3n+1}{3n-2}$ (*e*) $\lim_{n\to\infty}\frac{n}{n+1}$ (*h*) $\lim_{n\to\infty}\frac{2^n+1}{2^{n+1}+1}$

(*c*) $\lim_{n\to\infty}\frac{2n}{(n+1)(n+2)}$ (*f*) $\lim_{n\to\infty}\frac{(3n)!n}{(3n+1)!}$

Ans. (*a*) 2 (*b*) 1 (*c*) 0 (*d*) 2 (*e*) 1 (*f*) $\frac{1}{3}$ (*g*) 0 (*h*) $\frac{1}{2}$

26.30 Explain why each of the following has no limit:

(*a*) $1, 3, 5, 7, 9, \ldots$ (*c*) $1, -2, 4, -8, 16, -32, \ldots$

(*b*) $1, 0, 1, 0, 1, 0, \ldots$ (*d*) $\frac{1}{25}, \frac{4}{25}, \frac{9}{25}, \frac{16}{25}, \ldots$

26.31 Write out the next four terms given the recursive formula:

(*a*) $a_1 = -1, a_{n+1} = \frac{1}{3}a_n$

(*b*) $a_1 = 2, a_2 = 3, a_{n+2} = a_{n+1} + \frac{a_n}{2}$

Ans. (*a*) $-\frac{1}{3}, -\frac{1}{9}, -\frac{1}{27}, -\frac{1}{81}$

(*b*) $2 + \frac{3}{2}, 3 + \frac{7}{4}, \frac{7}{2} + \frac{19}{8}, \frac{19}{4} + \frac{47}{16}$

26.32 Investigate each of the following series for convergence or divergence:

(*a*) $\frac{1}{3}+\frac{1}{6}+\frac{1}{11}+\cdots+\frac{1}{2^n+n}+\cdots$ (*d*) $\frac{2}{1!}+\frac{2^2}{2!}+\frac{2^3}{3!}+\cdots+\frac{2^n}{n!}+\cdots$

(*b*) $\frac{1}{2}+\frac{1}{4}+\frac{1}{6}+\cdots+\frac{1}{2n}+\cdots$ (*e*) $2+\frac{1}{2}+\frac{8}{27}+\frac{1}{4}+\cdots+\frac{2^n}{n^3}+\cdots$

(*c*) $1+\frac{1}{3}+\frac{1}{5}+\cdots+\frac{1}{2n-1}+\cdots$ (*f*) $1+\frac{1}{\sqrt[3]{2}}+\frac{1}{\sqrt[3]{3}}+\cdots+\frac{1}{\sqrt[3]{n}}+\cdots$

(*g*) $\frac{1}{2}+\frac{1}{2\cdot 2^2}+\frac{1}{3\cdot 2^3}+\cdots+\frac{1}{n\cdot 2^n}+\cdots$

(*h*) $\frac{1}{2}+\frac{2}{3}+\frac{3}{4}+\cdots+\frac{n}{n+1}+\cdots$

(*i*) $\frac{1}{1\cdot 3}+\frac{1}{3\cdot 5}+\frac{1}{5\cdot 7}+\cdots+\frac{1}{(2n-1)(2n+1)}+\cdots$

(*j*) $1+\frac{2^2+1}{2^3+1}+\frac{3^2+1}{3^3+1}+\cdots+\frac{n^2+1}{n^3+1}+\cdots$

Ans. (*a*) Convergent (*c*) Divergent (*e*) Divergent (*g*) Convergent (*i*) Convergent
(*b*) Divergent (*d*) Convergent (*f*) Divergent (*h*) Divergent (*j*) Divergent

26.33 Investigate the following alternating series for convergence or divergence:

(*a*) $\frac{1}{4}-\frac{1}{10}+\frac{1}{28}-\frac{1}{82}+\cdots$

(*b*) $2-\frac{3}{2}+\frac{4}{3}-\frac{5}{4}+\cdots$

(*c*) $1-\frac{1}{2}+\frac{1}{3}-\frac{1}{4}+\cdots$

(*d*) $\frac{1}{2}-\frac{2}{3}+\frac{3}{4}-\frac{4}{5}+\cdots$

(*e*) $\frac{1}{4}-\frac{3}{6}+\frac{5}{8}-\frac{7}{10}+\cdots$

(*f*) $\frac{2}{3}-\frac{3}{4}\cdot\frac{1}{2}+\frac{4}{5}\cdot\frac{1}{3}-\frac{5}{6}\cdot\frac{1}{4}+\cdots$

(*g*) $\frac{2}{2\cdot 3}-\frac{2^2}{3\cdot 4}+\frac{2^3}{4\cdot 5}-\frac{2^4}{5\cdot 6}+\cdots$

(*h*) $2-\frac{2^3}{3!}+\frac{2^5}{5!}-\frac{2^7}{7!}+\cdots$

Ans. (*a*) Abs. Conv. (*c*) Cond. Conv. (*e*) Divergent (*g*) Divergent
(*b*) Divergent (*d*) Divergent (*f*) Cond. Conv. (*h*) Abs. Conv.

In Problems 26.34–26.46 find the interval of convergence including the end points.

26.34 $1+x^2+x^4+\cdots+x^{2n-2}+\cdots$

Ans. $-1 < x < 1$

26.35 $\frac{x}{1\cdot 2}+\frac{x^2}{2\cdot 3}+\frac{x^3}{3\cdot 4}+\cdots+\frac{x^n}{n(n+1)}+\cdots$

Ans. $-1 \leq x \leq 1$

26.36 $\frac{x}{1\cdot 3}+\frac{x^2}{2\cdot 3^2}+\frac{x^3}{3\cdot 3^3}+\cdots+\frac{x^n}{n\cdot 3^n}+\cdots$

Ans. $-3 \leq x < 3$

26.37 $\frac{x}{1^2+1}+\frac{x^2}{2^2+1}+\frac{x^3}{3^2+1}+\cdots+\frac{x^n}{n^2+1}+\cdots$

Ans. $-1 \leq x \leq 1$

26.38 $\frac{x^2}{4}-\frac{x^4}{8}+\frac{x^6}{16}-\cdots+(-1)^{n-1}\frac{x^{2n}}{2^{n+1}}+\cdots$

Ans. $-\sqrt{2} < x < \sqrt{2}$

26.39 $\frac{1\cdot x}{2\cdot 1}-\frac{1\cdot 3\cdot x^3}{2\cdot 4\cdot 3}+\frac{1\cdot 3\cdot 5\cdot x^5}{2\cdot 4\cdot 6\cdot 5}-\frac{1\cdot 3\cdot 5\cdot 7\cdot x^7}{2\cdot 4\cdot 6\cdot 8\cdot 7}+\cdots$

Ans. $-1 \leq x \leq 1$

26.40 $\frac{1 \cdot 1}{3 \cdot 2}x + \frac{3 \cdot 2}{5 \cdot 3}x^2 + \frac{5 \cdot 3}{7 \cdot 4}x^3 + \frac{7 \cdot 4}{9 \cdot 5}x^4 + \cdots$

Ans. $-1 < x < 1$

26.41 $(x-2) + \frac{1}{2}(x-2)^2 + \frac{1}{3}(x-2)^3 + \frac{1}{4}(x-2)^4 + \cdots$

Ans. $1 \le x < 3$

26.42 $\frac{x+1}{1 \cdot 2} - \frac{(x+1)^2}{3 \cdot 2^2} + \frac{(x+1)^3}{5 \cdot 2^3} - \frac{(x+1)^4}{7.2^4} + \cdots$

Ans. $-3 < x \le 1$

26.43 $\frac{x-a}{b} + \frac{(x-a)^2}{b^2} + \frac{(x-a)^3}{b^3} + \frac{(x-a)^4}{b^4} + \cdots$

Ans. $a - b < x < a + b$

26.44 $\frac{x-2}{x} + \frac{1}{2}\left(\frac{x-2}{x}\right)^2 + \frac{1}{3}\left(\frac{x-2}{x}\right)^3 + \frac{1}{4}\left(\frac{x-2}{x}\right)^4 + \cdots$

Ans. $x \ge 1$

26.45 $x + x^2/2^2 + x^3/3^3 + x^4/4^4 + \cdots + x^n/n^n + \cdots$

Ans. All values of x.

26.46 $x - 2^2x^2 + 3^3x^3 - 4^4x^4 + \cdots$

Ans. $x = 0$

Rectangular Coordinates in Space

RECTANGULAR COORDINATES IN SPACE. Consider the three mutually perpendicular planes of Fig. A-1. These three planes (the xy plane, the xz plane, the yz plane) are called the *coordinate planes*; their three lines of intersection are called the *coordinate axes* (the x axis, the y axis, the z axis); and their common point O is called the *origin*. Positive direction is indicated on each axis by an arrow-tip.

(NOTE: The coordinate system of Fig. A-1 is called a left-handed system. When the x and y axes are interchanged, the system becomes right-handed.)

The coordinate planes divide the space into eight regions, called *octants*. The octant whose edges are $\overrightarrow{Ox}, \overrightarrow{Oy}, \overrightarrow{Oz}$ is called the *first octant*; the other octants are not numbered.

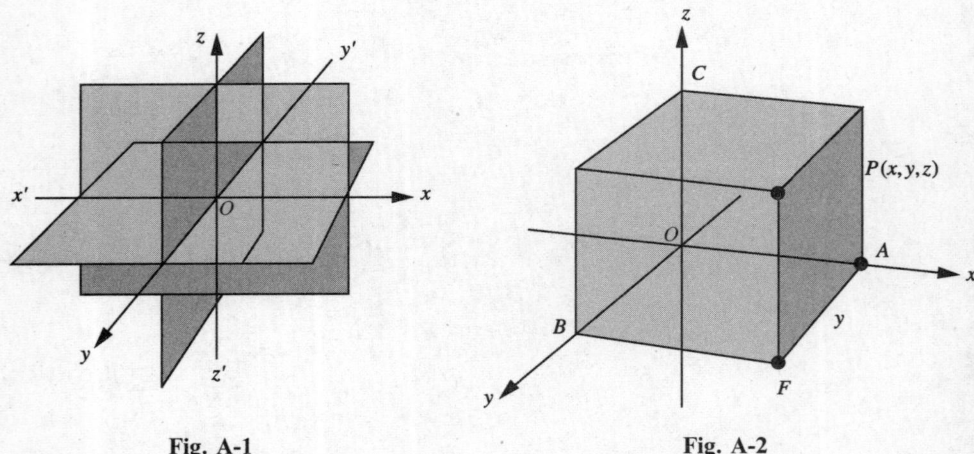

Fig. A-1

Fig. A-2

Let P be any point in space, not in a coordinate plane, and through P pass planes parallel to the coordinate planes meeting the coordinate axes in the points A, B, C and forming the rectangular parallelepiped of Fig. A-2. The directed distances $x = OA$, $y = OB$, $z = OC$ are called, respectively, the x coordinate, the y coordinate, the z coordinate of P and we write $P(x, y, z)$.

Since $\overline{AF} \cong \overline{OB}$ and $\overline{FP} \cong \overline{OC}$, it is preferable to use the three edges $\overline{OA}, \overline{AF}, \overline{FP}$ instead of the complete parallelepiped in locating a given point.

EXAMPLE 1. Locate the points:
(*a*) $(2, 3, 4)$ (*b*) $(-2, -2, 3)$ (*c*) $(2, -2, -3)$

As standard procedure in representing on paper the left-handed system, we shall draw $\angle xOz$ measuring 90° and $\angle xOy$ measuring 135°. Then distances on parallels to the x and z axes will be drawn to full scale while distances parallel to the y axis will be drawn about $\frac{7}{10}$ of full scale.

(*a*) From the origin move 2 units to the right along the x axis to $A(2, 0, 0)$, from A move 3 units forward parallel to the y axis to $F(2, 3, 0)$, and from F move 4 units upward parallel to the z axis to $P(2, 3, 4)$. See Fig. A-3(*a*).
(*b*) From the origin move 2 units to the left along the x axis to $A(-2, 0, 0)$, from A move 2 units backward parallel to the y axis to $F(-2, -2, 0)$, and from F move 3 units upward parallel to the z axis to $P(-2, -2, 3)$. See Fig. A-3(*b*).
(*c*) From the origin move 2 units to the right along the x axis to $A(2, 0, 0)$, from A move 2 units backward parallel to the y axis to $F(2, -2, 0)$, and from F move 3 units downward parallel to the z axis to $P(2, -2, -3)$. See Fig. A-3(*c*).

(See Problem A.1.)

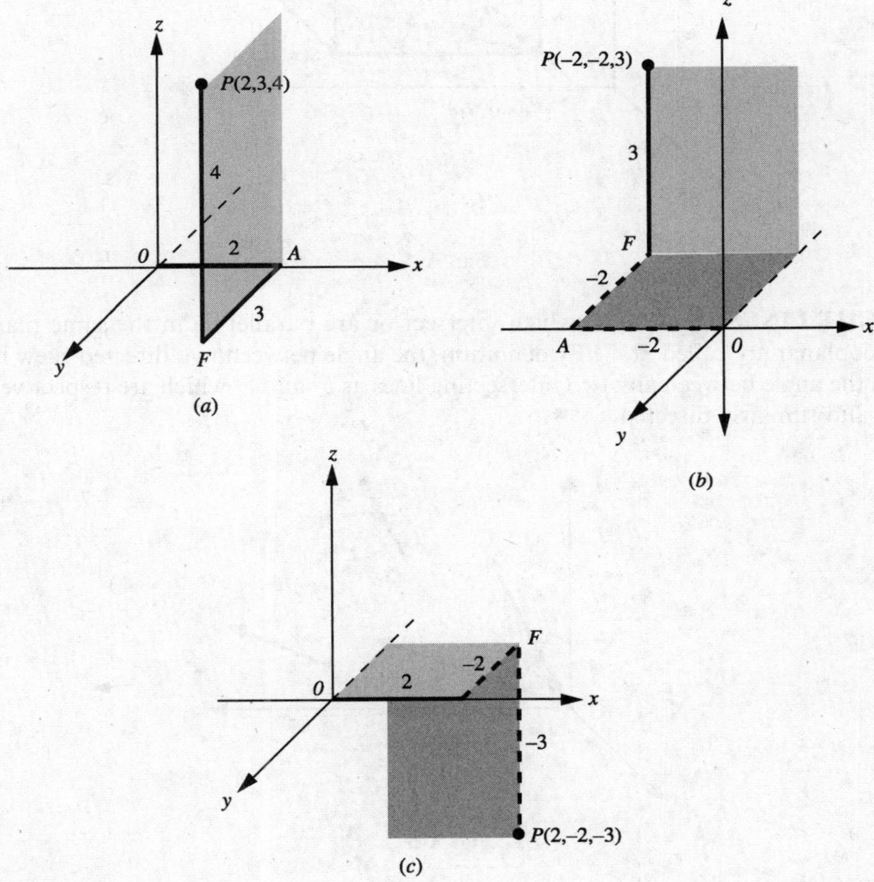

Fig. A-3

THE DISTANCE BETWEEN TWO POINTS $P_1(x_1, y_1, z_1)$ and $P_2(x_2, y_2, z_2)$ is from Fig. A-4

$$d = P_1P_2 = \sqrt{(P_1R)^2 + (RP_2)^2} = \sqrt{(P_1S)^2 + (SR)^2 + (RP_2)^2} = \sqrt{(x_2 - x_1)^2 + (y_2 - y_1)^2 + (z_2 - z_1)^2} \tag{A.1}$$

(See Problem A.2.)

IF $P_1(x_1, y_1, z_1)$ **AND** $P_2(x_1, y_2, z_2)$ are the end points of a line segment and if $P(x, y, z)$ divides the segment in the ratio $P_1P/PP_2 = r_1/r_2$, then

$$x = \frac{r_2x_1 + r_1x_2}{r_1 + r_2}, \quad y = \frac{r_2y_1 + r_1y_2}{r_1 + r_2}, \quad z = \frac{r_2z_1 + r_1z_2}{r_1 + r_2} \tag{A.2}$$

The coordinates of the midpoint of $\overline{P_1P_2}$ are $\left(\frac{1}{2}(x_1 + x_2),\ \frac{1}{2}(y_1 + y_2),\ \frac{1}{2}(z_1 + z_2)\right)$. (See Problems A.3–A.4.)

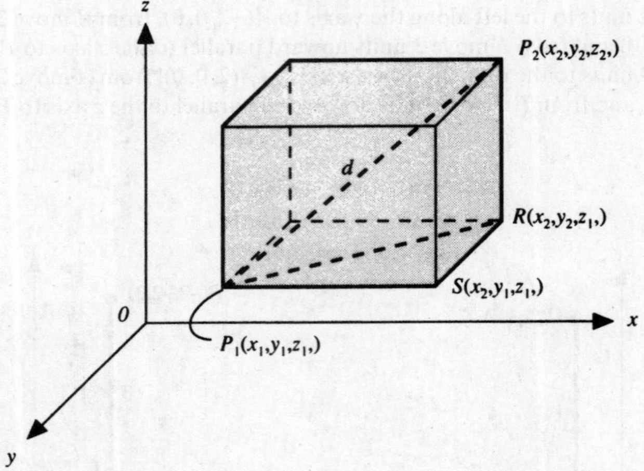

Fig. A-4

TWO STRAIGHT LINES IN SPACE which intersect or are parallel lie in the same plane; two lines which are not coplanar are called *skew*. By definition, the angle between two directed skew lines as b and c in Fig. A-5 is the angle between any two intersecting lines as b' and c' which are respectively parallel to the skew lines and similarly directed.

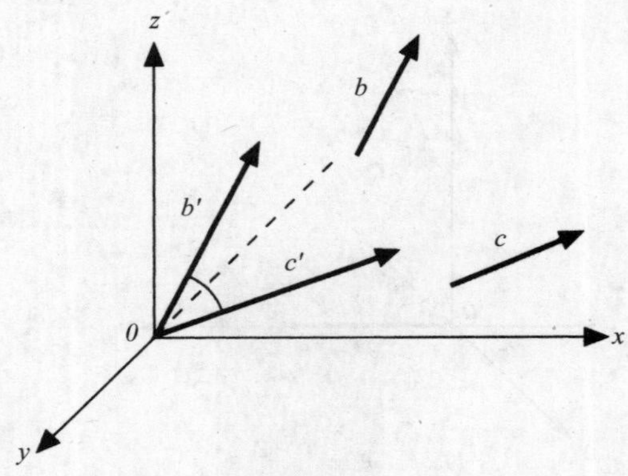

Fig. A-5

DIRECTION COSINES OF A LINE. In the plane a directed line l [positive direction upward in Figs. A-6(*a*) and (*b*)] forms the angles α and β with the positive directions on the x and y axes. However, in our study of the line in the plane we have favored the angle α over the angle β, calling it the angle of inclination of the line and its tangent the slope of the line.

In Fig. A-6(*a*), $\alpha + \beta = \frac{1}{2}\pi$ and $m = \tan\alpha = \dfrac{\sin\alpha}{\cos\alpha} = \dfrac{\sin(\frac{1}{2}\pi - \beta)}{\cos\alpha} = \dfrac{\cos\beta}{\cos\alpha}$, and in Fig. A-6(*b*), $\alpha = \frac{1}{2}\pi + \beta$ and $\tan\alpha = \dfrac{\sin(\frac{1}{2}\pi + \beta)}{\cos\alpha} = \dfrac{\cos\beta}{\cos\alpha}$. Now the angles α and β, called *direction angles* of the line, or their cosines, $\cos\alpha$ and $\cos\beta$, called *direction cosines* of the line, might have been used instead of the slope to give the direction of the line l. Indeed, it will be the direction cosines which will be generalized in our study of the straight line in space.

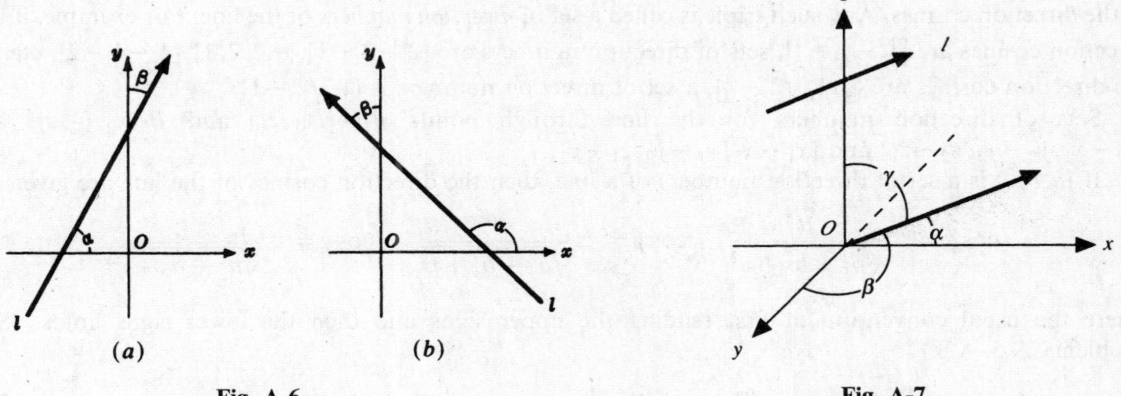

Fig. A-6 Fig. A-7

The direction of a line in space will be given by the three angles, called *direction angles* of the line, which it or that line through the origin parallel to it makes with the coordinate axes. If, as in Fig. A-7, the direction angles α, β, γ, where $O \leq \alpha, \beta, \gamma < \pi$, are the respective angles between the positive directions on the x, y, z axis and the directed line l (positive direction upward), the direction angles of this line when oppositely directed are $\alpha' = \pi - \alpha, \beta' = \pi - \beta, \gamma' = \pi - \gamma$. Thus, an undirected line in space has two sets of direction angles α, β, γ and $\pi - \alpha,\ \pi - \beta, \pi - \gamma$, and two sets of direction cosines $[\cos\alpha, \cos\beta, \cos\gamma]$ and $[-\cos\alpha, -\cos\beta, -\cos\gamma]$ since $\cos(\pi - \phi) = -\cos\phi$. To avoid confusion with the coordinates of a point, the triples of direction cosines of a line will be enclosed in a bracket. Thus we shall write $l : [A, B, C]$ to indicate the line whose direction cosines are the triple A, B, C.

The direction cosines of the l determined by points $P_1(x_1, y_1, z_1)$ and $P_2(x_2, y_2, z_2)$ and directed from P_1 to P_2 are (see Fig. A-4)

$$\cos\alpha = \cos\angle P_2P_1S = \frac{P_1S}{P_1P_2} = \frac{x_2 - x_1}{d}, \quad \cos\beta = \frac{y_2 - y_1}{d}, \quad \cos\gamma = \frac{z_2 - z_1}{d}.$$

when l is directed from P_2 to P_1, the direction cosines are

$$\left[\frac{x_1 - x_2}{d}, \frac{y_1 - y_2}{d}, \frac{z_1 - z_2}{d}\right]$$

Except for the natural preference for $[\frac{1}{3}, \frac{2}{3}, \frac{2}{3}]$ over $[-\frac{1}{3}, -\frac{2}{3}, -\frac{2}{3}]$, it is immaterial which set of direction cosines is used when dealing with an undirected line.

EXAMPLE 2. Find the two sets of direction cosines and indicate the positive direction along the line passing through the points $P_1(3, -1, 2)$ and $P_2(5, 2, -4)$.

We have $d = \sqrt{(2)^2 + (3)^2 + (-6)^2} = 7$. One set of direction cosines is

$$\left[\frac{x_2 - x_1}{d}, \frac{y_2 - y_1}{d}, \frac{z_2 - z_1}{d}\right] = \left[\frac{2}{7}, \frac{3}{7}, -\frac{6}{7}\right]$$

the positive direction being from P_1 to P_2. When the line is directed from P_2 to P_1, the direction cosines are $[-\frac{2}{7}, -\frac{3}{7}, \frac{6}{7}]$.

The sum of the squares of the direction cosines of any line is equal to 1; i.e.,

$$\cos^2 \alpha + \cos^2 \beta + \cos^2 \gamma = 1$$

It follows immediately that at least one of the direction cosines of any line is different from 0.

DIRECTION NUMBERS OF A LINE. Instead of the direction cosines of a line, it is frequently more convenient to use any triple of numbers, preferably small integers when possible, which are proportional to the direction cosines. Any such triple is called a set of *direction numbers* of the line. For example, if the direction cosines are $[\frac{2}{3}, -\frac{2}{3}, -\frac{1}{3}]$, sets of direction numbers are $[2, -2, -1]$, $[-2, 2, 1]$, $[4, -4, -2]$, etc.; if the direction cosines are $[\frac{1}{2}, 1/\sqrt{2}, -\frac{1}{2}]$, a set of direction numbers is $[1, \sqrt{2}, -1]$.

Sets of direction numbers for the line through points $P_1(x_1, y_1, z_1)$ and $P_2(x_2, y_2, z_2)$ are $[x_2 - x_1, y_2 - y_1, z_2 - z_1]$ and $[x_1 - x_2, y_1 - y_2, z_1 - z_2]$.

If $[a, b, c]$ is a set of direction numbers of a line, then the direction cosines of the line are given by

$$\cos\alpha = \pm\frac{a}{\sqrt{a^2 + b^2 + c^2}}, \quad \cos\beta = \pm\frac{b}{\sqrt{a^2 + b^2 + c^2}}, \quad \cos\gamma = \pm\frac{c}{\sqrt{a^2 + b^2 + c^2}} \qquad (A.3)$$

where the usual convention of first reading the upper signs and then the lower signs holds. (See Problems A.5–A.8.)

THE ANGLE θ BETWEEN TWO DIRECTED LINES

$$l_1: \ [\cos\alpha_1, \cos\beta_1, \cos\gamma_1] \quad \textit{and} \quad l_2: \ [\cos\alpha_2, \cos\beta_2, \cos\gamma_2]$$

is given by

$$\cos\theta = \cos\alpha_1 \cos\alpha_2 + \cos\beta_1 \cos\beta_2 + \cos\gamma_1 \cos\gamma_2 \qquad (A.4)$$

(For a proof see Problem A.9.)

If the two lines are parallel then $\theta = 0$ or π, according as the lines are similarly or oppositely directed, and $\cos\alpha_1 \cos\alpha_2 + \cos\beta_1 \cos\beta_2 + \cos\gamma_1 \cos\gamma_2 = \pm 1$. If the sign is $+$, then $\cos\alpha_1 = \cos\alpha_2, \cos\beta_1 = \cos\beta_2, \cos\gamma_1 = \cos\gamma_2$; if the sign is $-$, then $\cos\alpha_1 = -\cos\alpha_2, \cos\beta_1 = -\cos\beta_2, \cos\gamma_1 = -\cos\gamma_2$. Thus, two undirected lines are parallel if and only if their direction cosines are the same or differ only in sign. In terms of direction numbers, *two lines are parallel if and only if corresponding direction numbers are proportional.*

If the two lines are perpendicular, then $\theta = \frac{1}{2}\pi$ or $3\pi/2$, according as the lines are similarly or oppositely directed, and

$$\cos\alpha_1 \cos\alpha_2 + \cos\beta_1 \cos\beta_2 + \cos\gamma_1 \cos\gamma_2 = 0 \qquad (A.5)$$

In terms of direction numbers, *two lines with direction number $[a_1, b_1, c_1]$ and $[a_2, b_2, c_2]$, respectively, are perpendicular if and only if*

$$a_1 \cdot a_2 + b_1 \cdot b_2 + c_1 \cdot c_2 = 0. \qquad (A.5')$$

(See Problems A.10–A.12.)

THE DIRECTION NUMBER DEVICE. If l_1: $[a_1, b_1, c_1]$ and l_2: $[a_2, b_2, c_2]$ are two nonparallel lines, then a set of direction number $[a, b, c]$ of any line perpendicular to both l_1 and l_2 is given by

$$a = \begin{vmatrix} b_1 & c_1 \\ b_2 & c_2 \end{vmatrix}, \quad b = \begin{vmatrix} c_1 & a_1 \\ c_2 & a_2 \end{vmatrix}, \quad c = \begin{vmatrix} a_1 & b_1 \\ a_2 & b_2 \end{vmatrix}.$$

These three determinants can be obtained readily as follows:

(1) Write the two sets of direction numbers in three columns $\begin{matrix} a_1 & b_1 & c_1 \\ a_2 & b_2 & c_2 \end{matrix}$.

(2) Repeat the first two columns to obtain $\begin{matrix} a_1 & b_1 & c_1 & a_1 & b_1 \\ a_2 & b_2 & c_2 & a_2 & b_2 \end{matrix}$ and strike out the first column to have $\begin{matrix} \not{a_1} & b_1 & c_1 & a_1 & b_1 \\ \not{a_2} & b_2 & c_2 & a_2 & b_2 \end{matrix}$.

Then a is the determinant of the first and second columns remaining, b is the determinant of the second and third columns, and c is the determinant of the third and fourth columns. This procedure will be called the *direction number device*. Note, however, that it is a mechanical procedure for obtaining one solution of two homogeneous equations in three unknowns and thus has other applications.

EXAMPLE 3. Find a set of direction numbers $[a, b, c]$ of any line perpendicular to $l_1 : [2, 3, 4]$ and $l_2 : [1, -2, -3]$. Using the direction number device, we write $\begin{matrix} \not{2} & 3 & 4 & 2 & 3 \\ \not{1} & -2 & -3 & 1 & -2 \end{matrix}$. Then

$$a = \begin{vmatrix} 3 & 4 \\ -2 & -3 \end{vmatrix} = -1, \quad b = \begin{vmatrix} 4 & 2 \\ -3 & 1 \end{vmatrix} = 10, \quad c = \begin{vmatrix} 2 & 3 \\ 1 & -2 \end{vmatrix} = -7.$$

A set of direction numbers is $[-1, 10, -7]$ or, if preferred, $[1, -10, 7]$. (See Problem A.13.)

Solved Problems

A.1 What is the locus of a point:

(*a*) Whose z coordinate is always 0?

(*b*) Whose z coordinate is always 3?

(*c*) Whose x coordinate is always -5?

(*d*) Whose x and y coordinates are always 0?

(*e*) Whose x coordinate is always 2 and whose y coordinate is always 3?

(*a*) All points $(a, b, 0)$ lie in the xy plane; the locus is that plane.

(*b*) Every point is 3 units above the xy plane; the locus is the plane parallel to the xy plane and 3 units above it.

(*c*) A plane parallel to the yz plane and 5 units to the left of it.

(*d*) All points $(0, 0, c)$ lie on the z axis; the locus is that line.

(*e*) In locating the point $P(2, 3, c)$, the x and y coordinates are used to locate the point $F(2, 3, 0)$ in the xy plane and then a distance $|c|$ is measured from F parallel to the z axis. The locus is the line parallel to the z axis passing through the point $(2, 3, 0)$ in the xy plane.

A.2 (*a*) Find the distance between the points $P_1(-1, -3, 3)$ and $P_2(2, -4, 1)$.

(*b*) Find the perimeter of the triangle whose vertices are $A(-2, -4, -3)$, $B(1, 0, 9)$, $C(2, 0, 9)$.

(*c*) Show that the points $A(1, 2, 4)$, $B(4, 1, 6)$, and $C(-5, 4, 0)$ are collinear.

(*a*) Here $d = \sqrt{(x_2 - x_1)^2 + (y_2 - y_1)^2 + (z_2 - z_1)^2} = \sqrt{[2 - (-1)]^2 + [-4 - (-3)]^2 + (1 - 3)^2} = \sqrt{14}$.

(*b*) We find $AB = \sqrt{[1 - (-2)]^2 + [0 - (-4)]^2 + [9 - (-3)]^2} = 13$,

$BC = \sqrt{(2 - 1)^2 + (0 - 0)^2 + (9 - 9)^2} = 1$, and $CA = \sqrt{(-2 - 2)^2 + (-4 - 0)^2 + (-3 - 9)^2} = 4\sqrt{11}$.

The perimeter is $13 + 1 + 4\sqrt{11} = 14 + 4\sqrt{11}$.

(c) Here $AB = \sqrt{(3)^2 + (-1)^2 + (2)^2} = \sqrt{14}$, $BC = \sqrt{(-9)^2 + (3)^2 + (-6)^2} = 3\sqrt{14}$, and $CA = \sqrt{(6)^2 + (-2)^2 + (4)^2} = 2\sqrt{14}$. Since $BC = CA + AB$, the points are collinear.

A.3 Find the coordinates of the point P of division for each pair of points and given ratio. Find also the midpoint of the segment. (a) $P_1(3, 2, -4)$, $P_2(6, -1, 2)$; 1 : 2 (b) $P_1(2, 5, 4)$, $P_2(-6, 3, 8)$; $-3 : 5$.

(a) Here $r_1 = 1$ and $r_2 = 2$. Then

$$x = \frac{r_2x_1 + r_1x_2}{r_1 + r_2} = \frac{2 \cdot 3 + 1 \cdot 6}{1 + 2} = 4, \qquad y = \frac{r_2y_1 + r_1y_2}{r_1 + r_2} = \frac{2 \cdot 2 + 1(-1)}{1 + 2} = 1, \qquad z = \frac{r_2z_1 + r_1z_2}{r_1 + r_2} = -2$$

and the required point is $P(4, 1, -2)$. The midpoint has coordinates $(\frac{1}{2}(x_1 + x_2), \frac{1}{2}(y_1 + y_2), \frac{1}{2}(z_1 + z_2)) = (\frac{9}{2}, \frac{1}{2}, -1)$.

(b) Here $r_1 = -3$ and $r_2 = 5$. Then

$$x = \frac{5 \cdot 2 + (-3)(-6)}{-3 + 5} = 14, \quad y = \frac{5 \cdot 5 + (-3)3}{-3 + 5} = 8, \quad z = \frac{5 \cdot 4 + (-3)8}{-3 + 5} = -2$$

and the required point is $P(14, 8, -2)$. The midpoint has coordinates $(-2, 4, 6)$.

A.4 Prove: The three lines joining the midpoints of the opposite edges of a tetrahedron pass through a point P which bisects each of them.

Let the tetrahedron, shown in Fig. A-8, have vertices $O(0, 0, 0)$, $A(a, 0, 0)$, $B(b, c, 0)$, and $C(d, e, f.)$. The midpoints of $\overline{OB}$ and $\overline{AC}$ are, respectively, $D(\frac{1}{2}b, \frac{1}{2}c, 0)$ and $E(\frac{1}{2}(a + d), \frac{1}{2}e, \frac{1}{2}f)$, and the midpoint of DE is $P(\frac{1}{4}(a + b + d), \frac{1}{4}(c + e), \frac{1}{4}f)$. The midpoints of OA and BC are, respectively, $F(\frac{1}{2}a, 0, 0)$ and $G(\frac{1}{2}(b + d), \frac{1}{2}(c + e), \frac{1}{2}f)$, and the midpoint of FG is P. It is left for the reader to find the midpoints H and I of $\overline{OC}$ and $\overline{AB}$, and show that P is the midpoint of $\overline{HI}$.

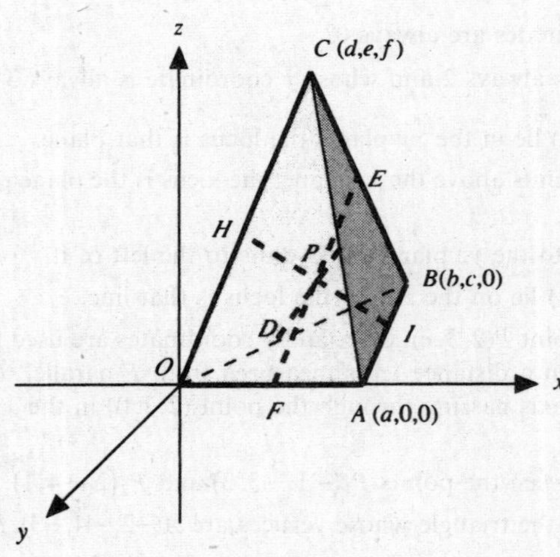

Fig. A-8

A.5 Find the direction cosines of the line:

(a) Passing through $P_1(3, 4, 5)$ and $P_2(-1, 2, 3)$ and directed from P_1 to P_2

(b) Passing through $P_1(2, -1, -3)$ and $P_2(-4, 2, 1)$ and directed from P_2 to P_1

(*c*) Passing through $O(0,0,0)$ and $P(a,b,c)$ and directed from O to P

(*d*) Passing through $P_1(4,-1,2)$ and $P_2(2,1,3)$ and directed so that γ is acute

(*a*) We have

$$\cos\alpha = \frac{x_2 - x_1}{d} = \frac{-4}{2\sqrt{6}}, \qquad \cos\beta = \frac{y_2 - y_1}{d} = \frac{-2}{2\sqrt{6}}, \qquad \cos\gamma = \frac{z_2 - z_1}{d} = \frac{-2}{2\sqrt{6}}.$$

The direction cosines are $\left[-\frac{2}{\sqrt{6}}, -\frac{1}{\sqrt{6}}, -\frac{1}{\sqrt{6}}\right]$.

(*b*)
$$\cos\alpha = \frac{x_1 - x_2}{d} = \frac{6}{\sqrt{61}}, \qquad \cos\beta = \frac{y_1 - y_2}{d} = \frac{-3}{\sqrt{61}}, \qquad \cos\gamma = \frac{z_1 - z_2}{d} = \frac{-4}{\sqrt{61}}.$$

The direction cosines are $\left[\frac{6}{\sqrt{61}}, -\frac{3}{\sqrt{61}}, -\frac{4}{\sqrt{61}}\right]$.

(*c*) $\cos\alpha = \frac{a-0}{d} = \frac{a}{\sqrt{a^2+b^2+c^2}}, \quad \cos\beta = \frac{b-0}{d} = \frac{b}{\sqrt{a^2+b^2+c^2}}, \quad \cos\gamma = \frac{c-0}{d} = \frac{c}{\sqrt{a^2+b^2+c^2}}$

The direction cosines are $\left[\frac{a}{\sqrt{a^2+b^2+c^2}}, \frac{b}{\sqrt{a^2+b^2+c^2}}, \frac{c}{\sqrt{a^2+b^2+c^2}}\right]$.

(*d*) The two sets of direction cosines of the undirected line are

$$\cos\alpha = \pm\tfrac{2}{3}, \quad \cos\beta = \mp\tfrac{2}{3}, \quad \cos\gamma = \mp\tfrac{1}{3}$$

one set being given by the upper signs and the other by the lower signs. When γ is acute, $\cos\gamma > 0$; hence the required set is $[-\frac{2}{3}, \frac{2}{3}, \frac{1}{3}]$.

A.6 Given the direction angles α measuring 120° and β measuring 45°, find γ if the line is directed upward.

$\cos^2\alpha + \cos^2\beta + \cos^2\gamma = \cos^2 120° + \cos^2 45° + \cos^2\gamma = (-\frac{1}{2})^2 + (1\sqrt{2})^2 + \cos^2\gamma = 1$. Then $\cos^2\gamma = \frac{1}{4}$ and $\cos\gamma = \pm\frac{1}{2}$. When the line is directed upward, $\cos\gamma = \frac{1}{2}$ and $\gamma = 60°$.

A.7 The direction numbers of a line l are given as $[2,-3,6]$. Find the direction cosines of l when directed upward.

The direction cosines of l are given by

$$\cos\alpha = \pm\frac{a}{\sqrt{a^2+b^2+c^2}} = \pm\frac{2}{7}, \quad \cos\beta = \pm\left(\frac{-3}{7}\right), \quad \cos\gamma = \pm\frac{6}{7}.$$

When γ is acute, $\cos\gamma > 0$, and the direction cosines are $[\frac{2}{7}, -\frac{3}{7}, \frac{6}{7}]$.

A.8 Use direction numbers to show that the points $A(1,2,4)$, $B(4,1,6)$, and $C(-5,4,0)$ are collinear. [See Problem A.2(*c*).]

A set if direction numbers of the line $\overleftrightarrow{AB}$ is $[3,-1,2]$, for BC is $[-9,3,-6]$. Since the two sets are proportional, the lines are parallel; since the lines have a point in common they are coincident and the points are collinear.

A.9 Prove: the angle θ between two directed lines l_t: $[\cos\alpha_1, \cos\beta_1, \cos\gamma_1]$ and l_2: $[\cos\alpha_2, \cos\beta_2, \cos\gamma_2]$ is given by $\cos\theta = \cos\alpha_1\cos\alpha_2 + \cos\beta_1\cos\beta_2 + \cos\gamma_1\cos\gamma_2$.

The angle θ is by definition the angle between two lines issuing from the origin parallel, respectively, to the given lines l_t and l_2 and similarly directed.

Consider the triangle OP_1P_2, in Fig. A-9, whose vertices are the origin and the points $P_1(\cos\alpha_1, \cos\beta_1, \cos\gamma_t)$ and $P_2(\cos\alpha_2, \cos\beta_2, \cos\gamma_2)$. The line segment OP_1 is of length 1 (why?) and is parallel to l_1; similarly, OP_2 is of length 1 and is parallel to l_2. Thus, $\angle P_1OP_2 = \theta$. By the Law of Cosines, $(P_1P_2)^2 = (OP_1)^2 + (OP_2)^2 - 2(OP_1)(OP_2)\cos\theta$ and $\cos\theta = \cos\alpha_1\cos\alpha_2 + \cos\beta_1\cos\beta_2 + \cos\gamma_1\cos\gamma_2$.

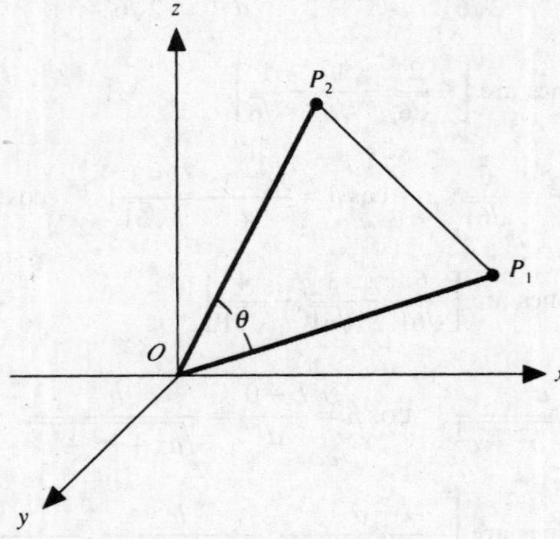

Fig. A-9

A.10 (*a*) Find the angle between the directed lines l_1: $[\frac{2}{7}, \frac{3}{7}, \frac{6}{7}]$ and l_2: $[\frac{2}{3}, -\frac{1}{3}, \frac{2}{3}]$.

(*b*) Find the acute angle between the lines l_1: $[-2, 1, 2]$ and l_2: $[2, -6, -3]$.

(*c*) The line l_1 passes through $A(5, -2, 3)$ and $B(2, 1, -4)$, and the line l_2 passes through $C(-4, 1, -2)$ and $D(-3, 2, 3)$. Find the acute angle between them.

(*a*) We have $\cos\theta = \cos\alpha_1\cos\alpha_2 + \cos\beta_1\cos\beta_2 + \cos\gamma_1\cos\gamma_2 = \frac{2}{7}\cdot\frac{2}{3} + \frac{3}{7}(-\frac{1}{3}) + \frac{6}{7}\cdot\frac{2}{3} = \frac{13}{21} = 0.619$ and $\theta = 51°50'$.

(*b*) Since $\sqrt{(-2)^2 + (1)^2 + (2)^2} = 3$, we take $[-\frac{2}{3}, \frac{1}{3}, \frac{2}{3}]$ as direction cosines of l_1. Since $\sqrt{(2)^2 + (-6)^2 + (-3)^2} = 7$, we take $[\frac{2}{7}, -\frac{6}{7}, -\frac{3}{7}]$ as direction cosines of l_2. Then $\cos\theta = -\frac{2}{3}\cdot\frac{2}{7} + \frac{1}{3}(-\frac{6}{7}) + \frac{2}{3}(-\frac{3}{7}) = -\frac{16}{21} = -0.762$, and $\theta = 139°40'$. The required angle is $40°20'$.

(*c*) Take $[3\sqrt{67}, -3\sqrt{67}, 7/\sqrt{67}]$ as direction cosines of l_1 and $[1/3\sqrt{3}, 1/3\sqrt{3}, 5/3\sqrt{3}]$ as direction cosines of l_2. Then

$$\cos\theta = \frac{1}{\sqrt{3\cdot 67}} - \frac{1}{\sqrt{3\cdot 67}} + \frac{35}{3\sqrt{3\cdot 67}} = \frac{35}{3\sqrt{201}} = 0.823 \quad \text{and} \quad \theta = 34°40'$$

A.11 (*a*) Show that the line joining $A(9, 2, 6)$ and $B(5, -3, 2)$ and the line joining $C(-1, -5, -2)$ and $D(7, 5, 6)$ are parallel.

(*b*) Show that the line joining $A(7, 2, 3)$ and $B(-2, 5, 2)$ and the line joining $C(4, 10, 1)$ and $D(1, 2, 4)$ are mutually perpendicular.

(*a*) Here $[9-5, 2-(-3), 6-2] = [4, 5, 4]$ is a set of direction numbers of AB and $[-1-7, -5-5, -2-6] = [-8, -10, -8]$ is a set of direction number of CD. Since the two sets are proportional, the two lines are parallel.

(*b*) Here $[9, -3, 1]$ is a set of direction numbers of AB and $[3, 8, -3]$ is a set of direction numbers of CD. Since [see Equation (*A.5*)] $9\cdot 3 + (-3)8 + 1(-3) = 0$, the lines are perpendicular.

A.12 Find the area of the triangle whose vertices are $A(4, 2, 3)$, $B(7, -2, 4)$ and $C(3, -4, 6)$.

The area of triangle ABC is given by $\frac{1}{2}(AB)(AC)\sin A$. We have $AB = \sqrt{26}$ and $AC = \sqrt{46}$.

To find $\sin A$, we direct the sides $\overline{AB}$ and $\overline{AC}$ away from the origin as in Fig. A-10. Then $\overline{AB}$ has direction cosines $[3/\sqrt{26}, -4/\sqrt{26}, 1/\sqrt{26}]$, $\overline{AC}$ has direction cosines $[-1/\sqrt{46}, -6/\sqrt{46}, 3/\sqrt{46}]$,

$$\cos A = \frac{3}{\sqrt{26}} \cdot \frac{-1}{\sqrt{46}} + \frac{-4}{\sqrt{26}} \cdot \frac{-6}{\sqrt{46}} + \frac{1}{\sqrt{26}} \cdot \frac{3}{\sqrt{46}} = \frac{24}{\sqrt{26}\sqrt{46}}$$

and
$$\sin A = \sqrt{1 - \cos^2 A} = \frac{2\sqrt{155}}{\sqrt{26}\sqrt{46}}$$

The required area is $\frac{1}{2}\sqrt{26} \cdot \sqrt{46} \cdot \dfrac{2\sqrt{155}}{\sqrt{26}\sqrt{46}} = \sqrt{155}$.

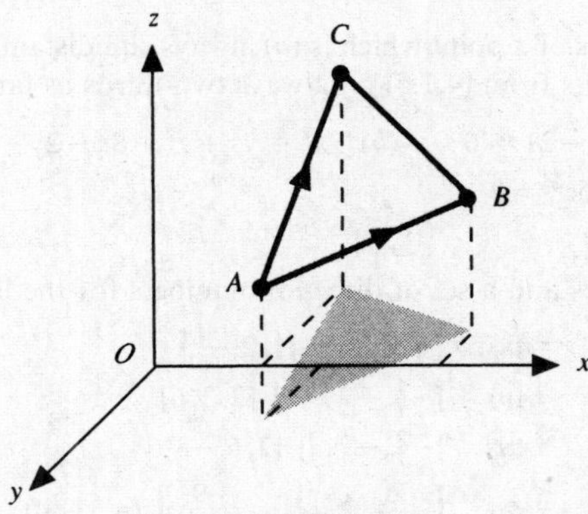

Fig. A-10

A.13 Find a set of direction numbers for any line which is perpendicular to

(*a*) l_1: $[1 - 2, -3]$ and l_2: $[4, -1, -5]$

(*b*) The triangle whose vertices are $A(4,2,3)$, $B(7,-2,4)$, and $C(3,-4,6)$

(*a*) Using the direction number device

$$\begin{matrix} \not{1} & -2 & -3 & 1 & -2 \\ \not{4} & -1 & -5 & 4 & -1 \end{matrix}, \quad \text{we obtain} \quad a = \begin{vmatrix} -2 & -3 \\ -1 & -5 \end{vmatrix} = 7, \quad b = \begin{vmatrix} -3 & 1 \\ -5 & 4 \end{vmatrix} = -7, \quad c = \begin{vmatrix} 1 & -2 \\ 4 & -1 \end{vmatrix} = 7.$$

Thus, a set of direction numbers is $[7, -7, 7]$; a simpler set is $[1, -1, 1]$.

(*b*) Since the triangle lies in a plane determined by the lines AB and AC, we seek direction numbers for any line perpendicular to these lines. For AB and AC, respective sets of direction numbers are $[3, -4, 1]$ and $[-1, -6, 3]$. Using the direction number device

$$\begin{matrix} \not{3} & -4 & 1 & 3 & -4 \\ \not{-1} & -6 & 3 & -1 & -6 \end{matrix}, \quad a = \begin{vmatrix} -4 & 1 \\ -6 & 3 \end{vmatrix} = -6, \quad b = \begin{vmatrix} 1 & 3 \\ 3 & -1 \end{vmatrix} = -10, \quad c = \begin{vmatrix} 3 & -4 \\ -1 & -6 \end{vmatrix} = -22.$$

Then $[-6, -10, -22]$ is a set of direction numbers and $[3,5,11]$ is a simpler set.

Supplementary Problems

A.14 Find the undirected distance between each pair of points:

(*a*) (4, 1, 5) and (2, −1, 4)	(*b*) (9, 7, −2) and (6, 5, 4)	(*c*) (9, −2, −3) and (−3, 4, 0).

Ans. (*a*) 3	(*b*) 7	(*c*) $3\sqrt{21}$

A.15 Find the undirected distance of each of the following points from (*i*) the origin, (*ii*) the *x* axis, (*iii*) the *y* axis, and (*iv*) the *z* axis: (*a*) (2,6,–3) (*b*) $(2,-\sqrt{3},3)$.

Ans. (*a*) 7, 2, 6, 3 (*b*) $4, 2, \sqrt{3}, 3$

A.16 For each pair of points, find the coordinates of the point dividing $\overline{P_1P_2}$ in the given ratio; find also the coordinates of the midpoint.

(*a*) $P_1(4,1,5), P_2(2,-1,4), 3:2$ *Ans.* $(\frac{14}{5},-\frac{1}{5},\frac{22}{5}),(3,0,\frac{9}{2})$

(*b*) $P_1(9,7,-2)$, $P_2(6,5,4)1:4$ *Ans.* $(\frac{42}{5},\frac{33}{5},-\frac{4}{5}),(\frac{15}{2},6,1)$

(*c*) $P_1(9,-2,-3), P_2(-3,4,0), -1:3$ *Ans.* $(15,-5,-\frac{9}{2}),(3,1,-\frac{3}{2})$

(*d*) $P_1(0,0,0), P_2(2,3,4), 2:-3$ *Ans.* $(-4-6,-8),(1,\frac{3}{2},2)$

A.17 Find the equation of the locus of a point which is (*a*) always equidistant from the points (4,1,5) and (2,–1,4) (*b*) always at a distance 6 units from (4,1,5) (*c*) always two-thirds as far from the *y* axis as from the origin.

Ans. (*a*) $4x+4y+2z-21=0$ (*b*) $x^2+y^2+z^2-8x-2y-10z+6=0$
(*c*) $5x^2-4y^2+5z^2=0$

A.18 Find a set of direction cosines and a set of direction numbers for the line joining P_1 and P_2, given

(*a*) $P_1(0,0,0), P_2(4,8,-8)$ *Ans.* $[\frac{1}{3},\frac{2}{3},-\frac{2}{3}],[1,2,-2]$

(*b*) $P_1(1,3,5), P_2(-1,0,-1)$ *Ans.* $[-\frac{2}{7},-\frac{3}{7},-\frac{6}{7}],[2,3,6]$

(*c*) $P_1(5,6,-3), P_2(1,-6,3)$ *Ans.* $[-\frac{2}{7},-\frac{6}{7},\frac{3}{7}],[2,6,-3]$

(*d*) $P_1(4,2,-6), P_2(-2,1,3)$ *Ans.* $\left[\frac{6}{\sqrt{118}},\frac{1}{\sqrt{118}},\frac{-9}{\sqrt{118}}\right],[6,1,-9]$

A.19 Find cos γ, given (*a*) $\cos\alpha=\frac{14}{15},\cos\beta=-\frac{1}{3}$ (*b*) $\alpha=60°,\beta=135°$

Ans. (*a*) $\pm\frac{2}{15}$ (*b*) $\pm\frac{1}{2}$

A.20 Find

(*a*) The acute angle between the line having direction numbers [–4, –1, –8] and the line joining the points (6,4,–1) and (4,0,3).

(*b*) The interior angles of the triangle whose vertices are *A*(2,–1,0), *B*(4,1,–1), *C*(5,–1,–4).

Ans. (*a*) 68°20′ (*b*) *A* = 48°10′, *B* = 95°10′, *C* = 36°40′

A.21 Find the coordinates of the point *P* in which the line joining *A*(5,–1,4) and *B*(–5, 7, 0) pierces the *yz* plane.

Hint: Let *P* have coordinates (0,*b*, *c*) and express the condition (see Problem A.8) that *A*, *B*, *P* be collinear.

Ans. *P*(0,3,2)

A.22 Find relations which the coordinates of $P(x,y,z)$ must satisfy if *P* is to be collinear with (2,3,1) and (1,–2,–5).

Ans. $x-2:y-3:z-1=1:5:6$ or $\frac{x-2}{1}=\frac{y-3}{5}=\frac{z-1}{6}$

A.23 Find a set of direction numbers for any line perpendicular to

(*a*) Each of the lines l_1: [1,2,–4] and l_2: [2, –1,3]

(*b*) Each of the lines joining *A*(2, –1, 5) to *B*(–1, 3, 4) and *C*(0, –5, 4)

Ans. (*a*) [2, –11, –5] (*b*) [8, 1, –20]

A.24 Find the coordinates of the point P in which the line joining the points $A(4,11,18)$ and $B(-1,-4,-7)$ intersects the line joining the points $C(3,1,5)$ and $D(5,0,7)$.

Hint: Let P divide $\overline{AB}$ in the ratio $1:r$ and $\overline{CD}$ in the ratio $1:s$, and obtain relations $rs - r - 4s - 6 = 0$, etc.

Ans. (1,2,3)

A.25 Prove that the four line segments joining each vertex of a tetrahedron to the point of intersection of the medians of the opposite face have a point G in common. Prove that each of the four line segments is divided in the ratio 1:3 by G.

(NOTE: G, the point P of Problem A.4, is called the centroid of the tetrahedron.)

Appendix B

Units and Dimensions

The **standard units of measurement** in physics are the System International (SI) units, which use meters (m) for length, kilograms (kg) for mass, and seconds (s) for time.

Some common **conversions** to and from alternative units of measurement, such as English units (using, for example, inches, feet, miles, pounds) are shown below.

These conversions may be accomplished by multiplying by various forms of "1".

For example, to convert 3 years into seconds, one multiplies by several forms of "1", as shown below:

$$3\ \text{y} = 3\ \text{y} \times \left(\frac{365.25\ \text{d}}{\text{y}}\right) \times \left(\frac{24\ \text{h}}{\text{d}}\right) \times \left(\frac{60\ \text{m}}{\text{h}}\right)\left(\frac{60\ \text{s}}{\text{m}}\right) = 9.47 \times 10^7\ \text{s}$$

Note that one cancels various units as if they were numbers.

Greek prefixes may be used to represent both positive and negative powers of 10 (see below and Appendix G).

In combining quantities with different units, one needs to check that the final quantity has the expected units. For example, the momentum p is equal to the product of the mass and the velocity (p = mv). In SI units, m has units of kg and v has units of m/s, and so p must have units of kg m/s.

Some useful conversions

Length

1 km = 1000 m
1 m = 100 cm = 1000 mm = 10^6 μm
1 cm = .01 m
1 mm = .001 m
1 μm = 10^{-6} m
1 km = .621 mi
1 m = 1.094 yd
1 cm = .3937 in.
1 yd = 36 in.
1 mi = 1760 yd
1 mi = 1.609 km
1 yd = .9144 m
1 in. = 2.54 cm

Mass, Force, and Weight

1 kg = 1000 g
1 kg = 2.20 lb
1 dyne = 10^{-5} N
1 pound = 4.448 N
1 lb = .454 kg

Energy

1 electron-volt (eV) = 1.602×10^{-19} J
1 erg = 10^{-7} J
1 calorie (cal) = 4.186 J
1 kilowatt-hour (kWh) = 3.6×10^6 J

Solving Physics Problems

While a vast diversity of physics problems exist, there are some general approaches that can be helpful. We choose to illustrate some of these approaches in the following problem, which involves multiple forces acting upon an object (Problem 18.9).

What is the acceleration of a 90 kg sled (including the rider) rolling down a frictionless incline at an angle of 30°? What is the normal force acting upon the sled?

We use this problem to illustrate a general method for addressing problems involving multiple forces on an object. Its solution is described below.

The situation is sketched in Fig. C-1.

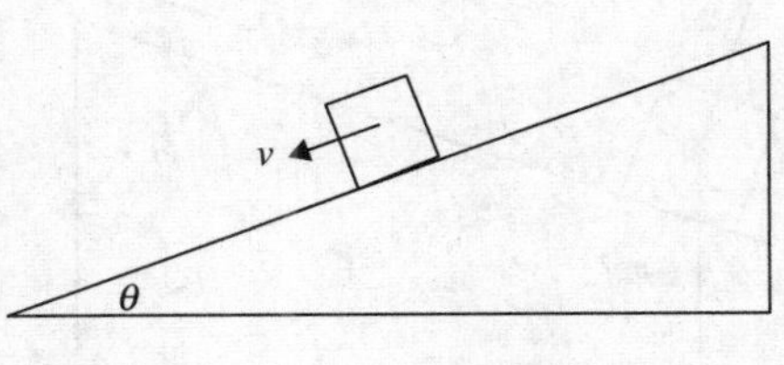

Fig. C-1

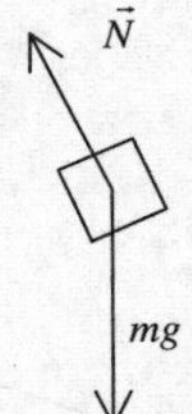

Fig. C-2

Problems of this sort—involving multiple forces and motion that is neither horizontal nor vertical—can often be effectively addressed through the following sequence of steps:

(1) Read the problem carefully—twice; and try to make an estimate (or even a wild guess) of what might be a rough answer;
(2) Draw a clear diagram;
(3) Indicate all forces acting upon the object of interest;
(4) Choose and label coordinate axes carefully;
(5) Decompose the forces into vector components along these axes;
(6) Apply Newton's Second Law (force = mass · acceleration) to each of the components;
(7) Solve the problem, avoiding the substitution of numbers as long as possible;
(8) Compare to your initial estimate or guess, and check that the units are correct.

Let us go through these steps for this case:

Step (1): Read the problem carefully—twice; and try to make an estimate (or even a wild guess) of what might be a rough answer.

This step could apply to any physics problem. Take your time to understand what you are being asked to find, what information is given and which physics principles may be relevant. Related concepts, examples from class, from the textbook or related problems can all be helpful. (Ultimately, there is no substitute for simply gaining experience in doing as many problems as possible—many problems will then begin to look familiar.)

In this case, we are asked to find *the acceleration of the sled down the incline.*

A rough guess as to the acceleration? Well, if it were falling vertically, the acceleration would be the acceleration of gravity, which is 9.8 m/s^2. It is moving down a slope instead, so the acceleration will be some fraction of that—let us guess about 3 m/s^2. We certainly do not expect the acceleration to be *greater* than if it were falling vertically!

Steps (2 and 3): Draw a clear diagram; indicate all forces of interest.

Draw a simple but clear diagram (Fig. C-1) to represent all the important features of the problem, including lengths, angles, velocities, etc.

It is often helpful to draw a *free-body diagram* (Fig. C-2) showing all of the forces acting on the sled. For our problem, this includes the force of gravity and the "normal force" of the incline against the sled (i.e., the force of the incline acting perpendicular to its surface).

Step (4): Choose and label coordinate axes carefully.

In situations like these, it is often convenient to choose coordinate axes that are, respectively, parallel and perpendicular to the direction of motion of the object; and to then label these axes in the diagram (Fig. 18-24).

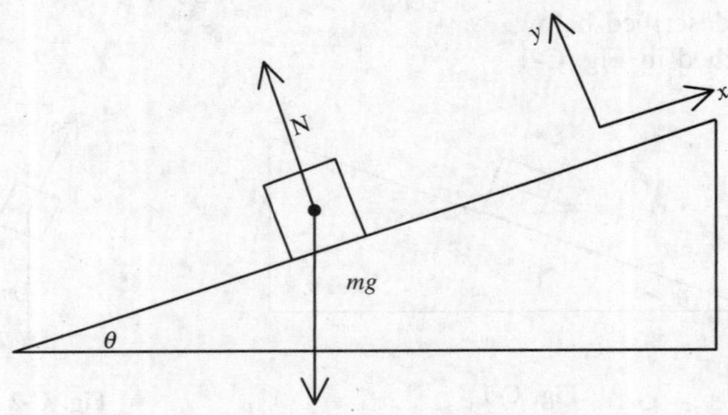

Fig. C-3

Step (5): Decompose forces along these coordinate axes.

Try to resolve each force vector into components acting along these axes. The normal force is already perpendicular to that motion, but the vertical force of gravity must be resolved into components as indicated in Fig. C-3.

In order to find these components of force along the chosen axes, one needs to recognize that the angle between the vertical and the downward going normal is equal to θ, the incline angle.* Recognizing this, and being able to draw the triangle in Fig. C-3 with the vertical force of gravity as the hypotenuse, is critical to the solution of many sloped incline problems.

---

* This is not difficult to see: the original angle θ is the angle between a horizontal line and the sloped surface. If one rotates these lines by 90°, then one gets the angle in question.

Step (6): Apply Newton's Second Law (force = mass · acceleration) to each of the components

We now apply Newton's Second Law of Motion, which states that the sum of the external forces acting upon a body is equal to the product of its mass and its acceleration. In equation form, this reads:

$$\vec{F}^{\text{total}} = m\vec{a}$$

where $\vec{F}^{\text{total}}$ is the sum of all external forces acting on the sled of mass m and $\vec{a}$ is the acceleration resulting from that force.

Resolving this equation into its components then yields†:

$$\vec{F}_x^{\text{total}} = ma_x$$
$$\vec{F}_y^{\text{total}} = ma_y$$

Recall that we are trying to find a_x—the acceleration of the sled down the incline. On the other hand, we know that a_y must equal zero, because the sled does not move in a direction perpendicular to the incline surface.

In the direction parallel to the incline, we have the component of gravity shown. In the direction perpendicular to the incline, we have both a component of gravity and the *normal force* of the incline on the ball, which, by definition, acts in a direction perpendicular to the incline.

Hence,

$$\vec{F}_x^{\text{total}} = -mg\sin\theta = ma_x$$
$$\vec{F}_y^{\text{total}} = \text{N} - mg\cos\theta = ma_y = 0$$

Step (7): Solve the problem, avoiding the substitution of numbers as long as possible.

$$-mg\sin\theta = ma_x \Rightarrow a_x = -g\sin\theta$$

$$\text{N} - mg\cos\theta = ma_y = 0 \Rightarrow \text{N} = mg\cos\theta$$

The acceleration of the sled down the incline, and the normal force on the sled, are therefore

$$a_x = g\sin\theta = \left(9.8\,\text{m/s}^2\right)(\sin 30°) = 4.9\,\text{m/s}^2$$

$$\text{N} = mg\cos\theta = (90\,\text{kg}\left(9.8\,\text{m/s}^2\right)\cos 30° = 763.8\,\text{N}$$

Step (8): Compare to your initial estimate or guess, and check that the units are correct.

The acceleration—the quantity we were asked to find—compares reasonably well to our initial guess of 3 m/s^2, and the SI units of acceleration and force are correctly indicated.

It is worth noting that the mathematics of this problem only required a few lines of algebra. The heart of the challenge was not in the mathematics but in recognizing the physical principles that apply (primarily Newton's Second Law in this case). Of course, being able to deal with the mathematics—the focus of this book—is also essential.

The reader is encouraged to review the steps presented in this problem.

† The force equation is also commonly expressed using $\sum \vec{F}_i = m\vec{a}$, where the summation symbol $\sum_i$ represents the sum over contributions from a total of "i" different forces.

Appendix D

Selected Physics Formulas

Density	$\rho = \text{mass}/\text{Volume}$
Average speed	$\bar{v} = \Delta x/\Delta t$
Instantaneous speed	dx/dt
Average acceleration	$\bar{a} = \Delta v/\Delta t$
Instantaneous acceleration	$a = d^2x/dt^2$
acceleration in uniform circular motion:	$a = v^2/r$
centripetal acceleration	$F = mv^2/r$
Gravity near Earth's surface	$F = mg$
Kinematic equations (with constant acceleration):	
	$x = x_0 + v_0 t + \frac{1}{2}at^2$
	$v = v_0 + at$
	$v^2 = v_0^2 + 2a(x - x_0)$
	$x = x_0 + \frac{(v_i + v)}{2}t$
Newton's Second law of Motion:	$\vec{F} = m\vec{a}$
Newton's Third Law of Motion	$\vec{F}_{12} = -\vec{F}_{21}$
Simple Harmonic Motion	$F = -kx$
Newton's Law of Gravitation	$F = G\frac{m_1 m_2}{r^2}$
Coulomb's Law	$F = \frac{1}{4\pi\varepsilon_0}\frac{q_1 q_2}{r^2}$
Current	$i = \frac{dq}{dt}$
Voltage across a resistor	$V = iR = R\frac{dq}{dt}$
Charge on a capacitor	$q = CV$
Potential difference across inductor	$V = -L\frac{di}{dt}$

Appendix E

Selected Physical Constants

Speed of light	c	2.99×10^8 m/s
Gravitational constant	G	$6.67 \times 10^{-11} \frac{\text{m}^3}{\text{kg s}^2}$
Electron mass	m_e	9.11×10^{-31} kg
Proton mass	m_p	1.67×10^{-27} kg
Neutron mass	m_n	1.67×10^{-27} kg
Electron charge	e	1.60×10^{-19} C
Planck's constant	h	6.63×10^{-34} J s
Permittivity constant	ε_0	$8.85 \times 10^{-12} \frac{\text{C}^2}{\text{N m}^2}$
Permeability constant	μ_0	$4\pi \times 10^{-7}$ T m/A
Bohr radius	a_0	$.529 \times 10^{-10}$ m

Some Astrophysical Constants

gravity at Earth's surface	g	$9.81 \frac{\text{m}}{\text{s}^2}$
equatorial radius of Earth	R_e	6.374×10^6 m
mass of Earth		5.98×10^{24} kg
mass of Moon		7.36×10^{22} kg
mass of Sun		1.99×10^{30} kg
Earth-moon distance		3.84×10^8 m
Earth-Sun distance (mean)		1.50×10^{11} m

Appendix F

Integration by Parts

Certain products and quotients of functions are easily integrated. Take, for example,

$$\int \frac{\ln x}{x} \mathrm{d}x$$

In this case, if we let

$$u = \ln x,$$

then $\mathrm{d}u = \frac{1}{x}\mathrm{d}x$.

Thus,

$$\int \frac{\ln x}{x} \mathrm{d}x = \int \frac{u}{x} x \, \mathrm{d}u = \int u \, \mathrm{d}u = \frac{u^2}{2} + C = \frac{\ln^2 x}{2} + C$$

However, note that this simple substitution method does not work for $\int x \ln x \, \mathrm{d}x$.

In this case, setting $u = \ln x$ simply makes the integrand more complicated. Clearly, a different technique is required. In the study of physics, the reader will come across many integrands containing products of functions. Consider the following statement of the Product Rule for Derivatives and the informal argument that follows:

$$\frac{\mathrm{d}}{\mathrm{d}x}(u \cdot v) = u\frac{\mathrm{d}v}{\mathrm{d}x} + v\frac{\mathrm{d}u}{\mathrm{d}x}$$

where u and v are differentiable functions of x. If one finds the antiderivative of each side of the above equation, one then finds that

$$u \cdot v = \int u\frac{\mathrm{d}v}{\mathrm{d}x} + \int v\frac{\mathrm{d}u}{\mathrm{d}x};$$

in other words,

$$\int u\frac{\mathrm{d}v}{\mathrm{d}x} = u \cdot v - \int v\frac{\mathrm{d}u}{\mathrm{d}x}$$

In simpler terms:

Integration by Parts

$$\int u\,\mathrm{d}v = uv - \int v\,\mathrm{d}u$$

where u and v are differentiable functions of x and $v\,\mathrm{d}u$ possesses an elementary antiderivative.

Let us turn to an example:

$$\int x\,e^x \mathrm{d}x$$

If we let $u = x$ and $\mathrm{d}v = e^x \mathrm{d}x$, then $\mathrm{d}u = \mathrm{d}x$ and $v = e^x$

Thus,

$$\begin{aligned}\int x\,e^x \mathrm{d}x &= uv - \int v\,\mathrm{d}u \\ &= x\,e^x - \int e^x\,\mathrm{d}x \\ &= x\,e^x - e^x + C \\ &= e^x(x-1) + C\end{aligned}$$

Another example in which the integrand appears to be quite simple is $\int \ln x\,\mathrm{d}x$.

Let $u = \ln x$ and $\mathrm{d}v = \mathrm{d}x$.

Then $\mathrm{d}u = \frac{1}{x}\mathrm{d}x$ and $v = x$;

Thus,

$$\begin{aligned}\int \ln x\,\mathrm{d}x &= uv - \int v\,\mathrm{d}u \\ &= x \ln x - x\left(\frac{1}{x}\right)\mathrm{d}x \\ &= x \ln x - x + C \\ &= x(\ln x - 1) + C\end{aligned}$$

This paves the way for the integrand that sparked this discussion:

$$\int x \ln x\,\mathrm{d}x$$

Let $u = x$ and $\mathrm{d}v = \ln x\,\mathrm{d}x$.

Then $du = dx$ and $v = \int \ln x\,\mathrm{d}x = x(\ln x - 1)$.

Thus, $\int x \ln x\,\mathrm{d}x = x \cdot x(\ln x - 1) - \int x(\ln x - 1)\mathrm{d}x$.

Then $\int x \ln x\,\mathrm{d}x = x^2(\ln x - 1) - \int x \ln x\,\mathrm{d}x + \int x\,\mathrm{d}x$

Hence, $\int x \ln x\,\mathrm{d}x = x^2(\ln x - 1) - \int x \ln x\,\mathrm{d}x + \frac{x^2}{2}$

$$2\int x \ln x \,\mathrm{d}x = x^2(\ln x - 1) + \frac{x^2}{2}$$

and finally,

$$\begin{aligned}\int x \ln x \,\mathrm{d}x &= \frac{x^2}{2}(\ln x - 1) + \frac{x^2}{4} + C \\ &= \frac{x^2}{2}\ln x - \frac{x^2}{2} + \frac{x^2}{4} + C \\ &= \frac{x^2}{2}\ln x - \frac{x^2}{4} + C \\ &= x^2\left(\frac{\ln x}{2} - \frac{1}{4}\right) + C\end{aligned}$$

The reader should differentiate this result to see that the result is, indeed, correct.

Solved Problems

1. Find the antiderivative of $\int \tan^{-1} x \,\mathrm{d}x$

Let $u = \tan^{-1} x$ and $\mathrm{d}v = \mathrm{d}x$

Then $v = x$ and $\mathrm{d}u = \frac{\mathrm{d}x}{1+x^2}$

Using $\int u\,\mathrm{d}v = uv - \int v\,\mathrm{d}u$, we then get

$$\int \tan^{-1} x \,\mathrm{d}x = x \tan^{-1} x - \int \frac{x}{1+x^2}\mathrm{d}x$$

To find $\int \frac{x}{1+x^2}\mathrm{d}x$, let $w = 1 + x^2$ and use a simple substitution. The final result is then

$$\int \tan^{-1} x = x \tan^{-1} x - \frac{1}{2}\ln(1 + x^2) + C$$

2. Find the antiderivative of $\int x e^x \mathrm{d}x$

Let $u = x$ and $\mathrm{d}v = e^x\,\mathrm{d}x$
Then $\mathrm{d}u = \mathrm{d}x$ and $v = e^x$
Using $\int u\,\mathrm{d}v = uv - \int v\,\mathrm{d}u$ with these substitutions, we then get

$$\begin{aligned}\int x e^x \mathrm{d}x &= x e^x - \int e^x \mathrm{d}x \\ &= x e^x - e^x + C \\ &= e^x(x - 1) + C\end{aligned}$$

Appendix G

The Greek Alphabet and Prefixes

The Greek Alphabet

Alpha	Α	α
Beta	Β	β
Gamma	Γ	γ
Delta	Δ	δ
Epsilon	Ε	ε
Zeta	Ζ	ς
Eta	Η	η
Theta	Θ	θ
Iota	Ι	ι
Kappa	Κ	κ
Lamda	Λ	λ
Mu	Μ	μ
Nu	Ν	ν
Xi	Ξ	ξ
Omicron	Ο	o
Pi	Π	π
Rho	Ρ	ρ
Sigma	Σ	σ
Tau	Τ	τ
Upsilon	ϒ	υ
Phi	Φ	ϕ
Chi	Χ	χ
Psi	Ψ	ψ
Omega	Ω	ω

Greek Prefixes

atto	10^{-18}
femto	10^{-15}
pico	10^{-12}
nano	10^{-9}
micro	10^{-6}
milli	10^{-3}
centi	10^{-2}
deci	10^{-1}
deka	10
hecto	10^{2}
kilo	10^{3}
mega	10^{6}
giga	10^{9}
tera	10^{12}
peta	10^{15}

INDEX

Abscissa, 13
Absolute inequality, 40
Absolute value, 151
Absolutely convergent series, 366
Acceleration, 167, 197–199, 241, 242, 245
  average change in velocity per unit time, 198
  first derivative, 198
  instantaneous, 198
  kinematic equations of motion, 199
  second derivative, 198
Activity, radioactive decay and, 138
Acute angle, 91–92
  adjacent side, 91–92
  hypotenuse, 91–92
  opposite side, 91–92
Addition
  algebraic vectors, 216
  complex numbers and, 149
  graphic representation of complex numbers, 150
Adjacent side, 91–92
Algebra, 1–7
  algebraic expression, 3
  elementary linear, 277–301
  positive integral exponents, 3
Algebraic
  additions, vectors and, 216
  expression, 3
  operations, 149–150
    addition, 149
    division, 150
    multiplication, 149
    subtraction, 149
  representation, vectors and, 215
  trigonometric functions and, 91
  solution
    elimination by addition, 24
    elimination by substitution, 24
    simultaneous linear equations and, 24
Alternating series 365–366
Amplitude, 97–98
Angles
  acute, 91–92
  between two directed lines, 382
  complementary, 92
  conterminal, 89, 93
Angles (*Cont.*):
  negative, 93
  quadrantal, 89, 91
  standard position, 89
    first quadrant angle, 89
  trigonometric functions and, 89–91
Antiderivatives, 200
  boundary, 200
  displacement function, 201
  indefinite integral, 200
  initial conditions, 200
  integrals and, 200
Approximation of irrational roots, 77–88
  Horner's method of approximation, 78–79
  method of successive linear approximations, 77–78
Area by summation, 169
  definite integral, 169
Area elements, 246
  Cartesian coordinates, 246
  polar coordinates, 246
Associativity, vectors and, 217
Asymptotes, 46–47
Average acceleration, 198
Average change in velocity per unit time, 198
Axis of symmetry, 45
  even, 46
  odd, 46
Azimuthal angle, 242

Base, positive integral exponents and, 3
Bijections, 9
Boundary conditions, 200, 320–321
  initial, 320

Calculators, 135
  graphing, 27
Calculus
  fundamental theorem of, 203
  multivariate, 264–276
  natural logarithm and, 135
  single variable functions, mathematics approach, 161–191
  vector, 302–318
Carbon dating, 139
Cartesian coordinates, 241, 246, 247

Cartesian coordinates (*Cont.*):
 cylindrical coordinate systems, transformation between, 243
 spherical coordinate systems, transformation between, 242
Center of circle, 61
Center of symmetry, 45
Circle, 61–69
 center, 61
 equation, 62
 general form of, 61–62
 locus, 61
 standard form of, 61–62
 tangent
  equation, 62
  length, 62
Circuit theory basics, 330
Closed surface, flux and, 315
Coefficients, 25–27
Cofactor of element, 280–281
Common logarithms, 135
Commutativity, vectors and, 217
Complementary angles, 92
Completing the square, 32–33
Complex conjugate, 73
Complex numbers, 149–160
 algebraic operations, 149–150
  addition, 149
  division, 150
  multiplication, 149
  subtraction, 149
 conjugate, 149
 De Moivre's theorem, 152
 graphic representation, 150
  addition, 150
  subtraction, 150
 polar form, 151
 pure imaginary
  numbers, 149
  part, 149
 real part, 149
 roots of, 153
Complex root, 73
Composition of sine curve, 98
Concavity, 166
Conditional inequality, 40
 solution of, 40–41
Conditionally convergent series, 366
Conjugate, complex number and, 149
Consistent and independent equations, determinant of
 order two, 277–278
 order three, 278–279
Consistent linear equations, 23
Conterminal angles, 89
Continuity
 closed interval, 162
Continuity (*Cont.*):
 equation of, 306–307
Continuous functions, 162
Convergence
 comparison test, 365
 ratio test, 365
Convergent condition, indicated sum and, 364–365
Coordinate axes, 378
Coordinate planes, 378
Coordinate systems, 235–263
 area elements, 246
 cylindrical, 243–245
 polar, 235–241
 rotating systems, 245–246
 spherical, 241–242
 translating systems, 245–246
 volume, 246–247
Cosecant θ, 90
Cosine θ, 90
Cosines, 380–382
 law of, 100
Cotangent θ, 90
Coterminal angles, 93
Cramer's rule, 283
Critical values, 165
Critically damped, 328
Cross products
 components, 221
 direction, 218–219
 magnitude, 218–219
 vectors and, 218–220
Curl of vector, 307–308
 cylindrical coordinates, 309
 spherical coordinates, 309
 Stokes' theorem, 307–308
Curve sketching, 237
 directions at the pole, 237–238
 extent, 237
 points on the locus, 238
 symmetry, 237
Cylindrical coordinates, 243–245, 247
 acceleration, 245
 Cartesian coordinates, transformation between, 243
 curl of vector and, 309
 divergence and, 309
 gradient and, 309
 vectors in, 243–244
 velocity, 245

Damping coefficient, 324
De Moivre's theorem, 152
Decay constant, 138
Decay rate, 138
Decay, radioactive, 138
Decreasing function, 164–165
Definite integral, 169

Definition, trigonometric functions and, 101–102
Degenerate, 45
Del operator, 302–305
Dependent events, mathematical probability and, 348
Dependent linear equations, 23
Dependent variable, 8, 264
  range of the function, 9
Derivative formulas, 195–197
  rules of, 196
Derivatives, 163
  difference quotient, 163
  displacement with respect to time, 194
  first, 192–195
  higher-order, 164
Descartes' Rule of Signs, 76, 78, 83, 84, 85
Detached coefficients, 25–27
Determinant of order two, 277
  consistent and independent equations, 277–278
  elements, 277
  principal diagonal, 277
  secondary diagonal, 277
  value of, 277
Determinants of order three, 278–279
  consistent and independent equations, 278–279
Determinants, 279–280
  cofactor of element, 280–281
  evaluation of, 281–282
  minor of given element, 279–280
  properties of, 281
  value of, 280
Difference quotient, 163
Differential elements, flux integration and, 314–315
Differential equations, 319–344
  boundary conditions, 320–321
  circuit theory basics, 330
    Kirchoff's law, 330
  critically damped, 328
  damping coefficient, 324
  electrical and mechanical systems, connection between, 330–331
  homogeneous, 319
  linear, 319
  non-homogeneous, 319
  order of, 319
  ordinary, 319
  overdamped, 328
  partial, 319, 331
  solving of, 321–330
  underdamped, 328
Differentials, single-variable functions and, 168
Differentiation
  formulas, 164
  multivariate functions and, 264–265
Diminishing roots of equation, 76–77
Direction angles, 381
Direction cosines of a line, 380–382
  direction angles, 381
Direction numbers, 382
  device, 382–383
Direction of motion, 167
Direction
  cross products and, 218–219
  vector and, 212
Directly vary, variation and, 20
Discriminant, 33
Displacement function, 201
Distance between two points, 380
Divergence, 305
  cylindrical coordinates, 309
  equation of continuity, 306–307
  positive, 305
  sinks, 305–306
  source, 305
  spherical coordinates, 309
Divergent condition, indicated sum and, 364–365
Division
  complex numbers and, 150
  polar form and, 152
Domain, functions and, 9
Dot product. *See* scalar product.

Elementary linear algebra, 277–301
  determinant, 279–280
    of order three, 278–279
    of order two, 277
  homogeneous equation, 283, 285
  nonhomogeneous equation, 283
  one linear and quadratic equation, 283–284
  systems of linear equations in unknowns, 282–283
    Cramer's rule, 283
  two quadratic equations, 284
Elementary probability, 345–361
  permutations, 345
  types of, 347–349
    empirical, 349
    mathematical, 347
Elements, determinants of order two and, 277
Empirical probability, 349
Equation of continuity, 306–307
Equation of tangent, circle and, 62
Equation
  definition of, 19
  locus, 45–50
  one linear and quadratic, 283–284
  polynomial, 73
  quadratic, 19
  straight line, 51
Even axis of symmetry, 46
Exponent, 3
  fractional, 4
  negative, 4

Exponent (*Cont.*):
  zero, 4
Exponential curve, 136
Exponential decay, 137
Exponential equation, logarithms and, 135
Exponential functions, 134–148
Exponential growth, 137
Extent, curve sketching and, 237
Extreme values, 193–194

Factoring, 32
Families of straight lines, 57–60
First derivative, acceleration and, 198
First derivatives, 192–195
First quadrant angle, 89
Flux, 313–315
  closed surface, 315
  integration over differential elements, 314–315
  sources, 315
    sinks, 316
Fractional exponent, 4
Functions, 8
  bijections, 9
  decreasing, 164–165
  definition of, 8
  dependent variable, 8
  domain, 9
  graphs of, 13–18
  increasing, 164–165
  independent variable, 8
  limit, 161–162
  multivalued, 8–9
  range of, 9
  single-valued, 8–9
  variable, 8
Fundamental theorem of calculus, 203
  applications of, 203–204

Gauss' law, 316
General angle, trigonometric functions and, 89–91
General form of circle, 61–62
General reduction formula, 94
General sine curve, 97–98
  amplitude, 97–98
  period, 97–98
General term of sequence, 362
General values, inverse trigonometric relations and, 104
Generalized ratio test, 366
Geometric representation of vectors, 212–213
Gradient
  cylindrical coordinates, 309
  operator, 302–305
    local density, 302
  spherical coordinates, 309
Graph representation, complex numbers and, 150
Graphical additions of vectors, 213–214
  parallelogram method, 214
Graphing calculator, 27
Graphs of functions, 13–18
  ordered pairs, 13
  rectangular cartesian coordinate system, 13
  roots, 15
Graphs of inverse trigonometric relations, 102
Graphs of trigonometric functions, 96–97
Graphs, simultaneous linear equations and, 24
Greek alphabet, 399
Greek prefixes, 399

Half-life, 138
Higher-order derivatives, 164
Homogeneous differential equation, 319
Homogeneous equation, 283, 285
  trivial solution, 283
Hooke's law, 325
Horner's method of approximation, 78–79, 85, 86
Hypotenuse, 91–92

Inconsistent linear equations, 23
Increasing function, 164–165
Increments, 162–163
Indefinite integral, 168, 200
Independent events, mathematical probability and, 348
Independent linear equations, 23
Independent variable, 8, 264
  domain, 9
Indicated sum, 364–365
  convergent condition, 364–365
  divergent condition, 364–365
Inequalities, 40–44
  absolute, 40
  conditional, 40
  same sense, 40
Infinite sequences, 362–364
  indicated sum, 364–365
  infinite series, 364–365
Infinite series of the form, 366–367
  interval of convergence, 366–367
  power series, 366–367
Infinite series, 362–377
  sequence, 362, 364–365
  series of
    positive terms, 365
    negative terms, 365–366
Inflection point of a curve, 166–167
Initial boundary conditions, 320
Initial conditions, 200
Instantaneous acceleration, 198
Instantaneous velocity, 193
  derivative of displacement with respect to time, 194
Integrals, 200
  antiderivatives and, 200
Integration by parts, 396–398

Integration, multivariate functions and, 264–268
Intercepts, 45
Intersections, 238–239
Interval of convergence, 366–367
Inverse functions, 101
Inverse trigonometric relations, 102–104
  general values, 104
  graphs, 102
Inversely vary, 20
Irrational root, 73
  approximation of, 77–88

Kinematic equations of motion, 199
Kirchoff's law, 330

Law of cosines, 100
Law of sines, 99
Length of tangent, circle and, 62
Limit theorems, 162
Limits, 73
  lower, 73
  upper, 73
Line representations, 94–95
Linear algebra, elementary, 277–301
Linear differential equations, 319
Linear equations, 19–22, 283–284
  directly vary, 20
  inversely vary, 20
  ratio and proportion, 19
  root, 19
  simultaneous, 23–31
  solution, 19
  three, 25
Local density, 302
Locus, 45–50
  asymptotes, 46–47
  circle, 61
  curve sketching and, 238
  definition, 45
  degenerate, 45
  intercepts, 45
  symmetry, 45–46
    axis of, 45
    center of, 45
Logarithmic curve, 136–137
  exponential decay, 137
  exponential growth, 137
  power law behavior, 139–140
  radioactive decay, 138
Logarithmic functions, 134–148
Logarithms
  calculators and, 135
  common, 135
  exponential
    curve, 136
    equation, 135
Logarithms, exponential (*Cont.*):
    functions, 136
  laws of, 134
  logarithmic curve, 136–137
  natural, 135
Lower limits, 73

Magnitude, 212
  cross products and, 218–219
Mathematical (theoretical) probability, 347–349
  dependent events, 348
  independent events, 348
  mutually exclusive, 347
Mathematics
  calculus of single variable functions and, 161–191
  single variable functions and
    acceleration, 167
    area by summation, 169
    concavity, 166
    continuity on closed interval, 162
    continuous functions 162
    critical values, 165
    decreasing function, 164–165
    derivative, 163
    differentials, 168
    differentiation formulas, 164
    direction of motion, 167
    function limit, 161–162
      higher-order derivatives, 164
    increasing function, 164–165
    increments, 162–163
    indefinite integral, 168
    inflection point of a curve, 166–167
    limit theorems, 162
    one-sided limit, 162
    relative maximum values, 165–166
    relative minimum values, 165–166
    speed, 167
    velocity, 167
Matrix, 25–27
Measurement, standard units of, 390
Method of successive linear approximations, 77–78
Minor of given element, 279–280
Modulus, 151
Multiplication
  complex numbers and, 149
  polar form and, 152
  vectors and, 216, 217
Multivalued function, 8–9
Multivariate calculus, 264–276
  functions of, 264
    dependent variable, 264
    independent variable, 264
Multivariate functions, 264–276
  differentiation, 264–265
  integration, 264–268

Mutually exclusive, probability and, 347

Natural logarithm, 135
calculus, 135
Negative angles, 93
Negative exponent, 4
Newton's Second Law, 227, 228, 229, 324, 325, 328, 329, 330, 393
Non-homogeneous differential equation, 319
Non-homogeneous equation, 283
Notation, 101
*N*th root, 3–4
principal, 4
Nuclear fission, 139–140
Null vector, 216

Oblique triangle, 98–99
Octants, 378
Odd axis of symmetry, 46
One parameter systems, 57
families of lines, 57
One-sided limit, 162
Opposite side angle 91–92
Order of differential equation, 319
Ordered pairs, 13
Ordinary differential equation, 319
Ordinate, 13
Origin, 13, 378
Overdamped, 328

Parabola, 32
Parallelogram method, 214
Parameter, straight line and, 57
one parameter systems, 57
Partial differential equations, 319, 331
separation of variables method, 331–333
Periodic functions, 97
Permutations, 345
calculating number of, 345–346
principles of, 345
Physical constants, 395
Physics formulas, 394
Physics problems
Newton's Second Law, 393
solving of, 391
Physics, single variable functions and, 192–211
acceleration, 197–199
antiderivatives, 200
derivative formulas, 195–197
first derivatives, 192–195
fundamental theorem of calculus, 203
integrals, 200
second derivatives, 197–199
velocity, 192–195
Polar angle, 242
Polar axis, 235
Polar coordinate systems, 235–241
acceleration, 241
curve sketching, 237
intersections, 238–239
slope of polar curve, 239–240
terminology, 235–236
polar axis, 235
pole, 235
radius vector, 235
vectorial angle, 235
vector representations, 240
velocity, 241
Polar coordinates
area elements and, 246
rectangular coordinates, transformations, 236
Polar curve, slope of, 238–240
Polar form (trigonometric form)
absolute value, 151
complex number, 151
division, 152
modulus, 151
multiplication, 152
rectangular form, 151
Pole, 235
Polynomial equations
approximation of irrational roots, 77–88
Descartes' Rule of Signs, 76
diminishing roots of, 76–77
rational roots, 74
root, 73
complex conjugate, 73
irrational, 73
limits, 73
rational, 74
simple, 73
standard form, 73
variation of sign, 75–76
Polynomial functions, 78–88
Positive divergence, 305
Positive integral exponents
base, 3
exponent, 3
*n*th roots, 3–4
principal, 4
Possible root, 74
Power law behavior, 139–140
nuclear fission, 139–140
Power series, 366–367
Principal diagonal, 277
Product identities, 220–221
Projection formulas, 99–100
Proportion, 19
Pure imaginary numbers, 149
Pure imaginary part, 149

Quadrantal angles, 89
trigonometric functions and, 91

Quadrants, 13
  angles, 89
Quadratic equation, 19, 32–44, 283–286
  completing the square, 32–33
  discriminant, 33
  factoring, 32
  graphing calculator, 34
  quadratic form, 33
  radicals, 33
  sum and products of roots, 33
Quadratic form, 33
Quadratic function, 32–44
  parabola, 32
  vertices, 32

Radial distance, 241
Radicals, 33
Radioactive decay, 138
  activity, 138
  carbon dating, 139
  decay constant, 138
  half-life, 138
  rate, 138
Radius vector, 235
Range of the function, 9
Range, 8
Ratio test
  convergence and, 365
  generalized, 366
Ratio, 19
Rational functions, 73–88
Rational root, 74
  possible, 74
Real number, 8
Real part complex number, 149
Reciprocal relations, 90
Rectangular Cartesian coordinate system, 13
  abscissa, 13
  ordinate, 13
  origin, 13
  quadrants, 13
Rectangular coordinates in space, 378–389
  angle between two directed lines, 382
  coordinate
    axes, 378
    planes, 378
  direction
    cosines of a line, 380–382
    number device, 382–383
  distance between two points, 380
  octants, 378
  origin, 378
  two straight lines in space, 380
Rectangular coordinates, polar coordinates and, transformations, 236
Rectangular form, 151
Recursive sequences, 364
Reduction formulas, 93–94
Relative maximum values, 165–166
  test for, 165–166
Roots, 15, 19, 73
  complex
    conjugate, 73
    numbers, 153
  irrational, 73
  limits, 73
  rational, 74
  simple, 73
  sum and product of, 33
Rotating systems, coordinate systems and, 245–246
Rule of Signs, 76

Same sense, 40
Scalar product (dot product), vectors and, 217–218
Scalar quantities, vector components and, 214
Schrodinger equation, 332–333
Secant θ, 90
Second derivatives, 197–199
  acceleration and, 198
Secondary diagonal, 277
Separation of variables method, 331–333
  Schrodinger equation, 332–333
Sequence
  general term, 362
  infinite, 362–364
  recursive, 364
  theorems, 364
Series of positive terms, 365
  convergence comparison test, 365
  convergence ratio test, 365
Series with negative terms, 365–366
  absolutely convergent, 366
  alternating series, 365–366
  conditionally convergent, 366
  generalized ratio test, 366
  infinite series of the form, 366–367
Simple root, 73
Simultaneous linear equations, 23–31
  algebraic solution, 24
    elimination by
      addition, 24
      substitution, 24
  consistent, 23
  dependent, 23
  detached coefficients, 25–27
  graphical solution, 24
  graphing calculator, 27
  inconsistent, 23
  independent, 23
  matrix, 25–27
  three linear equations, 25
Sine curve, 97–98
  composition of, 98

Sine, law of, 99
Single variable functions
  calculus of, mathematics approach, 161–191
  mathematics approach
    acceleration, 167
    area by summation, 169
    concavity, 166
    continuity on closed interval, 162
    continuous functions, 162
    critical values, 165
    decreasing function, 164–165
    derivative, 163
    differentials, 168
    differentiation formulas, 164
    direction of motion, 167
    function limit, 161–162
    higher-order derivatives, 164
    increasing function, 164–165
    increments, 162–163
    indefinite integral, 168
    inflection point of a curve, 166–167
    limit theorems, 162
    one-sided limit, 162
    relative maximum values, 165–166
    relative minimum values, 165–166
    velocity and speed, 167
    velocity, 167
  physics approach, 192–211
    acceleration, 197–199
    antiderivatives, 200
    derivative formulas, 195–197
    first derivatives, 192–195
    fundamental theorem of calculus, 203
    integrals, 200
    second derivatives, 197–199
    velocity, 192–195
Single-valued function, 8–9
Sinks
  divergence and, 305–306
  flux sources and, 316
Slope of polar curve, 238–240
Solution, 19
Sources
  divergence and, 305
  flux and, 315
    sinks, 316
Speed and velocity, 167
Spherical coordinate systems, 241–242, 247
  azimuthal angle, 242
  Cartesian coordinates, 241
    transformation between, 242
  curl of vector and, 309
  divergence and, 309
  gradient and, 309
  polar angle, 242
  radial distance, 241
Spherical coordinate systems (*Cont.*):
  unit vectors, 242
  velocity and acceleration, 242
Standard form of circle, 61–62
Standard position, angles in, 89
Standard units of measurement, 390
Stokes theorem, 307–308
Straight line, 51–56
  equation, 51
  families of, 57–60
  parameter, 57
    one-parameter systems, 57
Subtraction
  complex numbers and, 149
  graphic representation of complex numbers, 150
  vectors and, 216–217
Sum and product of roots, 33
Symmetry, 45–46
  axis of, 45
    even, 46
    odd, 46
  center of, 45
  curve sketching and, 237
Systems of linear equations in unknowns, 282–283
  Cramer's rule, 283

Tangent equation, circle and, 62
Tangent length, circle and, 62
Tangent, velocity and, 193
Theorems on limits, 162
Theoretical probability. *See* mathematical probability, 347–349
Three dimensions, vectors and, 217
Three linear equations in three unknowns, 25
Translating systems, coordinate systems and, 245–246
Triangle, oblique, 98–99
Trigonometric form. *See* polar form.
Trigonometric functions, 89–133
  acute angle, 91–92
    adjacent side, 91–92
    hypotenuse, 91–92
    opposite side, 91–92
  algebraic signs, 91
  angles in standard position, 89
  complementary angles, 92
  cosecant θ, 90
  cosine θ, 90
  cotangent θ, 90
  coterminal angles, 93
  definition, 101–102
  general
    angle, 89–91
    reduction formula, 94
    sine curve, 97–98
  graphs, 96–97
  inverse, 101

Trigonometric functions, inverse (*Cont.*):
trigonometric relations, 102–104
law of
cosines, 100
sines, 99
line representations, 94–95
negative angles, 93
notation, 101
oblique triangle, 98–99
periodic, 97
projection formulas, 99–100
quadrantal angles, 91
reciprocal relations, 90
reduction formulas, 93–94
secant θ, 90
sin θ, 90
tangent θ, 90
variations, 95–96
Trivial solution, 283
Two straight lines in space, 380

Underdamped, 328
Units and dimensions, 390
Unit vectors, 215, 242
Upper limits, 73

Value, determinant of order two and, 277
Variable,
dependent, 8, 264
independent, 8, 264
range, 8
real number, 8
Variation of sign, 75–76
Variations of trigonometric functions, 95–96
Vector calculus, 302–318
curl, 307–308
del operator, 302–305
divergence, 305
flux, 313–315
Gauss' law, 316
Vector calculus (*Cont.*):
gradient operator, 302–305
Vector representations, 240
Vectorial angle, 235
Vectors, 212–232
algebraic
additions, 216
epresentation, 215
associativity, 217
commutativity, 217
components of, 214–215
scalar quantities, 214
cross product, 218–220
components, 221
cylindrical coordinate systems and, 243–244
direction, 212
geometric representation, 212–213
graphical additions, 213–214
parallelogram method, 214
magnitude, 212
multiplication, 216, 217
Newton's Second Law, 227, 228, 229
null, 216
product identities, 220–221
scalar product, 217–218
subtraction of, 216–217
symbolic representation, 212
three dimensions, 217
unit, 215, 242
Velocity, 167, 192–195, 241, 242, 245
extreme values, 193–194
instantaneous, 193, 194
speed and, 167
tangent, 193
Vertices, 32
Volume elements, 246–247
Cartesian coordinates, 247
cylindrical coordinates, 247
spherical coordinates, 247

Zero exponent, 4